A Conceptual Approach to Genetics

A Conceptual Approach to Genetics

Editor: Rosanna Mann

RCALLISTO
REFERENCE

www.callistoreference.com

Callisto Reference,
118-35 Queens Blvd., Suite 400,
Forest Hills, NY 11375, USA

Visit us on the World Wide Web at:
www.callistoreference.com

ISBN: 978-1-64116-041-4 (Hardback)

Cataloging-in-Publication Data

A conceptual approach to genetics / edited by Rosanna Mann.
    p. cm.
Includes bibliographical references and index.
ISBN 978-1-64116-041-4
1. Genetics. 2. Genomes. 3. Biology. I. Mann, Rosanna.
QH430 .C66 2019
576.5--dc23

Table of Contents

Preface

Genetics is the study of variation in genes and heredity. The sub-disciplines of genetics include evolutionary genetics, epigenetics, population and conservation genetics among others. The field draws on the basic principles of trait inheritance and molecular inheritance. The emergence of gene editing and gene mapping tools has broadened the frontiers of genetics and facilitated the study of individual gene inheritance, expression and change on all stages of embryo growth. From theories to practical applications, all relevant topics in the field of genetics have been included in this book. It presents the complex subject of genetics in a comprehensive and easy to understand language. It will serve as a valuable reference guide for geneticists, academicians, researchers and students.

The information shared in this book is based on empirical researches made by veterans in this field of study. The elaborative information provided in this book will help the readers further their scope of knowledge leading to advancements in this field.

Finally, I would like to thank my fellow researchers who gave constructive feedback and my family members who supported me at every step of my research.

Editor

Study of the optimum haplotype length to build genomic relationship matrices

Mohammad H. Ferdosi[1,2]*, John Henshall[3,4] and Bruce Tier[2]*

Abstract

Background: As genomic data becomes more abundant, genomic prediction is more routinely used to estimate breeding values. In genomic prediction, the relationship matrix (**A**), which is traditionally used in genetic evaluations is replaced by the genomic relationship matrix (**G**). This paper considers alternative ways of building relationship matrices either using single markers or haplotypes of different lengths. We compared the prediction accuracies and log-likelihoods when using these alternative relationship matrices and the traditional **G** matrix, for real and simulated data.

Methods: For real data, we built relationship matrices using 50k genotype data for a population of Brahman cattle to analyze three traits: scrotal circumference (SC), age at puberty (AGECL) and weight at first corpus luteum (WTCL). Haplotypes were phased with hsphase and imputed with BEAGLE. The relationship matrices were built using three methods based on haplotypes of different lengths. The log-likelihood was considered to define the optimum haplotype lengths for each trait and each haplotype-based relationship matrix.

Results: Based on simulated data, we showed that the inverse of **G** matrix and the inverse of the haplotype relationship matrices for methods using one-single nucleotide polymorphism (SNP) phased haplotypes provided coefficients of determination (R^2) close to 1, although the estimated genetic variances differed across methods. Using real data and multiple SNPs in the haplotype segments to build the relationship matrices provided better results than the **G** matrix based on one-SNP haplotypes. However, the optimal haplotype length to achieve the highest log-likelihood depended on the method used and the trait. The optimal haplotype length (7 to 8 SNPs) was similar for SC and AGECL. One of the haplotype-based methods achieved the largest increase in log-likelihood for SC, i.e. from −1330 when using **G** to −1325 when using haplotypes with eight SNPs.

Conclusions: Building the relationship matrix by using haplotypes that comprise multiple SNPs will increase the accuracy of estimated breeding values. However, the optimum haplotype length that shows the correct relationship among individuals for each trait can be derived from the data.

Background

Advances in genotyping technologies have resulted in a substantial decrease in genotyping costs for many species. These advances have created a new era in livestock genetic evaluation by adding a new type of information to the traditional animal breeding techniques. In the past,

*Correspondence: mferdosi@myune.edu.au; btier@une.edu.au
[1] The Centre for Genetic Analysis and Applications, School of Environmental and Rural Science, University of New England, Armidale, Australia
[2] Animal Genetics and Breeding Unit, University of New England, Armidale, Australia
Full list of author information is available at the end of the article

genetic evaluation was based on phenotypic records and pedigree information with best linear unbiased prediction (BLUP) [1]. In spite of the complexity of the underlying biology, traditional genetic evaluation methods have had a large effect on the improvement of livestock production. Usually an animal model was used, i.e. a model that includes each animal's breeding value where the numerator relationship matrix (**A**) was used to define the genetic relationships among animals. Relationships in **A** are twice the co-ancestry of pairs of individuals and **A** is built by tracking the descent of founder genomes (from the base population, i.e. animals whose pedigrees are unknown and are assumed to be unrelated [2]) through the pedigree. Thus, elements in **A** are based on the idea of identity by descent (IBD).

The availability of cheap genomic data in large quantities allows the relationships among individuals to be defined directly and thus, more accurately. This led to the development of the genomic relationship matrix (**G**) [3–7], which can replace **A** in genetic evaluations. **G** was designed for use with large numbers of independent SNPs (single nucleotide polymorphisms).

Unlike **A**, **G** reflects identity by state (IBS) within the population, and thus relationships arise between pairs of individuals that were previously considered 'unrelated'. Hence **G** incorporates relationships that arise from unknown common ancestors. These ancestors predate the animals that are considered to be founders by pedigree information. Using **G** in place of **A** can increase the accuracy of parameter estimation and decrease the expense of progeny-testing [3, 8]. Furthermore, **G** allows us to estimate the breeding value for new individuals more accurately than **A** by using only the genotype data and the phenotype of ancestral individuals [9].

G considers complete linkage disequilibrium (LD) between SNPs and quantitative trait loci (QTL) but ignores LD between SNPs, especially in short regions [10]. Therefore, shuffling the order of SNPs has no effect on the final results in **G**. However, a desirable QTL allele in one sub-population may be in LD with the allele of one SNP in one strand of the haplotype, but with the other allele of the SNP in another haplotype in the other part of the population. Capturing this type of variable phase between SNPs and QTL requires using a group of SNPs that are joined together in the form of haplotypes [11].

Livestock populations usually consist of large numbers of half-sib and full-sib families. This population structure allows us to reconstruct (phase) the haplotypes accurately and rapidly [12, 13]. Combining IBD and LD information to describe relationships by using haplotypes can increase the accuracy of genetic evaluation and parameter estimation [4, 11, 14, 15]. Hickey et al. [15] used regional haplotype information (non-overlapping haplotype segments, i.e. distinct windows) by breaking haplotypes of all individuals into short segments of equal size (5 to 2000 SNPs) to estimate the relationships between individuals for each segment. The average relationships among all segments were calculated to estimate the total relationship between individuals using simulated data. As reported by Hickey et al. [15], this new method did not improve accuracy of prediction compared to the method based on unphased genotypes. However, they found a higher correlation between their diagonal and off-diagonal elements of the relationship matrix and the true relationship matrix (simulated data), than between those of the **G** matrix generated from individual SNPs. Simulated

data may not represent the real populations' genotypes and phenotypes because of the underlying biological complexity. The partitioning of the genome, as suggested by Hickey et al. [15], in non-overlapping haplotype segments [distinct windows (DW)] may not capture all the variation of the haplotype diversity across the entire genome, because the linkage between some haplotype segments may be ignored. This linkage can be accounted for, by partitioning the genome into segments that overlap (sliding windows (SW) [16]).

The challenge in choosing the optimal haplotype length is to model linkage between SNPs and QTL appropriately. When haplotypes are based on one-SNP (individual SNPs), there are only two possible alleles. When haplotypes are based on pairs of SNPs, four alleles are possible and as the number of SNPs in each segment increases, so does the number of possible alleles. Varying the length of haplotypes can assist in the modeling of the LD between SNPs and QTL. On the one hand, if haplotype similarity is the basis of determining relationships, increasing the number of haplotype alleles will generally result in lower relationships within any segment. On the other hand, using short haplotype segments will maintain remote relationships within the matrix, but may also capture different phases between SNPs and QTL by ignoring LD. The optimal haplotype length will balance the value of older relationships against the errors in LD that are assumed between SNPs and QTL across the whole population [4].

The aim of this study was to choose the best relationship matrix based on different ways of modeling LD between SNPs and QTL and identifying the optimal haplotype length. Three alternative relationship matrices, based on haplotypes of variable length, were considered and compared with the standard **G** matrix [3] for three traits. The underlying ideas in the construction of **G** and these alternative methods were explored with real and simulated data.

Methods
Data
Simulated data
A small dataset was simulated for the purpose of exploring **G** (VanRaden [3]—first method) when considering whether to include or exclude allele coding and allele frequencies on **G** and its inverse. In addition, our objective was to understand the effects of allele coding of the genotype and correcting allele frequencies in order to build **G** on the log-likelihood and variance components. Since the simulated **G** was based on one-SNP haplotypes, LD between markers was ignored. This dataset was based on a full-sib design of four males each mated to five females to produce one offspring per mating. The final population

included nine parents (four males and five females) and 20 offspring. A trait with a heritability of 0.55 and 99 SNPs was simulated. The phenotypes were simulated as:

$$\text{Phenotype} = \mathbf{q}\mathbf{X}_G + \mathbf{e}, \tag{1}$$

where $\mathbf{q}$ is a vector of SNP effects N(0,1), $\mathbf{X}_G$ is a genotype matrix with terms equal to the genotypes (defined as the number (0, 1, or 2) of second alleles of each animal at each SNP) and $\mathbf{e}$ is a vector of normally-distributed residuals. Genotypes were simulated at the gametic level with equally-spaced SNPs on 10 chromosomes, each 1 Morgan long.

Real data

A subset of the 50k SNP data obtained from the "Northern Breeding Project" resource Brahman population bred by the Cooperative Research Centre for Beef Genetic Technologies (BeefCRC) was used, with trait records for scrotal circumference (SC), age at puberty (AGECL), and weight at first corpus luteum (WTCL). The description and details of SC, WTCL and AGECL phenotypes were provided by Johnston et al. [17], Hawken et al. [18] and Zhang et al. [19]. Estimation of heritabilities was based on the single-trait animal model using $\mathbf{A}$ with the following fixed effects for each trait (see comments in Table 1):

SC: cohort, location, month of birth, operator, age and weight at 18 months;
AGECL: age of dam, cohort, origin, calving month, interaction of origin and calving month, interaction of cohort and origin, interaction of cohort and calving month;
WTCL: age of dam, cohort, line of origin and calving month.

Ethical approval

This experiment was approved by the JM. Rendel Laboratory Animal Experimental Ethics Committee (CSIRO, Queensland) as approvals TBC107 and RH225-06, respectively.

Table 1 Number of animals (N), mean (μ), standard deviation (SD) and heritability (h²) for different traits [17–19]

Trait	N	μ	SD	h²
SC (cm)	1007	26.6	2.94	0.75[a]
AGECL (days)	854	751	142.1	0.57[a]
WTCL (kg)	854	334	44.8	0.56[a]

SC scrotal circumference, AGECL age at puberty, WTCL weight at first corpus luteum

[a] BLUP using matrix A

Haplotypes

Overall haplotypes for the Brahman cattle were reconstructed for all the chromosomes using hsphase [12, 13] and missing genotypes were imputed by BEAGLE 3.3.2 [20]. hsphase and BEAGLE were run with default parameters. The whole genome was subsequently divided into segments of equal length (1, 2, 3, ..., 20, 40, 80 and 100) and the numbers of haplotype alleles in each segment were identified.

Relationship matrices

The following relationship matrices were built to determine the effect of centering and correcting allele frequency on the additive and residual variances.

Genomic relationship matrix using independent SNPs (centered)

In our work, $\mathbf{G}$ refers to VanRaden's first method [3] for calculating the genetic relationship matrix. To construct $\mathbf{G}$ in a population with 'a' animals genotyped for 'm' SNPs, the genotypes were centered so that the sum of each column was zero, $\mathbf{Z} = \mathbf{X}_G - \mathbf{P} = \mathbf{X}_G - \mathbf{J} - \mathbf{D}$, where $\mathbf{X}_G$ is the genotype matrix (a × m), with entries 0, 1 or 2, representing alleles AA, AB and BB, respectively; $\mathbf{D} = \mathbf{P} - \mathbf{J}$, where $\mathbf{P}$ is an (a × m) matrix with each row consisting of $2\mathbf{p}$ ($\mathbf{p}$ is the B allele frequency of each SNP) and $\mathbf{J}$ is a matrix of 1's with the same dimension as $\mathbf{P}$. Finally, $\mathbf{G}$ was calculated as:

$$\mathbf{G} = \frac{\mathbf{Z}\mathbf{Z}'}{2\sum \mathbf{p}(1 - \mathbf{p})}, \tag{2}$$

Because $\mathbf{G}$ was not positive definite, 0.001 was added to its diagonal elements, to allow inversion.

Genetic relationship matrix using independent SNPs (uncentered)

A matrix $\mathbf{M}$ was constructed from $\mathbf{X}_G$ by subtracting $\mathbf{1}$, via $\mathbf{M} = \mathbf{X}_G - \mathbf{J}$. This matrix included 1, 0 and −1 representing alleles AA, AB and BB.

Matrix $\mathbf{M}$ was used to calculate a matrix that is similar to $\mathbf{G}$ but uncentered for the allele frequencies ($\mathbf{G}_u$):

$$\mathbf{G}_u = \frac{\mathbf{M}\mathbf{M}'}{\mathrm{d}}, \tag{3}$$

where the denominator $\mathrm{d} = \mathrm{m}/2 = 2\sum \mathbf{p}(1 - \mathbf{p})$, assuming $\mathbf{p} = 0.5$. $\mathbf{G}_u$ was used to demonstrate the effect of centering on additive and residual variances. Alternatively, the same denominator that is used in $\mathbf{G}$ (i.e. calculating allele frequency after centering) could be used.

Relationship matrices using one-SNP haplotypes

Haplotypes of animals were used to create the one-SNP haplotype relationship matrix. Let $\mathbf{X}_H$ be a (h × m) matrix of haplotypes (h = 2a), with entries 0 or 1 indicating the number of copies of one of the two possible alleles. For a single locus, haplotypes were constructed without reference to the adjacent loci. Suppose that $K = I_{ah} \otimes [1\ 1]$ (I is an identity matrix, and $\otimes$ is the Kronecker product [21]). With X_H and K, the genotypes were reconstructed as $X_G = KX_H$. The allele frequencies for SNP in X_H were calculated as $p = 1X_H/h$.

The haplotype relationship matrix for one-SNP ($H_{*,1}$) can be calculated as follows:

$$H_{*,1} = K\Gamma K'/2, \tag{4}$$

such that:

$$\Gamma = (X_HX_H' + (X_H - J_{hm})(X_H - J_{hm})')/m. \tag{5}$$

Alternatively Γ can be computed as:

$$\Gamma = (J_{hm}J_{hm}' - (Q + Q'))/m, \tag{6}$$

where Q is $X_H(-(X_H - J_{hm})')$, which is similar to the method explained in [22].

Similarity of G and H_{*,1}

Expansion of the terms for the G (7), G_u (10) and $H_{*,1}$ (11) matrices helps to illustrate the differences between them.

$$G = ZZ'/d = ((X_G - J_{am} - D)(X_G - J_{am} - D)')/d, \tag{7}$$

$$G = \left(X_GX_G' - X_GJ_{am}' - X_GD' - J_{am}X_G' + J_{am}J_{am}' + J_{am}D' - DX_G' + DJ_{am}' + DD'\right)/d,$$

where

$$E = -X_GD' + JD' - DX_G' + DJ' + DD', \tag{8}$$

and

$$J_{am}J_{am}' = mJ_{aa}, \tag{9}$$

$$G = (X_GX_G' + mJ_{aa} - X_GJ_{am}' - J_{am}X_G' + E)/d.$$

$$G_u = (X_G - J_{am})(X_G - J_{am})'/d, \tag{10}$$

$$G_u = (X_GX_G' + J_{am}J_{am}' - X_GJ_{am}' - J_{am}X_G')/d,$$

$$G_u = (X_GX_G' + mJ_{aa} - X_GJ_{am}' - J_{am}X_G)/d, \tag{11}$$

$$H_{*,1} = ((K(X_HX_H' + (X_H - J_{hm})(X_H - J_{hm})')K')/2)/m,$$

$$H_{*,1} = ((KX_HX_H'K')/2 + (KX_HX_H'K')/2 + (KJ_{hm}J_{hm}'K')/2 - (KX_HJ_{hm}'K')/2 - (KJ_{hm}X_H'K')/2)/m,$$

$$H_{*,1} = KX_HX_H'K' + \left(KJ_{hm}J_{hm}'K'\right)/2 - \left(KX_HJ_{hm}'K'\right)/2 - \left(KJ_{hm}X_H'K'\right)/2/m,$$

and since $X_G = KX_H$ and $KJ_{hm} = 2J_{am}$,

$$H_{*,1} = \left(X_GX_G' + 2J_{am}J_{am}' - X_GJ_{am}' - J_{am}X_G'\right)/m,$$

$$H_{*,1} = \left(X_GX_G' + 2mJ_{aa} - X_GJ_{am}' - J_{am}X_G'\right)/m.$$

From Eqs. (10) and (11):

$$G_u = m(H - J_{aa})/d. \tag{12}$$

From Eqs. (7) and (10):

$$G = G_u + E/d. \tag{13}$$

And finally,

$$G_u + E/d = (m(H - J_{aa}) + E)/d. \tag{14}$$

As a result, the extension of $H_{*,1}$ (Eq. 11) produced the same result as the molecular coancestry suggested by Toro et al. [23].

Haplotype relationship matrices

Relationships among individuals were calculated in different ways. G [3] was calculated to provide the base to which the three methods were compared. The haplotype relationship matrices in the methods based on different lengths of haplotypes are designated as $H_{i,j}$ where i is method (1), (2) or (3) (see below and Fig. 1) and j is the length of the haplotypes (j = 1, 2, 3, ..., 20, 40, 80 and 100). The three methods used to calculate relationships based on haplotypes are illustrated in Fig. 1.

In method (1) or DW for distinct windows, which was used to construct $H_{1,*}$, each chromosome was divided into segments of length j. This method was similar to that described by Hickey et al. [15] for building H_1. A chromosome with k SNPs was divided into k /j segments so that each SNP appeared only once in any segment (Fig. 1b). Then, in the last segment of a chromosome that was shorter than the segment length, SNPs from the previous segment were included so that all segments had the same length (Fig. 1b).

The gametic relationship matrix ($\Gamma_{segment}$) among all pairs of haplotypes was determined for each segment by assuming that it was equal to 1 when two haplotypes were the same and 0 when they were not (Fig. 1b).

The gametic relationship matrices for each segment were summed to give a complete gametic relationship matrix (Fig. 1b). The relationship matrix for the whole genome was calculated as follows:

$$\Gamma = \sum_{i=1}^{n} \Gamma_i/n, \tag{15}$$

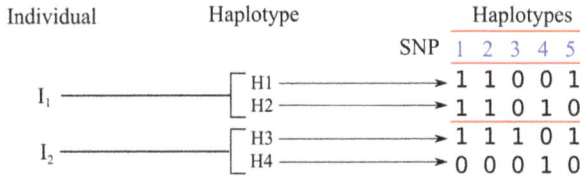

Fig. 1 Description of the methods used to build relationship matrices. **a** Haplotypes of five SNPs for two individuals. **b** Method DW, $\mathbf{H}_{1,2}$. Each haplotype segment has the same length ($j = 2$, $k/j = 5/2 = 3$, k SNPs and window of size j), with the last window potentially using SNPs that may have been used in the penultimate segment. $\boldsymbol{\Gamma}_i$ is the relationship matrix for each window and $\boldsymbol{\Gamma}$ is the final relationship matrix. **c** Method SW, $\mathbf{H}_{2,2}$. Haplotypes that are two SNPs long are constructed from adjacent pairs of SNPs with SNPs present in more than one segment ($j = 2$, $k - j + 1 = 5 - 2 + 1 = 4$). **d** Method TMS, $\mathbf{H}_{3,2}$. The total number of SNPs in contiguous segments that are identical in pairs of haplotypes. The segment size 2 defines the minimum number of SNPs in two contiguous haplotype segments to be considered as identical by descent (IBD)

where n is the number of segments.

This was converted to a relationship matrix at the animal level using (Fig. 1b):

$$\mathbf{H}_{1,1} = \mathbf{K}\boldsymbol{\Gamma}\mathbf{K}'/2. \tag{16}$$

Method (2) or SW for sliding windows was used to construct $\mathbf{H}_{2,*}$, which was similar to $\mathbf{H}_{1,*}$ but the genome was divided into segments in a different way. In this method, the genome was partitioned into $k - j + 1$ segments. The first segment had SNP 1 to j, the second segment had SNP 2 to $j + 1$ and, so on, to the last segment with SNP $k - j + 1$ to k (Fig. 1c).

In method (3) or TMS for total minimum similarity that was used to construct $\mathbf{H}_{3,*}$, haplotypes for whole chromosomes were considered. With this method, the number of SNPs in identical segments of length j or more in pairs of haplotypes were counted. These scores were divided by the numbers of SNPs on the chromosome (Fig. 1d).

Variance components

The model used to analyze the traits was as follows:

$$\mathbf{y} = \mathbf{X_v b} + \mathbf{Z_v u} + \mathbf{e},$$

where $\mathbf{y}$, $\mathbf{b}$, $\mathbf{u}$ and $\mathbf{e}$ are vectors of observations, fixed effects, breeding values and residuals, respectively, and $\mathbf{X_v}$ and $\mathbf{Z_v}$ are design matrices relating observations to effects. $\mathrm{Var}(\mathbf{u}) = \mathbf{W}\sigma_a^2$, where $\mathbf{W}$ is a relationship matrix which could be $\mathbf{G}, \mathbf{G_u}$ or $\mathbf{H_{i,j}}$. The residual variance was $\mathrm{Var}(\mathbf{e}) = \mathbf{I}\sigma_e^2$. Variance components and the log-likelihoods were estimated using ASReml-R version 3 [24].

Simulated data

Variance components were estimated for $\mathbf{G_u}$, which was similar to $\mathbf{G}$ but uncentered, and $\mathbf{H_{1,1}}$. In all cases, the only fixed effect ($\mathbf{b}$) was the mean.

Real data

Relationship matrices $\mathbf{G}$, and the three $\mathbf{H_{i,j}}$ with varying numbers of loci were used to model covariance between animals. For each method, genetic parameters (i.e. additive and residual variance) for SC, WTCL and AGECL were calculated using the standard single-trait animal models. The optimal length for haplotypes was found by using the profiled log-likelihood for each of the three haplotype-based methods.

Scaling the haplotype relationship matrix for comparison of additive variances

The additive variances (σ_a^2) of the haplotype relationship matrices were scaled as in Legarra [25]:

$$\left(\mathrm{tr}(\mathbf{H_{*,*}})/a - \left(\mathbf{J_{1h}H_{*,*}J_{h1}}\right)/a^2 \right)\sigma_a^2, \tag{17}$$

where 'tr' is the trace of the matrix and 'a' is the number of animals.

Cross-validation

Fivefold cross-validation was used to assess the accuracy of estimated breeding values (EBV). Individuals were grouped into five subsets of approximately equal size with all the progeny of a common sire in one group. EBV were estimated for each of the five subsets using data from the other subsets and compared with their adjusted phenotypes (phenotypes corrected for the fixed effects).

Results

Brahman haplotype diversity

Figure 2 shows boxplots for the number of haplotype alleles for all segments that explained 60 and 90 % of the observed haplotype alleles for chromosome 1 using the DW method (similar patterns were observed for the other chromosomes and the SW method—not shown). As the segment size increased, the number of haplotype alleles increased exponentially until the size of the population limited the number of unique haplotypes that could be found.

Simulated data

Table 2 shows the estimates of variance components when $\mathbf{G}$, $\mathbf{G_u}$, and $\mathbf{H_{1,1}}$ were used in the model. The value of the log-likelihood was similar for all three methods. The mean was somewhat different when $\mathbf{G}$ was used compared to that of the other two methods, which share the same mean. The residual variance components were the same when $\mathbf{G_u}$ and $\mathbf{H_{1,1}}$ were used to describe the covariance between animals, and very similar to the value obtained using $\mathbf{G}$. In spite of the differences in the relationship matrices as shown in Eqs. (7–14), the coefficient

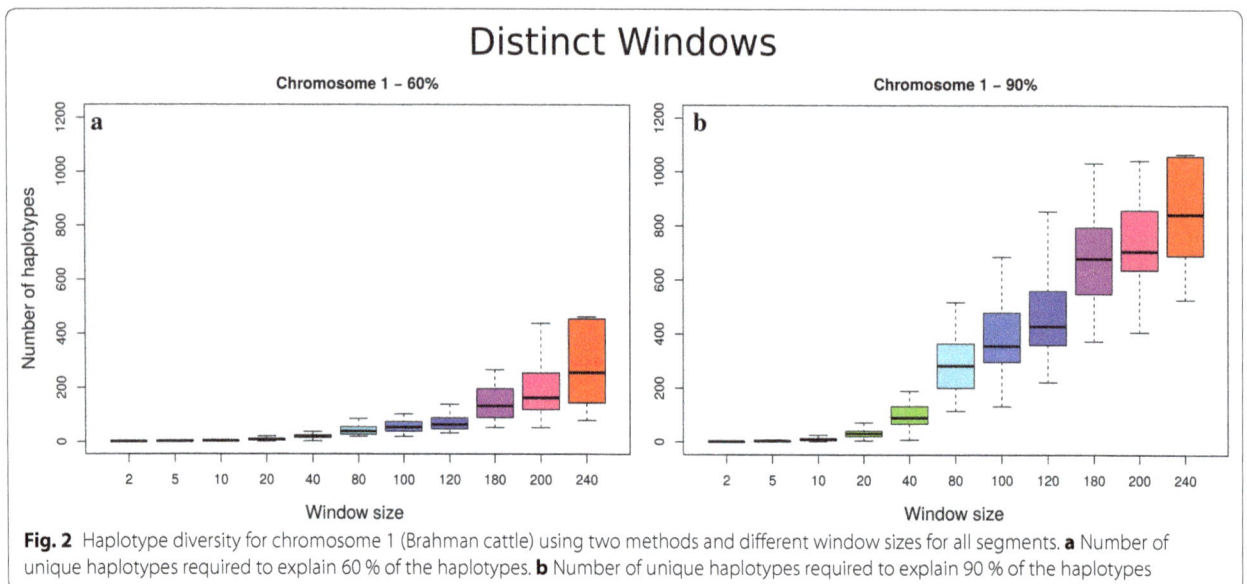

Fig. 2 Haplotype diversity for chromosome 1 (Brahman cattle) using two methods and different window sizes for all segments. **a** Number of unique haplotypes required to explain 60 % of the haplotypes. **b** Number of unique haplotypes required to explain 90 % of the haplotypes

of determination (R^2) between elements of the inverses of the different matrices was close to 1. Based on the slope of the regression, the values in $\mathbf{H}_{1,1}^{-1}$ were nearly two times higher than those in $\mathbf{G}^{-1}$ and $\mathbf{G}_u^{-1}$ (ignoring the intercept, see Table 3). Estimated genetic variances were similar

Table 2 Log-likelihood, residual variance (σ_e^2), additive variance (σ_a^2) and intercept (μ) using simulated data and different methods

Method	Log-likelihood	σ_e^2	σ_a^2	μ
G	−74.459	36.965	42.377	−5.050
$H_{1,1}$	−74.459	37.008	88.654	−5.106
G_u	−74.459	37.007	42.377	−5.106

Table 3 Intercept, slope and R^2 of linear regression of the elements of the inverse relationship matrices from models using different relationship matrices for the simulated data

Dependent variable	G^{-1}	G^{-1}	$H_{1,1}^{-1}$
Independent variable	$H_{1,1}^{-1}$	G_u^{-1}	G_u^{-1}
Intercept	34.482	34.470	−0.025
Slope	0.464	0.960	2.069
R^2	1.000	0.992	0.993

when $\mathbf{G}$ and $\mathbf{G}_u$ were used, but much greater when $\mathbf{H}_{1,1}$ was used [26]. The latter result was in agreement with Stranden et al. [27]. As shown in [26], the correlation between the breeding values were close to 1, as were the slopes when any one set of EBV was regressed on any other.

Real data

Haplotype relationship matrices and G

Figure 3 shows the scatter plots of the elements $\mathbf{H}_{1,1}$, $\mathbf{H}_{1,8}$, $\mathbf{H}_{1,17}$ and $\mathbf{H}_{1,100}$ against the corresponding elements of $\mathbf{G}$. The plots for $\mathbf{H}_{1,*}$ are in three groups across the X axis ($\mathbf{G}$), i.e. unrelated, half-sibs and diagonal. In $\mathbf{G}$, unrelated individuals have a mean close to 0, half-sibs around 0.23 and the diagonal elements around 1. In $\mathbf{H}$, the minimum was 1 in the diagonal elements and 0 in the off-diagonal elements, and the maximum, in both cases, was 2 (Table 4; Fig. 3). However, only the minimum of the diagonal and off-diagonal elements of the relationship matrices with large haplotype segments ($\mathbf{H}_{1,100}$) reached this minimum limit. The elements of $\mathbf{H}_{1,1}$ were much greater than these minimum limits. The mean for the off-diagonal and diagonal elements decreased as segment size increased (Table 4). However, the standard deviations for both off-diagonal and diagonal elements were higher for the intermediate segment sizes ($\mathbf{H}_{1,8}$ and $\mathbf{H}_{1,17}$) than for the very short and long segments ($\mathbf{H}_{1,1}$ and $\mathbf{H}_{1,100}$) (Table 4).

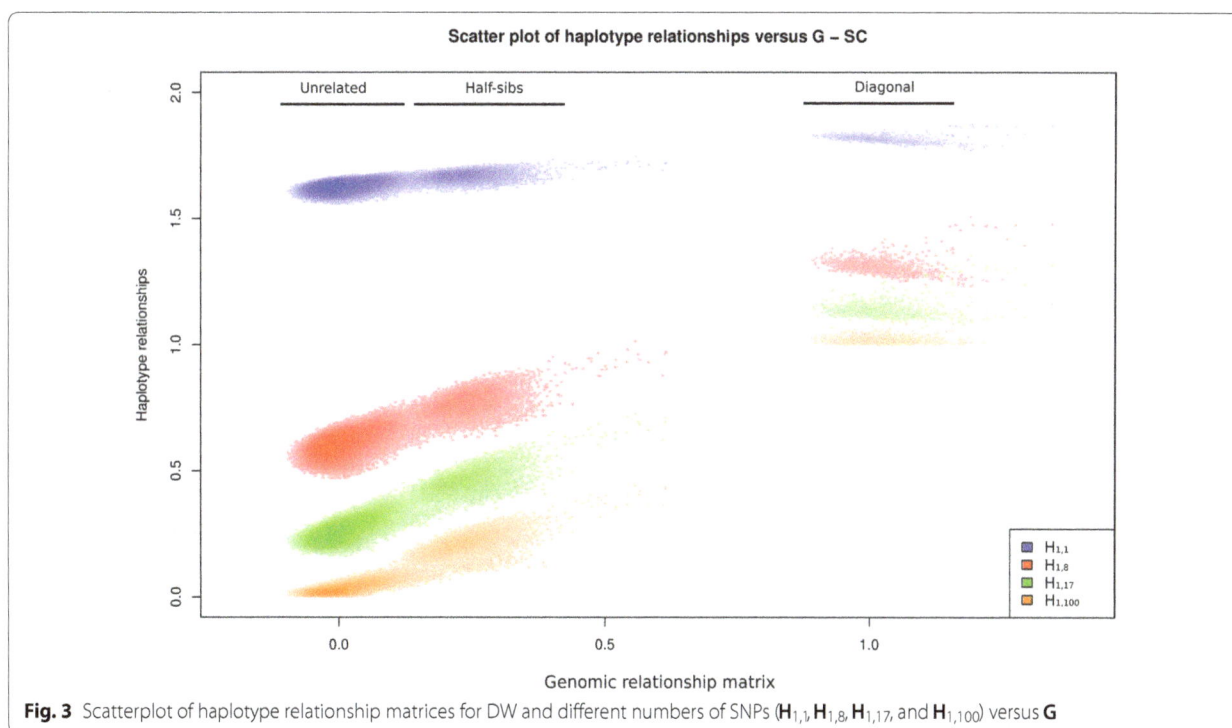

Fig. 3 Scatterplot of haplotype relationship matrices for DW and different numbers of SNPs ($\mathbf{H}_{1,1}$, $\mathbf{H}_{1,8}$, $\mathbf{H}_{1,17}$, and $\mathbf{H}_{1,100}$) versus $\mathbf{G}$

Correlations between the off-diagonal elements of **G** and $\mathbf{H}_{*,*}$ were positive. The correlation between off-diagonal elements increased as the segment size increased. However, only the diagonal elements of $\mathbf{H}_{1,17}$ and $\mathbf{H}_{1,100}$ were positively correlated with the elements of **G**. Although the elements of half-sib individuals were less correlated with **G** elements than the elements of unrelated individuals, there was a higher correlation between the elements of half-sibs individuals than between the diagonal elements and **G**.

Variance components

The log-likelihoods evaluated for SC, AGECL and WTCL using ASReml-R are in Fig. 4. For all traits, the log-likelihoods of the $\mathbf{H}_{*,*}$ methods were higher than that of **G** when haplotype length was longer than one-SNP. The three methods gave similar results for all traits. Regardless of the method used for dividing the haplotype, the log-likelihoods decreased as the segment size increased from 10 to 20 SNPs and the log-likelihoods were higher than that of **G** (black line), except for WTCL. However, the log-likelihood for WTCL increased slightly when the haplotype length was less than 10 SNPs. The best values for each trait are in Table 5.

The additive and residual variances for each trait are in Fig. 5. For short haplotypes, the additive variance estimated using **H** was much greater than that estimated using **G**. The additive variance component decreased substantially as the segment length increased to 20 SNPs, but stabilized as it became longer than 20 SNPs. The residual variance decreased considerably as the segment size increased, except for the TMS method. In contrast to the SW and DW methods, the residual component for the TMS method was larger when the segment size was less than 10 SNPs.

The additive variances generated from scaled relationship matrices [25] are in Fig. 6. Contrary to unscaled relationship matrices, the additive variance for the one-SNP relationship matrix was similar to that for **G** and as window size increased, additive variances increased.

Cross-validation

The correlation between adjusted phenotypes and EBV increased with the likelihood and number of SNPs per window. The SW method had the highest prediction accuracy for SC, and the TMS method had the highest accuracy for AGECL and WTLCL. However, when we looked at the standard deviations, the differences in accuracy could not be considered as significant. Similar to the log-likelihood, the best length of haplotype was trait-dependent (Fig. 7). The best accuracies for each trait are in Table 6. Except for SC, the window sizes that achieved the highest log-likelihood and accuracy were close to each other.

Discussion

Models based on haplotypes of optimum length to describe relationships among individuals were always better than models using **G**, however the optimum haplotype length depended on the trait. The improvement in

Table 4 Minimum, maximum, mean and standard deviation of G and $\mathbf{H}_{1,*}$, and the correlation of the elements of G with the elements of $\mathbf{H}_{1*}$

	Min	Max	Mean	SD	r_{GU}	r_{GH}	r_{GD}
G							
Off-diagonal	−0.01	0.61	0.00	0.05	1	1	1
Diagonal	0.89	1.34	1.02	0.06			
$\mathbf{H}_{1,1}$							
Off-diagonal	1.56	1.75	1.63	0.02	0.40	0.27	−0.34
Diagonal	1.77	1.88	1.82	0.01			
$\mathbf{H}_{1,8}$							
Off-diagonal	0.47	1.02	0.60	0.04	0.56	0.47	−0.09
Diagonal	1.23	1.51	1.31	0.03			
$\mathbf{H}_{1,17}$							
Off-diagonal	0.16	0.73	0.26	0.04	0.68	0.63	0.12
Diagonal	1.08	1.36	1.14	0.03			
$\mathbf{H}_{1,100}$							
Off-diagonal	0.00	0.44	0.03	0.03	0.71	0.73	0.30
Diagonal	1.00	1.19	1.02	0.02			

r_{GU}: correlation between the unrelated individuals (elements of **G** and $\mathbf{H}_{1,*}$); r_{GH}: correlation between the half-sibs individuals (elements of **G** and $\mathbf{H}_{1,*}$); r_{GD}: correlation between the diagonal elements of **G** and $\mathbf{H}_{1,*}$

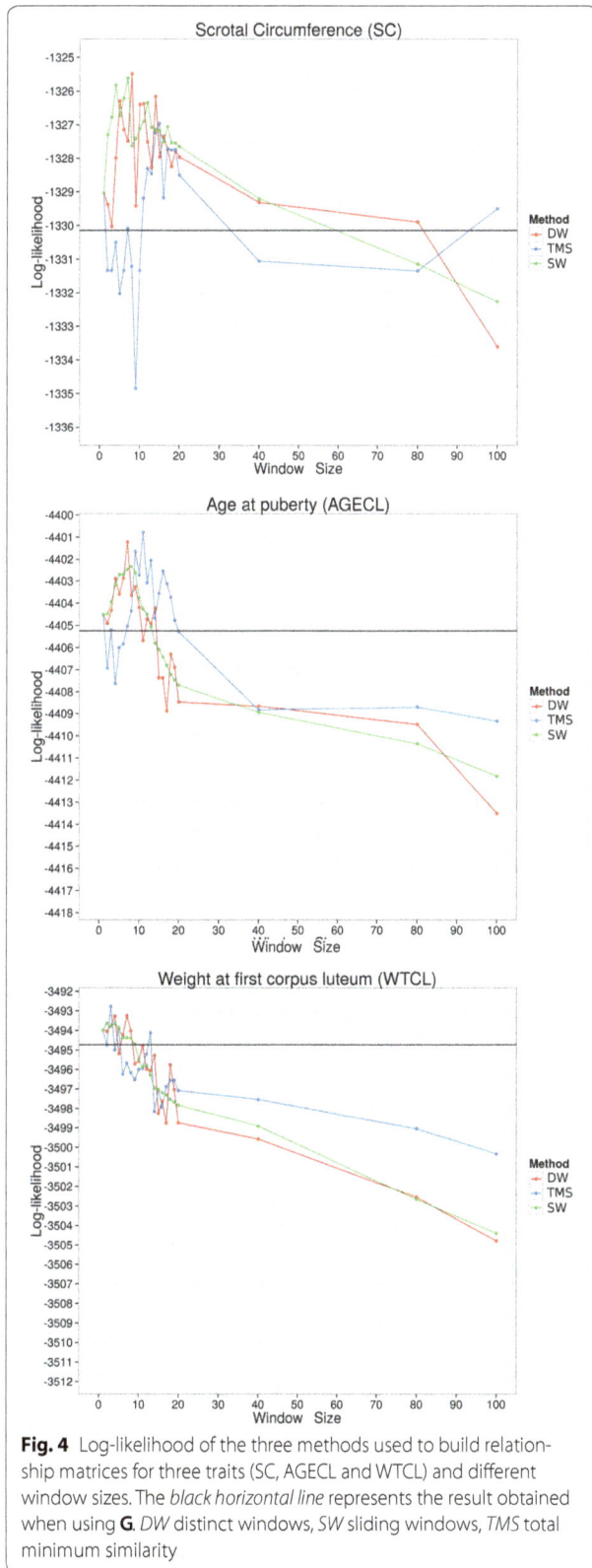

Fig. 4 Log-likelihood of the three methods used to build relationship matrices for three traits (SC, AGECL and WTCL) and different window sizes. The *black horizontal line* represents the result obtained when using **G**. *DW* distinct windows, *SW* sliding windows, *TMS* total minimum similarity

log-likelihood resulting from the use of haplotypes rather than one-SNP was most likely due to the LD between SNPs and QTL being better captured (at the intermediate age of the base population [4]), especially at short-range LD [10]. The other reason for the improvement in log-likelihood was that more genetic diversity was captured with haplotypes than with **G** [3]. Although these two datasets were relatively small, there is no reason to suspect that these results will not extend to the analysis of larger datasets. The results for different methods and traits suggest that haplotypes can make better use of genotype data for genomic prediction. With the real data, the optimum haplotype length was trait-dependent and could be estimated from the data.

Although there was considerable variation between the different haplotype-based methods both across and within the traits with longer haplotypes, the log-likelihood profiles and accuracies increased as the segment size increased up to window size 10, and then both decreased as the segment size increased further (Figs. 4, 7 for AGECL and SC). The decay in log-likelihood occurred because the use of large segments resulted in the relationship matrix tending towards an identity matrix as the variance of the relationships was reduced (Table 4). As a result, the relationships between individuals became closer to zero which makes it difficult to calculate the additive and residual variances. With very long haplotypes, relationships between parent and offspring or between full-sibs were less than 0.5, and between half-sibs less than 0.25. Therefore, using an appropriate method of haplotype partitioning is very important. As the segment size increased, only recent relationships between individuals could be captured and the optimum haplotype length may be an indicator of the optimum age of relationship between individuals [4]. However, for these analyses, there were only minor differences in optimum lengths of haplotype for each trait and method. Using this haplotype-based method in a multiple trait analysis may require the use of different relationship matrices for each trait. If so, then the blocks between animal and traits in the relationship matrix among all breeding values would need to be built and inverted explicitly, thus dramatically increasing the already difficult computational problem for these types of analyses. Nevertheless, there may be a suitable haplotype length that would permit the use of one genomic relationship matrix across all traits. For example, the method based on discrete windows had an optimum of seven SNPs per haplotype for AGECL and WTCL and eight for SC. However there was little difference between the results for SC when the $H_{1,8}$ was used compared to $H_{2,7}$ (Table 5).

Table 5 Window size, log-likelihood, residual variance (σ_e^2), additive variance (σ_a^2) and intercept (μ) with the best log-likelihood using real data and different methods

Trait	Method	DW	SW	TMS	G
Scrotal circumference	Window size	8	7	15	–
	Log-likelihood	−1325.48	−1325.60	−1326.95	−1330.15
	Additive variance	5.29	5.44	5.02	3.20
	Residual variance	2.03	2.11	2.32	2.44
Age at puberty	Window size	7	8	11	–
	Log-likelihood	−4401.20	−4402.31	−4400.77	−4405.25
	Additive variance	11273.73	11052.93	11853.68	6816.86
	Residual variance	4902.01	4694.35	5405.72	5729.85
Weight at first corpus luteum	Window size	7	2	3	–
	Log-likelihood	−3493.23	−3493.62	−3492.76	−3494.73
	Additive variance	1451.34	2758.13	3184.40	900.60
	Residual variance	750.88	804.50	845.43	833.38

A feature of all three haplotype-based methods was that the additive variance was much greater than that found when using **G**, simply because $\mathbf{H}_{*,*}$ and **G** have different scales (Table 2 and 3). However, the additive variance decreased rapidly as the number of SNPs that form the haplotypes increased (Fig. 5). Hence, it is important to estimate the genetic variance by using the appropriate relationship matrix.

Unlike additive variances, residual variances for $\mathbf{H}_{*,*}$ were generally smaller than those obtained when using **G**, except for the TMS method. Residual variances decreased as the window size increased for the same reason that the log-likelihood decreased, i.e. longer haplotypes resulted in a relationship matrix that was similar to an identity matrix. In the TMS method for small segments, the elements of the residual variances were greater than for **G** and the other methods. Consequently, this method may not be suitable for capturing the true relationships.

When only one-SNP haplotypes were used, all three methods provided the same $\mathbf{H}_{*,1}$ matrix and subsequent results [26]. As previously noted, the EBV obtained by using $\mathbf{H}_{*,1}$ to describe the relationships were the same as those estimated using **G**. The difference in their means was not important since it did not change relative merit as defined by differences in the breeding values. This occurred although the estimated genetic variances were much higher when $\mathbf{H}_{*,1}$ was used, than when **G** was used, to model the relationships. Clearly, the effects of using **G** and $\mathbf{H}_{*,1}$ for estimating breeding values were similar, as were their inverses (Table 3). However, the elements in $\mathbf{H}_{*,1}$ appeared to be on a different scale, being much higher than those observed in **G**. These very high coefficients suggest that the individuals were highly related and inbred, compared to the implied founder population.

The scale of the relationship matrices based on genomic data is very important for the computation of heritability and combining genotyped and ungenotyped individuals in the so-called single-step analysis. The EBV [26] clearly indicated that the genotypic information was used in the same way in all methods for prediction. We have demonstrated how a change to **G**, $\mathbf{G}_u$ and $\mathbf{H}_{*,*}$ can be directly related to one another, as demonstrated in Eqs. (7–14).

An alternative method for appropriate scaling of the relationships among individuals is necessary. There are three possible methods for scaling. One was developed for scaling **G** based on the pedigree [28]. A second method, since $\mathbf{H}_{1,1}$ demonstrates the molecular coancestry that can be rescaled to genealogical coancestry with the formula in [23], uses a similar formula with a slight modification to rescale $\mathbf{H}_{*,*}$ with segment sizes larger than 1. A third scaling method was suggested by Legarra [25] for scaling the relationship matrices in order to compare their additive variances. However, further research is required to identify which of these methods provide the most accurate scaling of the haplotype relationship matrix.

The optimum haplotype lengths to achieve the highest accuracy and log-likelihood were similar for AGECL and WTCL whereas for SC the optimum haplotype length for each method varied considerably. This may be caused by the high heritability of the SC trait, although the difference in improvement of accuracy for both optimum lengths was not significant (Table 6).

Only three methods for building haplotype-based relationships were used in this paper. Other methods to create the haplotypes or relationships may improve the accuracy. Two obvious methods that were not tested in this paper are based on the physical position of markers or the linkage maps of the genome.

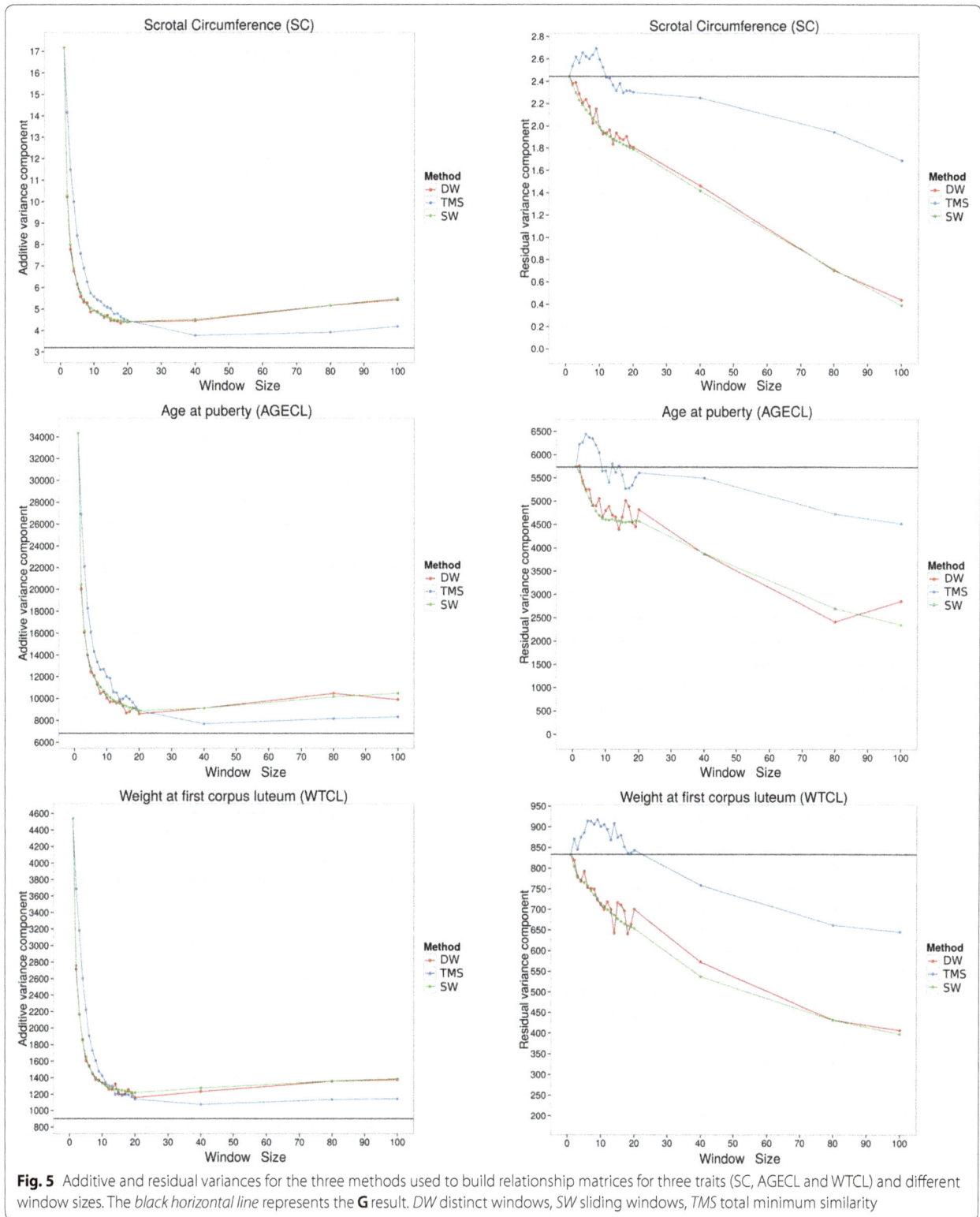

Fig. 5 Additive and residual variances for the three methods used to build relationship matrices for three traits (SC, AGECL and WTCL) and different window sizes. The *black horizontal line* represents the **G** result. *DW* distinct windows, *SW* sliding windows, *TMS* total minimum similarity

Alternatively, a more complete approach to modeling relationships between haplotypes within each segment would include non-zero correlations between haplotypes. Such correlations would be based on methods that estimate the evolutionary relationships among haplotypes.

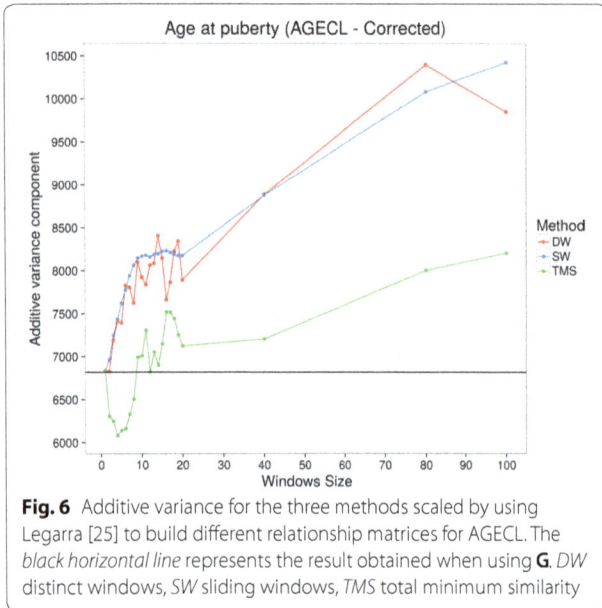

Fig. 6 Additive variance for the three methods scaled by using Legarra [25] to build different relationship matrices for AGECL. The *black horizontal line* represents the result obtained when using **G**. *DW* distinct windows, *SW* sliding windows, *TMS* total minimum similarity

In addition, the effect of heritability and genotyping errors should be considered when comparing the improvement in accuracy and log-likelihood of different traits. Simulation studies have shown that using haplotype segments can increase the accuracy of genomic selection for traits with a high heritability [11]. However, the effect of heritability on the accuracy should be checked with real data. In the current study, the effect of heritability on the increase in accuracy and log-likelihood profile was observed for SC with a heritability of 0.75, which led to a high accuracy even when large haplotype segments were used (Table 6). Moreover, genotyping and haplotype reconstruction errors should be considered when building the relationship matrix. These errors may be one of the reasons that explain the fluctuation in accuracy and log-likelihood observed in this paper for different window sizes, especially with the DW method, which is more sensitive to this kind of issue. In addition, the rate of the decrease in prediction accuracy as segment size increases would be affected by these errors, i.e. genotyping errors will cause more problems for large segments than for small segments.

Conclusions

In this article, three strategies to build relationship matrices using haplotype segments were evaluated. When one-SNP haplotypes are used, we showed and proved that the current methods and the **G** matrix of VanRaden [3] were the same but on different scales. In addition, using more

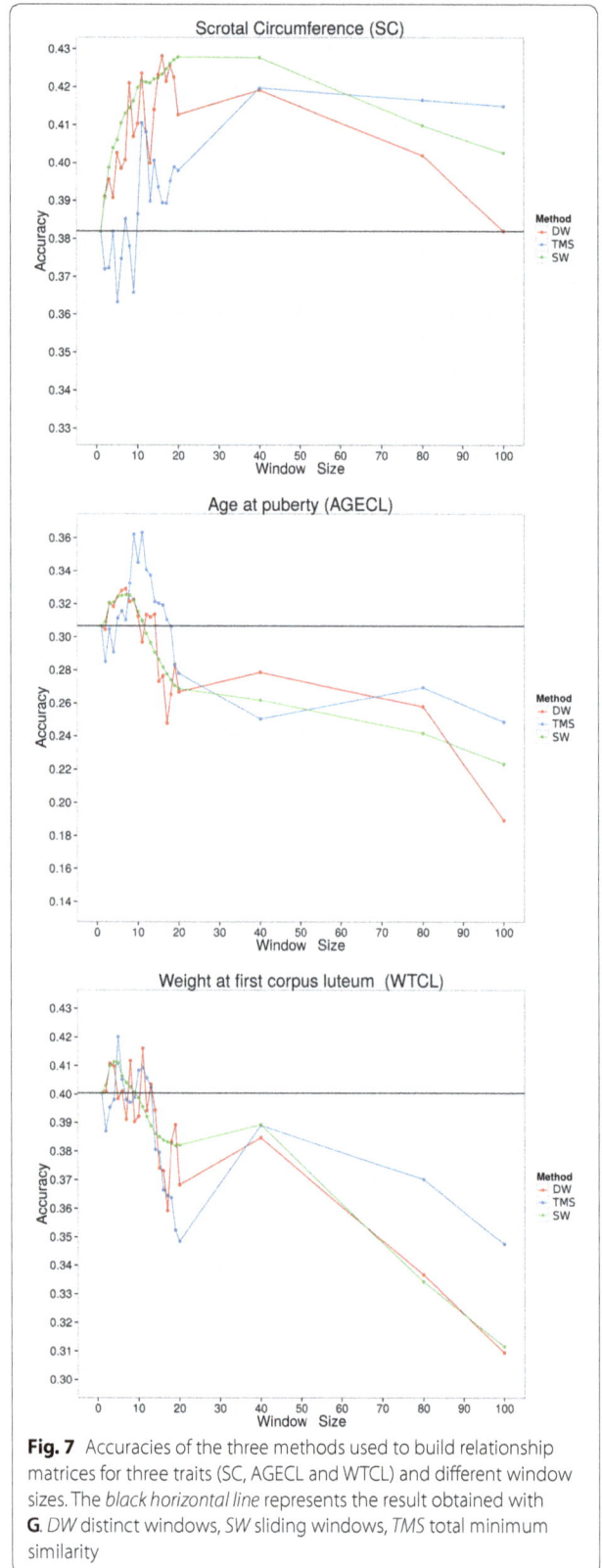

Fig. 7 Accuracies of the three methods used to build relationship matrices for three traits (SC, AGECL and WTCL) and different window sizes. The *black horizontal line* represents the result obtained with **G**. *DW* distinct windows, *SW* sliding windows, *TMS* total minimum similarity

Table 6 Window sizes for the best log-likelihood and accuracies using real datasets (mean ± SD)

Trait	Method	DW-AC	DW-LL	SW-AC	SW-LL	TMS-AC	TMS-LL
Scrotal circumference	Window size	16	8	20	7	40	15
	Accuracy	0.43 ± 0.07	0.42 ± 0.07	0.43 ± 0.09	0.41 ± 0.06	0.42 ± 0.07	0.39 ± 0.06
	G accuracy	0.38 ± 0.07					
Age at puberty	Window size	7	7	7	8	11	11
	Accuracy	0.33 ± 0.17	0.33 ± 0.17	0.33 ± 0.18	0.33 ± 0.19	0.36 ± 0.17	0.36 ± 0.17
	G accuracy	0.31 ± 15					
Weight at first corpus luteum	Window size	11	7	4	2	5	3
	Accuracy	0.42 ± 0.15	0.40 ± 0.15	0.41 ± 0.13	0.40 ± 0.14	0.42 ± 0.11	0.37 ± 0.14
	G accuracy	0.40 ± 14					

DW distinct windows, *SW* sliding windows, *TMS* total minimum similarity

AC: window size for the best accuracy, LL: accuracy and windows size for the best log-likelihood

than one-SNP as a haplotype segment can improve the log-likelihood of genomic selection. For example, the log-likelihood of SC with $H_{1,8}$ was increased by 4.67 in comparison to that with the **G** matrix of VanRaden [3] which was equal to -1330. The optimum haplotype length varied and depended on the methods used for creating relationship matrices, as well as the traits studied, and varied also across datasets. Hence, other methods for haplotype partitioning based on the linkage map or smooth correlation between haplotype segments may improve the prediction accuracy.

Authors' contributions
MHF, JH and BT designed the experiments. MHF wrote the program to build the relationship matrices efficiently and performed the data analysis. MHF derived equations to build the relationship matrix with one-SNP. BT supervised the overall analysis and expanded the equations. MHF drafted the manuscript and BT completed the major revision. All authors read and approved the final manuscript.

Author details
[1] The Centre for Genetic Analysis and Applications, School of Environmental and Rural Science, University of New England, Armidale, Australia. [2] Animal Genetics and Breeding Unit, University of New England, Armidale, Australia. [3] Cobb-Vantress, Siloam Springs, AR, USA. [4] CSIRO Agriculture Flagship, FD McMaster Laboratory Chiswick, Armidale, Australia.

Acknowledgements
The authors wish to thank Yuandan Zhang for providing the edited genotype and phenotype data, and Cooperative Research Centre for Beef Genetic Technologies (BeefCRC) for providing real genotype and phenotype data used in this work. In addition, we want to thank the editors and two anonymous reviewers for their comments and advice.

Competing interests
The authors declare that they have no competing interests.

References
1. Henderson CR. Best linear unbiased estimation and prediction under a selection model. Biometrics. 1975;31:423–47.
2. Wright S. Coefficients of inbreeding and relationship. Am Nat. 1922;56:330–8.
3. VanRaden PM. Efficient methods to compute genomic predictions. J Dairy Sci. 2008;91:4414–23.
4. Meuwissen TH, Odegard J, Andersen-Ranberg I, Grindflek E. On the distance of genetic relationships and the accuracy of genomic prediction in pig breeding. Genet Sel Evol. 2014;46:49.
5. Luan T, Woolliams JA, Odegard J, Dolezal M, Roman-Ponce SI, Bagnato A, et al. The importance of identity-by-state information for the accuracy of genomic selection. Genet Sel Evol. 2012;44:28.
6. Yang J, Benyamin B, McEvoy BP, Gordon S, Henders AK, Nyholt DR, et al. Common SNPs explain a large proportion of the heritability for human height. Nat Genet. 2010;42:565–9.
7. Nejati-Javaremi A, Smith C, Gibson JP. Effect of total allelic relationship on accuracy of evaluation and response to selection. J Anim Sci. 1997;75:1738–45.
8. Schaeffer LR. Strategy for applying genome-wide selection in dairy cattle. J Anim Breed Genet. 2006;123:218–23.
9. Calus MPL. Genomic breeding value prediction: methods and procedures. Animal. 2010;4:157–64.
10. Habier D, Fernando RL, Garrick DJ. Genomic BLUP decoded: a look into the black box of genomic prediction. Genetics. 2013;194:597–607.
11. Calus MPL, Meuwissen THE, de Roos APW, Veerkamp RF. Accuracy of genomic selection using different methods to define haplotypes. Genetics. 2008;178:553–61.
12. Ferdosi MH, Kinghorn BP, van der Werf JH, Gondro C. Detection of recombination events, haplotype reconstruction and imputation of sires using half-sib SNP genotypes. Genet Sel Evol. 2014;46:11.
13. Ferdosi MH, Kinghorn BP, van der Werf JH, Lee SH, Gondro C. hsphase: an R package for pedigree reconstruction, detection of recombination events, phasing and imputation of half-sib family groups. BMC Bioinformatics. 2014;15:172.
14. Villumsen TM, Janss L, Lund MS. The importance of haplotype length and heritability using genomic selection in dairy cattle. J Anim Breed Genet. 2009;126:3–13.
15. Hickey JM, Kinghorn BP, Tier B, Clark SA, van der Werf JH, Gorjanc G. Genomic evaluations using similarity between haplotypes. J Anim Breed Genet. 2013;130:259–69.
16. Johansson AM, Pettersson ME, Siegel PB, Carlborg O. Genome-wide effects of long-term divergent selection. PLoS Genet. 2010;6:e1001188.
17. Johnston DJ, Barwick SA, Corbet NJ, Fordyce G, Holroyd RG, Williams PJ, et al. Genetics of heifer puberty in two tropical beef genotypes in northern Australia and associations with heifer- and steer-production traits. Anim Prod Sci. 2009;49:399–412.
18. Hawken RJ, Zhang YD, Fortes MR, Collis E, Barris WC, Corbet NJ, et al. Genome-wide association studies of female reproduction in tropically adapted beef cattle. J Anim Sci. 2012;90:1398–410.
19. Zhang YD, Johnston DJ, Bolormaa S, Reverter A, Fortes MRS, Tier B. Using male performance to improve genomic selection for female fertility in Brahman cattle. In: Proceedings of the 20th conference of the Association

for the Advancement of Animal Breeding and Genetics, 20–23 Oct 2013, Napier. 2013: p. 224–8.

20. Browning SR, Browning BL. Rapid and accurate haplotype phasing and missing-data inference for whole-genome association studies by use of localized haplotype clustering. Am J Hum Genet. 2007;81:1084–97.

21. Smith SP. Dominance relationship matrix and inversion for an inbred population. Edited by Department of Dairy Science; Ohio State University, Columbus. 1984.

22. Ferdosi MH, Boerner V. A fast method for evaluating opposing homozygosity in large SNP data sets. Livest Sci. 2014;166:35–7.

23. Toro MA, Garcia-Cortes LA, Legarra A. A note on the rationale for estimating genealogical coancestry from molecular markers. Genet Sel Evol. 2011;43:27.

24. Gilmour A, Gogel B, Cullis B, Thompson R. ASReml-R reference manual. Hemel Hempstead: VSN International; 2009.

25. Legarra A. Comparing estimates of genetic variance across different relationship models. Theor Popul Biol. 2016;107:26–30.

26. Tier B, Meyer K, Ferdosi MH. Which genomic relationship matrix. In: Proceedings of the 21st conference of the Association for the Advancement of Animal Breeding and Genetics, 28–30 Sept 2015, Lorne. 2015.

27. Stranden I, Christensen O. Allele coding in genomic evaluation. Genet Sel Evol. 2011;43:25.

28. Forni S, Aguilar I, Misztal I. Different genomic relationship matrices for single-step analysis using phenotypic, pedigree and genomic information. Genet Sel Evol. 2011;43:1.

2

Influence of epistasis on response to genomic selection using complete sequence data

Natalia S. Forneris[1,2]*, Zulma G. Vitezica[3], Andres Legarra[3] and Miguel Pérez-Enciso[1,4,5]* (iD)

Abstract

Background: The effect of epistasis on response to selection is a highly debated topic. Here, we investigated the impact of epistasis on response to sequence-based selection via genomic best linear prediction (GBLUP) in a regime of strong non-symmetrical epistasis under divergent selection, using real Drosophila sequence data. We also explored the possible advantage of including epistasis in the evaluation model and/or of knowing the causal mutations.

Results: Response to selection was almost exclusively due to changes in allele frequency at a few loci with a large effect. Response was highly asymmetric (about four phenotypic standard deviations higher for upward than downward selection) due to the highly skewed site frequency spectrum. Epistasis accentuated this asymmetry and affected response to selection by modulating the additive genetic variance, which was sustained for longer under upward selection whereas it eroded rapidly under downward selection. Response to selection was quite insensitive to the evaluation model, especially under an additive scenario. Nevertheless, including epistasis in the model when there was none eventually led to lower accuracies as selection proceeded. Accounting for epistasis in the model, if it existed, was beneficial but only in the medium term. There was not much gain in response if causal mutations were known, compared to using sequence data, which is likely due to strong linkage disequilibrium, high heritability and availability of phenotypes on candidates.

Conclusions: Epistatic interactions affect the response to genomic selection by modulating the additive genetic variance used for selection. Epistasis releases additive variance that may increase response to selection compared to a pure additive genetic action. Furthermore, genomic evaluation models and, in particular, GBLUP are robust, i.e. adding complexity to the model did not modify substantially the response (for a given architecture).

Background

The relation between the genotype and phenotype of an individual can be extremely complex. Nevertheless, quantitative genetics is able to predict breeding values and response to selection with surprising accuracy based on highly simplified assumptions. Among all potential complexities, epistasis is one of the most widely studied and controversial [1]. In the physiological sense, recently reviewed experimental evidence [2] suggests that functional epistatic gene action is common, and

that additivity can be an emergent property of underlying genetic interaction networks. In the statistical sense, functional epistasis makes the statistical additive effects of alleles depend on the current genetic background, and their contribution to the total genetic variance depends on the allele frequencies [3].

The effect of epistasis on response to selection is a highly debated topic, both from the perspective of short-term response and from an evolutionary perspective [4]. Although some authors claim that its effect on the long term may be substantial, others argue that interaction effects contribute very little to the total genetic variance of a population, and consequently to its short-term response, because most of the variance is additive [5].

*Correspondence: forneris@agro.uba.ar; miguel.perez@uab.es
[1] Centre for Research in Agricultural Genomics (CRAG), CSIC-IRTA-UAB-UB Consortium, 08193 Bellaterra, Barcelona, Spain
Full list of author information is available at the end of the article

When epistasis is present, the level of additive variance can be sustained or even increased compared to that expected under a strict additive model and the genetic gain may be sustained for longer [4]. Moreover, if epistasis is not symmetrical, that is, if alleles with a positive marginal effect interact positively on average (or negatively), the rate of evolution will accelerate (or decelerate) [4].

In the presence of functional epistasis, allele frequency drift and changes in frequency of causal alleles due to selection will cause the response to artificial selection from the same base population to differ among replicate lines as well as within the same line over time [2]. Unless interacting loci are identified and co-introgressed, a favorable allele at one locus may be detrimental in a different genetic background. Although it was shown that an additive model may explain a major part of the genetic variance in different datasets [5], this model does not explicitly capture any kind of interaction which may be present in biochemical pathways that connect gene expression with the ultimate target phenotype. Therefore, statistical models that incorporate interactions between loci have been viewed as potentially beneficial for genomic prediction [2, 6–9].

So far, genomic selection (GS) has been mainly performed with manufactured genotyping arrays based on single nucleotide polymorphisms (SNPs), but the drop in sequencing costs should enable GS programs to routinely use genome sequencing instead of genotyping arrays in the near future. Since causative variants are themselves (potentially) included in the sequence data, the accuracy that can be achieved when sequence data is used instead of SNP arrays is expected to be no longer limited by linkage disequilibrium (LD) between SNPs and causal mutations [10]. Nevertheless, a few empirical studies on breeding schemes and recent simulations agree on the fact that full sequence data will probably not make SNP arrays obsolete for predicting genetic merit [11, 12]; yet, a modest increase (~4%) in genomic best linear unbiased prediction (GBLUP) accuracy, compared to SNP arrays, can be expected under some genetic architectures [13–15]. Using sequence and SNP data, VanRaden [16] obtained an average increase of 2.5% in accuracy in US Holstein cattle compared to using SNP data only.

The aim of this study was to quantify the impact of epistasis on the response to GS in an extreme regime of non-symmetrical epistasis in diploid genomes, as well as to study the possible advantage of including non-additive effects in the prediction model. Since large amounts of epistasis have been reported in Drosophila [2], in this study, we mimicked this genome and used real Drosophila sequence data as starting population.

Methods

We conducted an in silico divergent genomic selection experiment in Drosophila using the sequence based virtual breeding (SBVB) software [15]. SBVB can use real sequence data for founder animals in a simulated population and simulates the genomes and phenotypes of offspring in a very efficient and flexible manner according to specified genetic architectures. Although SBVB is essentially a gene-dropping algorithm, it allows implementing selection by efficiently reading and writing haplotypes (see Additional file 1: Figure S1).

Data

We downloaded the public SNP data from the 205 inbred lines of the *Drosophila melanogaster* Genetic Reference Panel v2 (DGRP Freeze 2.0, http://dgrp2. gnets.ncsu.edu/ [17]). SNPs from chromosome 4 and indels were removed, resulting in 3,954,651 SNPs that were used for analyses. Chromosome 4 in Drosophila is normally ignored in population genetic studies since it is very small, does not recombine and is mostly heterochromatic. Missing values were imputed with Beagle4 [18]. SBVB allows the specification of variable recombination rates along the genome and between sexes as well as sex chromosomes. We used the genetic map from Flybase (www.flybase.org) and allowed for the fact that no recombination occurs in male Drosophila.

Genetic architecture

Four hundred causal SNPs, i.e. quantitative trait nucleotides (QTN) were considered in the analysis. We used the 103 SNPs and their estimated additive effects on the phenotype "chill coma recovery time" which are reported by [19] in their supplementary Table 3. For those SNPs that were identified in both sexes (12 out of 103), we used the average between-sex effect size as QTN effect. Otherwise, the specific sex effect was taken as the additive genetic effect in both sexes. In addition, we used 297 randomly chosen SNPs with their additive effects that were simulated by using an exponential distribution with rate parameter equal to 5. The purpose of this was to generate a larger number of loci with smaller effects than those detected by the association study in [19], since the quantitative trait loci (QTL) that are reported as significant are typically those with the largest effects, leading to a marked upward bias [20]. The sign (− or +) of the simulated additive effect sizes was sampled with equal probability.

Genotypic values were simulated according to two extreme architectures.

Epistatic architecture

All loci ($400 = 103 + 297$) were randomly grouped in 200 epistatic pairs. Second-order epistasis followed the complementary model [5], where the genotypic values for the nine possible two-locus genotype combinations are equal to:

	C_1C_1	C_1C_2	C_2C_2
B_1B_1	z	z	0
B_1B_2	z	z	0
B_2B_2	0	0	0

This architecture assumes equal values for the additive effect of locus B and locus C and complete dominance ($a = d$), with B_1 and C_1 being the dominant alleles. The genotypic value 'z' was computed as the arithmetic mean of the effects of the two original loci, that is, either the effect reported by [19] or that simulated from the exponential distribution. In this architecture, the double mutation B_1C_1 yields the same phenotype as either one alone. The genotypic value of an individual was computed as the sum of the two-locus genotypic value at each of the 200 epistatic pairs. We chose the complementary model because it is simple to interpret and allows for substantial non-additive genetic effects (~25% of the genetic variance is non-additive with a 'U' shape distribution of allele frequencies) [5].

Additive architecture

The genotypic value of an individual was obtained by summing all additive genotypic effects across loci, according to each individual's genotypes (i.e., a, 0 or $-a$ for B_1B_1, B_1B_2 and B_2B_2, respectively). Here, the value 'a' of each locus is equal to the value 'z' used for the corresponding epistatic pair previously described.

Each individual's phenotype was calculated from its genotypic value adding an environmental effect taken from a normal distribution with mean 0 and variance σ_e^2. In both architectures, σ_e^2 was adjusted so that the broad sense heritability (H^2) was 0.5, before selection started. Narrow sense heritability was approximately 0.25 for the epistatic architecture.

Evaluation model

Breeding values were predicted using GBLUP [21]. Briefly, GBLUP uses SNPs to build genomic relationship matrices ($\mathbf{G}$). Four alternative models were used to evaluate individuals.

Additive model using sequence data (A-SEQ)

The evaluation model included a mean plus additive values distributed as $\mathbf{a} \sim N(0, \mathbf{G}\sigma_a^2)$, where $\mathbf{G}$ was obtained using all SNPs. Computation of $\mathbf{G}$ is described below.

Additive model using causal SNPs (A-QTN)

As above except that $\mathbf{G}$ was obtained by using only causal SNPs (QTN).

Full epistatic model using sequence data (E-SEQ)

The evaluation model included a mean, (statistical) additive values, distributed as $\mathbf{a} \sim N(0, \mathbf{G}\sigma_a^2)$, a dominant random deviation $\mathbf{d} \sim N(0, \mathbf{D}\sigma_d^2)$, and an (statistical) epistatic effect distributed as $\mathbf{h} \sim N(0, \mathbf{G} \neq \mathbf{G}\sigma_p^2)$, where $\neq$ denotes the Hadamard product [22, 23]; $\mathbf{G}$ and $\mathbf{D}$ were obtained by using all SNPs. Computation of $\mathbf{D}$ is described below.

Full epistatic model using causal SNPs (E-QTN)

As above except that $\mathbf{G}$ and $\mathbf{D}$ were obtained by using only the QTN.

For a given set of markers (all or only causal SNPs), $\mathbf{G}$ was obtained from $\mathbf{MM}' / \sum_{j=1}^{k} 2p_jq_j$, where the elements of the $\mathbf{m}$ vectors for each individual are equal to $-2p_j$, $1 - 2p_j$, and $2 - 2p_j$ for genotypes $B_{1j}B_{1j}$, $B_{1j}B_{2j}$ and $B_{2j}B_{2j}$, respectively [21], p_j is the frequency of allele B_{1j} for the genotyped individuals of the population, $q_j = 1 - p_j$, and k is the number of SNPs. $\mathbf{D}$ was obtained as in [7]:

$$\mathbf{D} = \frac{\mathbf{M_d}\mathbf{M_d}'}{\sum_{j=1}^{k} 4p_j^2q_j^2},$$

where the elements of the $\mathbf{m}_d$ vectors for each individual are equal to $-2p_j^2$, $2p_jq_j$ and $-2q_j^2$ for genotypes $B_{1j}B_{1j}$, $B_{1j}B_{2j}$ and $B_{2j}B_{2j}$, respectively.

Breeding values were predicted with the Bayesian generalized linear regression (BGLR) package [24]. For prediction purposes, variance components were assumed unknown and, thus, estimated simultaneously (marginalized). BGLR implements various Bayesian regression models that were developed for genomic applications, including the GBLUP model. An eigenvalue decomposition of the covariance matrices ($\mathbf{G}$ and $\mathbf{D}$) was used, given its good convergence properties [25]. Default prior parameters and 10,000 iterations plus 2000 burn-in cycles were used in the Markov chain Monte Carlo (MCMC) method, resulting in 100 to 150 effective iterations [26], depending on the parameter and replicate. To verify whether 10,000 iterations were sufficient for our purposes, we compared the prediction of breeding values obtained in chains with 10k and 200k iterations, which was on average equal to 0.98 so we used 10k for computational speed.

The linear predictor included an intercept plus a linear regression on additive effects. We also included both a linear regression on dominant effects and a linear regression on epistatic effects for both E-SEQ and E-QTN evaluation models. Gaussian prior densities were used for all the linear regressions. The residual variance prior was assigned a scaled-inverse Chi-square density and the intercept is assigned a flat prior by default. For the other variances, we also used scaled-inverse Chi-squared densities with hyperparameters set by using the default rules in BGLR (see appendix of [24] at http://www.genetics.org/content/suppl/2014/07/09/genetics.114.164442.DC1/164442SI.pdf). In short, the number of degrees of freedom was 5 (which provides a rather uninformative prior) and scale parameter S such that the R^2 of the model is matched. BGLR was run at each generation of selection, and predictions were obtained using all the phenotypes and genotypes of all the animals of the preceding generations including the present generation.

Selection scheme

We generated the base population starting with the 205 sequenced lines of DGRP2 and performing 10 generations of random mating to decrease LD and to generate heterozygous individuals, since parents were homozygous at most sites. The size of the generated base population was N = 500. Selection intensity was 10% in both sexes. At each generation, 25 males and 25 dams were chosen based on genomic breeding values that were predicted using different genomic models (A-SEQ, E-SEQ, A-QTN and E-QTN) and randomly mated; each mating produced 20 offspring with an equal sex ratio. At each generation, breeding values were predicted using all molecular and phenotypic information up to that generation. This scheme was continued for seven discrete generations. Both upward and downward selections were performed. We ran 10 replicates of each of the 16 experiments (two genetic architectures × four evaluation models × two directions of selection). In each replicate, the same set of SNP effects was used, but different base populations with different haplotype structures were generated, although all were initiated with the same real Drosophila data.

Response to selection, prediction accuracy and additive variance over generations

We investigated the influence of the genetic architecture, the direction of selection and the evaluation model on response to selection, genomic prediction accuracy and the narrow sense heritability over generations. For the total cumulative response to selection, we computed the phenotypic mean per generation (N = 500), averaged across replicates and expressed in standard deviation (SD) units of the base population phenotypic distribution. Prediction accuracy was computed as the Pearson correlation between true and predicted breeding values. Under random mating, the breeding value of an individual is defined as twice the phenotypic mean of its offspring since it

deviated from the phenotypic population mean. Calculating true breeding values is straightforward under an additive architecture, since they are equal to the simulated genotypic values. However, this equality does not hold for the epistatic architecture. From the simulations, we have the true total genetic values but not the breeding values. In this case, we used the original definition and empirically estimated the 'true' breeding value of an individual that generates 1000 offspring, which result from mating the individual to randomly chosen individuals from the same generation. Additive genetic variance was computed as the variance of 'true' breeding values among the individuals in the generation of interest.

Linkage disequilibrium and inbreeding

Long-range LD between causal SNPs was assessed at the beginning (t_0) and at the end (t_7) of each selection experiment and replicate. A pair-wise r^2 estimation implemented in PLINK [27], defined as the squared correlation coefficient of genotypes at two loci, was used to measure LD between all causal SNP pairs within a chromosome. The curve of the decay of r^2 with physical distance was fitted for each experiment by nonlinear regression, using Hill and Weir's [28] expectation of r^2. Genomic inbreeding coefficient estimates (F_h) were obtained for each individual using PLINK's—het function [27], which is based on excess SNP homozygosity, as $F_{h_i} = (O_i - E)/(k - E)$, where O_i is the number of observed homozygous genotype counts for individual i, $E = \sum_{j=1}^{k} 1 - 2p_j(1 - p_j)$ is the number of homozygous genotype counts expected by chance for the base population, p_j is the frequency of B_1 in the base population, and k is the number of SNPs.

Results

The distributions of QTN allele frequencies and additive effect sizes in the base population (t_0) are in Fig. 1. The distribution of absolute effect sizes is bimodal, the result of the empirically identified QTL [19] plus the effects simulated following an exponential distribution (Fig. 1a). The site frequency spectra of the alleles that increase the trait value was U-shaped, although the frequency of alleles that decreased the mean was higher (Fig. 1b). The correlation between effect size and frequency was negative, $\rho = -0.42$ (Fig. 1c). Note that a negative correlation between effect and allele frequency is expected under some directional or stabilizing selection [17, 29], so our data can mimic a trait that has been under continuous selection.

Response to selection

Phenotypic means in SD units along generations and for all scenarios considered are in Fig. 2. Response to selection was clearly asymmetric: for the additive

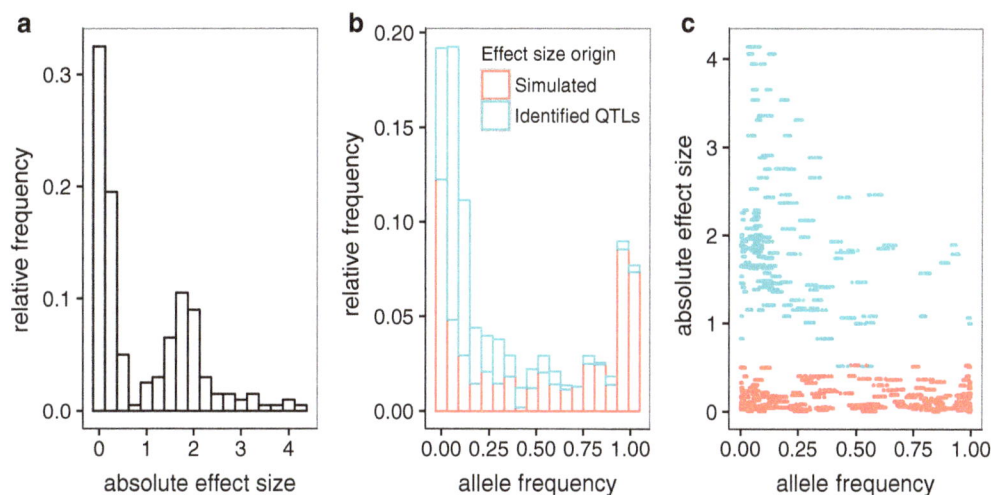

Fig. 1 a Distribution of absolute additive effects used in the simulations. **b** QTN site frequency spectrum of the allele that increases the trait value in the base population. Sites for which effect size was simulated following an exponential distribution are shown in *red* and those identified in Ober et al. [19] are shown in *blue*. **c** Absolute effect size versus allele frequency of the allele that increases the value of the trait

architecture, the upward response was equal to ~8 SD but the downward response was barely 2 SD; epistasis resulted in an even larger asymmetry (10 vs. 1 SD). As discussed below, response was almost exclusively due to changes in allele frequency at loci with a large effect, which were initially at low frequency (Fig. 1b). The frequency of alleles increasing the trait value was 0.17 on average. The complementary model of epistasis actually shrinks the absolute genotypic value of the double homozygotes while it gives the double heterozygote the same advantage as the double dominant homozygote, compared to an additive architecture. In our experiment, epistasis was not symmetrical overall because, for all SNPs, the alternative allele in the Drosophila genome was assigned to be the recessive allele, which happened to be the one that increased the trait value in most cases. Thus, under upward selection, alleles with a positive effect systematically tended to couple (in other words, for double mutants to become favorably selected), which accelerated the rate of response. In contrast, under downward selection, alleles with a positive effect tended to interact negatively on average, which decelerated the rate of response.

The second relevant result is that response to selection depended mainly on the underlying genetic architecture, and not so much on the statistical model used to perform the evaluation (to predict breeding values). For instance, upward response was almost two SD larger in the epistasis architecture than in the additive scenario, irrespective of the evaluation model. Remarkably, knowing the QTN would not have made a big difference. For the additive architecture in particular (Fig. 2a, c), response to

selection was rather insensitive to the evaluation model used, even if QTN were known. For instance, upward response using A-QTN was only ~6% larger than that obtained with A-SEQ, E-QTN or E-SEQ. In the presence of epistasis, the situation was somewhat more complex. In this case, knowing the causal QTN improved response to downward selection by about ~24% (A-QTN vs. A-SEQ) or ~37% (E-QTN vs. E-SEQ). In upward selection (Fig. 2b), the advantage of knowing the QTN was only substantial in the short term (~20% higher for t < 5) but was only ~2% onwards. The E-SEQ strategy resulted in an increase in response compared to A-SEQ (on average ~10%) in the medium term and comparable to that with the E-QTN model. Overall, results in Fig. 2b suggest that accounting for epistasis in the evaluation model may have a positive effect if epistasis exists, but mainly in the medium term. Otherwise, in an additive scenario, response is quite insensitive to the evaluation model used to predict breeding values and even to the knowledge of QTN positions.

Evolution of additive genetic variance and of allele frequencies

The evolution of the additive genetic variance (i.e., the variance of the 'true' breeding values, see "Methods") over generations was examined for each experiment (Fig. 3). Downward selection rapidly eroded additive variance. In contrast, additive variance increased in the first generations of upward selection because alleles with a large effect (Fig. 1b), which were initially at low frequencies, were favored by selection, thus the minor allele frequency increased at those sites. Subsequently, upward

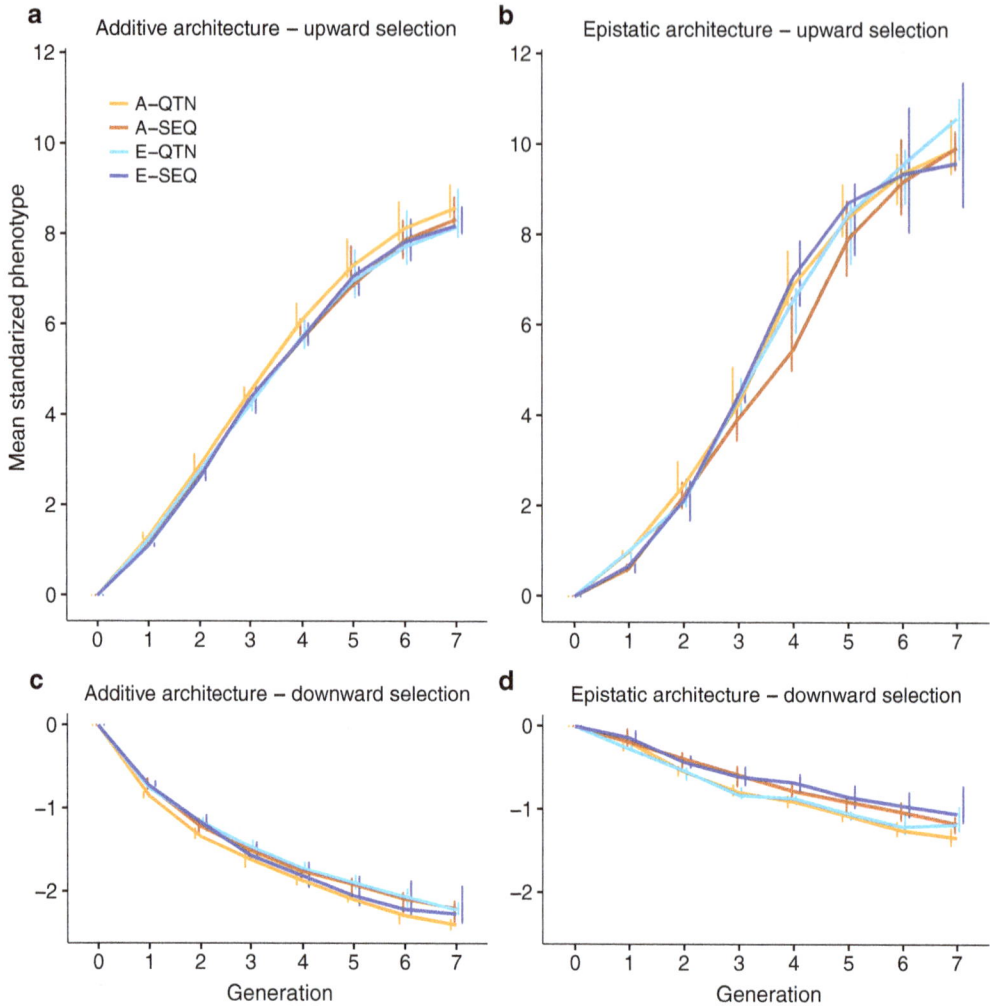

Fig. 2 Mode and 75th percentile (*bars*) of the phenotypic mean per generation (N = 500 individuals per generation, 10 replicates). A-QTN: additive model using causal SNPs; A-SEQ: additive model using sequence data; E-QTN: full epistatic model using causal SNPs; E-SEQ: full epistatic model using sequence data

selection also exhibited a marked decay in the additive genetic variance. This effect was specially marked in the additive architecture. Importantly, epistasis sustained additive genetic variance for longer, especially if QTN were known and epistasis was included in the model (E-QTN).

For each QTN, we computed the difference in allele frequency between the last and first generation; the distribution of these changes in allele frequencies across the genome is in Fig. 4. Clearly, the evaluation model used for prediction did not have a substantial impact on the change in allele frequency. Yet, the direction of selection and architecture did. First, for downward selection, patterns between additive and epistatic architectures were small. Average changes in allele

frequencies were -0.05 and -0.04 for additivity and epistasis, respectively. Again, this is due to the extreme allele frequency that we observed for loci with a large effect in the base population. Second, the pattern of changes in frequency for upward selection was clearly distinct from that for downward selection (Fig. 4a, b). Here, a small percentage of QTN (~7 and 4% for additivity and epistasis, respectively) went to fixation or near fixation starting from very low frequencies, while the frequencies of the other QTN did not change much. In the light of the results in, e.g., Fig. 2a, b, it seems that most of the response was due to these very few loci with a large effect. Note that fewer loci were affected by large changes in frequency with epistasis than with additivity, which again is coherent with the larger decrease in

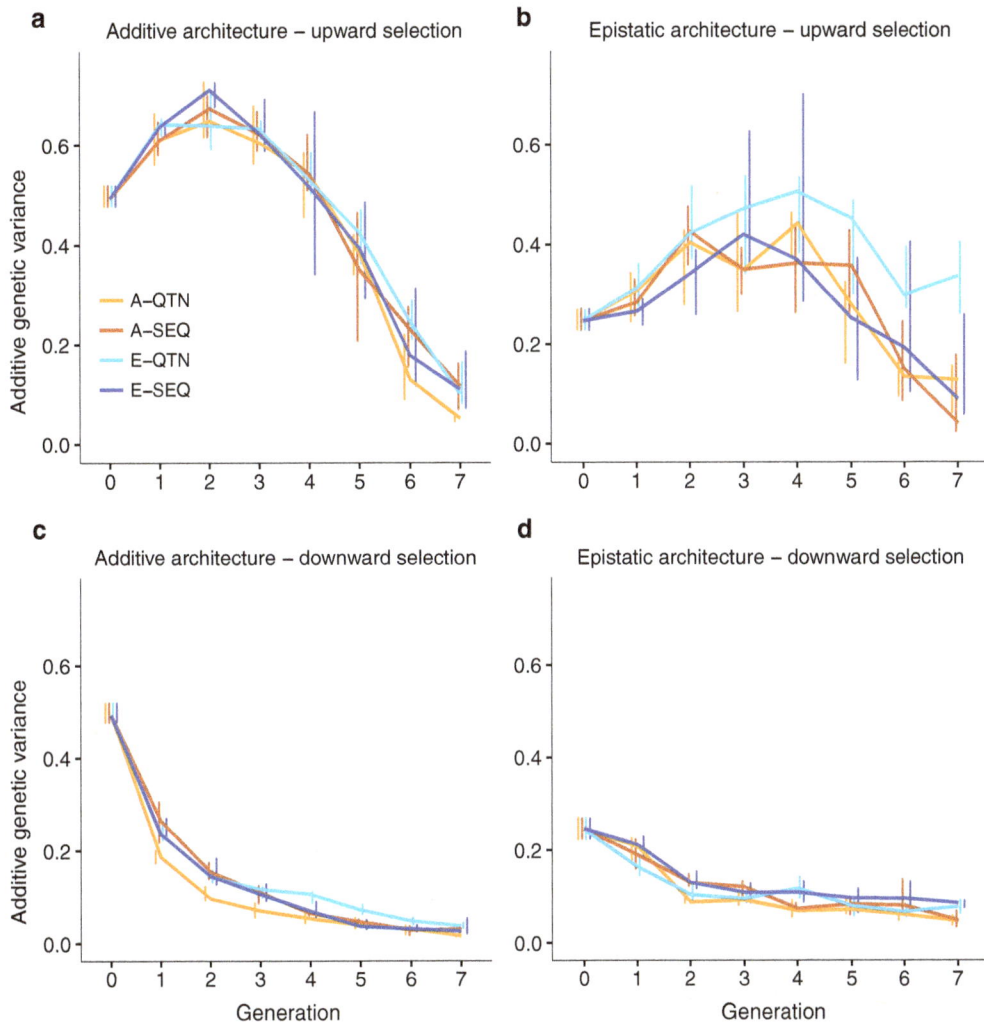

Fig. 3 Mode and 75th percentile (*bars*) of the proportion of phenotypic variance due to additive variance (or variance of true breeding values) per generation. A-QTN: additive model using causal SNPs; A-SEQ: additive model using sequence data; E-QTN: full epistatic model using causal SNPs; E-SEQ: full epistatic model using sequence data

additive variance observed in the additive than in the epistatic architecture (Fig. 4a, b).

To study this furthermore, we computed the expected contribution in the absence of disequilibrium for each locus j (i.e. $2p_j q_j \alpha_j^2$, α being the substitution effect [30]). Cumulative distributions both in the base population and in the last generation are in Figure S2 [see Additional file 1: Figure S2]. For all criteria and genetic architectures, selection resulted in fewer loci explaining a given percentage of variance (although note that the absolute value of additive variance decreased as selection proceeded, Fig. 3). About 110 (28%) loci explained 90% of the additive variance in the base population versus 20 to 40 (5 to 10%) loci in the last generation.

Prediction accuracy

In contrast to response to selection, prediction accuracy was affected by both genetic architecture and selection method. For the additive architecture and upward selection (Fig. 5a), accuracies obtained with the A-QTN model were high and remained relatively constant over generations; they were only slightly higher than with the A-SEQ model (ca. 4%). If epistasis was absent but accommodated in the model (E-QTN and E-SEQ), accuracies were initially comparable to those of the additive models but decreased eventually. Thus, in the long term, including epistasis in the model when there is none, affected negatively the GS performance, and this was observed in both upward and downward selection (Fig. 5a, c). In contrast, there were marked

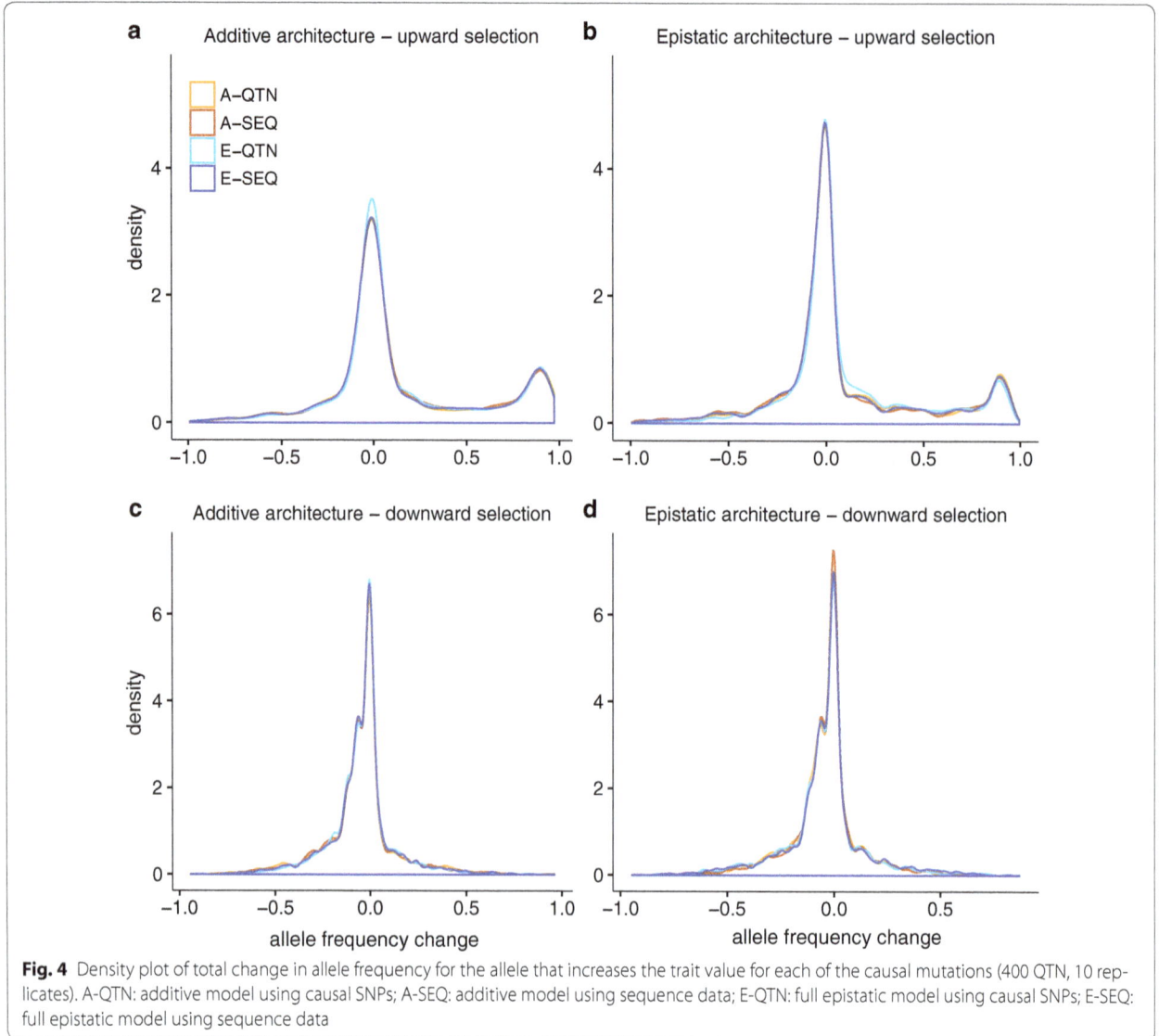

Fig. 4 Density plot of total change in allele frequency for the allele that increases the trait value for each of the causal mutations (400 QTN, 10 replicates). A-QTN: additive model using causal SNPs; A-SEQ: additive model using sequence data; E-QTN: full epistatic model using causal SNPs; E-SEQ: full epistatic model using sequence data

differences between downward and upward selection accuracies with epistasis (Fig. 5b, d). First, for downward selection, accuracies remained stable with A-QTN and A-SEQ but steadily declined when epistasis was included in the evaluation model. Second, results were clearly non-linear for upward selection. In this case, all accuracies decreased in the long term although there was an advantage of including epistasis in the model in the short term.

Linkage disequilibrium and inbreeding

We computed decay of LD (r^2) with physical distance at the beginning (t_0) and at the end (t_7) of each selection experiment. The level of long-range LD in the base population was very low (Fig. 6, black line), with average r^2 of 0.06 and 0.00 for SNPs within 1 and 5 Mb, respectively.

LD increased significantly after selection. Direction of selection seems to be the main factor that affects the extent of LD: upward selection experiments had an average r^2 of ~0.27 between QTN pairs up to 10 Mb apart, whereas average LD (in r^2) was ~0.16 for downward selection. This could be due to the fact that most favorable alleles in downward selection were nearly fixed in the initial generation. Genetic architecture did not affect the extent of LD strongly; yet, under the epistatic architecture, knowing the causal mutations and including epistasis in the prediction model hindered the buildup of LD (r^2 was ~0.21 for E-QTN vs. ~0.28 for the rest of the models).

There were no substantial differences between genetic architectures in terms of genomic inbreeding over

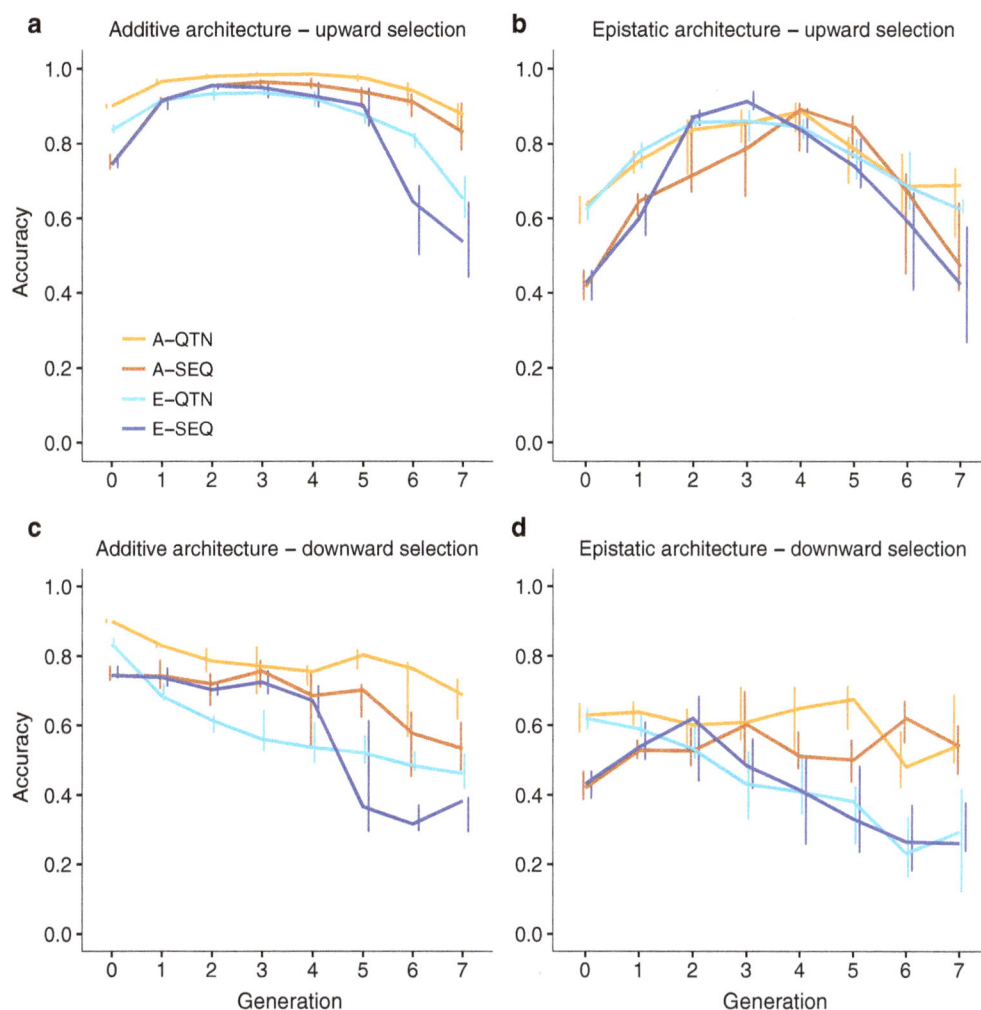

Fig. 5 Mode and 75th percentile (*bars*) of the accuracy of prediction of the selection candidates per generation. A-QTN: additive model using causal SNPs; A-SEQ: additive model using sequence data; E-QTN: full epistatic model using causal SNPs; E-SEQ: full epistatic model using sequence data

generations, yet the pattern of inbreeding differed with the direction of selection (Fig. 7). After seven generations of upward selection, the final genomic inbreeding coefficient was ~70% on average, compared to ~55% from downward selection experiments. Moreover, an initial reduction of the genomic inbreeding coefficient was observed in upward selection, which is likely related to an initial increase in minor allele frequencies. Knowing the causal mutations can favorably affect genomic inbreeding (on average, 54% with E-QTN versus ~65% for the other evaluation models). Considering that GS is expected to reduce the rates of inbreeding per generation, compared with traditional BLUP, because it provides additional information on Mendelian sampling terms of selection candidates, it is not surprising that the lowest inbreeding

corresponded to the model that better reflects the architecture of the trait.

Discussion

This study examined the impact of epistasis on the short-to-medium-term response to GS in an extreme scenario of non-symmetrical epistasis under divergent selection, as well as the possible advantage of including non-additive effects in the prediction model. To date, there is no previous work on the evolution of selection response to full-sequence-based GS over generations under epistasis.

As in any simulation study of this kind, our study aimed at computational feasibility and made some guesses on a likely genetic architecture. First, although GS is certainly most relevant for mammalian and avian genomes,

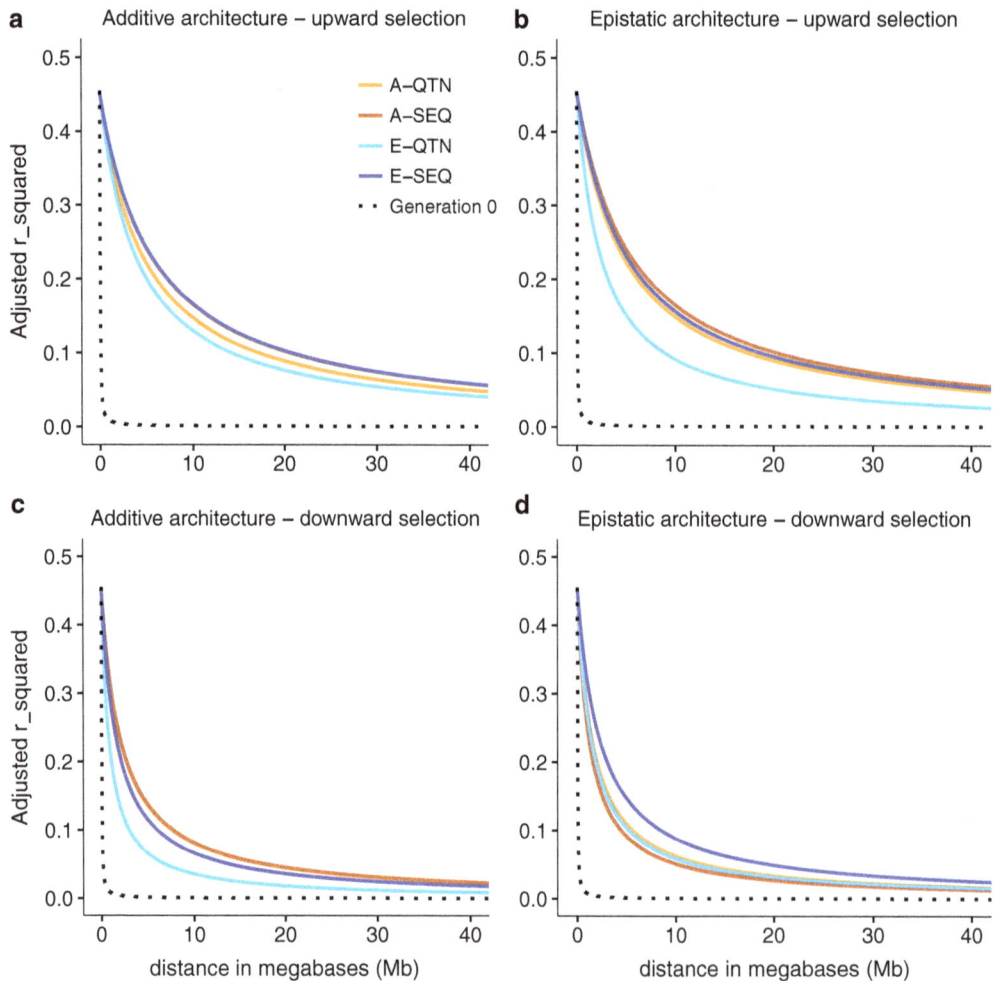

Fig. 6 Decay of linkage disequilibrium with physical distance after seven generations of selection. The *dotted line* corresponds to the linkage disequilibrium in the base population. A-QTN: additive model using causal SNPs; A-SEQ: additive model using sequence data; E-QTN: full epistatic model using causal SNPs; E-SEQ: full epistatic model using sequence data

simulating these large genomes would have added a large computational burden. We chose the Drosophila genome instead because it is about 15 times smaller than those of mammals and because this species has been traditionally used extensively in numerous selection experiments, their chromosome genetic lengths are comparable to those in mammals, and whole-genome sequences are available for a reasonably sized population [17]. Furthermore, starting with a high level of nucleotide variability and a low LD, the intensive selection process that we simulated induced strong LD (Figs. 6, 7), which mimics that of domestic species. Second, we used an extreme epistatic architecture, where all 200 QTN pairs showed interaction. This was done to set an upper limit on the influence of epistasis, for which real effects in comparison to complete additivity are likely weaker than those found here.

Third, we used a mixture of estimated and small simulated gene effects. This was done to compensate for the fact that most estimated effects are likely overestimated and result in a negative correlation between effect and frequency, which is expected in traits under directional selection [29].

We used the traditional encoding {0, 1, 2}, where 0 and 2 are for the homozygous genotypes and 1 is for the heterozygous genotype, together with a Hadamard product between **G** to account for epistasis. It should be recalled that epistasis analyses are coding-dependent [31], because the multiplications of different encodings differ. Recently, we showed that this coding and Hadamard product are equivalent to a model with an explicit effect for interactions [32], and that this model is orthogonal under Hardy–Weinberg conditions. Nevertheless,

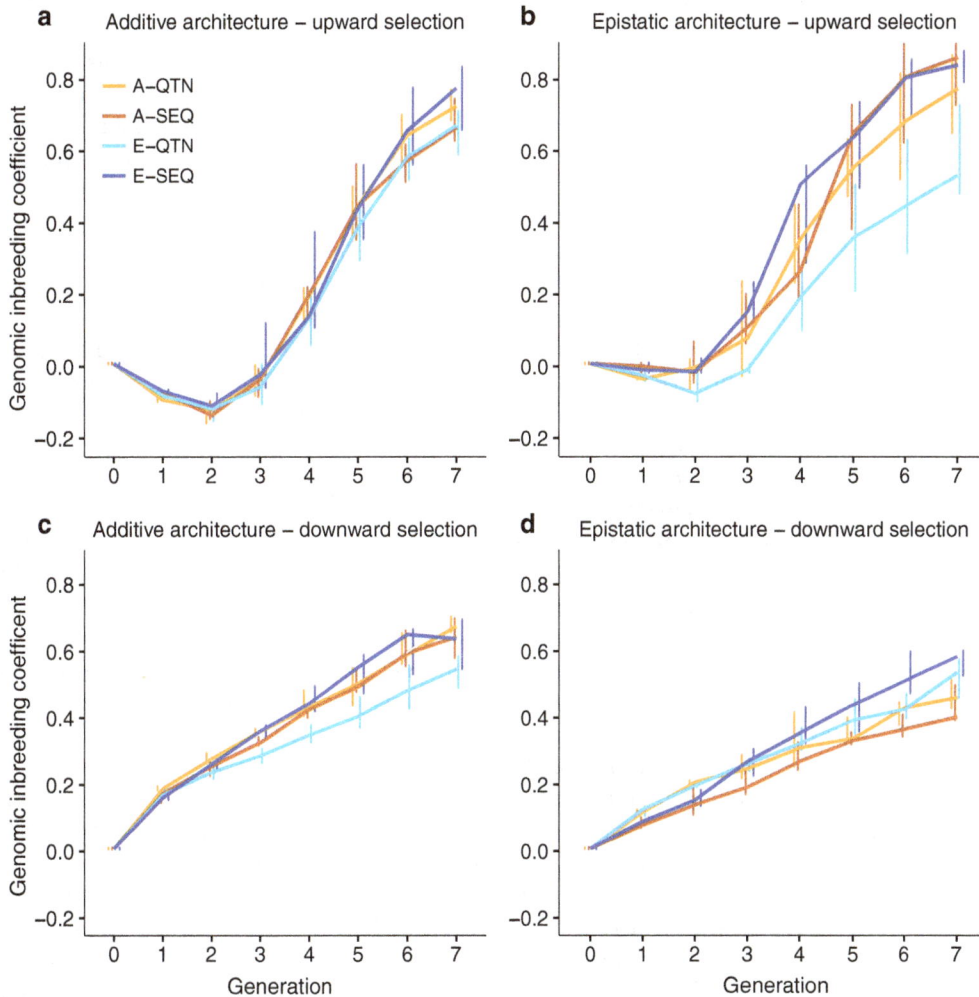

Fig. 7 Genomic inbreeding coefficient over generations

because we did not build a matrix of dominance-by-dominance or additive-by-dominance epistasis, only part of the dominance is accounted for. Theoretically, more realistic modeling options exist than the one used here but they would require more variance components to be estimated and, in the light of our results, are probably similar to the model used in our study.

In our study, the response to selection was highly asymmetric (Fig. 2) due to the skewed distribution of allele frequencies in the base population. Alleles that decreased the mean were at high frequency in the base population, thus dampening the efficacy of downwards artificial selection [33]. This asymmetric pattern is usually observed in traits that are closely associated with fitness, as is chill coma recovery time [34–37]. The asymmetry was accentuated under epistasis. This can be explained as follows. In the complementary epistatic

architecture, our simulated epistasis shrinks the absolute genotypic value of the double homozygotes, compared to an additive architecture (see "Epistatic architecture" in "Methods"). Because the dominant allele happened to be, in most loci, the one that decreased the value of the trait, epistasis is not symmetrical. Under upward selection, the dominant allele will be deleterious and, thus, the double dominant genotype will usually have a fitter phenotype than expected from pure additivity, protecting this positive interaction against the negative effects and causing a less severe fitness drop [38]. In contrast, under downward selection, the dominant allele will usually be the beneficial one and thus the double dominant will usually have a less fit phenotype than expected from an additive action, causing smaller than expected increments in the trait mean. As a result, the rate of response was systematically accelerated (decelerated) under upward (downward) selection, compared to an additive architecture, as predicted by Paixão and Barton [4]. In fact, one of the proposed evolutionary advantages for the presence of epistasis is past selection for resilience to environmental or genetic perturbations [4, 39].

In addition, epistatic interactions affect the response to selection mainly by modulating the additive genetic variance. Here, epistasis sustained additive genetic variance in GS for longer than with pure additivity (Fig. 3b, d). According to Paixão and Barton [4] and Mackay [2], alleles that are initially deleterious or near-neutral may acquire favorable effects as the genetic background changes, "converting" epistatic variance into additive variance, and thus prolonging the response to selection (Fig. 2b, d). In our simulations, downward selection rapidly eroded the additive genetic variance (Fig. 3c, d). This is likely due to the small size of the effective population (although population size was constant and equal to 500 individuals, only 10% in each sex was used to breed the next generation), shift in frequencies of alleles with a large effect, and to the Bulmer effect [40], which induces negative LD (i.e., negative correlation between frequencies of the beneficial alleles) [41–43] and, as a consequence, accuracies also decay rapidly (Fig. 5c, d).

GBLUP genomic evaluation models were robust since adding complexity to the model did not modify substantially the genetic gain for a given architecture. For instance, response was rather insensitive to the evaluation model used in an additive scenario (Fig. 2a, c). Including epistasis when there was none led to similar genomic predictions because the estimated non-additive variance components from GBLUP in this scenario were very close to zero (results not shown). If epistasis existed, including it into the evaluation model had a positive effect, but only in the medium term. This was clearly observed with the E-SEQ model, which achieved response values comparable to those when the causal mutations are known under upward selection (Fig. 2b). Under downward selection and epistasis (Fig. 2d), no substantial effect on the response was observed by adding complexity to the evaluation model, likely because the variation on which selection can work is already very small (Fig. 3d).

Adding complexity to the model had a noticeable impact on prediction accuracy (Fig. 5). For instance, including epistasis in the model when there is none eventually led to lower accuracies as selection proceeded (Fig. 5a, c). Yet, accounting for epistasis when it did exist, did not always lead to higher accuracies; in fact, for upward selection, including epistasis in the model was advantageous only in the medium term when using sequence data compared to using a strictly additive model (Figs. 2b, 5b). This is not too surprising, since most

of the genetic variance is additive even in the presence of epistasis, unless allele frequencies are intermediate [2, 5]. (In our scenario of extreme epistasis, narrow sense heritability in the base population was about 0.25, i.e., half the total genetic variance). Accuracies even declined under downward selection when epistasis existed and was included in the evaluation model (Fig. 5d), whereas strictly additive models exhibited relatively constant accuracy values. This may happen because there is not enough power to capture epistatic variance due to most favorable variants being already close to fixation in the base population. Although some studies predict gains in accuracy when accounting for epistasis in GS [19], our results show that this advantage can be eroded as selection proceeds, likely because the substitution effects change over time.

This study also examined the performance of sequence data compared to when the causal mutations are known (i.e., when only the causal SNPs were used to build the genomic relationship matrices). Surprisingly, there was not much gain in response when causal mutations were known (Fig. 2), except at the beginning of the experiment (Fig. 5). With epistasis, the E-QTN model may be beneficial in the long term, because it resulted in significantly lower levels of LD (Fig. 6b) and inbreeding (Fig. 7b) and thus it may sustain additive variance for longer (Fig. 3b). Nevertheless, the small advantage of knowing the causal loci contrasts with previous analyses [19, 44]. In [19], genomic prediction improved when adding SNPs or SNP combinations that were selected based on genome-wide association analyses. In [20], the improvement in prediction accuracy with only-QTN models was larger under an epistatic scenario due to a lower proportion of the total genetic variance being additive in this architecture. However, in our simulations, the initial advantage of knowing the causal mutations on prediction accuracy did not persist. First, this small advantage of knowing the causal mutations may be attributed to the high heritability of the trait, large LD as a result of recent selection and the small number of chromosomes (three in Drosophila), and to the fact that phenotypic information on candidates was available for genomic prediction at each generation.

Conclusions

This study shows that epistatic interactions affect the response to genomic selection by modulating the additive genetic variance used for selection. Depending on the kind of epistatic action and on the distribution of allele frequencies, epistasis releases additive variance that may increase selection response compared to a pure additive genetic action. Furthermore, genomic evaluation models and, in particular, GBLUP are robust, since adding complexity to the model did not substantially modify the response (for a given architecture). Nevertheless, the evaluation model did affect accuracy but in a nonlinear way, which suggests that complex architectures may require updating the evaluation model as selection proceeds. In practice, of course, the problem would be to have sufficient data to estimate parameters reliably. Finally, even if knowing the causal mutation can in principle boost accuracy, its impact on response to selection can be less impressive than anticipated, especially if the generated LD is very large, the effective population size is small and phenotypes are available on the candidates.

Additional file

> **Additional file 1: Figure S1.** Implementation of selection in SBVB. **(a)** In a given generation t, files containing pedigree, phenotypic (Y) and molecular information of individuals up to generation t are available, and these are used to perform genomic evaluation of candidates via an external program. **(b)** As a result, estimated breeding values (GEBV) are obtained, and the user selects the sires and dams that will be the parents of generation $t + 1$ (highlighted lines in pedigree file). **(c)** Next, the user needs to expand the pedigree file and generate the pedigree of the next generation, this simply requires adding ids to the pedigree file of as many offspring as desired per each selected couple. In the next round of SBVB, the program will simulate the phenotypes and genotypes (dotted lines) of the new offspring using the parents' genotypes and the architecture of the trait. **Figure S2.** Cumulative fraction of additive variance across loci. Theoretical individual loci contribution to additive variance, $2p(1 - p)$ a^2, in the base population (black dashed line) and in the last generation for each of the selection criteria and genetic architecture (colored lines). A-QTN: additive model using causal SNPs; A-SEQ: additive model using sequence data; E-QTN: full epistatic model using causal SNPs; E-SEQ: full epistatic model using sequence data.

Authors' contributions
MPE, ZV and AL conceived and designed the research. MPE and NF developed the software. NF carried out the simulations. NF drafted the manuscript with the help from the rest of authors. All authors read and approved the final manuscript.

Author details
[1] Centre for Research in Agricultural Genomics (CRAG), CSIC-IRTA-UAB-UB Consortium, 08193 Bellaterra, Barcelona, Spain. [2] Departamento de Producción Animal, Facultad de Agronomía, Universidad de Buenos Aires, C1417DSE Buenos Aires, Argentina. [3] GenPhySE, INRA, INPT, ENVT, Université de Toulouse, 31326 Castanet-Tolosan, France. [4] Departament de Ciència Animal i dels Aliments, Universitat Autònoma de Barcelona, 08193 Bellaterra, Barcelona, Spain. [5] ICREA, Passeig de Lluís Companys 23, 08010 Barcelona, Spain.

Acknowledgements
We thank the Genotoul bioinformatics platform Toulouse Midi-Pyrenees for providing help and computing resources.

Competing interests
The authors declare that they have no competing interests.

Funding
This work was funded by AGL2013-41834-R and AGL2016-78709-R grants (MINECO, Spain) to MPE, INRA SELGEN metaprogram - project 'EpiSel' to ZV and AL. The project was partly supported by Toulouse Midi-Pyrénées bioinformatics platform. We also acknowledge the support of MINECO for the 'Centro de Excelencia Severo Ochoa 2016-2019' award SEV-2015-0533 and of the CERCA program (Generalitat de Catalunya, Spain) to CRAG.

References
1. Mäki-Tanila A, Hill WG. Influence of gene interaction on complex trait variation with multilocus models. Genetics. 2014;198:355–67.
2. Mackay TFC. Epistasis and quantitative traits: using model organisms to study gene-gene interactions. Nat Rev Genet. 2014;15:22–33.
3. Cheverud JM, Routman EJ. Epistasis and its contribution to genetic variance components. Genetics. 1995;139:1455–61.
4. Paixão T, Barton NH. The effect of gene interactions on the long-term response to selection. Proc Natl Acad Sci USA. 2016;113:4422–7.
5. Hill WG, Goddard ME, Visscher PM. Data and theory point to mainly additive genetic variance for complex traits. PLoS Genet. 2008;4:e1000008.
6. Hu Z, Li Y, Song X, Han Y, Cai X, Xu S, et al. Genomic value prediction for quantitative traits under the epistatic model. BMC Genet. 2011;12:15.
7. Vitezica ZG, Varona L, Legarra A. On the additive and dominant variance and covariance of individuals within the genomic selection scope. Genetics. 2013;195:1223–30.
8. Wang D, Salah El-Basyoni I, Stephen Baenziger P, Crossa J, Eskridge KM, Dweikat I. Prediction of genetic values of quantitative traits with epistatic effects in plant breeding populations. Heredity (Edinb). 2012;109:313–9.
9. Wittenburg D, Melzer N, Reinsch N. Including non-additive genetic effects in Bayesian methods for the prediction of genetic values based on genome-wide markers. BMC Genet. 2011;12:74.
10. Meuwissen T, Goddard M. Accurate prediction of genetic values for complex traits by whole-genome resequencing. Genetics. 2010;185:623–31.
11. van Binsbergen R, Calus MPL, Bink MCAM, van Eeuwijk FA, Schrooten C, Veerkamp RF. Genomic prediction using imputed whole-genome sequence data in Holstein Friesian cattle. Genet Sel Evol. 2015;47:71.
12. Calus MPL, Bouwman AC, Schrooten C, Veerkamp RF. Efficient genomic prediction based on whole-genome sequence data using split-and-merge Bayesian variable selection. Genet Sel Evol. 2016;48:49.
13. Hayes BJ, MacLeod I, Daetwyler HD, Bowman PJ, Chamberlain AJ, Vander Jagt CJ. Genomic prediction from whole genome sequence in livestock: the 1000 bull genomes project. In: Proceedings of the 10th world congress on genetics applied to livestock production, 17–22 August 2014. Vancouver; 2014.
14. Heidaritabar M, Calus MPL, Megens HJ, Vereijken A, Groenen MAM, Bastiaansen JWM. Accuracy of genomic prediction using imputed whole-genome sequence data in white layers. J Anim Breed Genet. 2016;133:167–79.
15. Pérez-Enciso M, Forneris N, de los Campos G, Legarra A. Evaluating sequence-based genomic prediction with an efficient new simulator. Genetics. 2016;205:939–53.
16. VanRaden PM, Tooker ME, O'Connell JR, Cole JB, Bickhart DM, Brent S. Selecting sequence variants to improve genomic predictions for dairy cattle. Genet Sel Evol. 2017;49:32.
17. Mackay TFC, Richards S, Stone EA, Barbadilla A, Ayroles JF, Zhu D, et al. The *Drosophila melanogaster* genetic reference panel. Nature. 2012;482:173–8.
18. Browning BL, Browning SR. Improving the accuracy and efficiency of identity-by-descent detection in population data. Genetics. 2013;194:459–71.
19. Ober U, Huang W, Magwire M, Schlather M, Simianer H, Mackay TFC. Accounting for genetic architecture improves sequence based genomic prediction for a Drosophila fitness trait. PLoS One. 2015;10:e0126880.
20. Allison DB, Fernandez JR, Heo M, Zhu S, Etzel C, Beasley TM, et al. Bias in estimates of quantitative-trait–locus effect in genome scans: demonstration of the phenomenon and a method-of-moments procedure for reducing bias. Am J Hum Genet. 2002;70:575–85.
21. VanRaden PM. Efficient methods to compute genomic predictions. J Dairy Sci. 2008;91:4414–23.
22. Su G, Christensen OF, Ostersen T, Henryon M, Lund MS. Estimating additive and non-additive genetic variances and predicting genetic merits using genome-wide dense single nucleotide polymorphism markers. PLoS One. 2012;7:e45293.
23. Henderson CR. Best linear unbiased prediction of nonadditive genetic merits in noninbred populations. J Anim Sci. 1985;60:111–7.

24. Pérez P, de los Campos G. Genome-wide regression and prediction with the BGLR statistical package. Genetics. 2014;198:483–95.

25. De los Campos G, Gianola D, Rosa GJM, Weigel K, Crossa J. Semi-parametric genomic-enabled prediction of genetic values using reproducing kernel Hilbert spaces methods. Genet Res (Camb). 2010;92:295–308.

26. Geyer CJ. Practical Markov chain Monte Carlo. Stat Sci. 1992;7:473–511.

27. Purcell S, Neale B, Todd-brown K, Thomas L, Ferreira MAR, Bender D, et al. PLINK: a tool set for whole-genome association and population-based linkage analyses. Am J Hum Genet. 2007;81:559–75.

28. Hill WG, Weir BS. Variances and covariances of squared linkage disequilibria in finite populations. Theor Popul Biol. 1988;33:54–78.

29. Caballero A, Tenesa A, Keightley PD. The nature of genetic variation for complex traits revealed by GWAS and regional heritability mapping analyses. Genetics. 2015;201:1601–13.

30. Falconer D, Mackay T. Introduction to quantitative genetics. Essex: Longman Publishing Group; 1996.

31. He D, Wang Z, Parida L. Data-driven encoding for quantitative genetic trait prediction. BMC Bioinformatics. 2015;16:S10.

32. Vitezica ZG, Legarra A, Toro MA, Varona L. Orthogonal estimates of variances for additive, dominance and epistatic effects in populations. Genetics. 2017;206:1297–307.

33. Frankham R. Are responses to artificial selection for reproductive fitness characters consistently asymmetrical? Genet Res. 1990;56:35–42.

34. Edwards AC, Rollmann SM, Morgan TJ, Mackay TFC, Leips J. Quantitative genomics of aggressive behavior in *Drosophila melanogaster*. PLoS Genet. 2006;2:e154.

35. Gerken AR, Mackay TFC, Morgan TJ. Artificial selection on chill-coma recovery time in *Drosophila melanogaster*: direct and correlated responses to selection. J Therm Biol. 2016;59:77–85.

36. Krebs RA, Loeschcke V. Estimating heritability in a threshold trait: heat-shock tolerance in *Drosophila buzzatii*. Heredity (Edinb). 1997;79:252–9.

37. Mackay TFC, Heinsohn SL, Lyman RF, Moehring AJ, Morgan TJ, Rollmann SM. Genetics and genomics of Drosophila mating behavior. Proc Natl Acad Sci USA. 2005;102:6622–9.

38. He X, Qian W, Wang Z, Li Y, Zhang J. Prevalent positive epistasis in *Escherichia coli* and *Saccharomyces cerevisiae* metabolic networks. Nat Genet. 2010;42:272–6.

39. Barton NH. How does epistasis influence the response to selection? Heredity (Edinb). 2017;118:96–109.

40. Bulmer MG. The effect of selection on genetic variability. Am Nat. 1971;105:201–11.

41. Zhang XS, Hill WG. Predictions of patterns of response to artificial selection in lines derived from natural populations. Genetics. 2005;169:411–25.

42. Wei M, Caballero A, Hill WG. Selection response in finite populations. Genetics. 1996;144:1961–74.

43. Walsh B. Population and quantitative genetic models of selection limits. Plant Breed Rev. 2004;24:177–226.

44. Perez-Enciso M, Rincón JC, Legarra A. Sequence- vs. chip-assisted genomic selection: accurate biological information is advised. Genet Sel Evol. 2015;47:43.

3

Tag SNP selection for prediction of tick resistance in Brazilian Braford and Hereford cattle breeds using Bayesian methods

Bruna P. Sollero[1*], Vinícius S. Junqueira[2], Cláudia C. G. Gomes[1], Alexandre R. Caetano[3] and Fernando F. Cardoso[1,4]

Abstract

Background: Cattle resistance to ticks is known to be under genetic control with a complex biological mechanism within and among breeds. Our aim was to identify genomic segments and tag single nucleotide polymorphisms (SNPs) associated with tick-resistance in Hereford and Braford cattle. The predictive performance of a very low-density tag SNP panel was estimated and compared with results obtained with a 50 K SNP dataset.

Results: BayesB ($\pi = 0.99$) was initially applied in a genome-wide association study (GWAS) for this complex trait by using deregressed estimated breeding values for tick counts and 41,045 SNP genotypes from 3455 animals raised in southern Brazil. To estimate the combined effect of a genomic region that is potentially associated with quantitative trait loci (QTL), 2519 non-overlapping 1-Mb windows that varied in SNP number were defined, with the top 48 windows including 914 SNPs and explaining more than 20% of the estimated genetic variance for tick resistance. Subsequently, the most informative SNPs were selected based on Bayesian parameters (model frequency and t-like statistics), linkage disequilibrium and minor allele frequency to propose a very low-density 58-SNP panel. Some of these tag SNPs mapped close to or within genes and pseudogenes that are functionally related to tick resistance. Prediction ability of this SNP panel was investigated by cross-validation using K-means and random clustering and a BayesA model to predict direct genomic values. Accuracies from these cross-validations were 0.27 ± 0.09 and 0.30 ± 0.09 for the K-means and random clustering groups, respectively, compared to respective values of 0.37 ± 0.08 and 0.43 ± 0.08 when using all 41,045 SNPs and BayesB with $\pi = 0.99$, or of 0.28 ± 0.07 and 0.40 ± 0.08 with $\pi = 0.999$.

Conclusions: Bayesian GWAS model parameters can be used to select tag SNPs for a very low-density panel, which will include SNPs that are potentially linked to functional genes. It can be useful for cost-effective genomic selection tools, when one or a few key complex traits are of interest.

Background

Bovine ticks are endemic throughout some of the most productive livestock farming regions in the world [1]. In Brazil, the *Rhipicephalus* (*Boophilus*) *microplus* tick is one of the main causes of economic losses in cattle production and affects negatively the performance of their hosts both directly by blood sucking and indirectly as a vector of viral, bacterial and protozoal diseases [2].

Resistance to ticks is known to be under genetic control and the utility of genetic evaluations to classify cattle as resistant or susceptible based on natural tick infestations has already been demonstrated [3]. In addition, it is now well established that several biological mechanisms control host genetic resistance within and among breeds [4, 5]. Therefore, understanding the precise biological mechanisms that underlie vector–host–pathogen interactions is essential to develop innovative and sustainable tick management strategies [6].

The use of genome-wide single nucleotide polymorphism (SNP) panels of varying densities to detect

*Correspondence: bruna.sollero@embrapa.br
[1] Embrapa Pecuária Sul, Caixa Postal 242 - BR 153 - Km 633, Bagé, Rio Grande do Sul 96.401-970, Brazil
Full list of author information is available at the end of the article

statistical associations between phenotypes of interest and SNPs is a powerful method to identify the major genes that are involved in the control of complex traits. However, confounding factors, such as multicollinearity and estimability, which are embedded within multidimensional genotypic and/or phenotypic complex datasets must be considered, since it is necessary to weight the rate of false associations for the interpretation of results [7].

To date, several genomic regions associated with tick burden in dairy and/or beef cattle have been identified through association studies based on different regression methods [2, 8–14]. However, to estimate a greater proportion of the genetic variance explained by SNPs and to identify more complex relationships between SNPs, a shift to models that fit multiple SNPs simultaneously was proposed [15].

Bayesian methods provide a flexible approach to solve high-dimensional problems and enable simultaneous estimation of the effects of high-density SNPs [16]. The application of Bayesian inference methods in genome-wide association studies (GWAS) may improve the mapping of regions across the genome that contain causal variants, especially in the case of complex traits for which the majority of the SNPs each explain a small proportion of the total observed variance. Identification of the most informative SNPs associated with complex traits may contribute to the design of a low-density SNP panel with high predictive performance. This would be highly desirable since cost-effective solutions are needed for genomic selection to be implemented in most animal production sectors [17].

In this study, Bayesian methods were used on 50 K SNP panel data from Hereford and Braford cattle to identify genomic regions and tag SNPs associated with tick resistance. The predictive performance of the very low-density panel based on a selected subset of significant SNPs was estimated and compared with results obtained with the full SNP panel.

Methods
Animal sampling and data analyzed
All Hereford (HH) and Braford (BO) samples were derived from eight herds associated with the Delta G Connection breeding program (Rio Grande do Sul, Brazil). A subset of 3455 phenotyped animals was genotyped with the Illumina BovineSNP50 BeadChip. Total tick counts from one side of the body were recorded for each animal born between 2008 and 2011, two to three times consecutively, during the post-weaning period. In total, 10,673 tick counts were available for analyses. Variance components and breeding values were generated from log-transformed tick counts [3, 18] with the BLUPf90

family of programs [19] and estimated breeding values were used for GWAS analyses.

Quality control analysis
SNP data quality control (QC) was performed using the R version 3.0.2/snpStats package [20] and the following criteria (thresholds): individuals for which the call rate was lower than 90%, heterozygosity deviations were above or below three standard deviations from the mean of the genotyped animals, with sex misidentifications and those that showed near-perfect collinearity with other individuals (>99.5%) were removed. Expected heterozygosity deviations were checked to identify individuals with either an excessive or reduced proportion of heterozygous genotypes, which may be indicative of DNA sample contamination or inbreeding, respectively. Individual SNPs were excluded from further analysis if their call rate was lower than 98%, their minor allele frequency (MAF) was lower than 3%, if they deviated significantly from Hardy–Weinberg equilibrium (Chi square test, $P < 10^{-7}$) and if identical genotypes were found with other SNPs in neighboring positions. Moreover, only the SNP with the highest MAF was retained within groups of SNPs at the same position or that were highly correlated (>98%). A total of 41,045 SNPs (78%) and 3455 animals (98%) were retained for further analyses; these 3455 animals included 2803 BO and 652 HH and comprised yearling bulls, steers and heifers with respective phenotypes for tick count. Sporadically missing genotypes were imputed using FImpute software [21].

Bayesian GWAS
Estimated breeding values (EBV) were obtained by adjusting a pedigree-based repeatability animal model to the tick count data. This model considered fixed effects for contemporary groups, regression coefficients with the linear additive effect for the zebu breed proportion, zebu–HH dominance effect, zebu–HH additive by additive epistatic effect [22], and linear and quadratic coefficients for animal age. Breed composition coefficients were derived from pedigree data [18]. Subsequently, deregressed estimated breeding values (DEBV) for tick resistance were calculated according to Garrick et al. [23], in order to remove parent average values and account for heterogeneous variance. It should be mentioned that the Hereford and Braford population studied here is evaluated and selected as a single breed-type with common breeding objectives and variance components by the Delta G Connection Breeding Program [24]. Moreover, as demonstrated by Biegelmeyer et al. [25], correlation of marker phase between these two breeds was estimated at 0.92 for SNPs less than 50 kb apart, which further supports the assumption that the initial detection analyses

based on the 50 K SNP panel was suitable [18]. Therefore, we carried out a joint analysis that accounted for breed differences and heterosis to calculate DEBV. These pseudophenotypes, which did not include breed effects, were then analyzed with a model that includes random SNP allele substitution effects using the GenSel software version 4.0 [26]. Different Bayesian methods were applied to analyze DEBV data using genotypes as explanatory variables: BayesA, BayesB [16] and BayesCπ [27]. In BayesA and BayesB, each SNP is considered to have a locus-specific variance, which is derived from a scale inverted Chi square distribution X^{-2} (v, S) with $v = 4$ degrees of freedom and a scale $S = 0.0091$. In addition, a prior distribution for the residual variance was also considered as X^{-2} (v, S), but with $v = 10$ and scale $S = 0.0572$. Prior expected values of these Chi square distributions for the dispersion parameters that were equal to 0.0182 and 0.0715, respectively for the genetic and residual variances, were based on estimates previously obtained for tick counts in these BO and HH populations [18].

Prior specification for SNP effects in BayesB allows a proportion of the SNPs to have a zero effect, with a fixed probability π, while the remaining SNPs have normally distributed effects with a locus-specific variance and a probability 1-π. Conversely, in BayesA all SNP covariates are fitted, i.e., π = 0, for each Markov chain Monte Carlo (MCMC) cycle. The statistical model used for Bayesian analyses was: $\mathbf{y} = \sum_{i=1}^{k=41,045} \delta_i \mathbf{z}_i a_i + \mathbf{e}$, where $\mathbf{y}$ is a vector of phenotypes (DEBV); k is the total number of SNPs; δ_i indicates whether SNP i is included in ($\delta_i = 1$) or excluded ($\delta_i = 0$) from the model for a given iteration of the MCMC; $\mathbf{z}_i$ is a vector of genotypes of the fitted SNP i, coded $-10/0/10$; a_i is the random substitution effect of the fitted SNP i with its own variance $\sigma^2_{a_i}$ and an a priori zero effect with probability π or a non-zero effect with probability 1-π, and $\mathbf{e}$ is the vector of normally distributed random residuals. In BayesCπ, the probability that a SNP has a zero effect was treated as unknown and a common effect variance was assumed for all the SNPs having a non-zero effect, while for BayesA δ_i was always equal to 1. Initial SNP effects were estimated for all individuals with BayesCπ (setting π to 0.5 a priori and as starting value) as proposed by Sun et al. [28] and de Oliveira et al. [29]. Subsequent analyses with BayesB tested the posterior mean of π obtained with BayesCπ and π = 0.99. A total of 41,000 chain iterations was used, of which the first 1000 were discarded as burn-in. Convergence of MCMC chains was verified by the Geweke test [30] using the boa (Bayesian output analysis) R package [31].

Top windows and tag SNPs
SNPs were allocated to 2519 non-overlapping 1-Mb genome windows that contained different numbers of

SNPs based on the physical map order derived from the bovine genome assembly UMD3.1 [32]. Genetic variance explained jointly by each SNP subset, considered as window variance, was estimated and subsequently converted into the proportion of total genetic variance explained by the window [28, 33].

Genome regions that potentially contained quantitative trait loci (QTL) associated with tick resistance, referred to as top windows, were identified based on a threshold that is defined in terms of genetic variance contribution as described by Schurink et al. [34]. Top windows were identified in the GWAS by considering all 3455 animals and 41,045 SNPs and by applying the BayesB method (π = 0.99). Assuming an equal contribution of all genomic regions, the expected proportion of genetic variance explained by each of the 2519 windows was equal to 0.04%. Hence, 1-Mb size windows that explained at least 0.2% of the genetic variance, which corresponds to five times the expected variance (0.04% × 5 = 0.2%), were considered as putative QTL [35, 36] and selected for further analyses.

To identify potential SNPs to construct a low-density panel, a tag SNP selection strategy was tested within the top windows by considering model frequency (MF), t-like statistic (TL), linkage disequilibrium (LD) and minor allele frequency (MAF) parameters. In GenSel, MF reflects the proportion of post-burn-in iterations that included that particular covariate (SNP) in the model, while TL is the absolute value of posterior mean effects (for only those chains that included the SNP in the model) divided by the respective standard deviations of those effects. The R/snpStats package [20] was used to obtain LD values and the R/LDheatmap package [37] was applied to generate plots of LD in relation to physical distances.

We selected SNPs with the maximum MF within each top window as top SNPs. Then, we also selected all SNPs within top windows that had MF values above the minimum observed MF value for top SNPs. This step aimed at selecting SNPs that were not at the top of their own windows, but that had sufficiently large MF to exceed the MF value of the top SNPs located in other selected windows. A similar approach was used to evaluate consistency of SNP effects by considering TL. Within those pre-selected SNPs based on MF, the minimum TL value was determined and set as the threshold to select the remaining SNPs within top windows that exceeded this minimum TL value. The final step to construct the tag SNP panel aimed at removing redundant SNPs due to observed high LD among subsets of SNPs pre-selected by MF and TL. Thus, when two SNPs were observed with r^2 values higher than 0.4 [38], only the SNP with the highest MAF was retained.

Prediction ability of selected tag SNP panels

To check the effectiveness of choosing only the most informative SNPs for genetic prediction of tick resistance, genotypic and tick count data from 3455 animals were divided into five sub-groups based on two strategies: K-means clustering according to SNP relationship distance, or randomly, using the R 3.0.2/base package. Cross-validation was carried out within the grouping strategy by selecting subsets of SNPs as described above using data from four of the five groups and then testing the derived tag SNP panel for genomic prediction in the group that was not included in the selection process.

For individuals within each testing group, direct genomic values (DGV) were calculated based on their tag SNP genotypes and corresponding allele substitution effects estimated from training data, which consisted in data on tick counts and genotypes from the four other groups. In this step, we used the BayesA method, such that all selected SNPs had non-zero effects. For the jth individual:

$$\widehat{DGV}_j = \sum_{i=1}^{K} z_{ji}\hat{a}_i,$$

where the estimated SNP effect, $\hat{a}_i$, is represented by its posterior mean obtained by the BayesA method, and z_{ji} represents the genotype for the ith SNP from the total K SNPs included in the very low-density panel.

Pooled prediction accuracies of DGV were derived from their genetic correlations with tick count data in a bivariate analysis using a within-group pedigree-based numerator relationship matrix ($\mathbf{A}^*$; [39]) and were computed using the Gibbs2f90 software [19]. For our fivefold cross-validation:

$$\mathbf{A}^* \begin{bmatrix} \mathbf{A}_{11} & \cdots & 0 \\ \vdots & \ddots & \vdots \\ 0 & \cdots & \mathbf{A}_{55} \end{bmatrix},$$

$\mathbf{A}_{cc}$ is the numerator relationship matrix within cluster c.

Prediction accuracies were also estimated within each cluster c, as proposed by Legarra et al. [40]. Additional details of this cross-validation approach have been described by Cardoso et al. [18] for the full set of 41,045 SNPs.

To further check the effectiveness of our selection process, prediction accuracies of DGV were also obtained with the same fivefold cross-validation with BayesB considering all 41,045 SNPs and $\pi = 0.999$. With this model, the built-in selection process fits, within each cycle, a number of SNPs that is comparable to that included in our proposed panel ($\pi \approx 1 - n_{\text{tagSNP}}/41,045$).

Functional analysis

To map tag SNPs to genes and genomic regions, the BED-Tools software [41] was used to relate SNP data with the *Bos taurus* genome information provided by the Ensembl database [42]. Alternatively, for the SNPs that were not mapped within any known gene within ±100 kb on the *Bos taurus* genome, the package NCBI2R [43] was used to search for the closest known genes in the genome of other species. Using DAVID bioinformatics resources [44], the biological meaning of the genes mapped to tag SNPs was extracted. The online software STRING v9.1 [45] was used to identify potential protein–protein interactions related to the identified genes.

Results and discussion

Groups of animals

Genomic relationships between the five groups of animals based on K-means clustering and the number of individuals in each group are in Table 1. Each of the five groups that were obtained from random distribution contained 691 animals and displayed similar relatedness within and across groups.

Choice of π

The BayesCπ analysis that included all animals and SNPs simultaneously resulted in a posterior (π) of 0.9999 and therefore, only approximately four SNPs (0.01%) were fitted in each iteration of the MCMC chain. Using π = 0.9999 in a BayesB analysis resulted in a very low estimated heritability ($h^2 = 0.02$), which corresponded to a small fraction of the pedigree-based heritability ($h^2 = 0.19$) obtained with the same dataset [18], and was similar to the lower-bound heritability estimates recently reported for cattle tick resistance [14] in a GWAS that analyzed *A. hebraeum* tick counts on the tail of South African Nguni cattle (0.02). Some cycles contained no fitted SNPs when an extremely high value of π (0.9999) was used in BayesB, which resulted in the absence of any predictive SNPs, and thus this model contributed mostly to the estimated residual variance. These results suggest

Table 1 Number of individuals (N) and average (±SD) zebu proportions, and within- and between-group genomic relationships (Gij) for the K-means clustering groups

Group	N	Zebu proportion	Within-group Gij	Between-group Gij
1	629	0.02	0.140 ± 0.04	−0.030 ± 0.04
2	230	0.37	0.070 ± 0.05	0.005 ± 0.05
3	1211	0.35	0.004 ± 0.03	0.003 ± 0.03
4	471	0.34	0.010 ± 0.04	0.002 ± 0.03
5	914	0.35	0.020 ± 0.03	0.010 ± 0.04

The majority of the Hereford breed animals were clustered into Group 1 that BayesCπ could not estimate π appropriately based on the present data.

Subsequently, more SNPs were fitted in the BayesB model by setting $\pi = 0.99$ for the GWAS including all animals simultaneously and for each group in the cross-validation process. With this new π value and the full GWAS data, the proportion of variance explained by SNPs increased to 0.1132, which corresponds to 58% of the estimated heritability based on pedigree-based analysis of this dataset on tick resistance [18]. This reduced genomic heritability may result from incomplete linkage disequilibrium between the SNPs studied and the QTL affecting the trait [46], when only 1% of the markers were fit in each chain cycle ($\pi = 0.99$). Alternatively, the proportion of phenotypic variance explained by SNPs when fitting BayesA with the full SNP panel (0.1755) was much closer to that based on pedigree analysis (0.19). These BayesA and pedigree estimated heritabilities were higher than that reported by Porto Neto et al. [13] for the analysis of tick burden in Brahman cattle (0.09). Setting π at 0.99, in spite of the lower estimated heritability compared to BayesA or pedigree analysis, has the advantage of fitting only the regions in strong association with the trait [33, 35, 47]. According to Fernando and Garrick [48], higher values of π can be more discriminating for the identification of the largest QTL, which is an important factor for selecting tag SNPs. Moreover, it was shown that the SNP-specific variances in BayesB led to less shrinkage for SNPs with the largest effects compared to BayesC [27].

All Bayesian GWAS analyses were visually checked and passed the Geweke's test for convergence.

Top windows and QTL detection
The proportion of genetic variance explained by each of the 2519 1-Mb windows including all 41,045 SNPs across

the genome is shown in Fig. 1. The number of SNPs included in the windows varied from 1 (only 10 windows) to 30. Forty-eight windows represented by 914 SNPs were found to jointly explain more than 20% of the genetic variance and were considered as top windows containing QTL (Table 2).

Some of the detected windows coincided with previously reported QTL from linkage analyses and GWAS for tick burden (Cattle QTL database, [49]), i.e. on BTA2 (BTA for *Bos taurus* chromosome) top windows number 163 located at 4 Mb (identified according to the first SNP position in the window) and number 214 at 55 Mb, top windows number 364 at 68 Mb on BTA3, number 553 at 14 Mb on BTA5, number 794 at 13 Mb on BTA7, number 1190 at 77 Mb on BTA10, number 1283 at 65 Mb on BTA11, and number 1553 at 54 Mb on BTA14 (Table 2).

The first 12 top windows jointly explained more than 10% of the genetic variance and three genomic regions (top three windows) individually explained more than 1% of the genetic variance for tick resistance (Table 2). In these three regions (BTA15 at 37 Mb, BTA11 at 101 Mb and BTA10 at 51 Mb), within ±100 kb on each side of the SNPs included in the respective top windows, four SNPs (rs110197574 and rs41665212, rs29019899 and rs110144789) were mapped to annotated genes or pseudogenes in the bovine or human genomes (see Additional file 1). Two SNPs on BTA15, were located at ~40 kb apart from each other (rs110197574 and rs41665212) and mapped to HSA5 (HSA for *Homo sapiens* chromosome) close to the *RPS15P8* pseudogene (*ribosomal protein S15, pseudogene 8*). Other positional candidate genes close to SNP rs110144789 (BTA11) are *LAMC3* (*laminin, gamma 3*), *ABL1* (*ABL proto-oncogene 1, non-receptor tyrosine kinase*), *FIBCD1* (*fibrinogen C domain containing 1*), *QRFP* (*pyroglutamylated RFamide peptide*); and

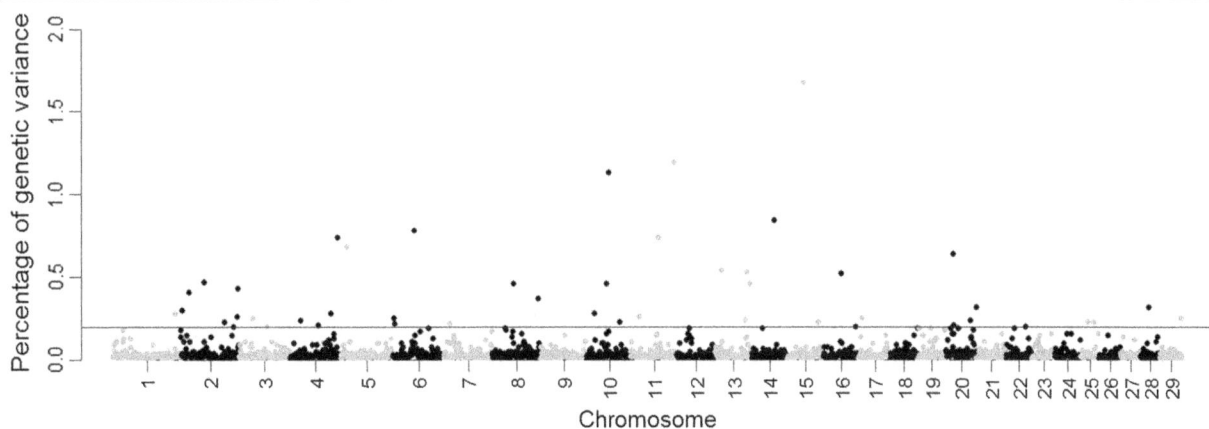

Fig. 1 Manhattan plot displaying Bayesian genome-wide association estimates (BayesB, $\pi = 0.99$) for tick resistance. The *Y-axis* represents the proportion of the total genetic variance explained by 1-Mb windows across the bovine genome and the *X-axis* represents the chromosomal location of windows (2519 non-overlapping windows). Windows explaining more than 0.2% of the genetic variance are above the *grey line*

Table 2 Windows explaining the largest percentages of tick resistance genetic variance in Hereford and Braford cattle breeds

Obs	Window	Start SNP name	End SNP name	N SNP	%Var	Cum Var	chr_Mb	Top SNP name	ModelFreq	t.like	Stand_effect
1	1621	ARS-BFGL-BAC-27751	ARS-BFGL-NGS-115263	17	1.67	1.67	15_37	ARS-BFGL-NGS-5811	0.7574	1.393	1.4036
2	1319	ARS-BFGL-NGS-112243	ARS-BFGL-NGS-12954	17	1.19	2.86	11_101	ARS-BFGL-NGS-111179	0.6731	1.243	1.1350
3	1164	ARS-BFGL-NGS-1854	ARS-BFGL-NGS-24556	27	1.13	3.99	10_51	Hapmap58695-rs29019899	0.5048	1.076	0.8828
4	1553	ARS-BFGL-NGS-115527	ARS-BFGL-NGS-1112	15	0.84	4.83	14_54	BTB-00915241	0.3045	0.982	0.4614
5	710	BTB-01688071	ARS-BFGL-NGS-58275	23	0.78	5.6	6_49	BTB-02002785	0.3176	0.986	0.4781
6	1283	ARS-BFGL-NGS-44192	Hapmap60779-rs29022104	14	0.74	6.35	11_65	Hapmap60779-rs29022104	0.3697	1.005	0.6143
7	531	Hapmap57291-ss46526771	Hapmap22875-BTA-155031	14	0.74	7.09	4_113	Hapmap22875-BTA-155031	0.4834	1.06	0.8291
8	553	Hapmap36482-SCAF-FOLD163485_1458	BTA-87049-no-rs	12	0.68	7.77	5_14	Hapmap52967-rs29017027	0.4047	1.022	0.6210
9	1974	ARS-BFGL-NGS-13160	BTB-00774670	20	0.64	8.4	20_17	Hapmap34041-BES1_Con-tig298_838	0.327	0.988	0.5321
10	1429	Hapmap48542-BTA-97857	ARS-BFGL-NGS-27497	11	0.54	8.95	13_14	Hapmap44228-BTA-34185	0.3342	0.99	0.4960
11	1488	ARS-BFGL-NGS-107401	ARS-BFGL-NGS-4602	18	0.53	9.48	13_73	Hapmap40517-BTA-33731	0.3514	0.997	0.5201
12	1709	Hapmap54735-ss46526095	Hapmap56619-rs29009970	16	0.52	10	16_40	ARS-BFGL-NGS-40365	0.2408	0.956	0.3808
13	214	Hapmap25908-BTA-160304	Hapmap60963-rs29015781	26	0.47	10.47	2_55	ARS-BFGL-NGS-111213	0.1665	0.937	0.2428
14	1159	ARS-BFGL-NGS-113665	ARS-BFGL-NGS-10383	21	0.46	10.93	10_46	ARS-BFGL-NGS-60054	0.2224	0.953	0.3154
15	944	Hapmap41647-BTA-81135	ARS-BFGL-NGS-111988	20	0.46	11.39	8_50	BTB-01398754	0.0986	0.902	0.1449
16	1495	ARS-BFGL-NGS-83969	Hapmap41120-BTA-99310	20	0.46	11.85	13_80	Hapmap41120-BTA-99310	0.1838	0.942	0.2532
17	293	ARS-BFGL-NGS-22691	ARS-BFGL-NGS-17681	22	0.43	12.28	2_134	ARS-BFGL-NGS-113378	0.2608	0.956	0.4564
18	179	ARS-BFGL-NGS-39206	BTB-01168392	21	0.41	12.69	2_20	BTB-00082871	0.2273	0.953	0.3302
19	1001	Hapmap25843-BTA-146186	ARS-BFGL-NGS-20859	18	0.37	13.07	8_107	Hapmap40677-BTA-121871	0.1692	0.936	0.2455
20	2440	BTB-00980670	Hapmap34915-BES7_Con-tig278_1082	17	0.32	13.39	28_20	BTB-01129090	0.1208	0.923	0.1530
21	2028	Hapmap44700-BTA-34998	ARS-BFGL-NGS-118166	15	0.32	13.71	20_71	ARS-BFGL-NGS-13702	0.2361	0.947	0.3888
22	163	Hapmap43083-BTA-86781	BTA-47785-no-rs	17	0.3	14.01	2_4	BTB-00077766	0.1948	0.944	0.2567
23	1132	ARS-BFGL-NGS-94247	ARS-BFGL-NGS-32828	24	0.28	14.29	10_18	ARS-BFGL-NGS-107048	0.2059	0.947	0.3001
24	147	Hapmap30204-BTA-124882	BTB-01761180	28	0.28	14.57	1_147	BTB-01301015	0.1687	0.934	0.2423
25	515	Hapmap25270-BTA-142450	ARS-BFGL-NGS-12738	11	0.28	14.85	4_97	Hapmap25270-BTA-142450	0.1796	0.931	0.2960
26	1239	Hapmap43962-BTA-86597	Hapmap42711-BTA-87541	24	0.26	15.11	11_21	BTB-00464454	0.142	0.926	0.2100
27	1238	ARS-BFGL-NGS-20053	BTB-00464777	19	0.26	15.37	11_20	BTA-104373-no-rs[a]	0.1533	0.922	0.2280
28	290	ARS-BFGL-NGS-54356	ARS-BFGL-NGS-77887	26	0.26	15.63	2_131	BTB-00117780	0.1469	0.931	0.1938
29	664	Hapmap30828-BTA-143720	Hapmap30881-BTA-159706	23	0.25	15.88	6_3	Hapmap30881-BTA-159706	0.171	0.937	0.2465
30	2515	Hapmap24672-BTA-140771	ARS-BFGL-NGS-29493	27	0.25	16.13	29_48	BTA-66199-no-rs	0.211	0.951	0.2818
31	329	ARS-BFGL-NGS-117560	Hapmap51025-BTA-67309	19	0.25	16.38	3_33	ARS-BFGL-NGS-119309	0.0962	0.91	0.1189

Table 2 continued

Obs	Window	Start SNP name	End SNP name	N SNP	%Var	Cum Var	chr_Mb	Top SNP name	ModelFreq	t.like	Stand_effect
32	1758	ARS-BFGL-NGS-5880	BTA-122662-no-rs	23	0.25	16.63	17_7	ARS-BFGL-BAC-27352	0.1857	0.944	0.2496
33	2013	ARS-BFGL-NGS-55465	Hapmap51244-BTA-50863	19	0.24	16.87	20_56	Hapmap43377-BTA-85612	0.1385	0.927	0.1717
34	442	ARS-BFGL-NGS-113848	BTB-00169886	16	0.24	17.12	4_24	Hapmap45129-BTA-72713	0.1427	0.925	0.2042
35	1485	ARS-BFGL-NGS-118627	ARS-BFGL-BAC-15769	17	0.24	17.35	13_70	Hapmap57013-rs29019369	0.1502	0.934	0.1073
36	1656	ARS-BFGL-BAC-18252	Hapmap45825-BTA-25376	16	0.23	17.58	15_72	Hapmap45825-BTA-25376	0.1828	0.943	0.2411
37	261	Hapmap48190-BTA-114376	BTA-48503-no-rs	11	0.23	17.81	2_102	Hapmap36094-SCAF-FOLD96944_22403	0.1323	0.928	0.1683
38	1190	ARS-BFGL-NGS-38839	Hapmap50492-BTA-86239	16	0.23	18.04	10_77	ARS-BFGL-NGS-111871	0.1883	0.945	0.2477
39	2294	ARS-BFGL-NGS-26313	ARS-BFGL-NGS-34801	17	0.23	18.26	25_15	ARS-BFGL-NGS-84660	0.1403	0.933	0.1543
40	2309	ARS-BFGL-BAC-3777	ARS-BFGL-BAC-47171	21	0.23	18.49	25_30	ARS-BFGL-BAC-37178	0.1584	0.911	0.3051
41	665	ARS-BFGL-NGS-56212	BTB-01468045	20	0.22	18.71	6_4	BTB-01280976	0.0959	0.922	0.1089
42	794	ARS-BFGL-NGS-93802	Hapmap57279-ss46526160	19	0.22	18.93	7_13	ARS-BFGL-NGS-111257	0.0898	0.91	0.1073
43	484	ARS-BFGL-NGS-26541	ARS-BFGL-NGS-44674	20	0.21	19.14	4_66	ARS-BFGL-NGS-36591	0.1501	0.932	0.2019
44	1976	BTB-00775794	Hapmap49633-BTA-50009	16	0.21	19.35	20_19	Hapmap28040-BTA-134983	0.1529	0.935	0.2100
45	1743	BTB-00661933	ARS-BFGL-NGS-99802	17	0.2	19.56	16_74	Hapmap48746-BTA-40116	0.1664	0.938	0.2172
46	282	ARS-BFGL-NGS-102874	ARS-BFGL-NGS-15468	22	0.2	19.76	2_123	ARS-BFGL-NGS-102874	0.1163	0.915	0.1622
47	2146	ARS-BFGL-NGS-118471	Hapmap38075-BTA-54630	18	0.2	19.96	22_45	BTB-00849206	0.1658	0.922	0.2876
48	364	BTA-68264-no-rs	Hapmap44273-BTA-68311	24	0.2	20.15	3_68	ARS-BFGL-NGS-33433	0.1601	0.912	0.2923

Obs sequence number of the top 48 1-Mb non-overlapping windows, *Window* window coded number by GenSel according to physical map order, *Start SNP name* name of the first SNP flanking the window, *End SNP name* name of the last SNP flanking the window, *N SNP* Number of SNPs within the window, *%Var* percentage of genetic variance explained by the window, *Cum Var* cumulative percentage of genetic variance, *chr_Mb* BTA autosome and position of the window in Mb pairs, *Top SNP name* name of the top SNP in the window (in terms of *ModelFreq* and or *t.like* statistics) and respective values of *ModelFreq, t.like* and standard effect (*Stand_effect*) of each top SNP

[a] This SNP was not included in the low-density panels proposed because it is in LD with BTB-00464454 and it has a low MAF

to rs29019899 (BTA10) are *ADAM10* (*metallopeptidade domain 10*), *LIPC* (*lipase, hepatic*) and the gene *5S_rRNA* (*ENSBTAG00000037226, 5S ribosomal RNA*); all these genes are annotated on the bovine genome sequence.

The SNP with the largest effect on tick count (rs110197574) was mapped to the *RPS15P8* gene, which in humans encodes a ribosomal protein that is a component of the 40S subunit [50]. Analysis of the genes that encode components of the ribosome or proteins involved in ribosome biosynthesis is very complex, and considering the wide range of biological processes in which ribosomal genes may be involved, the potential role of *RPS15P8* in tick resistance needs to be further investigated. Barendse [10] reported a polymorphism in the *RPS13* (*ribosomal protein S13*) gene that is associated with increased tick resistance in cattle. The *ADAM10* (*ADAM metallopeptidase domain 10*) gene (BTA10) encodes a characterized member of the *ADAM*-family of metalloproteases, which has a prominent role in inflammation [51]. Furthermore, different inflammatory responses can activate ADAM10-mediated proteolysis of E-cadherin, which is a prime mediator of epithelial cell-to-cell interactions, in primary human keratinocytes and in diseased human skin [52]. According to Porto Neto et al. [53] at approximately 15 Mb on BTA10, some locus-haplotypes that include SNPs in the *ITGA11* (*integrin alpha 11*) gene are associated with tick burden in dairy cattle breeds (Australian Red, Brown Swiss, Channel Isle, Holstein and composites) and Brahman beef cattle. Although this gene is functionally described as related with cellular adhesion control, these authors suggested that it had a role in modulating cellular immune responses. Both of these genes (*ADAM10* and *ITGA11*) are on BTA10 and may be involved in the control of cellular adhesion and migration during the process of skin infection caused by tick burden. Other studies based on microsatellite whole-genome scans [2, 54] and a GWAS with a low-density SNP panel [10] also reported QTL associated to tick burden on BTA10. In agreement with Regitano et al. [54], we identified potential QTL at 18 Mb on BTA10, as well as on BTA4 (97 Mb). On BTA10, beyond the region that contains the *ADAM10* gene (~50 Mb), three other top windows (Table 2) were detected as potential QTL in agreement with Machado et al. [2].

Anaplasmosis is an infectious rickettsial disease (*Anaplasma marginale*) that is mainly transmitted by ticks [55] and negatively impacts cattle production in tropical and subtropical areas [56]. In humans, *A. phagocytophilum*, an obligatory intracellular parasite of human granulocytes, causes a similar disease and was shown to activate the ABL1 signaling pathway during cell invasion. This protein is critical for intracellular invasion

and infection establishment. Thus, a novel strategy for the treatment of human granulocytic anaplasmosis was proposed through inhibition of the host cell Abl-1 signaling pathway [57]. In addition to being possibly directly associated with tick resistance, results that we obtained from the analysis with STRING suggest the occurrence of interactions between the *ABL1* gene and other genes that are associated with the most informative SNPs found to affect tick count (*LAMC3* or *PLCG1, phospholipase C, gamma 1* on BTA13; *CDC42, cell division cycle 42* on BTA2; *SDC3, syndecan 3* on BTA2 and *EPS8L3, epidermal growth factor receptor kinase substrate 8-like protein 3* on BTA3), which indicates that a gene network may be involved in cattle resistance to ticks. Two other top windows on BTA11 with putative QTL were also reported by Machado et al. [2].

Other genomic regions on BTA17 at 7 Mb flanked by SNPs ARS-BFGL-NGS-5880 and BTA-122662-no-rs (top window 1758, Table 2) also include two SNPs (rs43499108 and rs29011077), which have been reported to be associated with *R. evertsi evertsi* tick count in African cattle [14]. The top SNP in this window (rs109822497) was mapped to the *double cortin-like kinase 2* (*DCLK2*) gene, near the *LRBA* gene, which is suggested by these authors to be associated with protein kinase A that supports the secretion and/or membrane deposition of immune effector molecules.

Selecting tag SNPs

Based on the model frequency (MF) and *t*-like statistic (TL) provided by GenSel, in the strategy used for tag SNP selection, a minimum MF value of 0.0898 was determined among all top SNPs representing each of the 48 top windows. Nine additional SNPs with an MF above this threshold were selected from the list of 914 SNPs within the top windows. Within those 57 (48 + 9) preselected SNPs, the minimum observed TL of 0.902 was set as another threshold to select SNPs within the 914-SNP list that exceeded this lower bound TL value. The subset of SNPs that were pre-selected based on MF and TL contained 63 SNPs, which were subsequently analyzed in terms of LD and MAF, resulting in a final list of 58 SNPs. These selected SNPs were distributed on most of the bovine chromosomes, except BTA9, 12, 18, 19, 21, 23, 24, 26 and 27. It is interesting to mention that nine of the 58 SNPs were located on BTA2. Previous studies identified significant allele effects associated with tick burden in a GWAS analysis, as well as positional candidate genes on chromosome BTA2 [2, 10, 11, 58]. Our proposed panel included SNPs that represented 47 of the 48 top windows, because SNP rs43669951 was included in two adjacent windows on BTA11. The resulting minimum MF value among the SNPs in this panel was equal to 0.0744 and only two windows included three SNPs on BTA28 at ~20 Mb and BTA13 at ~80 Mb, while all other windows included only one or two SNPs.

This strategy to select more informative SNPs that are uniquely linked to QTL related to cattle tick resistance within top windows favored those that were more often included in the Bayesian mixture model (greater MF) and with a more consistent effect (greater TL), but avoided redundancy due to LD. Based on that, our goal was to retain SNPs that had a suitable prediction ability to build very low-density panels for cost-effective genomic selection of tick resistance in cattle.

The proportion of fitted models that included a SNP and used it to infer associations with the phenotype under study, represented by MF [28, 59], was highly correlated ($r = 0.99$) to the SNP adjusted effect, $(\hat{a}_i)^2/\mathrm{var}(\hat{a}_i)$, in the subset of 914 SNPs. In contrast, a moderate correlation was found between MF and TL ($r = 0.46$). Since TL is an alternative measure of SNP effect (i.e. $|\hat{a}_i|/\mathrm{sd}(\hat{a}_i)$ is calculated by considering only the cycles in which SNP i was included in the model) and due to its incomplete correlation with MF, we were able to combine both parameters, MF and TL, to select informative SNPs for which the estimated effects were consistent [26].

According to some authors [33, 36], SNPs with an MF higher than 0.90 are deemed significant in a Bayesian GWAS analysis, and those with an MF lower than 0.10 represent false positives. In the current study, the highest MF for a top SNP was 0.7574 for rs110197574/ ARS-BFGL-NGS-5811 located on BTA15 within the 1-Mb window that explained the greatest proportion of the genetic variance (Table 2). Therefore, this particular SNP had a non-zero effect in 75% of MCMC samples. Considering all SNPs with the highest MF within each of the 48 top windows according to our BayesB analysis (Table 2), the average MF was equal to 0.23 ± 0.14. This result indicates that there are no major genes affecting tick resistance and that most of the SNPs each explained a small proportion of the phenotypic variation for this trait. Similar results were reported by other authors who concluded that selection programs must use SNP panels rather than single SNPs with high predictive value [60]. This emphasizes the fact that it cannot be expected to find a very small number of genes with a large effect, which would lead to accurate prediction for tick resistance. In this regard, our approach was to identify a minimal set of informative SNPs that would still yield useful predictions compared to those derived from high-density SNP panels, but potentially reducing genotyping costs.

Figure 2 shows LD-heatmaps and respective MF and TL values for two distinct windows, which highlight SNPs in the proposed list of 58 SNPs. The first window,

on BTA3 at ~33 Mb (Fig. 2a), represents a chromosomal region that contains the SNP ARS-BFGL-NGS-119309 (rs110043221) selected as its tag SNP. Figure 2a also illustrates the case of the ARS-BFGL-NGS-77834 (rs110132430) SNP that has TL and MF values higher than the selection threshold, but that was excluded because it was in LD ($r^2 > 0.4$) with another SNP with a higher MF (rs110043221). SNPs within this window were mapped to a bovine genomic region that contains three genes, *EPS8L3* (*EPS8-like 3*), *GSTM1/3* (*glutathione S-transferase mu 1 and 3*), *CSF1* (*colony stimulating factor 1—macrophage*) and a microRNA bta-mir-2413. De Rose et al. [61] showed that cytokines, such as the granulocyte and macrophage colony stimulating factor (GM-CSF) or interleukin (IL)-1b have increased vaccine effectiveness by enhancing the immune response against *Rhipicephalus* (*Boophilus*) *microplus* in sheep. The second top window with an effect on tick resistance is located at ~54 Mb on BTA14 (Fig. 2b), contains 15 SNPs and includes an LD block represented by SNP BTB-00915241 (rs42075995). In this case, SNP BTA-60194-no-rs (rs41587782) was in high LD with the representative tag SNP and thus, was excluded in the final step of the selection strategy. The differential pattern of MF and TL variation of SNPs was critical to effective tag SNP selection, since the top SNPs were clearly distinct in the histograms of those windows (Fig. 2). Therefore, most of the SNPs with low MF/TL were excluded in the first two selection steps and the remaining ones were evaluated in terms of LD/MAF in a final step with only a few additional exclusions.

The genes that map to the regions containing the 58 SNPs that were selected to compose the proposed very low-density panel are listed in Additional file 1. One hundred and three genes are located in the genome regions that are on either side of 52 of these SNPs, based on the information derived from the bovine (43 SNPs) and human (9 SNPs) genomes. Gene ontologies and biological pathways which may be related to the biological processes that underlie vector-host-pathogen interactions, such as pathways involved in inflammation mediated by chemokine and cytokine signaling, cell receptor signaling and calcium signaling, were identified for these genes. Also, enrichment analysis identified genes that are associated with biological processes such as regulation of adaptive immune response (e.g. *ADA*), activation of immune response (e.g. *ABL-1*), positive regulation of macrophage derived from cell differentiation (e.g. *CSF1*), regulation of inflammatory response and leukocyte chemotaxis (e.g. *ADAM10*), cell–cell junction organization (e.g. *CDC42*) and leukocyte activation involved in immune response (e.g. *ADA, ABL-1*).

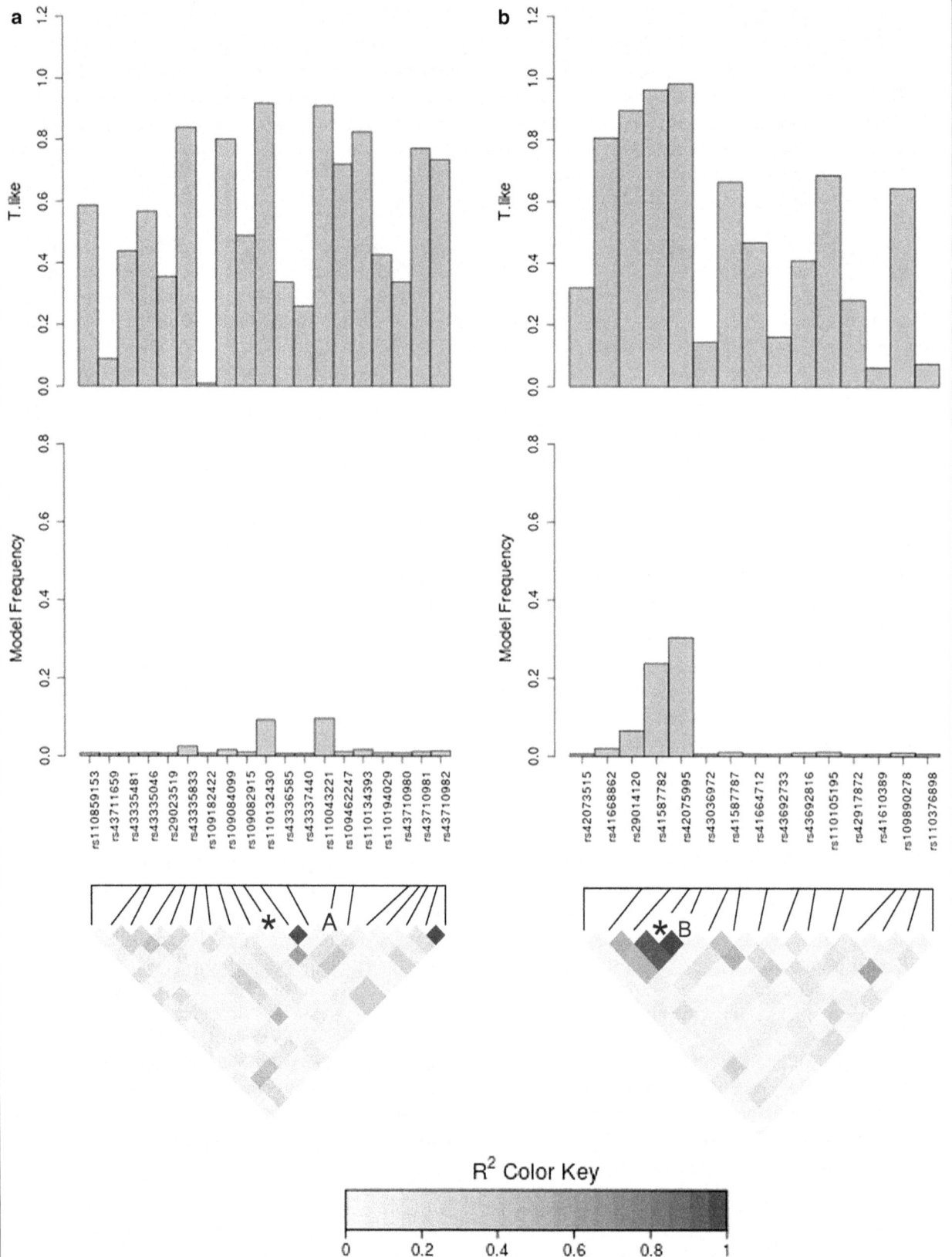

Fig. 2 MF and TL estimates and LD heatmaps, for neighboring SNPs in two windows (1 Mb) according to physical map order. **a** Top window on BTA3. *Markers excluded by LD parameter. "A" Marker selected as tag SNP in the low-density panel. **b** Top window on BTA14. *Markers excluded by LD parameter. "B" Marker selected as tag SNP in the low-density panel

Prediction ability of tag SNP panels

The proposed BayesB ($\pi = 0.99$) GWAS and tag SNP selection strategy was applied to each of five K-means and five random cross-validation subsets and generated 10 alternative SNP panels, which included 47 to 86 SNPs (Table 3). Three hundred and fifty unique SNPs were selected based on the combination of all 10 tag SNP panels derived by the cross-validation analyses. The number of times that each of the original 58 tag SNPs (our proposed panel using the whole data) was represented in those 10 cross-validation subsets is presented in Additional file 1.

The posterior proportions of the phenotypic variance, which was explained by the SNPs, i.e. the SNP-heritabilities (h^2), that were estimated with BayesA using the very low-density panel derived for each of the 10 cross-validation groups ranged from 0.09 to 0.12, and were very similar to the h^2 estimated with BayesB ($\pi = 0.99$) using the full set of 41,045 SNPs (Table 3). Conversely, the h^2 obtained with BayesB also using all 41,045 SNPs but with $\pi = 0.999$ (~1–58/41,045) were lower and ranged from 0.04 to 0.06 (Table 3). These results demonstrate that a very small number of SNPs selected based on the tag-method explains more variation than a similar number of SNPs chosen with the Bayes-B method (on average 0.14% of the total number of available 41,045 SNPs).

Bayesian approaches generally combine shrinkage procedures to consider different variances for individual SNPs and mixture models, in which the prior information about the distribution of SNP effects is used to coerce negligible effects towards zero. In the case of tag

SNP panels, BayesA was chosen because these panels are expected to include only the most significant SNPs each with a detectable effect ($\pi = 0$), while allowing for SNPs to have specific variances and consequently different effect sizes [16].

The effectiveness of the applied strategy for selecting more informative SNPs for genomic prediction of cattle tick resistance was assessed by pooled breeding value prediction accuracies measured as the genetic correlation between cross-validation DGV and tick count data, which were equal to 0.27 ± 0.09 for the K-means clustering groups and 0.30 ± 0.09 for the random groups.

Accuracies within each cluster were also obtained using the method of Legarra et al. [38] and substantial differences between groups were observed with values ranging from 0.08 to 0.41 (Fig. 3). The lowest values were observed for K-means group 1, which was the most distinct cluster that included mainly Hereford animals (the zebu proportion was near zero) and showed the largest genetic distance to the other groups (Table 1). Therefore, this result is consistent with the fact that a reference population that includes only Braford cattle would not result in suitable accuracies for Hereford selection candidates [18]. Although all random groups had the same number of animals and the same genetic distance within and between clusters, accuracy for Group 3 was considerably lower (0.16) compared to the other random clusters. The highest accuracies were observed for K-means Group 5 (0.40) and random Group 2 (0.41).

Using the full set of 41,045 SNPs, pooled cross-validation accuracies for K-means and random clustering,

Table 3 Posterior mean proportion of variance explained by markers (h^2) using different Bayesian methods, and number of chromosome segments and SNPs involved in the very low-density panel selection by K-means and random cross-validation group

Group	h^2				SNP panel selection		
	BayesB $\pi = 0.99$	BayesB $\pi = 0.999$	BayesA full	BayesA tag	Top windows[a]	Top SNPs[b]	Tag SNPs[c]
K-means 1	0.13	0.06	0.19	0.10	41	741	47
K-means 2	0.10	0.04	0.17	0.09	46	878	57
K-means 3	0.12	0.05	0.18	0.12	39	727	67
K-means 4	0.12	0.05	0.18	0.11	48	941	79
K-means 5	0.11	0.04	0.18	0.10	43	799	55
Random 1	0.12	0.05	0.18	0.11	42	778	57
Random 2	0.11	0.04	0.18	0.12	53	956	70
Random 3	0.11	0.05	0.18	0.13	52	1008	86
Random 4	0.12	0.05	0.18	0.11	48	900	78
Random 5	0.11	0.04	0.18	0.12	55	1005	79

[a] Top windows represents the number of windows that explained above 0.2% of the genetic variance in the BayesB ($\pi = 0.99$) GWAS analysis

[b] Top SNPs represents the number of SNPs included in those top windows

[c] Tag SNPs represents the number of SNPs selected as more informative according to the criteria based on model frequency and t.like statistics, linkage disequilibrium and minor allele frequency

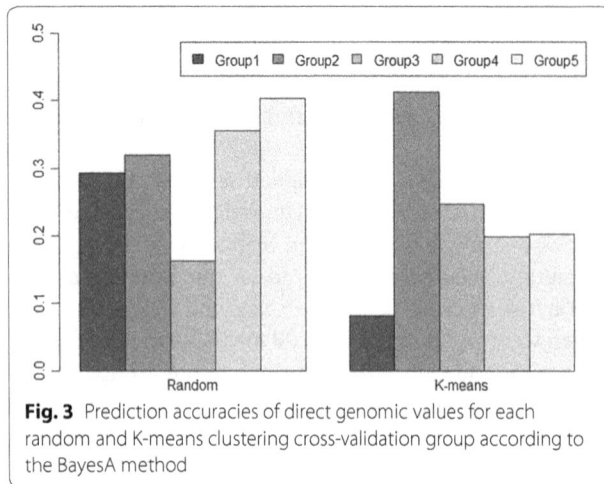

Fig. 3 Prediction accuracies of direct genomic values for each random and K-means clustering cross-validation group according to the BayesA method

respectively, were equal to 0.37 ± 0.08 and 0.43 ± 0.08 for BayesA, 0.37 ± 0.08 and 0.43 ± 0.08 for BayesB with $\pi = 0.99$, and 0.28 ± 0.07 and 0.40 ± 0.08 for BayesB with $\pi = 0.999$. When compared to the above results, accuracies that are derived using the proposed very low-density tag SNP panel with 58 SNPs would represent at least 68% of the accuracies of predictions obtained using all 41,045 SNPs with BayesB or BayesA methods. These results demonstrate that tag SNP panels may be used in commercial applications for genomic predictions in beef cattle as an alternative to more costly high-density panels. Nevertheless, the decision about the most suitable SNP density should be trait- and population-specific, depending on the relative accuracy and cost of the alternative SNP panels.

Cardoso et al. [18] reported pooled cross-validation accuracies of 0.39 and 0.44 for K-means and random clustering, respectively, for BayesB ($\pi = 0.95$) predictions obtained for tick count with the same population. These preview results obtained with a 50 K SNP panel represented accuracy gains of 50.0 and 51.7% when compared, respectively, to pedigree best linear unbiased prediction (PBLUP) accuracies of 0.26 for K-means and 0.29 for random groups obtained by the same authors. Compared to these results, the very low-density SNP panel that we propose here shows very similar accuracies to predictions based on conventional PBLUP. This would be the case for animals that are closely related to the reference population as in Cardoso et al. [18]. Even with similar accuracies, the very low-density panel predictions have the advantage of being applicable in the absence of historical tick count data, when phenotypes on ancestors may not be available, thus avoiding the need of population parasite burden. Moreover, blending strategies to combine tag SNP panel predictions with historical data from non-genotyped animals deserves further investigation, since

they could improve prediction accuracies of selection candidates using, for example, single-step methodologies [62–64]. For these Hereford and Braford tick resistance datasets and the full set of SNPs, accuracy gains of using blended historical data by single-step genomic BLUP compared to Bayes B ($\pi = 0.95$) DGV were equal to 23 and 27% respectively for K-means and random cross-validation groups [18].

Bayesian approaches have been proposed for predicting genomic breeding values with high-density SNP panels, but in practice they may be more useful for low-density panels [65]. Decreased predictive abilities are expected for very low-density in comparison to high-density SNP panels, due to the expected reduced LD between SNPs and highly dispersed QTL affecting a particular trait. However, some studies have demonstrated the superiority of Bayesian methods to capture this LD between SNPs and QTL [66, 67]. Cleveland et al. [65] compared Bayesian prediction accuracies of different scenarios including high- and low-density SNP panels. These authors found similar accuracies when SNPs were selected based on the size of their additive effects, even when SNP coverage was extremely low, which corroborates our results. Dynamic schemes to successfully apply genomic selection technology for genetic improvement of livestock, invariably aim at minimizing genotyping costs while maximizing genetic gains and overall profits. Genotype data that are generated at lower costs from small subsets of highly informative SNPs could be used to genotype most of the animals in a herd and generate genomic breeding value predictions based on SNP effects that are estimated from high-density training datasets [68]. Moreover, our results show that recalculation of genetic effects for the most informative SNPs that were originally chosen from the full dataset resulted in the reduction of redundancy and/or confounding effects, which might have been included in estimates obtained during the original discovery using the 41,045 SNPs, as a result of multicollinearity among SNP effects. The applied strategy appears promising, since the obtained DGV retained about 70% of the accuracy of DGV derived from the full high-density panel, with only about 0.14% of the SNP density (58 out of 41,045). Similar strategies have already been proposed for predicting breeding values in young dairy cattle seedstock by using panels of about 3000 SNPs (larger than the panel proposed here) and resulted in accuracies representing 80 to 90% of those obtained with high-density panels with [69].

Genome-wide association studies allow for a much finer description of the genome and genomic selection results in increased genetic gains because early and accurate selection decisions are made possible, for traits that were previously ignored because of high associated phenotyping costs. Observed trends of decreasing genotyping costs in contrast to increasing expenses for phenotyping are expected to lead various livestock sectors to widely adopt genomic technology. However, the development of beef cattle training populations has been generally conducted by private companies and at a significantly slower pace compared to the dairy industry [17]. For a worldwide adoption of genomic selection in beef cattle breeding, it is still necessary to develop cost-effective strategies and robust training populations for more economically-relevant traits. Very low-density panels including informative SNPs may represent a viable alternative for including only one or a very few key complex traits of high economic value, such as tick resistance, that are not yet considered in traditional genetic evaluation, because they are too difficult or too costly to measure. These additional trait tag-SNP predictions could be combined with pedigree/phenotype-based breeding values that are regularly derived for production traits using selection index theory [70]. However, if genomic predictions require high prediction accuracies for many traits in a complex breeding goal, the tag-SNP panel strategy may not be effective due to a likely large number of SNPs in the panel when adding tag-SNPs across various traits.

Fine-mapping investigation using next-generation sequencing could also be used to target flanking regions around the currently identified tag SNPs, which may be involved in the biological mechanisms of tick resistance in Hereford and Braford cattle. The identification of causal mutations, along with the availability of a larger training population or suitable blending with historical data, would be decisive to propose cost-effective genomic evaluations based on very low-density marker panels to improve tick resistance in commercials herds.

Conclusions

BayesB appears to be a suitable method for selecting tag SNPs based on Bayesian model frequency and t-like statistics. The resulting very low-density panel included SNPs that are potentially linked to functional gene networks and accounted for most of the genetic variance in tick resistance. The accuracy of genomic predictions derived from the proposed very low-density SNP panel using BayesA was moderate and may be useful for delivering cheaper genomic tests to the industry and for further studies related to fine-mapping for causal variants discovery.

Additional file

Additional file 1. SNP name, reference sequence, chromosome and position (Chr_Pos), window coded number by GenSel in physical map order, description of the genes (symbol, database search, name, HGNC_id, genome which were mapped) mapped to the 58 SNPs selected from the GWAS analysis.

Authors' contributions

BPS was involved in drafting the manuscript, made substantial contributions to conception, analysis and interpretation of data; VSJ made substantial contributions to the design and analysis of data; CCGG made substantial contributions to conception and acquisition of data; ARC made substantial contributions to the interpretation of data and was involved in revising the manuscript; FFC was involved in revising the manuscript, made substantial contributions to analyses and interpretation of data, and gave final approval of the version to be published. All authors read and approved the final manuscript.

Author details

[1] Embrapa Pecuária Sul, Caixa Postal 242 - BR 153 - Km 633, Bagé, Rio Grande do Sul 96.401-970, Brazil. [2] Departamento de Zootecnia, Universidade Federal de Viçosa, Avenida Peter Henry Rolfs, s/n - Campus Universitário, Viçosa, Minas Gerais 36.570-000, Brazil. [3] Embrapa Recursos Genéticos e Biotecnologia, Parque Estacao Biologica Final Av. W/5 Norte, Brasilia-DF, C.P. 02372, Brasília, Distrito Federal 70770-917, Brazil. [4] Universidade Federal de Pelotas, Capão do Leão, Rio Grande do Sul 96.000-010, Brazil.

Acknowledgements

The authors acknowledge the Delta G Connection for providing data for this research and Dr. John B. Cole for reviewing the manuscript and providing a critical contribution to its final version.

Competing interests

The authors declare that they have no competing interests.

Funding

Research supported by CNPq—National Council for Scientific and Technological Development Grant 478992/2012-2, CAPES—Coordination for the Improvement of Higher Level Personnel Grant PNPD 02645/09-2 and Embrapa-Brazilian Agricultural Research Corporation Grants 02.13.10.002 and 01.11.07.002.

References

1. Mapholi NO, Marufu MC, Maiwashe A, Banga CB, Muchenje V, MacNeil MD, et al. Towards a genomics approach to tick (Acari: Ixodidae) control in cattle: a review. Ticks Tick Borne Dis. 2014;5:475–83.
2. Machado MA, Azevedo ALS, Teodoro RL, Pires MA, Peixoto MGCD, de Freitas C, et al. Genome wide scan for quantitative trait loci affecting tick resistance in cattle (Bos taurus x Bos indicus). BMC Genomics. 2010;11:280.
3. Biegelmeyer P, Nizoli LQ, da Silva SS, dos Santos TRB, Dionello NJL, Gulias-Gomes CC, et al. Bovine genetic resistance effects on biological traits of Rhipicephalus (Boophilus) microplus. Vet Parasitol. 2015;208:231–7.
4. Piper EK, Jonsson NN, Gondro C, Lew-Tabor AE, Moolhuijzen P, Vance ME, et al. Immunological profiles of Bos taurus and Bos indicus cattle infested with the cattle tick, Rhipicephalus (Boophilus) microplus. Clin Vaccine Immunol. 2009;16:1074–86.
5. Wambura PN, Gwakisa PS, Silayo RS, Rugaimukamu E. Breed-associated resistance to tick infestation in Bos indicus and their crosses with Bos taurus. Vet Parasitol. 1998;77:63–70.
6. Brake DK, Pérez de León AA. Immunoregulation of bovine macrophages by factors in the salivary glands of Rhipicephalus microplus. Parasit Vectors. 2012;5:38.

7. Chan EKF, Hawken R, Reverter A. The combined effect of SNP-marker and phenotype attributes in genome-wide association studies. Anim Genet. 2009;40:149–56.

8. Martinez ML, Machado MA, Nascimento CS, Silva MV, Teodoro RL, Furlong J, et al. Association of BoLA-DRB3. 2 alleles with tick (*Boophilus microplus*) resistance in cattle. Genet Mol Res. 2006;5:513–24.

9. Gasparin G, Miyata M, Coutinho LL, Martinez ML, Teodoro RL, Furlong J, et al. Mapping of quantitative trait loci controlling tick [*Riphicephalus* (*Boophilus*) *microplus*] resistance on bovine chromosomes 5, 7 and 14. Anim Genet. 2007;38:453–9.

10. Barendse W. Assessing tick resistance in a bovine animal for selecting cattle for tick resistance by providing a nucleic acid from the bovine animal and assaying for the occurrence of a single nucleotide polymorphism (SNP). Patent application WO2007051248-A1. 2007; 1–146.

11. Turner LB, Harrison BE, Bunch RJ, Porto Neto LR, Li Y, Barendse W. A genome-wide association study of tick burden and milk composition in cattle. Anim Prod Sci. 2010;50:235–45.

12. Porto Neto LR, Bunch RJ, Harrison BE, Barendse W. DNA variation in the gene ELTD1 is associated with tick burden in cattle. Anim Genet. 2011;42:50–5.

13. Porto-Neto LR, Reverter A, Prayaga KC, Chan EKF, Johnston DJ, Hawken RJ, et al. The genetic architecture of climatic adaptation of tropical cattle. PLoS One. 2014;9:e113284.

14. Mapholi NO, Maiwashe A, Matika O, Riggio V, Bishop SC, MacNeil MD, et al. Genome-wide association study of tick resistance in South African Nguni cattle. Ticks Tick Borne Dis. 2016;7:487–97.

15. Moore JH, Asselbergs FW, Williams SM. Bioinformatics challenges for genome-wide association studies. Bioinformatics. 2010;26:445–55.

16. Meuwissen THE, Hayes BJ, Goddard ME. Prediction of total genetic value using genome-wide dense marker maps. Genetics. 2001;157:1819–29.

17. Van Eenennaam AL, Weigel KA, Young AE, Cleveland MA, Dekkers JCM. Applied animal genomics: results from the field. Annu Rev Anim Biosci. 2014;2:105–39.

18. Cardoso FF, Gomes CCG, Sollero BP, Oliveira MM, Roso VM, Piccoli ML, et al. Genomic prediction for tick resistance in Braford and Hereford cattle. J Anim Sci. 2015;93:2693–705.

19. Misztal I, Tsuruta S, Strabel T, Auvray B, Druet T, Lee DH. BLUPF90 and related programs (BGF90). In: Proceedings of the world congress on genetics applied to livestock production: 19–23 August 2002; Montpellier. Communication № 28-07; 2002.

20. Clayton D. snpStats: SnpMatrix and XSnpMatrix classes and methods. R Package version 1.26.0. 2015.

21. Sargolzaei M, Chesnais JP, Schenkel FS. A new approach for efficient genotype imputation using information from relatives. BMC Genomics. 2014;15:478.

22. Cardoso FF, Tempelman RJ. Hierarchical Bayes multiple-breed inference with an application to genetic evaluation of a Nelore-Hereford population. J Anim Sci. 2004;82:1589–601.

23. Garrick DJ, Taylor JF, Fernando RL. Deregressing estimated breeding values and weighting information for genomic regression analyses. Genet Sel Evol. 2009;41:55.

24. Piccoli ML, Braccini J, Cardoso FF, Sargolzaei M, Larmer SG, Schenkel FS. Accuracy of genome-wide imputation in Braford and Hereford beef cattle. BMC Genet. 2014;15:157.

25. Biegelmeyer P, Gulias-Gomes CC, Caetano AR, Steibel JP, Cardoso FF. Linkage disequilibrium, persistence of phase and effective population size estimates in Hereford and Braford cattle. BMC Genet. 2016;17:32.

26. Fernando RL, Garrick DJ. GenSel-user manual for a portifolio of genomic selection related analyses. Ames: Iowa State University; 2009.

27. Habier D, Fernando RL, Kizilkaya K, Garrick DJ. Extension of the Bayesian alphabet for genomic selection. BMC Bioinformatics. 2011;12:186.

28. Sun X, Habier D, Fernando RL, Garrick DJ, Dekkers JC. Genomic breeding value prediction and QTL mapping of QTLMAS2010 data using Bayesian methods. BMC Proc. 2011;5(Suppl 3):S13.

29. de Oliveira PSN, Cesar ASM, do Nascimento ML, Chaves AS, Tizioto PC, Tullio RR, et al. Identification of genomic regions associated with feed efficiency in Nelore cattle. BMC Genet. 2014;15:100.

30. Geweke J. Evaluating the accuracy of sampling-based approaches to the calculation of posterior moments. Federal Reserve Bank of Minneapolis: Research Department Staff Report, Minneapolis; 1991.

31. Smith BJ. boa: an R package for MCMC output convergence assessment and posterior inference. J Stat Softw. 2007;21:1–37.

32. Zimin AV, Delcher AL, Florea L, Kelley DR, Schatz MC, Puiu D, et al. A whole-genome assembly of the domestic cow. *Bos taurus*. Genome Biol. 2009;10:R42.

33. Wolc A, Arango J, Settar P, Fulton JE, O'Sullivan NP, Preisinger R, et al. Genome-wide association analysis and genetic architecture of egg weight and egg uniformity in layer chickens. Anim Genet. 2012;43(Suppl 1):87–96.

34. Schurink A, Wolc A, Ducro BJ, Frankena K, Garrick DJ, Dekkers JCM, et al. Genome-wide association study of insect bite hypersensitivity in two horse populations in the Netherlands. Genet Sel Evol. 2012;44:31.

35. Onteru SK, Gorbach DM, Young JM, Garrick DJ, Dekkers JCM, Rothschild MF. Whole genome association studies of residual feed intake and related traits in the pig. PLoS One. 2013;8:e61756.

36. Zare Y, Shook GE, Collins MT, Kirkpatrick BW. Genome-wide association analysis and genomic prediction of Mycobacterium avium subspecies paratuberculosis infection in US Jersey cattle. PLoS One. 2014;9:e88380.

37. Shin JH, Blay S, McNeney B, Graham J. LDheatmap: an R function for graphical display of pairwise linkage disequilibria between single nucleotide polymorphisms. J Stat Softw. 2006;16:1–10.

38. Badke YM, Bates RO, Ernst CW, Schwab C, Steibel JP. Estimation of linkage disequilibrium in four US pig breeds. BMC Genomics. 2012;13:24.

39. Saatchi M, Ward J, Garrick DJ. Accuracies of direct genomic breeding values in Hereford beef cattle using national or international training populations. J Anim Sci. 2013;91:1538–51.

40. Legarra A, Robert-Granié C, Manfredi E, Elsen JM. Performance of genomic selection in mice. Genetics. 2008;180:611–8.

41. Quinlan AR, Hall IM. BEDTools: a flexible suite of utilities for comparing genomic features. Bioinformatics. 2010;26:841–2.

42. Hammond MP, Birney E. Genome information resources–developments at Ensembl. Trends Genet. 2004;20:268–72.

43. Melville S, Fuchsberger C. NCBI2R—an R package to navigate and annotate genes and SNPs. R Package version 1; 2012.

44. Huang DW, Sherman BT, Zheng X, Yang J, Imamichi T, Stephens R, et al. Extracting biological meaning from large gene lists with DAVID. Curr Protoc Bioinformatics. 2009;Chapter 13:Unit 13.11.

45. Franceschini A, Szklarczyk D, Frankild S, Kuhn M, Simonovic M, Roth A, et al. STRING v9. 1: protein–protein interaction networks, with increased coverage and integration. Nucleic Acids Res. 2013;41:D808–15.

46. Yang J, Benyamin B, McEvoy BP, Gordon S, Henders AK, Nyholt DR, et al. Common SNPs explain a large proportion of the heritability for human height. Nat Genet. 2010;42:565–9.

47. Boddicker NJ, Bjorkquist A, Rowland RR, Lunney JK, Reecy JM, Dekkers JC. Genome-wide association and genomic prediction for host response to porcine reproductive and respiratory syndrome virus infection. Genet Sel Evol. 2014;46:18.

48. Fernando RL, Garrick D. Bayesian methods applied to GWAS. In: Gondro C, van der Werf J, Hayes B, editors. Genome-wide association studies and genomic prediction. Berlin: Springer; 2013. p. 237–74.

49. Cattle QTL Database. QTL for trait Tick resistance in the cattle genome. 2017. http://www.animalgenome.org/cgi-bin/QTLdb/BT/traitmap?trait_ID=123&traitnm=Tick%20resistance. Accessed 13 Jan 2017.

50. Robledo S, Idol RA, Crimmins DL, Ladenson JH, Mason PJ, Bessler M. The role of human ribosomal proteins in the maturation of rRNA and ribosome production. RNA. 2008;14:1918–29.

51. Dreymueller D, Pruessmeyer J, Groth E, Ludwig A. The role of ADAM-mediated shedding in vascular biology. Eur J Cell Biol. 2012;91:472–85.

52. Maretzky T, Scholz F, Köten B, Proksch E, Saftig P, Reiss K. ADAM10-mediated E-cadherin release is regulated by proinflammatory cytokines and modulates keratinocyte cohesion in eczematous dermatitis. J Invest Dermatol. 2008;128:1737–46.

53. Porto Neto LR, Bunch RJ, Harrison BE, Prayaga KC, Barendse W. Haplotypes that include the integrin alpha 11 gene are associated with tick burden in cattle. BMC Genet. 2010;11:55.

54. Regitano LCA, Ibelli AMG, Gasparin G, Miyata M, Azevedo ALS, Coutinho LL, et al. On the search for markers of tick resistance in bovines. Dev Biol. 2008;132:225–30.

55. Connell ML. Transmission of Anaplasma marginale by the cattle tick *Boophilus microplus*. Qld J Agric Anim Sci. 1974;31:185–93.

56. Jonsson NN, Bock RE, Jorgensen WK. Productivity and health effects of anaplasmosis and babesiosis on *Bos indicus* cattle and their crosses, and the effects of differing intensity of tick control in Australia. Vet Parasitol. 2008;155:1–9.

57. Lin M, den Dulk-Ras A, Hooykaas PJJ, Rikihisa Y. *Anaplasma phagocytophilum* AnkA secreted by type IV secretion system is tyrosine phosphorylated by Abl-1 to facilitate infection. Cell Microbiol. 2007;9:2644–57.

58. Porto Neto LR, Piper EK, Jonsson NN, Barendse WGC, Gondro C. Meta-analysis of genome wide association and gene expression studies to identify candidate genes for tick burden in cattle. In: Proceedings of the 9th world congress of genetic applied to livestock production: 1–6 August 2010; Leipzig; 2010.

59. Yi N, Yandell BS, Churchill GA, Allison DB, Eisen EJ, Pomp D. Bayesian model selection for genome-wide epistatic quantitative trait loci analysis. Genetics. 2005;170:1333–44.

60. Porto Neto LR, Jonsson NN, D'Occhio MJ, Barendse W. Molecular genetic approaches for identifying the basis of variation in resistance to tick infestation in cattle. Vet Parasitol. 2011;180:165–72.

61. De Rose R, McKenna RV, Cobon G, Tennent J, Zakrzewski H, Gale K, et al. Bm86 antigen induces a protective immune response against *Boophilus microplus* following DNA and protein vaccination in sheep. Vet Immunol Immunopathol. 1999;71:151–60.

62. Aguilar I, Misztal I, Johnson DL, Legarra A, Tsuruta S, Lawlor TJ. Hot topic: a unified approach to utilize phenotypic, full pedigree, and genomic information for genetic evaluation of Holstein final score. J Dairy Sci. 2010;93:743–52.

63. Christensen OF, Lund MS. Genomic prediction when some animals are not genotyped. Genet Sel Evol. 2010;42:2.

64. Fernando RL, Dekkers JCM, Garrick DJ. A class of Bayesian methods to combine large numbers of genotyped and non-genotyped animals for whole-genome analyses. Genet Sel Evol. 2014;46:50.

65. Cleveland MA, Forni S, Deeb N, Maltecca C. Genomic breeding value prediction using three Bayesian methods and application to reduced density marker panels. BMC Proc. 2010;4(Suppl 1):S6.

66. Habier D, Fernando RL, Dekkers JCM. The impact of genetic relationship information on genome-assisted breeding values. Genetics. 2007;177:2389–97.

67. Zhong S, Dekkers JCM, Fernando RL, Jannink JL. Factors affecting accuracy from genomic selection in populations derived from multiple inbred lines: a barley case study. Genetics. 2009;182:355–64.

68. Habier D, Fernando RL, Dekkers JCM. Genomic selection using low-density marker panels. Genetics. 2009;182:343–53.

69. Moser G, Khatkar MS, Hayes BJ, Raadsma HW. Accuracy of direct genomic values in Holstein bulls and cows using subsets of SNP markers. Genet Sel Evol. 2010;42:37.

70. Reis ÂP, Boligon AA, Yokoo MJ, Cardoso FF. Design of selection schemes to include tick resistance in the breeding goal for Hereford and Braford cattle. J Anim Sci. 2017;95:572–83.

4

Interaction of direct and social genetic effects with feeding regime in growing rabbits

Miriam Piles[1]*[iD], Ingrid David[2], Josep Ramon[1], Laurianne Canario[2], Oriol Rafel[1], Mariam Pascual[1], Mohamed Ragab[1,3] and Juan P. Sánchez[1]

Abstract

Background: Most rabbit production farms apply feed restriction at fattening because of its protective effect against digestive diseases that affect growing rabbits. However, it leads to competitive behaviour between cage mates, which is not observed when animals are fed ad libitum. Our aim was to estimate the contribution of direct (d) and social (s) genetic effects (also known as indirect genetic effects) to total heritable variance of average daily gain (ADG) in rabbits on different feeding regimens (FR), and the magnitude of the interaction between genotype and FR (G × FR).

Methods: A total of 6264 contemporary kits were housed in cages of eight individuals and raised on full (F) or restricted (R) feeding to 75% of the ad libitum intake. A Bayesian analysis of weekly records of ADG (from 32 to 60 days of age) in rabbits on F and R was performed with a two-trait model including d and s.

Results: The ratio between total heritable variance and phenotypic variance (T^2) was low (<0.10) and did not differ significantly between FR. However, the ratio between h^2 (i.e. variance of d relative to phenotypic variance) and T^2 was ~0.52 and 0.86 for animals on R and F, respectively, thus s contributed more to the heritable variance of animals on R than on F. Feeding regimen also affected the sign and magnitude of the correlation between d and s, i.e. −0.5 and ~0 for animals on R and F, respectively. The posterior mean (posterior sd) of the correlation between estimated total breeding values (ETBV) of animals on R and F was 0.26 (0.20), indicating very strong G × FR interactions. The correlations between d and s in rabbits on F and R ranged from −0.47 (d on F and s on R) to 0.64.

Conclusions: Our results suggest that selection of rabbits for ADG under F may completely fail to improve ADG in rabbits on R. Social genetic effects contribute substantially to ETBV of rabbits on R but not on F. Selection for ADG should be performed under production conditions regarding the FR, by accounting for s if the amount of food is limited.

Background

Feed efficiency is a key factor of profitability, productivity and sustainability of rabbit meat production. However, direct selection for this trait is difficult to implement because it requires individual recording of feed intake (FI), which is expensive and time consuming for animals housed in individual cages, and not possible for animals housed in groups since automatic feeding systems are still not available for this species. As a consequence, selection for feed efficiency has been performed either by indirect selection for average daily gain (ADG) of animals fed ad libitum and housed in groups [1], or by direct selection for residual feed intake (RFI) or for ADG under restricted feeding [2] with a limited number of selection candidates kept in individual cages. Results that compare the production performance of young rabbits selected for ADG and RFI and bred under different feeding regimens (FR) suggest there is an interaction effect between genotype and FR (G × FR) on ADG but not on other traits such as body weight (BW), FI or feed efficiency [2]. However, to date, variance estimates due to the G × FR interaction or its components (i.e. difference in genetic variances and genetic correlation between different conditions) for production traits in rabbit have not been

*Correspondence: miriam.piles@irta.es
[1] Institute for Food and Agriculture Research and Technology, Torre Marimon s/n, 08140 Caldes de Montbui, Barcelona, Spain
Full list of author information is available at the end of the article

reported. This interaction effect could be relevant when animals are bred in collective cages, which is the most common practice on commercial rabbit farms and elicits competition for feed intake between cage mates.

Social effects might be particularly important when feed restriction is applied at fattening, which is a common practice on production farms to reduce mortality associated to digestive disorders that are caused by some diseases, such as epizootic rabbit enteropathy [3]. By restricting the amount of food to 75% of the ad libitum intake and providing it once a day, Dalmau et al. [4] observed that signs of antagonistic behaviour such as biting, displacement and animals jumping one on top of each other occurred during the whole growing period. However, they did not find any effect of feed restriction on the coefficient of variation in body weight, which could indicate that, in spite of the competition, all kits faced the same level of feed restriction [5].

The benefits of selection for feed efficiency in individual cages could be lost when animals are kept in collective cages if substantial G × FR interaction effects exist. In the presence of those effects, phenotypic differences among individuals are not the same under different management conditions, with possible re-ranking of individuals [6]. In addition, ignoring the existence of social interaction effects in a breeding program could have negative consequences on the magnitude and sign of response to selection, which depend on the genetic parameters for direct and social genetic effects. Selection for individual performance may lead to strong competition when the covariance between direct and social effects is negative. Then, response to selection, which is determined by the sign of the covariance between an individual's phenotypic trait value and its total breeding value [7], could take the opposite direction to that desired [8, 9].

The current study aimed at estimating the genetic parameters for direct and social effects on ADG of growing rabbits that were raised on an ad libitum or restricted FR, and the interaction effect between the individual genotype [i.e. total breeding value (TBV)] and FR.

Methods
Animals and housing conditions
The experiment was carried out between July 2012 and June 2014 on the experimental farm of IRTA in Spain. We used 7864 kits, which were produced from a rabbit sire line (Caldes line [10]) selected for ADG in kits fed ad libitum during the fattening period (from 32 to 60 days of age) and housed in cages of eight individuals on a nucleus farm. All animals in the experiment were bred under the same management conditions except for the FR at fattening (5 weeks long in this experiment), which was either (1) full feeding i.e. ad libitum (F) or (2) restricted feeding

(R) to 75% of the ad libitum feed intake, with in both cases, the same standard diet. After weaning at 32 days of age, kits were randomly assigned to one of these two FR. In order to obtain homogeneous groups regarding animal size, kits under a FR were assigned to two groups based on their BW: big size kits (BS, i.e. with a BW >700 g) and small size kits (SS, i.e. with a BW ≤700 g), which is a common management practice on rabbit farms to obtain homogeneous growth and body weight at slaughter. Animals from a same litter were distributed to both FR. A maximum of two kits per litter were allocated to the same cage in order to minimize the effect of maternal and pre-weaning environmental effects on behaviour and growth performance at fattening. Kits were housed on a farm close to the selection nucleus (6.2 km) in 969 cages, each containing eight rabbits. Cages assigned to each group were interleaved on the farm.

The fattening period of the experiment lasted 5 weeks and food was supplied once per day in a feeder with three places and in the form of commercial pellets for rabbits that contain antibiotics to control gut disorders. At the last week of fattening, it was changed to a standard food without drugs. Data from this period were not considered for analysis due to the possible impact of the change in diet on the results. Details on the composition of the food for the analysed period are in Table 1. Water was available ad libitum (one nipple drinker per cage). The surface of the cage was 0.38 m^2. All these housing conditions are considered as standard conditions on commercial farms.

To obtain a feed restriction of 75% of the ad libitum feed intake, the amount of food supplied during week i was computed as 0.75 times the average feed intake of kits on F in a specific group j (j = BS or SS) during the week before (i.e., $i - 1$), plus 10% corresponding to the estimated increase in FI as the animals grows, i.e.:

$$FI_{R,ji} = (0.75 + 0.10) \times FI_{F,j(i-1)} \quad \text{for } i = 1, 2, 3, 4$$
$$\text{and 5 and } j = \text{BS or SS.}$$

This amount of food was multiplied by the number of animals alive in each cage at that time to determine feed requirements of the group. The amount of food for week

Table 1 Feed composition on a wet basis

Component	Amount
Crude fibre (%)	18.70
Crude protein (%)	15.02
Ashes (%)	8.97
Ether extract (%)	3.28
Oxytetracycline (ppm)	400
Valnemulin (ppm)	30
Colistin (ppm)	100

1 was computed from data that were recorded in previous experiments on the same line with animals raised in the same season. Actual feed restriction was on average 75 and 74.1% of the ad libitum intake in BS and SS kits, respectively.

Individual BW and total FI of kits in the same cage were weekly recorded after weaning (32 days). Average daily gain for a specific week was calculated as the difference in BW at the beginning and end of the week divided by the number of days elapsed (i.e., 7 ± 1 day). On control days, food was supplied after the kits were weighed. Information on sick animals was recorded. For each week, groups with animals that presented symptoms of an infectious disease, which was not caused a priori by antagonist behaviour (e.g. epizootic rabbit enteropathy, ERE, or respiratory problems), were discarded from the analyses, so that group size was always equal to 8. The average number of weeks with records per cage was 3.45 for animals on F and 3.41 for animals on R. The distribution of the data for each FR, BW class and week, after data filtering, is in Table 2. The final set of data for analysis included information on 6264 individuals born from 1303 litters housed in 783 cages. The pedigree included information on 7701 individuals, tracing back 5 generations from that corresponding to the animals of the experiment. The average relatedness coefficient within a cage was equal to 0.16.

Models and statistical analyses

Preliminary analyses of the data were performed using mixed linear models that were implemented with "lme4" and "lsmeans" packages of the R software [11]. Weekly records of BW, ADG and their coefficient of variation (CV) within a cage were analysed. Analysis of BW and ADG included the random effects of animal, cage and litter and the systematic effects of FR (two levels: F and R), week of fattening (four levels), body weight at weaning (two levels: BS and SS), batch (14 levels), parity order (four levels: 1, 2, 3, >3), number of kits born alive in the litter (seven levels: <6, 6, 7, 8, 9, 10, >10) and the interaction between week and all other systematic effects in the final model. Triple interactions between all systematic effects were initially

included but finally discarded because they were not significant. Models for the analysis of weekly CV in BW and ADG included the random effect of cage and the systematic effects of week of fattening, FR, BW at weaning, batch and the interactions between week and other systematic effects.

The genetic analysis was performed in two steps. In a first step, in order to estimate the G × FR interaction, ADG of animals on F and R (ADG_F and ADG_R, respectively) were considered as different but correlated traits and analysed with a two-trait model. The following repeatability animal model was fitted to weekly records of the same trait:

$$\mathbf{y} = \mathbf{X}\boldsymbol{\beta} + \mathbf{Z}_D\mathbf{d} + \mathbf{Z}_P\mathbf{p} + \mathbf{Z}_L\mathbf{l} + \mathbf{Z}_G\mathbf{g} + \mathbf{e}, \quad \text{(Model 1)}$$

where $\mathbf{y}$ is the vector of ADG_F or ADG_R, $\boldsymbol{\beta}$ is the vector of systematic effects with the corresponding incidence matrix $\mathbf{X}$, $\mathbf{d}$ is the vector of additive direct genetic effects with the corresponding incidence matrix $\mathbf{Z}_D$, $\mathbf{p}$ is the vector of permanent animal effects (6264 levels) with the corresponding incidence matrix $\mathbf{Z}_P$, $\mathbf{l}$ is the vector of litter birth effects (1303 levels) with the corresponding incidence matrix $\mathbf{Z}_L$, $\mathbf{g}$ is the vector of non-genetic group effects (783 levels) with the corresponding incidence matrix $\mathbf{Z}_G$, and $\mathbf{e}$ is the vector of residuals. Systematic effects were the same as those that were finally included in the preliminary analysis of ADG except FR and its interaction with week of fattening.

In a second step, the previous model was extended to include social genetic effects between cage mates [12]. This model can be written as:

$$\mathbf{y} = \mathbf{X}\boldsymbol{\beta} + \mathbf{Z}_D\mathbf{d} + \mathbf{Z}_S\mathbf{s} + \mathbf{Z}_P\mathbf{p} + \mathbf{Z}_L\mathbf{l} + \mathbf{Z}_G\mathbf{g} + \mathbf{e},$$

$$\text{(Model 2)}$$

where $\mathbf{d}$ and $\mathbf{s}$ are the vectors of the direct and social genetic effects, respectively and $\mathbf{Z}_D$ and $\mathbf{Z}_S$ are their corresponding incidence matrices. All other terms are identical to Model 1.

Bayesian methodology was used to estimate model parameters. The prior distribution of the additive genetic values was $\mathbf{d}|\mathbf{G}_1 \sim N(\mathbf{0}, \mathbf{A} \otimes \mathbf{G}_1)$ for Model 1, where $\mathbf{A}$ is the matrix of coefficients of relatedness between individuals, $\otimes$ denotes the Kronecker product and $\mathbf{G}_1$ is the

Table 2 Distribution of the number of records

Feeding regimen	Class according to body weight at weaning	Week of fattening			
		1	2	3	4
Restricted	Small	124	108	97	76
	Big	212	220	188	167
Full	Small	122	109	83	78
	Big	258	233	182	161

2×2 additive genetic covariance matrix for ADG of animals on F and R:

$$\mathbf{G}_1 = \begin{bmatrix} \sigma_{dF}^2 & \sigma_{dF,dR} \\ \sigma_{dR,dF} & \sigma_{dR}^2 \end{bmatrix}.$$

For Model 2, the prior distribution of the additive genetic values was $\begin{bmatrix} \mathbf{d} \\ \mathbf{s} \end{bmatrix} \Big| \mathbf{G}_2 \sim N(\mathbf{0}, \mathbf{G}_2 \otimes \mathbf{A})$, where all the terms are defined as before except $\mathbf{G}_2$ which, in this case, is the 4×4 additive genetic covariance matrix of direct and social effects for ADG of animals on F and R:

$$\mathbf{G}_2 = \begin{bmatrix} \sigma_{dF}^2 & \sigma_{dF,dR} & \sigma_{dF,sF} & \sigma_{dF,sR} \\ \sigma_{dR,dF} & \sigma_{dR}^2 & \sigma_{dR,sF} & \sigma_{dR,sR} \\ \sigma_{sF,dF} & \sigma_{sF,dR} & \sigma_{sF}^2 & \sigma_{sF,sR} \\ \sigma_{sR,dF} & \sigma_{sR,dR} & \sigma_{sR,sF} & \sigma_{sR}^2 \end{bmatrix},$$

where, σ_{ij}^2 is the variance of direct ($i = d$) or social ($i = s$) additive genetic effects for ADG of animals on full ($j = F$) or restricted ($j = R$) feeding; and $\sigma_{ij,i'j'}$ is the covariance terms for $ij \neq i'j'$ with (i and $i' = d$ or s) and (j and $j' = F$ or R).

The prior distribution of the litter effects ($l_i, i = 1, \ldots, N_l$), group effects ($g_i, i = 1, \ldots, N_g$) and permanent effects ($p_i, i = 1, \ldots, N_p$) in both models were $\mathbf{l}|\mathbf{L} \sim N(\mathbf{0}, \mathbf{I} \otimes \mathbf{L})$, $\mathbf{g}|\mathbf{G} \sim N(\mathbf{0}, \mathbf{I} \otimes \mathbf{Gr})$, and $\mathbf{p}|\mathbf{p} \sim N(\mathbf{0}, \mathbf{I} \otimes \mathbf{P})$, respectively, where $\mathbf{l}$, $\mathbf{g}$ and $\mathbf{p}$ are the corresponding vectors of litter, group and permanent effects, respectively, and $\mathbf{L}$, $\mathbf{Gr}$ and $\mathbf{P}$ are the corresponding 2×2 covariance matrices defined as: $\mathbf{L} = \begin{bmatrix} \sigma_{lF}^2 & \sigma_{lF,lR} \\ \sigma_{lR,lF} & \sigma_{lR}^2 \end{bmatrix}$, $\mathbf{Gr} = \begin{bmatrix} \sigma_{gF}^2 & 0 \\ 0 & \sigma_{gR}^2 \end{bmatrix}$ and

$\mathbf{P} = \begin{bmatrix} \sigma_{pF}^2 & 0 \\ 0 & \sigma_{pR}^2 \end{bmatrix}$, where σ_{ij}^2 is the variance of litter ($i = l$) or group effects ($i = g$) of animals on F or R ($j = F$ or R) and $\sigma_{lF,lR}$ is the covariance between litter effects of animals on F and R. $\mathbf{I}$ is the identity matrix. N_l, N_g and N_p are the number of litters, groups and animals with records, respectively. The residual variance matrix was defined as $\mathbf{R} = \begin{bmatrix} \sigma_{eF}^2 & 0 \\ 0 & \sigma_{eR}^2 \end{bmatrix}$, where σ_{eF}^2 and σ_{eR}^2 are the residual variances for animals on F and R, respectively.

Flat priors were used for systematic effects and variance components of the animal mixed models. The marginal posterior distributions of all the unknowns were approximated by Gibbs sampling [13] using the gibbs2f90 software [14].

Two sampling processes of 1,500,000 iterations each were run. The first 100,000 iterations were discarded as burn-in. One sample of the parameters of interest was saved every 100 iterations. The sampling variance of the

chains was obtained by computing Monte Carlo standard errors [15]. Statistics for the marginal posterior distributions were calculated directly from the samples.

Ratios of the phenotypic variance under FR j ($j = F$ or R), σ_{Pj}^2, were computed from the variance components of the different models. In all models, the contribution of the additive genetic variance to the total phenotypic variance for each trait was computed as [16]:

$$\sigma_{dj}^2 + (n-1)\sigma_{sj}^2 + r_j \times (n-1) \times \left[2\sigma_{dj,sj} + (n-2)\sigma_{sj}^2\right], \tag{1}$$

where σ_{dj}^2 and σ_{sj}^2 are the variances of direct and social additive genetic effects, respectively; $\sigma_{dj,sj}$ is the covariance between σ_{dj}^2 and σ_{sj}^2; n is the number of cage mates (8 in our case); and r_j is the average relationship coefficient between all pairs of individuals within a cage.

The total heritable contribution of the genes of a single individual on the mean trait value, defined as the individual's total breeding value (TBV) [17], was computed for individual i under FR j as:

$$\text{TBV}_{ij} = d_{ij} + (n-1)s_{ij}, \tag{2}$$

the total heritable variance available for selection σ_{TBV}^2 [18] was calculated as follows:

$$\sigma_{TBV_j}^2 = \sigma_{dj}^2 + 2(n-1)\sigma_{dj,sj} + (n-1)^2\sigma_{sj}^2, \tag{3}$$

and the ratio between total heritable variance and phenotypic variance [19] was calculated as follows:

$$T_j^2 = \frac{\sigma_{TBV_j}^2}{\sigma_{P_j}^2}. \tag{4}$$

The covariance of TBV between FR was computed as:

$$\begin{aligned} Cov(\text{TBV}_F, \text{TBV}_R) = {} & \sigma_{dF,dR} + (n-1)\sigma_{dF,sR} \\ & + (n-1)\sigma_{sF,dR} + (n-1)^2\sigma_{sF,sR}, \end{aligned} \tag{5}$$

where $\sigma_{dF,dR}$ is the covariance between direct genetic effects for animals on F and R, $\sigma_{dF,sR}$ is the covariance between direct genetic effects of animals on F and social genetic effects of animals on R, $\sigma_{sF,dR}$ is the covariance between social genetic effects of animals on F and direct genetic effects of animals on R, $\sigma_{sF,sR}$ is the covariance between social genetic effects of animals on F and R, and n is the number of kits in a cage (8 in our analysis).

The deviance information criterion (DIC [20]) was used to compare the models and assess which one yielded the best fit (considering a penalty for model complexity) to the data.

Results

Phenotypic analysis of BW and ADG

In this experiment, we confirmed that feed restriction has a protective effect on the health of growing rabbits with the mortality rate being 14.6% for animals on F and 9.5% for animals on R. The FR also had an important effect on BW and ADG, as expected, with slightly different values between large and small kits at weaning. Thus, the overall means for BW were equal to 1773 and 1487 g for BS kits on F and R, respectively, and 1460 and 1164 g for SS kits on F

and R, respectively. The overall means for ADG were equal to 51.9 and 36.9 g/day for BS kits on F and R, respectively, and 49.6 and 34.4 g/day for SS kits on F and R, respectively.

The pattern of growth also differed between animals on F or R (Fig. 1). Post-weaning growth decelerated after the first week for animals on F whereas it accelerated until week 3 and then remained constant for animals on R (Fig. 1c). The mean ADG was 106.9% higher for animals on F than on R for week 1 [difference in LSmeans $\pm$ standard error (se) $= 26.8 \pm 0.3$ g/

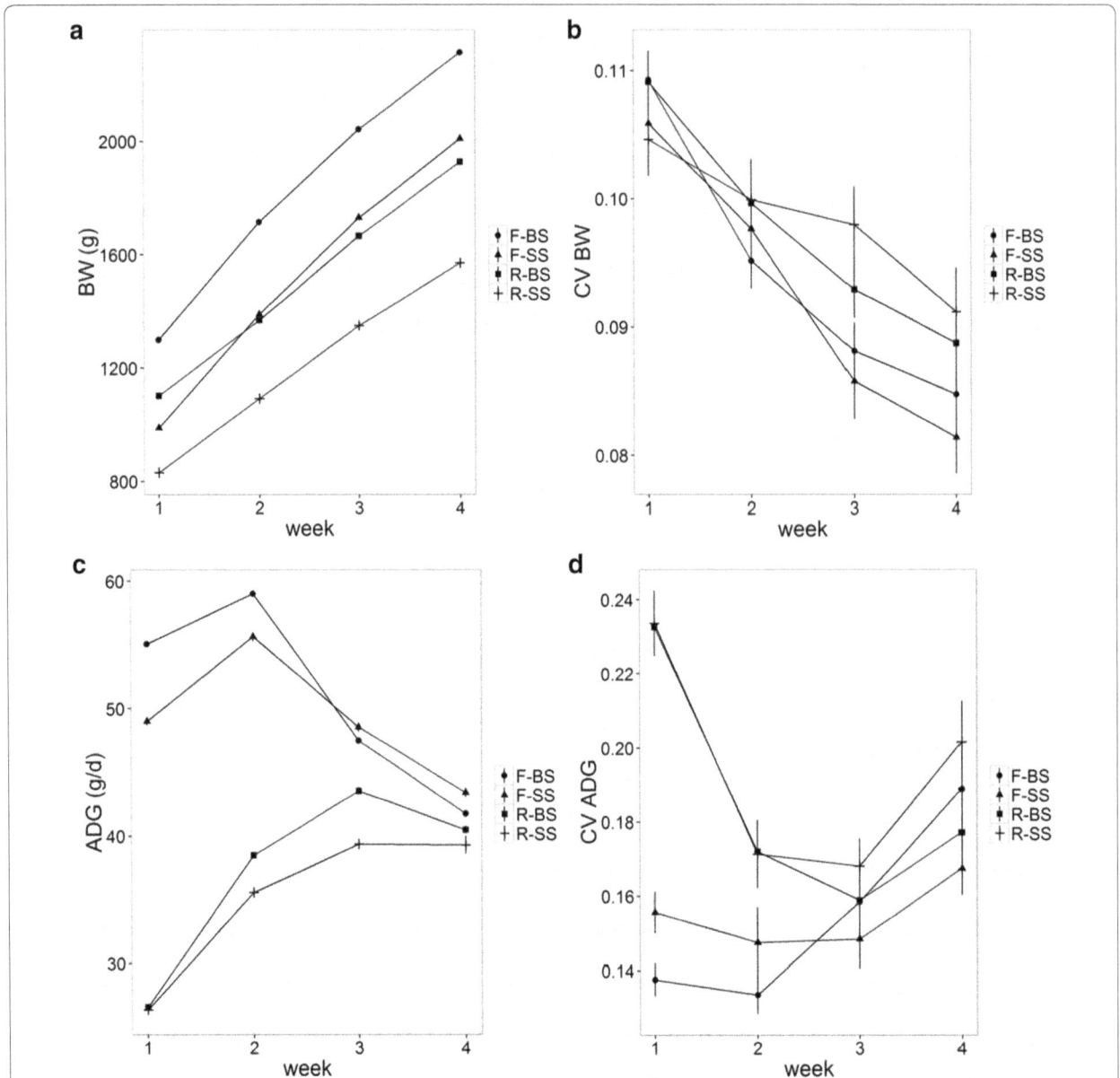

Fig. 1 Mean (**a**, **c**) and coefficient of variation within cage (CV; **b**, **d**) in body weight at the end of each week (BW) and average daily gain (ADG) for growing rabbits of big (BS) or small (SS) weaning body weight on full (F) or restricted (R) feeding regimen during the fattening period

day] but it decreased to 4.5% for week 4 (difference in LSmeans $\pm$ se $= 1.8 \pm 0.4$ g/day).

Variation in growth rate between cage mates was larger for animals on R than on F for the first 2 weeks after weaning. The magnitude of the difference was 0.091 ± 0.006 and 0.034 ± 0.006 for weeks 1 and 2, respectively (Fig. 1d). On the contrary, no differences between the CV in BW of the four groups of animals were observed for weeks 1 and 2 but small differences were found for weeks 3 and 4, and these were also larger for animals on R than on F (difference $= 5.7e{-}3 \pm 2.4e{-}3$ and $5.1e{-}3 \pm 2.5e{-}3$ for weeks 3 and 4, respectively).

Genetic analysis of ADG when social effects were ignored

The DIC was equal to 231,366.50 for the model that ignored social effects. Ratios of phenotypic variance from the classical repeatability model for ADG_F and ADG_R are in Table 3. We found a 40% larger phenotypic variance for animals on F than on R. This difference was mainly explained by a difference in additive genetic variance which was 4.7 times larger for animals on F than on R. Residual variance also differed between FR and was 1.3 times larger for animals on F than on R, whereas the variance due to permanent animal effects, litter or group effects did not differ between FR.

Differences in variance components between FR are due in part to the reduced magnitude of ADG due to limited feed intake. When the ratios of phenotypic variance were computed for each FR, we found similar values for residual variance (around 0.8), permanent, litter, and group effects but not for additive genetic effects which were quite different. Thus, posterior means (posterior sd) of heritability were 0.08 (0.02) and 0.02 (0.01) for ADG_F and ADG_R, respectively. Posterior means (posterior sd) of additive genetic and litter correlations between FR were 0.81 (0.16) and 0.92 (0.07), respectively (Table 4).

Table 4 Correlations between components of average daily gain of growing rabbits between full and restricted feeding regimens from the classical animal model

Effect	Mean	Lower and upper limits of HPD 95%[a]		MCse[b]
Additive	0.807	0.504	1.000	0.01212
Litter	0.924	0.790	1.000	0.00282

[a] HPD 95%: highest posterior density interval at 95%

[b] MCse: Monte Carlo standard error

Genetic analysis of ADG when social effects were included

The DIC was equal to 231,062.12 for the model that accounted for social effects, which was 304.38 units less than for Model 1. Results of this analysis are in Table 5, Figs. 2 and 3. The marginal posterior distributions of the proportion of phenotypic variance due to group, litter and permanent effects were low and very close for both FR (2, 7 and 1% for group, litter and permanent effects respectively; Table 5). These values did not differ from those obtained when social genetic effects were ignored (Table 3).

When estimating social genetic effects, it is essential to account for non-genetic effects between group mates [8]. Variance between groups takes the covariance among individuals within a group into account when it is positive. In a preliminary analysis, we checked that there was a positive correlation between individuals within a group. This correlation was estimated per week and per FR using a model that included social effects but not non-genetic group effects and that considered a uniform correlation between residuals for animals in the same cage. Models were fitted using ASReml [21]. The results obtained showed that the correlation between residuals was positive for all weeks and FR, ranging from 0.07 to 0.20 for ADG_F and 0.11 to 0.19 for ADG_R. Therefore, the

Table 3 Genetic parameters of average daily gain under different feeding regimens from the classical animal repeatability model

Parameter[a]	Restricted feeding				Full feeding			
	Mean	HPD 95%[b]		MCse[c]	Mean	HPD 95%[b]		MCse[c]
h^2	0.023	0.009	0.041	0.00047	0.078	0.049	0.108	0.00038
g^2	0.032	0.020	0.044	0.00007	0.024	0.012	0.036	0.00006
p^2	0.013	0.001	0.027	0.00029	0.008	0.000	0.021	0.00036
l^2	0.072	0.053	0.091	0.00026	0.073	0.052	0.096	0.00027
σ_p^2	56.014	54.308	57.773	0.00835	78.265	75.813	80.738	0.01194

[a] h^2: heritability; g^2: variance of group effects relative to phenotypic variance; p^2: variance of permanent animal effects relative to phenotypic variance; l^2: variance of litter effects relative to phenotypic variance; σ_p^2: phenotypic variance

[b] HPD 95%: highest posterior density interval at 95% with lower and upper limits

[c] MCse: Monte Carlo standard error

Table 5 Genetic parameters for average daily gain under different feeding regimens from the model that includes social genetic effects

Parameter[a]	Restricted feeding				Full feeding			
	Mean	HPD 95%[b]		MCse[c]	Mean	HPD 95%[b]		MCse[c]
TBV	3.616	0.663	6.981	0.07878	7.484	3.786	11.300	0.08072
h^2	0.033	0.017	0.051	0.00029	0.082	0.053	0.111	0.00041
T^2	0.064	0.012	0.123	0.00139	0.095	0.050	0.144	0.00102
s^2	0.0017	0.0005	0.0030	2.9e−05	0.0003	5.7e−05	0.0006	6.5e−06
$\rho_{d,s}$	−0.505	−0.912	−0.072	0.01059	−0.030	−0.553	0.495	0.01286
g^2	0.025	0.011	0.038	0.00017	0.023	0.011	0.036	0.00012
p^2	0.010	0.001	0.021	0.00027	0.008	0.000	0.020	0.00029
l^2	0.066	0.048	0.084	0.00021	0.071	0.049	0.092	0.00031
σ_p^2	56.296	54.590	58.144	0.01309	78.415	75.998	81.045	0.01620

[a] TBV: total breeding value; h^2: variance of direct genetic effects relative to phenotypic variance; T^2: variance of TBV relative to phenotypic variance; s^2: variance of social genetic effects relative to phenotypic variance; $\rho_{d,s}$: correlation between direct and social genetic effects; g^2: variance of group effects relative to phenotypic variance; p^2: variance of permanent animal effects relative to phenotypic variance; l^2: variance of litter effects relative to phenotypic variance; σ_p^2: phenotypic variance

[b] HPD 95%: highest posterior density interval at 95%

[c] MCse: Monte Carlo standard error

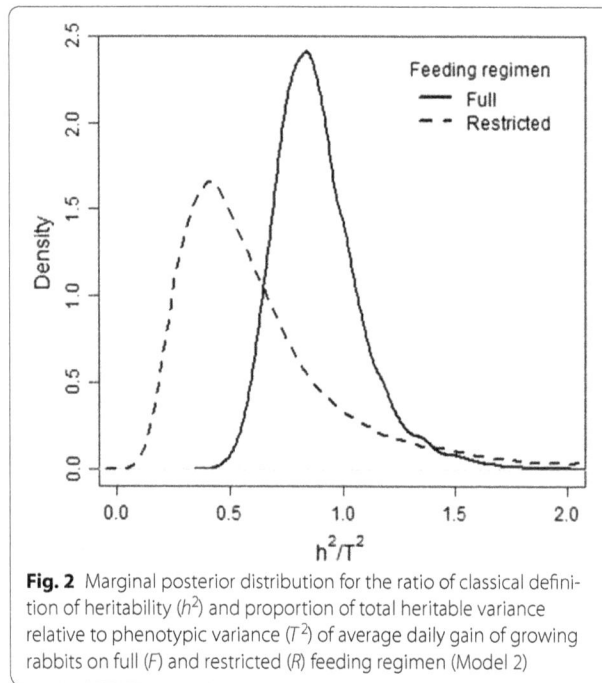

Fig. 2 Marginal posterior distribution for the ratio of classical definition of heritability (h^2) and proportion of total heritable variance relative to phenotypic variance (T^2) of average daily gain of growing rabbits on full (F) and restricted (R) feeding regimen (Model 2)

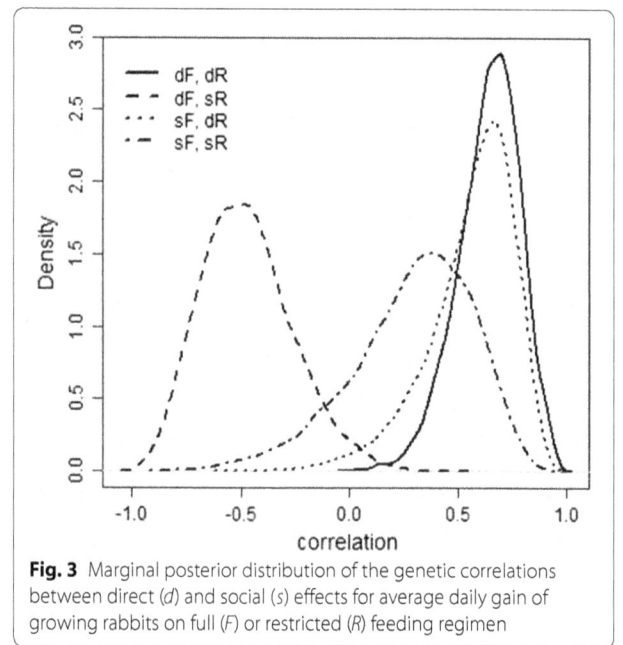

Fig. 3 Marginal posterior distribution of the genetic correlations between direct (d) and social (s) effects for average daily gain of growing rabbits on full (F) or restricted (R) feeding regimen

covariance between cage mates was estimated from the non-genetic group effects of Model 2.

Estimates of the variance of direct genetic effects did not significantly change when social genetic effects were included in the model (Tables 3, 5). Posterior means (posterior sd) were 6.14 (1.21) and 6.42 (1.20) g/day for ADG_F with Models 1 and 2, respectively, and 1.31 (0.50) and 1.86 (0.50) g/day for ADG_R with Models 1 and 2,

respectively. Variances of social genetic effects were estimated at 0.025 (0.012) and 0.096 (0.038) g/day for ADG_F and ADG_R, respectively.

Heritability (i.e. h^2, the ratio between variance of direct genetic effects and phenotypic variance) was low (0.08) for ADG_F and very low (0.03) for ADG_R. The ratio between total heritable variance and phenotypic variance (T^2) was also low (<0.10) and did not significantly differ between FR (Table 5). Thus, there is no evidence that the potential of the population to respond to selection depends on FR. However, the contribution of social genetic effects to the heritable variance was higher for animals on R than on F. Thus, the ratio between h^2 and T^2 was around 0.52 for animals on R and 0.86 for animals on F (Fig. 2). Feeding regimen also affected the sign and magnitude of the correlation between direct and social genetic effects which was negative and moderate for animals on R (posterior mean = −0.51; posterior sd = 0.22) and did not significantly differ from 0 for animals on F (Table 5).

Figure 3 shows the correlations between direct (d) and social (s) genetic effects for animals on F and R which all differ significantly from 1. The correlation between direct genetic effects for animals on R and direct and social genetic effects for animals on F were both clearly positive, with posterior means of (posterior sd) 0.64 (0.14) and 0.56 (0.20), respectively. The correlation between direct genetic effects for animals on F and social genetic effects for animals on R was high and clearly negative (posterior mean = −0.47; posterior sd = 0.21), whereas the correlation between social genetic effects for animals on both feeding regimens did not significantly differ from 0.

Relationship between models

The posterior mean (posterior sd) of the correlation between litter effects from Models 1 and 2 was very high, i.e. 0.94 (0.06). Regarding the genetic correlations, Table 6 shows the Pearson (upper diagonal) and Spearman (lower diagonal) correlations between the posterior means of individual EBV from Model 1, which ignores social genetic effects, and the posterior means of individual estimated total breeding values (ETBV) and its components from Model 2, which includes those effects. Pearson and Spearman correlations were very similar for all combinations of genetic effects. Therefore, all the comments provided below refer to Pearson correlations.

Regarding the interaction between genotype and FR, when social genetic effects were ignored, the correlation between direct genetic effects (i.e. EBV) for animals on F and R was high and positive (0.94), which indicated that there was no G × FR interaction apart from a scale effect that originated from the effect of FR on the genetic variance. On the contrary, when social genetic effects were taken into account, the correlation between ETBV for animals on F and R was null (posterior mean = 0.23; posterior sd = 0.26), which indicated that a strong interaction between genotype and FR exists. This result originates from the structure of the correlations between direct and social genetic effects on FR as described above.

The correlation between EBV and ETBV for animals on F was very high and positive (0.98). This was due to the high and positive (0.99) correlation between direct genetic effects from Models 1 and 2, the low and positive (0.12) correlation between EBV and social effects from Model 2, and the small contribution of social effects to the total heritable variance under these conditions. However, for animals on R, the correlation between EBV and ETBV was almost negligible (0.11) as a result of a high and positive correlation between EBV and direct genetic effects from Model 2 (0.93), which is counteracted by a negative and moderate to high (−0.60) correlation between EBV and social genetic effects from Model 2. This is indicative of a sizeable contribution of social genetic effects to ETBV.

The correlation between EBV for animals on F and ETBV for animals on R was negative and very small (−0.05), which was mainly due to the negative and moderate to high (−0.64) correlation between EBV for animals on F and social genetic effects for animals on R (which account for 48% of the total heritable variance), since the correlation between direct genetic effects for animals on F and R was high and positive (0.83).

On the contrary, the correlation between EBV for animals on R and ETBV for animals on F was high and

Table 6 Pearson (upper diagonal) and Spearman (lower diagonal) correlations between estimated breeding value (EBV) from Model 1 and estimated total breeding value (ETBV), direct genetic effects (d) and social genetic effects (s) from Model 2 for average daily gain of growing rabbits under full (F) or restricted (R) feeding regimen

	EBV_F	EBV_R	ETBV_F	ETBV_R	d_F	d_R	s_F	s_R
EBV_F		0.943	0.98	−0.047	0.985	0.834	0.119	−0.636
EBV_R	0.938		0.972	0.105	0.905	0.93	0.391	−0.602
ETBV_F	0.977	0.969		0.007	0.967	0.918	0.266	−0.66
ETBV_R	−0.028	0.101	0.013		−0.195	−0.014	0.761	0.689
d_F	0.983	0.898	0.964	−0.168		0.829	0.01	−0.733
d_R	0.817	0.917	0.909	−0.035	0.819		0.463	−0.734
s_F	0.115	0.362	0.252	0.741	0.018	0.424		0.181
s_R	−0.595	−0.572	−0.628	0.675	−0.696	−0.726	0.172	

positive (0.97). This was the result of the high and positive correlation between EBV for animals on R and direct genetic effects for animals on F (0.91), the low to moderate and positive correlation (0.39) between EBV for animals on R and social genetic effects for animals on F, and the small contribution of social effects to total heritable variance for animals on F.

Discussion

Our findings indicate that although the ratio between total heritable variance and phenotypic variance did not differ significantly between FR, there was a higher contribution of social interaction effects to the heritable variance of animals on R than on F. Feeding regimen also affected the sign and magnitude of the correlation between direct and social effects, which was negative and moderate for animals on R and not significantly different from 0 for animals on F. In addition, the correlation between ETBV for animals under both FR was null, which indicates very strong G × FR interactions. This correlation results from the correlations between direct and social effects for animals on F and R that ranged from −0.47 (direct effects for animals on F and social effects for animals on R) to 0.64. Therefore, selection of rabbits for ADG under F may be completely ineffective to improve ADG in rabbits raised on R when animals are housed in groups.

Feed restriction has been demonstrated to have good effects on the health and productive performance of growing rabbits (see review by Gidenne et al. [5]). Although weight gain is reduced during the period of feed restriction, a limitation of the amount of food reduces the risk of digestive diseases and improves feed efficiency both during restriction and especially after it, when the amount of food is gradually increased. As a consequence, feeding costs and the use of antibiotics are reduced, which result in economic profit and a reduced environmental impact of rabbit meat production. Thus, feed restriction has become a common management technique on commercial farms in Europe.

In our study, feed restriction was performed by supplying 75% of the ad libitum intake of a pelleted feed, once per day (~8 a.m.) in a feeder with three places, to breed rabbits that were housed in cages of eight individuals of similar size at weaning. These are standard conditions for rabbit meat production. According to Gidenne et al. [5], this technique of feed distribution is appropriate to achieve a good control of post-weaning intake and health status of the animals. This was confirmed by the 5.1% lower mortality rate for kits on R than on F that we observed in our study, which means an improvement of 35% over F conditions. This result is in agreement with

other findings in rabbits by Boisot et al. [22], Gidenne et al. [3] or Romero et al. [23].

Feed restriction modifies the feeding behaviour of growing rabbits who adapt to it very quickly. Rabbits on F eat 30 to 40 meals throughout the day [24]. Conversely, rabbits on R eat ~40% of the ad libitum intake within 2 h after feed distribution and complete their total intake within 10 h [25]. In our experiment, we also observed this feeding behaviour, which led to a high competition for food at distribution time with clear signs of antagonistic behaviour [4]. This could be a matter of concern for animal well-being. In spite of this feeding behaviour, Tudela and Lebas [26] pointed out that within-group variability of individual weight was not affected by feed limitation in collective cages because of the limited volume of the rabbit's stomach, which limits the amount of feed an animal can eat in one meal. This means that all kits have access to food at some time of the day and therefore, roughly the same level of feed restriction is applied to all cage mates. However, in our experiment, we observed significant differences between FR in the within-group homogeneity of ADG during the first 2 weeks and in the within-group homogeneity of BW during the last 2 weeks. Some animals grew at different rates and this could be related to differences in the amount of feed ingested, which could be caused or not by competition for feed, or to differences in feed efficiency, which could be more visible when the amount of feed is limited. Therefore, it was necessary to explore the role that social interactions between cage mates may have on the genetic determinism of growth and feed efficiency when the amount of food is limited. It was also important to explore the consequences of selection for growth and feed efficiency under specific conditions of feeding regimen and housing on the productive performance of young rabbits bred under different conditions. For example, it was necessary to evaluate the productive performance of growing rabbits on R that are housed in collective cages on commercial farms when breeding animals come from a nucleus herd in which selection for ADG is performed on animals bred on F in collective cages or on R in individual cages.

In order to achieve this objective, ADG of animals under different FR were assumed to be different traits following a character state model [6] for the analysis of G × E interactions. Two models were fitted to each trait: a classical repeatability animal model and an extension of that model including social genetic effects between cage mates. It was not possible to better consider the covariance structure of the longitudinal data by fitting a random regression model because of convergence problems of the sampling procedure. Since the animals were mixed, this model would have allowed the assessment of how social genetic and environmental effects change over time in growing rabbits. In fact, in a study on pigs, Canario et al. [27] and Camerlink et al. [28] highlighted that the association between social genetic effects and agonistic interactions, which is observed after mixing animals, decreased after several weeks. However, according to Piles et al. [29], the repeatability model could be a proper approximation for selection if there is no need to change the pattern of growth over time.

The models used in this study assumed that the residual variance was homogeneous with time and with body weight at weaning. In a preliminary analysis (results not shown), we demonstrated that residual variance was homogeneous with body weight at weaning but not with week (DIC heteroscedastic Model 1 = 231,366). However, since the estimates of other variance components of the models did not differ between the homoscedastic and heteroscedastic models, we present only the results of the homoscedastic models for the sake of simplicity.

The results from the classical repeatability model indicated that heritability of ADG_F was low (0.08) compared to published estimates of this parameter in the same (0.15; [30]) or other rabbit populations (0.11 [31]; 0.18 [32] and [33]; 0.29 [34]). However, it is important to note that the definition of the trait and the model used for analysis were different in our study. In the previously published studies, ADG was computed as the difference between body weight at the beginning and end of the fattening period divided by the number of days elapsed (ranging from 28 to 42 days), whereas we analysed weekly measures of ADG. It is well known that heritability increases with the length of the period measured because the residual variance is reduced by averaging the observations over a longer time period [35]. Therefore, estimates from those studies are not directly comparable with ours. The limited control of the environmental conditions on the experimental farm might have contributed to the large residual variance that we found.

The effect of the interaction between genotype and FR on ADG has been documented in several species such as mouse [36–38], pig [39], mink [40] and rabbit [2] by comparing, in most of the studies, the performance of a small number of individuals raised under different feeding regimens. However, the magnitude of the variance of this interaction effect and its components was not previously estimated in rabbit.

In rabbits, Drouilhet et al. [2] compared BW at 63 days of age, ADG, FI, feed conversion ratio and RFI of growing rabbits from two lines that were selected for feed efficiency following different strategies (selection for ADG_R and selection for RFI) and bred on F and R. The authors observed a significant interaction between FR and line effects only for ADG but not for the other traits.

Both lines had similar ADG when raised on R but the line selected for ADG_R had a higher ADG when raised on F than that selected for RFI (48.51 ± 0.72 vs. 45.29 ± 0.68 g/day, respectively). In addition, FR affected the phenotypic variance of the traits in a different way in the two selected lines depending on the trait. ADG was less variable in the line selected for ADG_R than in the line selected for RFI when raised on R (sd = 1.95 vs. 2.76 g/day, respectively) but was more variable in the line selected for ADG_R than in the line selected for RFI when raised on F (sd = 4.48 vs 3.55 g/day, respectively). However, contrary to our results, ADG_R was moderately to highly heritable (0.22 ± 0.06), which suggests that the animals were able to express their genetic potential for growth even when the amount of food was limited. However, unlike in our experiment, animals were housed in individual cages and therefore, competition for feed did not occur. Taken together, the different results on the genetic determinism of ADG_R that were obtained between the experiment performed by Drouilhet et al. [2] and our experiment and the results that are reported by Dalmau et al. [4] on the differences in kit behaviour under different FR indicate that it is likely that social interactions played an important role in our experiment.

The social effect of an individual refers to its effect on the trait value of a social partner. It can be of environmental or genetic origin and generates an additional level of heritable variation, which is not part of the observed phenotypic variance of the individual trait. As a result, the heritable variance of socially affected traits can exceed the phenotypic variance and, thus, ignoring these effects could lead to the absence of an optimal response or even to a response in the opposite direction to that of the selection objective [17, 18].

In our experiment, the proportion of variance between groups relative to the phenotypic variance was equivalent under both feeding regimens and small (around 2% of the phenotypic variance). In spite of this, the environmental component of social effects may still be large if the covariance between direct and social environmental effects is strongly negative, as may occur for animals on R.

The ratio between total heritable variance and phenotypic variance was the same irrespective of the FR. However, the contribution of the social genetic effects to the heritable variance was higher for animals on R than on F. The estimated values indicated that almost 50% of the heritable variance would be hidden when the classical model is used for selection to increase ADG in growing rabbits on R, but only 14% when they are on F. The strategy used to reduce the initial variation in body weight may have reduced competition between cage mates. However, the effect of maternal and pre-weaning environmental effects on the estimated direct and social interaction effects for ADG at fattening is expected to be minimized. In pigs, Bergsma et al. [19] estimated direct and social genetic effects for ADG during fattening by distinguishing between a sub-population of animals fed ad libitum and the ~90% of animals that were fed on a restricted regimen. In their case, the residual variance for ADG differed significantly between FR and was larger under R. Although the FR were less contrasted in their study than in ours, estimates for T^2 did not differ between FR.

Interestingly, in our study on rabbit, the sign and magnitude of the correlation between direct and social genetic effects differed between the two FR. It was negative and moderate for ADG_R, whereas it did not statistically differ from 0 for ADG_F. This means that, if social genetic effects are ignored, selection for increased ADG_R could lead to more competitive animals, which would have a negative effect on the growth of their cage mates, whereas when feed is not limited this is not expected to happen. Conversely, Bergsma et al. [19] obtained similar non-significant correlations between direct and social genetic effects in pigs under two FR.

In addition, the consequences of ignoring social effects on selection decisions to increase ADG in growing rabbits would be minimal when they are raised on F (the rank correlation between EBV from Model 1 and ETBV from Model 2 for animals on F was equal to 0.98) but they could be very important when they are raised on R given the absence of correlation between estimated genetic values of a selection candidate from both models (the rank correlation between EBV and ETBV for animals on R was equal to 0.11). Therefore, response to selection could be impacted if social effects are not taken into account in a breeding program for increasing ADG_R when animals are kept in collective cages. This result stresses the importance of considering social genetic effects in selection programs, as a potential method to improve production and eventually reduce harmful behaviours in livestock [18, 27].

In a study on Japanese quail that were raised under feed restriction and housed in 16-bird cages, Muir [16] found a moderate to large and negative (-0.56) genetic correlation between direct and social effects on weight at 6-weeks, which is comparable to that obtained in our experiment. In pigs fed ad libitum in groups of 6 to 12 pen mates of the same sex, Bergsma et al. [19] found a positive but low value for this parameter, which also agrees with our results. However, Chen et al. [41] reported positive to negative social interactions for growth in pig populations from the United States housed in groups of 15 pen-mates, with a maximum negative value at -0.37. Canario et al. [42] found that direct and social effects were independent in pigs raised under feed restriction. In

mice and pigs, social genetic effects are known to favour antagonistic behaviours when mixing animals [27, 43].

The magnitude of the interaction between genotype and feeding regimen is mainly due to the covariance structure of the direct and social genetic effects under different FR, which results in a null correlation between total heritable effects for animals on R and F. All the correlations between combinations of direct and social genetic effects for animals on F and R differed significantly from 1. The most remarkable result may be the correlation between direct genetic effects for animals on F and social genetic effects for animals on R, which was high and clearly negative. This result suggests that animals with high genetic potential for growth on F could be those that display a more competitive behaviour when they are feed restricted. Thus, the environment modulates the expression of social skills. Competition for food favours animals that have a less altruistic attitude towards cage mates. As a consequence, the productive performance of growing rabbits raised under feed restriction and housed in collective cages may differ greatly from that expected when selection for ADG_F is performed in a nucleus herd.

The consequences of ignoring social genetic effects would be minimal if animals were selected on R and produced on F (rank correlation between EBV for animals on R and ETBV for animals on F was 0.97) because of the low contribution of social genetic effects to ADG_F. However, if animals were selected on F and produced on R, no genetic response would be achieved on production farms because of the null rank correlation between EBV for animals on F and ETBV for animals on R (-0.028).

To date, few selection experiments that include social genetic effects have been performed in livestock species. In Japanese quail [16], two lines were selected for increased weight at 6 weeks under feed restriction (i.e. feeding was limited to once per day and access to the feeder was restricted). One of the lines was selected only for direct effects while the other was selected for ETBV using an index. After 23 cycles of selection (i.e. six generations), selection based on ETBV led to a positive response of 0.52 ± 0.25 g/hatch, whereas selection based only on direct genetic effects led to no response on weight at 6 weeks and to an increase in mortality of 0.32 ± 0.15 deaths/hatch. Therefore, ignoring social genetic effects was detrimental not only to response to selection but also to animal well-being. Selection for direct genetic effects only worsened the social genetic effects, which explains the lack of response in this line because responses in direct and social effects were in opposite directions.

These results were confirmed in a later experiment that was carried out on the same population under the same management and environmental conditions [44]. In this case, multilevel selections with birds housed in either kin or random groups were compared. Selection was based on best linear unbiased predictions (BLUP) of breeding values (EBV) which, when relatives are in the same group, is equivalent to multilevel selection since BLUP also weights the group performance. On the contrary, if relatives are in different groups, EBV do not include any weight in group performance. Results over 18 selection cycles indicated that response with multilevel selection in kin groups was 1.30 g/hatch, which was greater than that obtained with multilevel selection in random groups (1.13 g/hatch) and also significantly greater than that with selection based on TBV (0.52 ± 0.25 g/hatch). In addition, the mortality rate was also significantly lower (6 vs. 8% in kin and random groups, respectively). In our current analysis, the consequences on rabbit survival were not evaluated since it was observed that mortality was mainly due to some bouts of ERE on the farm. This is an infectious disease, which clearly impairs animal growth, but it is not considered to be caused by antagonistic behaviour between cage mates (animals in individual cages also have ERE). However, the impact that individuals have on each other is a crucial factor for the prevalence of infectious diseases in animals that are housed in groups [45, 46].

Conclusions

Social genetic effects contribute largely to the total heritable variance of average daily gain when growing rabbits are raised under restricted feeding but not when they are fed ad libitum. Ignoring those effects in a breeding program for increasing rabbit growth is likely to have negative consequences on the productive performance of young rabbits and eventually on animal well-being when the amount of food is limited. The interaction between genotype and feeding regimen will lead to a substantial re-ranking of the selection candidates under different conditions because of the null correlation between total heritable variance for animals on R and F. This is mainly due to the null and negative genetic correlations between direct and social genetic effects on full and restricted feeding regimen, respectively. Therefore, we recommend to select animals under the same conditions of feeding and housing as those applied on production farms for rabbit meat production, especially when feed restriction is applied on commercial farms.

Authors' contributions
MP designed the experiment, analysed the data and was the major contributor in writing the manuscript. JR and OR performed the experiment and data recording. ID contributed to data analysis and to writing and editing the manuscript. LC, MPa, and MR discussed the results and contributed to writing and editing the manuscript. JPS also participated in the design of the experiment and contributed to data analysis, discussion of results and writing and editing of the manuscript. All authors read and approved the final manuscript.

Author details
[1] Institute for Food and Agriculture Research and Technology, Torre Marimon s/n, 08140 Caldes de Montbui, Barcelona, Spain. [2] GenPhySE, INRA, Université de Toulouse, INPT, ENVT, 31326 Castanet Tolosan, France. [3] Poultry Production Department, Kafr El-Sheikh University, Kafr El-Sheikh 33516, Egypt.

Acknowledgements
The authors are grateful to the staff of Unitat de Cunicultura, IRTA (Oscar Perucho, Carmen Requena, Jaume Salinas and Juan Vicente) for their invaluable contribution to data recording and animal care during the experiment. MP thanks Dr. Manuel Ramon for his help in programming and the two anonymous reviewers for their comments.

Competing interests
The authors declare that they have no competing interests.

Funding
This research was supported by the Instituto Nacional de Investigación y Tecnología Agraria y Alimentaria (INIA, Madrid, Spain) Project RTA2011-00064-00-00 and the Feed-a-Gene Project funded by the European's Union H2020 Programme under Grant Agreement EU 633531.

References
1. Piles M, Gomez EA, Rafel O, Ramon J, Blasco A. Elliptical selection experiment for the estimation of genetic parameters of the growth rate and feed conversion ratio in rabbits. J Anim Sci. 2004;82:654–60.
2. Drouilhet L, Achard CS, Zemb O, Molette C, Gidenne T, Larzul C, et al. Direct and correlated responses to selection in two lines of rabbits selected for feed efficiency under ad libitum and restricted feeding: I. Production traits and gut microbiota characteristics. J Anim Sci. 2016;94:38–48.
3. Gidenne T, Combes S, Feugier A, Jehl N, Arveux P, Boisot P, et al. Feed restriction strategy in the growing rabbit. 2. Impact on digestive health, growth and carcass characteristics. Animal. 2009;3:509–15.
4. Dalmau A, Abdel-Khalek AM, Ramon J, Piles M, Sanchez JP, Velarde A, et al. Comparison of behaviour, performance and mortality in restricted and ad libitum-fed growing rabbits. Animal. 2015;9:1172–80.
5. Gidenne T, Combes S, Fortun-Lamothe L. Feed intake limitation strategies for the growing rabbit: effect on feeding behaviour, welfare, performance, digestive physiology and health: a review. Animal. 2012;6:1407–19.
6. Kolmodin R. Reaction norms for the study of genotype by environment interaction in animal breeding. PhD thesis, Swedish University of Agricultural Sciences. 2003.
7. Griffing B. Selection in reference to biological groups. I. Individual and group selection applied to populations of unordered groups. Aust J Biol Sci. 1967;20:127–39.
8. Bijma P, Muir WM, Ellen ED, Wolf JB, Van Arendonk JAM. Multilevel selection 2: estimating the genetic parameters determining inheritance and response to selection. Genetics. 2007;175:289–99.
9. Ellen ED, Rodenburg TB, Albers GAA, Bolhuis JE, Camerlink I, Duijvesteijn N, Knol EF, Muir WM, Peeters K, Reimert I, Sell-Kubiak E, van Arendonk JAM, Visscher J, Bijma P. The prospects of selection for social genetic effects to improve welfare and productivity in livestock. Front Genet. 2014;5:377.

10. Gómez EA, Rafel O, Ramon J. The Caldes strain Rabbit genetic resources in mediterranean countries. Opt Méditerr Ser B Etudes Rech. 2002;38:189–98.

11. R Core Team. R: a language and environment for statistical computing. Vienna: R Foundation for Statistical Computing. 2016. https://www.R-project.org/.

12. Muir WM, Schinckel A. Incorporation of competitive effects in breeding programs to improve productivity and animal well-being. In: 7th World Congress on Genetics Applied Livestock Production, Montpellier, France; 2002. p. 35–6.

13. Sorensen D, Gianola D. Likelihood, Bayesian, and MCMC methods in quantitative genetics. New York: Springer; 2002.

14. Misztal I, Tsuruta S, Strabel T, Auvray B, Druet T, Lee DH. BLUPF90 and related programs (BGF90). Montpellier: Institut National de la Recherche Agronomique (INRA); 2002. p. 1–2.

15. Geyer CJ. Practical Markov chain Monte Carlo. Stat Sci. 1992;7:473–511.

16. Muir WM. Incorporation of competitive effects in forest tree or animal breeding programs. Genetics. 2005;170:1247–59.

17. Bijma P, Muir WM, Van Arendonk JAM. Multilevel selection 1: quantitative genetics of inheritance and response to selection. Genetics. 2007;175:277–88.

18. Bijma P. A general definition of the heritable variation that determines the potential of a population to respond to selection. Genetics. 2011;189:1347–59.

19. Bergsma R, Kanis E, Knol EF, Bijma P. The contribution of social effects to heritable variation in finishing traits of domestic pigs (Sus scrofa). Genetics. 2008;178:1559–70.

20. Spiegelhalter DJ, Best NG, Carlin BP, van der Linde A. Bayesian measures of model complexity and fit. J R Stat Soc B. 2002;64:583–616.

21. Gilmour AR, Gogel BJ, Cullis BR, Welham SJ, Thompson R. ASReml. User guide release 4.1 structural specification. Hemel Hempstead: VSN International Ltd.; 2015.

22. Boisot P, Licois D, Gidenne T. Une restriction alimentaire réduit l'impact sanitaire d'une reproduction expérimentale de l'entéropathie épizootique (EEL) chez le lapin en croissance. In: Proceedings of the 10th Journées de la Recherche Cunicole, Paris, 19–20 Nov 2003. p. 267–70.

23. Romero C, Cuesta S, Astillero JR, Nicodemus N, De Blas C. Effect of early feed restriction on performance and health status in growing rabbits slaughtered at 2 kg live-weight. World Rabbit Sci. 2010;18:211–8.

24. Gidenne T, Lebas F. Feeding behaviour in rabbits. In: Bels V, editor. Feeding in domestic vertebrates. From structure to behaviour. Wallingford: CABI Publishing; 2006. p. 179–209.

25. Martignon MH, Combes S, Gidenne T. Effect of the feed distribution mode in a strategy of feed restriction: effect on the feed intake pattern, growth and digestive health in the rabbit. In: Proceedings of the 13th Journées des Recherches Cunicoles, Le Mans; 2009. p. 39–42.

26. Tudela F, Lebas F. Modalités du rationnement des lapins en engraissement. Effets du mode de distribution de la ration quotidienne sur la vitesse de croissance, le comportement alimentaire et l'homogénéité des poids. Cuniculture. 2006;33:21–7.

27. Canario L, Turner S, Roehe R, Lundeheim N, D'Eath R, Lawrence A, et al. Genetic associations between behavioral traits and direct-social effects of growth rate in pigs. J Anim Sci. 2012;90:4706–15.

28. Camerlink I, Turner SP, Bijma P, Bolhuis JE. Indirect genetic effects and housing conditions in relation to aggressive behaviour in pigs. PLoS One. 2013;8:e65136.

29. Piles M, Garcia ML, Rafel O, Ramon J, Baselga M. Genetics of litter size in three maternal lines of rabbits: repeatability versus multiple-trait models. J Anim Sci. 2006;84:2309–15.

30. Piles M, Tusell L. Genetic correlation between growth and female and male contributions to fertility in rabbit. J Anim Breed Genet. 2012;129:298–305.

31. Piles M, Blasco A. Response to selection for growth rate in rabbits estimated by using a control cryopreserved population. World Rabbit Sci. 2003;11:53–62.

32. McNitt JI, Lukefahr SD. Genetic and environmental parameters for post-weaning growth traits of rabbit using an animal model. In Proceedings of the 6th World Rabbit Congress, Toulouse; 1996. p. 325–9.

33. Lavara R, Vicente JS, Baselga M. Genetic parameter estimates for semen production traits and growth rate of a paternal rabbit line. J Anim Breed Genet. 2011;128:44–51.

34. Larzul C, Gondret F, Combes S. Rochambeau deH. Divergent selection on 63-day body weight in the rabbit: response on growth, carcass and muscle traits. Genet Sel Evol. 2005;37:105–22.

35. Wetten M, Ødegård J, Vangen O, Meuwissen THE. Simultaneous estimation of daily weight and feed intake curves for growing pigs by random regression. Animal. 2012;6:433–9.

36. Timon VM, Eisen EJ. Comparison of ad libitum and restricted feeding of mice selected and unselected for postweaning gain. 1. Growth, feed consumption and feed efficiency. Genetics. 1970;64:41–57.

37. Nielsen BVH, Andersen S. Selection for growth on normal and reduced protein diets in mice. Genet Res (Camb). 1987;50:9.

38. Urrutia MS, Hayes JF. Selection for weight gain in mice at two ages and under ad libitum and restricted feeding. 1. Direct and correlated responses in weight gain and body weight. Theor Appl Genet. 1988;75:415–23.

39. Cameron ND, Curran MK. Genotype with feeding regime interaction in pigs divergently selected for components of efficient lean growth-rate. Anim Sci. 1995;61:123–32.

40. Nielsen V, Møller SH, Hansen BK, Berg P. Response to selection and genotype by environment interaction in mink (Neovison vison) selected on ad libitum and restricted feeding. Can J Anim Sci. 2011;91:231–7.

41. Chen CY, Johnson RK, Newman S, Kachman SD, van Vleck LD. Effects of social interactions on empirical responses to selection for average daily gain of boars. J Anim Sci. 2009;87:844–9.

42. Canario L, Lundeheim N, Bijma P. The early-life social experiences of a pig shape the phenotypes of its social partners in adulthood. Heredity (Edinb). 2017;118:534–41.

43. Wilson AJ, Gelin U, Perron M-C, Réale D. Indirect genetic effects and the evolution of aggression in a vertebrate system. Proc Biol Sci. 2009;276:533–41.

44. Muir WM, Bijma P, Schinckel A. Multilevel selection with kin and non-kin groups, experimental results with japanese quail (Coturnix japonica). Evolution. 2013;67:1598–606.

45. Lipschutz-Powell D, Woolliams JA, Bijma P, Doeschl-Wilson AB. Indirect genetic effects and the spread of infectious disease: are we capturing the full heritable variation underlying disease prevalence? PLoS One. 2012;7:e39551.

46. Anacleto O, Garcia-Cortes LA, Lipschutz-Powell D, Woolliams JA, Doeschl-Wilson AB. A novel statistical model to estimate host genetic effects affecting disease transmission. Genetics. 2015;201:871–84.

5

Multiple-trait QTL mapping and genomic prediction for wool traits in sheep

Sunduimijid Bolormaa[1,7]*, Andrew A. Swan[2,7], Daniel J. Brown[2,7], Sue Hatcher[3,7], Nasir Moghaddar[4,7], Julius H. van der Werf[4,7], Michael E. Goddard[1,5] and Hans D. Daetwyler[1,6,7]

Abstract

Background: The application of genomic selection to sheep breeding could lead to substantial increases in profitability of wool production due to the availability of accurate breeding values from single nucleotide polymorphism (SNP) data. Several key traits determine the value of wool and influence a sheep's susceptibility to fleece rot and fly strike. Our aim was to predict genomic estimated breeding values (GEBV) and to compare three methods of combining information across traits to map polymorphisms that affect these traits.

Methods: GEBV for 5726 Merino and Merino crossbred sheep were calculated using BayesR and genomic best linear unbiased prediction (GBLUP) with real and imputed 510,174 SNPs for 22 traits (at yearling and adult ages) including wool production and quality, and breech conformation traits that are associated with susceptibility to fly strike. Accuracies of these GEBV were assessed using fivefold cross-validation. We also devised and compared three approximate multi-trait analyses to map pleiotropic quantitative trait loci (QTL): a multi-trait genome-wide association study and two multi-trait methods that use the output from BayesR analyses. One BayesR method used local GEBV for each trait, while the other used the posterior probabilities that a SNP had an effect on each trait.

Results: BayesR and GBLUP resulted in similar average GEBV accuracies across traits (~0.22). BayesR accuracies were highest for wool yield and fibre diameter (>0.40) and lowest for skin quality and dag score (<0.10). Generally, accuracy was higher for traits with larger reference populations and higher heritability. In total, the three multi-trait analyses identified 206 putative QTL, of which 20 were common to the three analyses. The two BayesR multi-trait approaches mapped QTL in a more defined manner than the multi-trait GWAS. We identified genes with known effects on hair growth (i.e. *FGF5*, *STAT3*, *KRT86*, and *ALX4*) near SNPs with pleiotropic effects on wool traits.

Conclusions: The mean accuracy of genomic prediction across wool traits was around 0.22. The three multi-trait analyses identified 206 putative QTL across the ovine genome. Detailed phenotypic information helped to identify likely candidate genes.

Background

Merino sheep are traditionally bred for wool. The value of a sheep's fleece depends on many characteristics including fleece weight, fibre diameter, staple strength and length, crimp (or curvature), wool color, and dust penetration [1]. Flystrike, particularly around the breech region, is an important disease of Australian sheep, which costs the industry $280 million annually [2]. There are attempts to select for resistance to breech flystrike but direct selection for this trait is not easy to implement in ram breeding flocks because breeding animals are valuable and managed to reduce incidence of flystrike. Several easily assessed or measured indicator traits associated with breech flystrike are available, namely breech wool cover, breech skin wrinkle, dags, wool colour and fleece rot [3, 4].

Genetic variation for these traits is well documented. Estimated heritabilities and correlations for wool traits in Merino sheep are reported in the literature [5, 6]. Genetic correlations between many wool traits including greasy

---

*Correspondence: bolormaa.sunduimijid@ecodev.vic.gov.au
[1] Agriculture Victoria Research, AgriBio Centre, Bundoora, VIC 3083, Australia
Full list of author information is available at the end of the article

fleece weight and staple length are positive and moderate to high [7] but there are significant antagonisms between some traits. For instance, unfavourable correlations exist between fleece weight and fibre diameter, fibre diameter and staple strength, and wrinkle score and fleece weight [8]. Thus, individual causal polymorphisms are likely to have pleiotropic effects on multiple traits, possibly because they operate through physiological mechanisms that affect wool growth generally. Genetic correlations between measurements at yearling and adult ages for the same trait are moderate to high and range from 0.6 to 0.9 for fleece weights and are higher for fibre diameter [9].

Genomic prediction is an attractive approach for sheep breeders because estimated breeding values (EBV) can be calculated from DNA marker data (genomic selection [10]) when animals are too young to be measured for some phenotypes. In several cases, non-linear Bayesian methods, such as BayesR, result in more accurate EBV than genomic best linear unbiased prediction (GBLUP) because they give more weight to markers that are close to the causal polymorphisms [11].

Genome-wide association studies (GWAS), which use similar data to genomic selection, have been widely used to map causal variants in livestock and humans [12, 13]. Both genomic selection and GWAS are usually performed one trait at a time, which limits their power to detect single nucleotide polymorphisms (SNPs) that are associated with multiple traits and hence to study patterns of pleoitropy. Bolormaa et al. [14] showed that a novel multi-trait analysis, which combines the results from a single-trait GWAS for 56 individual body composition traits in sheep, increased the power to detect pleiotropic quantitative trait loci (QTL). However, precise mapping of QTL may be difficult in such GWAS studies because SNPs that are at a long distance from the QTL can be identified as associated with the QTL due to long-range linkage disequilibrium (LD) between SNPs in livestock. Genomic selection models, especially non-linear models, which fit all SNPs simultaneously, tend to be less affected by this problem, and map causal variants or QTL more precisely than GWAS [11]. Ideally, we would like to find a common set of SNPs with a maximum power to map QTL, to show their pleiotropic effects and to estimate breeding values.

Although multi-trait BLUP genomic selection methods are available, they become cumbersome when the number of traits is very large. Here, we present three approximate multi-trait analyses that use the results from single-trait BayesR and GWAS analyses as data to identify the SNPs that are closest to pleiotropic QTL. We applied these methods for 22 traits (each measured at two ages) that describe wool production and quality (i.e. measured and visually assessed wool traits), and indicator traits associated with susceptibility to breech flystrike on 5726 sheep with genotypes for 510,174 SNPs.

Methods

Phenotype data and traits

The Merino and Merino crossbred animals used in this study were sourced from the Information Nucleus (IN) flock of Cooperative Research Centre for Sheep Industry Innovation (Sheep CRC) [15, 16]. In total, 7191 animals were available with phenotype records on 22 traits each measured at two ages ("yearling"; 150 < days < 550 days, and "adult"; ≥550 days), including wool production and quality traits and breech flystrike indicator traits. Trait definitions, numbers of records for each trait, raw means and standard deviations based on the phenotyped animals and number of genotyped animals are in Table 1. A complete description of the design, methods and analyses of wool production and quality assessments is in Hatcher et al. [17].

Prior to shearing at each IN site, the sheep were assessed for a series of visual wool scores, including staple structure (SSTRC), staple weathering (WEATH), wool character (CHAR), fleece rot (FLROT), dust penetration (DUST), and greasy colour (GCOL), and visual breech traits including breech cover (BCOV), crutch cover (CCOV), and dag (DAG) [18]. Each assessment was based on a five-point system in which low scores represent desirable attributes and high scores represent undesirable attributes. A mid-side wool sample (75 to 85 g) was taken from the right side of each animal using an electric handpiece. The samples were measured in a commercial laboratory (AWTA Limited, Melbourne, Vic., Australia) for a range of wool traits. Ten staples from each mid-side sample were randomly sub-sampled to measure staple length (SL) and staple strength (SS). The remainder of each mid-side sample was weighed, washed in hot water with detergent, rinsed in cold water twice, spun and oven-dried at 105 °C. The oven-dried weight was recorded and the 16% regain used to calculate the washing yield (YLD). A Shirley Analyser (AWTA Limited, Melbourne, Vic., Australia) was used to card the dried scoured sample before conditioning at 20 °C and 65% relative humidity for 24 h, after which 2-mm snippets were sampled via mini-coring. The snippets were measured for mean fibre diameter (FD), FD coefficient of variation (FDCV) and mean fibre curvature (CURV) by Sirolan™ Laserscan (AWTA Limited). The washed carded sample was further subsampled and measured for various tristimulus values (T units) that are routinely used to describe aspects of clean colour (X, Y, Z and Y–Z), where X refers to reflected red light, Y to reflected green light and Z to reflected blue light. With wool, the Y value indicates

Table 1 Number of records, their mean, standard deviation (SD), estimated heritabilities (h²), and variance explained by sire-by-flock interaction for each trait at yearling (Y) and adult (A) ages based on the animals with phenotypic measurements

Trait	Phenotyped animals					Genotyped animals	Trait	Phenotyped animals					Genotyped animals
	Nb	Mean	SD	h²	s.f.			Nb	Mean	SD	h²	s.f.	
YGFW	5840	3.6	1.06	0.41	0.11	5365	AGFW	4446	5.4	1.77	0.54	0.06	4428
YYLD	5807	71.2	6.46	0.46	0.06	5334	AYLD	4460	74.0	6.02	0.44	0.06	4442
YSL	3859	85.3	15.7	0.62	0.03	3403	ASL	3405	98.9	16.8	0.66	0.04	3399
YSS	3862	30.9	11.6	0.38	0.02	3398	ASS	3397	34.8	10.9	0.38	0.04	3391
YFD	4375	17.3	1.92	0.84	0.04	3915	AFD	3389	18.8	2.70	0.98	0.02	3383
YFDCV	4460	19.3	3.07	0.60	0.02	3999	AFDCV	3425	18.0	2.83	0.55	0.01	3419
YCURV	5353	72.0	12.0	0.63	0.03	5335	ACURV	4437	72.4	12.7	0.72	0.00	4419
YBRWR	5127	2.2	0.95	0.46	0.03	4981	ABRWR	3899	2.2	0.90	0.35	0.06	3884
YBCOV	3826	3.5	0.90	0.22	0.04	3826	ABCOV	2389	3.2	0.98	0.04	0.12	2381
YCCOV	3726	3.5	0.85	0.26	0.05	3724	ACCOV	2418	3.3	0.87	0.37	0.07	2407
YDAG	3956	1.8	1.01	0.09	0.09	3955	ADAG	2762	1.7	0.92	0.04	0.06	2748
YSSTRC	5138	2.7	0.86	0.17	0.14	5127	ASSTRC	3174	2.7	0.94	0.34	0.11	3161
YWEATH	5137	3.1	1.04	0.02	0.16	5126	AWEATH	3174	3.0	1.12	0.13	0.07	3161
YCHAR	5138	2.7	0.86	0.26	0.07	5127	ACHAR	3174	2.7	0.91	0.28	0.05	3161
YFLROT	5027	1.8	1.26	0.20	0.06	5016	AFLROT	3396	1.9	1.43	0.16	0.06	3381
YDUST	5137	3.1	0.98	0.18	0.07	5126	ADUST	3174	2.9	1.15	0.04	0.13	3161
YGCOL	5138	2.5	0.79	0.29	0.08	5127	AGCOL	3174	2.6	0.86	0.20	0.05	3161
YCOLZ	2740	65.6	2.54	0.32	0.00	2738	ACOLZ	2700	65.2	2.42	0.26	0.06	2695
YCOLYZ	2728	8.1	0.78	0.50	0.04	2726	ACOLYZ	2697	8.4	0.77	0.40	0.07	2692
YCOLY	2740	73.8	2.43	0.21	0.00	2738	ACOLY	2708	73.5	2.17	0.19	0.05	2703
YCOLX	2739	69.6	2.24	0.22	0.00	2737	ACOLX	2708	69.4	1.99	0.20	0.04	2703
YSKINQ	1972	2.9	0.72	0.25	0.07	1972	ASKINQ	1798	2.6	0.76	0.07	0.07	1785

GFW = greasy fleece weight; YLD = wool yield; SL = staple length; SS = staple strength; FD = mean fibre diameter; FDCV = fibre diameter coefficient of variation; mean fibre curvature (CURV); BRWR = breech wrinkle; BCOV = breech cover; CCOV = crutch cover; DAG = dag; SSTRC = staple structure, WEATH = staple weathering; CHAR = wool character; FLROT = fleece rot; DUST = dust penetration; GCOL = greasy colour; COL(Z, YZ,Y, and X) = wool clean colour: Z = reflected blue light; YZ = yellowness; Y = brightness; X = reflected red light; SKINQ = skin quality; s.f. = proportion of phenotypic variance explained by sire-by-flock interaction

brightness, with increasing values indicating increasing brightness, and the difference between the Y and Z values (Y–Z) indicating wool yellowness [19]. An additional visual breech score, i.e. breech wrinkle (BRWR), was scored post shearing [18]. Not all sheep were measured for all traits.

Genotype data

This study used the Ovine Infinium® HD SNP BeadChip that was developed under the auspices of the International Sheep Genomics Consortium (http://www.farmiq.co.nz/) and includes 606,006 high-density (HD) SNPs, and the Illumina 50 k Ovine SNP chip (Illumina Inc., San Diego, CA, USA) that includes 54,241 (50 k) SNPs. All SNPs were mapped to the OAR 3.1 build of the ovine genome sequence using SNPchiMp v.3 [20]. Sporadic missing genotypes for the HD SNP chip were filled using FImpute [21]. Quality control of genotypes, imputation of sporadic missing genotypes within each SNP chip, and imputation of the 50 k SNP genotypes to HD SNPs are

described in [14, 22]. The details of the quality control are summarised below.

Stringent quality control procedures were applied to the SNP data, i.e. SNPs were excluded if the call rate per SNP (which is the proportion of SNP genotypes that have a GC (Illumina GenCall) score above 0.6) was less than 95%, the minor allele frequency was lower than 0.01 or if departure from Hardy–Weinberg equilibrium ($P < 10^{-5}$) was extreme. These criteria were applied on each batch of genotypes separately rather than to the whole dataset. Furthermore, if the average call rate per individual was less than 90%, those animals were removed from the SNP data. The final set of genotyped animals used in this study included 5726 animals with phenotypic records for at least one trait: 690 animals were genotyped for 510,174 SNPs, and the remaining 5036 animals were genotyped with the 50 k SNPs, which were imputed to 510,174 SNPs using a multi-breed population of 1735 animals. Cross-validation within these 1735 HD genotypes revealed an average accuracy of imputation (correlation of imputed empirical non-50 k genotypes) of 0.9871. Most sires of phenotyped animals were genotyped with the HD SNP chip.

Single-trait genome-wide association studies

Mixed models that fit fixed and random effects simultaneously were used to estimate heritabilities and the effects of SNPs associated with each of the traits were studied. Pedigree heritabilities were estimated based on all animals for which genotype and phenotype data were available. The pedigree file included 10,360 animals (including 785 sires and 3891 dams). The analysis was performed using the ASReml software [23]. The same mixed model was used for the GWAS, except that each SNP (SNPi, $i = 1, 2, 3, ..., 510,174$) was added to the model as a fixed effect, one at a time, and tested for association with the trait:

$$\mathbf{y} = \mathbf{1}_n\mu + \mathbf{Xb} + \mathbf{s}_i\alpha_i + \mathbf{Z_1a} + \mathbf{Z_1Qq} + \mathbf{Z_2s.f} + \mathbf{e}, \quad (1)$$

where $\mathbf{y}$ is the vector of observed phenotypic values of the animals, $\mathbf{1}_n$ is an n × 1 vector of 1s (n = number of animals with phenotypes), μ is the overall mean, $\mathbf{X}$, $\mathbf{Z_1}$, and $\mathbf{Z_2}$ are design matrices relating observations to the corresponding fixed and random effects, $\mathbf{b}$ is a vector of fixed effects (described below), $\mathbf{a}$ is a vector of polygenic additive genetic effects sampled from the distribution $N(0, \mathbf{A}\sigma_a^2)$, where σ_a^2 is additive genetic variance and $\mathbf{A}$ is the additive relationship matrix constructed from the pedigree of the animals and their ancestors, $\mathbf{q}$, $\mathbf{s.f}$, and $\mathbf{e}$ are the vectors of random effects of breed (including Merino strains), sire-by-flock interaction, and residual error, respectively. $\mathbf{Q}$ is a matrix with breed and strain proportions calculated from pedigree ($q \sim N(0, \mathbf{I}\sigma_q^2)$) [24]; $\mathbf{s}_i$ is a vector of genotypes for each animal at the ith SNP and α_i is the corresponding SNP fixed effect. The sire-by-flock interaction effect being significant ($P < 0.05$), it was retained in the model. All models included contemporary group, flock, drop year, sex, birth type, and rearing type as fixed effects and age of measurement, age of dam and its squared value as covariates. Birth type (BT: single = 1, twin = 2, triplet = 3, and quadruplet = 4) and rearing type (RT: single = 1, twin = 2, and triplet = 3) were grouped together (BTRT). The significant interactions between fixed effects (including flock by BTRT, sex by BTRT, and BTRT by age of dam) were fitted in the model. SNPs were tested for significant association with particular traits at several significance thresholds, and the false discovery rate (FDR) [25] was calculated to account for the thousands of significance tests performed. Based on FDR, we chose stringent significance thresholds ($P < 10^{-5}$ and $P < 5 \times 10^{-7}$) to minimise false discoveries (Table 2).

Genomic prediction

Genomic prediction analyses were performed using two methods: GBLUP [26, 27], with the genomic relationship matrix (GRM) constructed as described by [28] and BayesR [29]. Genomic EBV (GEBV) were estimated in GBLUP directly and calculated from SNP effects in BayesR.

Validation populations used for BayesR and GBLUP

The same validation populations were used for both methods. All 5726 genotyped and phenotyped animals were assigned to two groups (straightbred MER and crossbred MER) according to their breed using the breed proportions ($\mathbf{Q}$) of animals, which were derived from pedigree [24]. Of the 5726 individuals, 3883 were straightbred MER ($Q_{MER} > 0.90$) and 1843 were crossbred MER ($0.25 < Q_{MER} \leq 0.90$). All crossbred MER animals and the straightbred animals that had sires in common with crossbred MER were in all training sets but not in the validation set. The remaining straightbred MER individuals were split into five sets by allocating all offspring of randomly selected sires to one of the five datasets (fivefold cross-validation approach). In this way, no animal used for validation had paternal half-sibs in the training population. Thus, the analysis was performed five times using each data fold in turn as a validation group and the remaining fourfolds as the training population (i.e. fourfolds plus crossbred MER from above).

GBLUP

GEBV were predicted using Model (1), but no single SNP effect ($\mathbf{s}_i$) was fitted and $\mathbf{a}$ was replaced by $\mathbf{g}$, where $\mathbf{g}$ is a vector of GEBV $\sim N(0, \mathbf{G}\sigma_g^2)$, where σ_g^2 is the genetic variance and $\mathbf{G}$ is the genomic relationship matrix (GRM). For a SNP to be included in the GRM, its minor allele frequency had to be higher than 0.005, once genotypes (real and imputed) were combined in the whole dataset. Validation animals were included in the GRM but had unknown phenotypes in the calculation of GEBV.

BayesR

The BayesR method [29] assumes that SNP effects are from a mixture of four normal distributions with the variance of each distribution equal to 0, 0.01, 0.1 or 1% of the genetic variance. Gibbs sampling was used to sample from the posterior distributions of the parameters, running 40,000 iterations with 20,000 iterations of burn-in, which were averaged across five parallel chains. Since the BayesR software used in this study does not allow fitting a full-model, residuals were calculated by adjusting the phenotypes for fixed effects, breed effects, and the sire-by-flock interaction effect using ASReml [23]. These residuals were then used as phenotypes in the

Table 2 Number of significant SNPs ($P < 10^{-5}$ and $P < 5 \times 10^{-7}$) and their false discovery rates (FDR, %) for each trait from the single-trait GWAS

Trait[a]	$P < 1 \times 10^{-5}$		$P < 5 \times 10^{-7}$		Trait[a]	$P < 1 \times 10^{-5}$		$P < 5 \times 10^{-7}$	
	Nb SNPs	FDR	Nb SNPs	FDR[b]		Nb SNPs	FDR	Nb SNPs	FDR[b]
YGFW	69	7.4	7	3.6	AGFW	304	1.7	157	0.2
YYLD	122	4.2	75	0.3	AYLD	113	4.5	59	0.4
YSL	68	7.5	26	1.0	ASL	39	13.1	3	8.5
YSS	56	9.1	31	0.8	ASS	39	13.1	15	1.7
YFD	202	2.5	36	0.7	AFD	295	1.7	69	0.4
YFDCV	97	5.3	31	0.8	AFDCV	109	4.7	47	0.5
YCURV	105	4.9	22	1.2	ACURV	103	5.0	8	3.2
YBRWR	50	10.2	23	1.1	ABRWR	15	34.0	2	12.8
YBCOV	3		2	12.8	ABCOV	6	85.0	1	25.5
YCCOV	19	26.9	1	25.5	ACCOV	45	11.3	19	1.3
YDAG	11	46.4	0		ADAG	5		1	25.5
YSSTRC	41	12.4	2	12.8	ASSTRC	19	26.9	1	25.5
YWEATH	5		0		AWEATH	8	63.8	2	12.8
YCHAR	46	11.1	15	1.7	ACHAR	28	18.2	5	5.1
YFLROT	22	23.2	1	25.5	AFLROT	75	6.8	12	2.1
YDUST	15	34.0	0		ADUST	10	51.0	1	25.5
YGCOL	25	20.4	2	12.8	AGCOL	21	24.3	0	
YCOLZ	28	18.2	5	5.1	ACOLZ	25	20.4	0	
YCOLYZ	7	72.9	1	25.5	ACOLYZ	28	18.2	0	
YCOLY	22	23.2	3	8.5	ACOLY	20	25.5	0	
YCOLX	24	21.3	4	6.4	ACOLX	23	22.2	0	
YSKINQ	9	56.7	0		ASKINQ	7	72.9	0	

[a] Trait names see Table 1

[b] For empty cells, FDR are not available or are higher than 100%

analysis. The BayesR analysis fits the effects of all SNPs and a residual polygenic effect, the latter with a covariance structure that is proportional to the numerator relationship matrix (**A**). The SNP effects from BayesR were then used to calculate GEBV for animals in the validation sets.

Estimation of the accuracy of GEBV

For each validation population, the accuracy of genomic prediction was calculated as the correlation between GEBV and the adjusted phenotype corrected for fixed effects, which was divided by the square root of the heritability of the trait (h^2) that was estimated by using the 8-generation pedigree of all recorded animals. Thus, we report accuracy as the estimated correlation between GEBV and true breeding values.

Multi-trait analyses to identify pleiotropic QTL

To identify the pleiotropic genomic regions that control a wide range of wool traits, we used three approximate multi-trait analyses: (1) multi-trait GWAS (multi-GWAS) following the procedure described by Bolormaa et al.

[30]; (2) approximate BayesR posterior probability of SNP effects across traits (multi-PP); and (3) the linear combination of local GEBV that were derived from BayesR estimates of SNP effects within 250-kb windows across all traits for a total of 9813 windows (multi-LGEBV).

Multi-GWAS

Multi-trait analyses were performed following the procedure in Bolormaa et al. [30] based on SNP effects that were estimated from 44 individual single-trait GWAS. The multi-trait χ^2 statistic was calculated as: multi-trait $\chi^2 = t_i' V^{-1} \mathbf{t}_i$, where $\mathbf{t}_i$ is a vector of the signed t-values of the effects of the ith SNP for the 44 traits and V^{-1} is the inverse of the 44×44 correlation matrix where the correlation is calculated over the 510,174 estimated SNP effects (signed t-values) between each pair of traits. The power of QTL detection was investigated by comparing the FDR [25] from the multi-trait test with the FDR from the single-trait GWAS. To avoid testing a large number of closely-linked SNPs, only the SNPs with the most stringent P values ($P < 5 \times 10^{-7}$) within each 1-Mb window were selected from the multi-trait analysis. These SNPs

$(P < 5 \times 10^{-7})$ were assumed to be near QTL that affect wool traits and were also examined as likely candidate gene positions.

Multi-PP

The posterior probability that a SNP had no effect on any trait was calculated as a product of posterior probabilities that a SNP had no effect on any individual trait. We use 1 minus this probability to approximate the probability that a SNP has an effect on at least one trait i.e.: $PP_{effect \neq 0} = 1 - \Pi(pp_{effect=0})$. We retained SNPs with a $PP_{effect \neq 0}$ higher than 0.3.

Multi-LGEBV

GEBV in the 9813 non-overlapping 250-kb windows (local GEBV) for each animal were calculated based on the BayesR effects of all SNPs in the window. A high variance of local GEBV in a window, means that the window includes a QTL for that trait [11]. For each of the 250-kb windows (segments), the covariance of local GEBV between each pair of the 44 traits was standardized for the variability of each trait as follows:

$$t_{(y,x)} = \frac{cov_{LGEBV_{(y,x)}}}{\sigma_y \sigma_x},$$

where $cov_{LGEBV_{(y,x)}}$ is the covariance of local GEBV between trait y and trait x and σ_y and σ_x are the phenotypic standard deviations of trait y and x. If a window contains a single QTL, we expect this covariance matrix to be dominated by one linear combination of traits representing the QTL effect. Therefore, we carried out a principal component (PC) analysis of the 44×44 covariance matrix and examined the highest eigenvalue. This eigenvalue is the variance, in phenotypic standard deviation units, of the linear combination of traits with the largest variance. Across all 9813 windows, the 120 windows with the highest eigenvalues of the first PC (PC1) were arbitrary selected as containing a QTL.

After selecting the 120 top 250-kb windows, we further investigated which SNP, in each of the selected windows, was most highly associated with the PC1 of the (co)variance matrix (the 'best' SNP). To identify the best SNP, a pseudo trait (S_{LC}), which consisted of the linear combination of local standardised GEBV and the PC1 eigenvector across the 44 traits, was calculated for each animal at each of the selected 250-kb windows using the following formula: $S_{LC} = y'x$, where y' is the transpose of a vector of the local GEBV that are standardised (divided) by the phenotypic standard deviation of each trait (1×44) in the corresponding 250-kb window, and x is the eigenvector of the PC1 (44×1), which was calculated based on the covariance matrix of the standardised local GEBV among 44 traits. Since not all

animals were measured for all traits, the missing local GEBV were replaced by the mean of local GEBV across all animals, for which measurements for that particular trait were available in order to calculate a linear combination of the 44 traits. This resulted in a S_{LC} for each animal in each of the 120 chosen windows, which were now used as a pseudo-trait in GWAS within each window to identify the "best" SNP that tagged variation within the segment. The model used in this GWAS was as follows: $S_{LC} \sim mean + SNP_i + animal + error$, where animal and error were fitted as random effects and SNP_i was fitted as a fixed effect, one at a time (the phenotypes used to calculate S_{LC} had already been corrected for other fixed effects, breed effect, and sire-by-flock interaction effect prior to BayesR). After performing this new GWAS, the SNP with the highest F value within each of the corresponding 250-kb windows was chosen as the best SNP to tag the QTL.

Validation of SNP effects

Predicting missing phenotypes

The multi-trait validation of SNP effects (i.e. using the linear index approach [30]) requires complete data for all traits at the individual level. For animals without records for a particular trait, missing phenotypes were predicted using a multiple regression approach. This multiple regression used phenotypes that were already corrected for fixed effects, breed effect, and sire-by-flock interaction effect. The multiple regression procedure uses the phenotypic (co)variance matrix between the 44 traits based on all animals (training and validation population), which was estimated using the available phenotypic values. Next, the phenotypic (co)variance matrix was inverted. Then, separately for each animal, traits with phenotypes were ordered before traits with missing phenotypes. Again, for each animal, the missing phenotypes (y_n) were then predicted using the following formula:

$$\hat{y}_n = -(U^{nn})^{-1} U^{nm} y_m,$$

where y_m is a vector of the traits measured on a particular animal, U^{nn} is the inverse of phenotypic covariance matrix between 44 traits with a missing record, and U^{nn} is the inverse of phenotypic covariance matrix of the traits with and without a missing record.

Validation populations for single-trait and multi-trait analyses

To enable validation of SNP effects in independent animals, the 5726 animals with full phenotypic data (including the predicted phenotypic values) were divided into training and validation populations. The same cross-validation approach was used as described in the genomic prediction section above, except that crossbred animals were not excluded from validation sets. Then, one of the five divisions was randomly used as a validation population and the other four divisions as the training population. Only one 4:1 division (i.e. 4649 training animals: 1077 validation animals, out of 5726 genotyped animals) was tested for each of the traits studied.

Validation of SNP effects from the single-trait analysis

The GWAS for all traits were performed separately in the training and validation populations. SNPs with a significant effect in the training population were validated in the validation population for the five traits that displayed the largest number of significant associations. We counted the number of SNPs for which the effect was in the same direction in both the validation and training populations.

Selection of top SNPs from each of the three multi-trait methods in the training population

The single-trait BayesR analyses for all traits were also performed using only the training population (4649 animals). Then, the three multi-trait analyses described in the previous section (multi-GWAS, multi-PP, and multi-LGEBV) were repeated in the validation set to validate the top SNPs from each multi-trait analysis in the training population.

Use of a linear index in multi-trait validation

The top SNPs in each of the three multi-trait analyses (i.e., three different lists of SNPs) in the training population were validated separately in an independent set of validation animals. A linear index (y_I) was calculated for each putative QTL (from each of the three methods) and for each animal. It summarised the information across the 44 traits (22 traits at two ages) and was calculated using the following formula [30]: $y_I = b'\mathbf{C}^{-1}\mathbf{y}$, where b' is the transpose of a vector of the estimated SNP effects (not t values) on the 44 traits (1×44), which were estimated from only the training population, $\mathbf{C}^{-1}$ is the inverse of the 44×44 (co)variance matrix among the 44 traits calculated from the estimated effects of the 510,174 SNPs, only in the training population, and $\mathbf{y}$ is a 44×1 vector of the phenotype values (adjusted for fixed, breed and sire-by-flock effects) for 44 traits for each animal in the validation sample. This resulted in a linear index (y_I) for all putative QTL where, for each QTL, each animal had a linear index summary "phenotype". Then, GWAS, in which each y_I was treated as a new trait, were performed using the following model:

$$y_I = \text{mean} + \text{SNP}_i + \text{animal} + \text{error},$$

where animal and error were fitted as random effects and SNP$_i$ was fitted as a fixed effect, one at a time, for significant SNPs ($P < 10^{-5}$ for each 1 Mbp) discovered in the training population. After performing the GWAS, the significance ($P < 0.05$) and the consistency of the direction of effects (positive or negative) for the selected significant SNPs were compared between the training and validation populations.

Identification of the most likely candidate genes

The genes that were located 50 kb upstream and downstream of the best SNP were identified using UCSC Genome Bioinformatics (http://genome.ucsc.edu/) and Ensembl (http://www.ensembl.org/biomart/). If there was more than one gene, we retained only the gene that was located nearest to the SNP or the particular gene with a known effect on wool or hair.

Results
Single-trait GWAS

The number of significant SNPs for each trait is in Table 2.

Genomic prediction

Using BayesR, mean accuracies of genomic prediction of 0.21 and 0.23 were obtained across wool traits at yearling and adult ages, respectively (Table 3). Accuracy tended to increase with Th^2, with T being the number of phenotypes in the training set (Fig. 1, $R^2 = 0.34$). GBLUP provided very similar mean accuracies (0.21 and 0.22 at yearling and adult ages, respectively). However, BayesR tended to result in higher accuracies than GBLUP for traits that had a large number of significant SNPs in the single-trait GWAS (Tables 2, 3; Fig. 2).

Multi-trait analyses for the identification of pleiotropic QTL
Multi-GWAS

The multi-trait analyses using GWAS identified many narrow regions that contained more than one significant SNP (e.g. on chromosomes OAR3, 6, 7, 13, 19, and 25, OAR for *Ovis aries* chromosome; Fig. 3a). Combining the single-trait GWAS in a multi-trait analysis resulted in 563 and 263 significant SNPs at significance thresholds of $P < 10^{-5}$ and $P < 5 \times 10^{-7}$, respectively. This corresponded to a FDR of 0.9 and 0.1% (respectively), which was lower than for any individual trait tested in the single-trait GWAS (Table 2). In order to avoid testing a large number of closely-linked SNPs, we identified 64 SNPs that were significant at $P < 5 \times 10^{-7}$ and which were separated from each other by at least 1 Mb. Figure 4a compares the multi-GWAS with five single-trait GWAS for a region around 59.0 Mb on OAR3. These five traits

Table 3 Average accuracies of GEBV of the fivefold cross-validation populations using BayesR and GBLUP methods for each trait at yearling (Y) and adult (A) ages

Trait[a]	Accuracy (SE)		Trait[a]	Accuracy (SE)	
	BayesR	GBLUP		BayesR	GBLUP
YGFW	0.28 (0.011)	0.24 (0.027)	AGFW	0.39 (0.024)	0.33 (0.015)
YYLD	0.33 (0.026)	0.30 (0.021)	AYLD	0.42 (0.042)	0.37 (0.029)
YSL	0.27 (0.024)	0.23 (0.012)	ASL	0.23 (0.033)	0.22 (0.040)
YSS	0.24 (0.026)	0.16 (0.029)	ASS	0.28 (0.028)	0.28 (0.039)
YFD	0.35 (0.015)	0.31 (0.015)	AFD	0.41 (0.033)	0.28 (0.021)
YFDCV	0.23 (0.029)	0.19 (0.021)	AFDCV	0.30 (0.029)	0.22 (0.048)
YCURV	0.24 (0.015)	0.20 (0.009)	ACURV	0.27 (0.018)	0.21 (0.011)
YBRWR	0.27 (0.027)	0.24 (0.018)	ABRWR	0.19 (0.030)	0.23 (0.021)
YBCOV	0.13 (0.035)	0.14 (0.046)	ABCOV	0.10 (0.079)	0.16 (0.054)
YCCOV	0.22 (0.037)	0.25 (0.041)	ACCOV	0.22 (0.042)	0.19 (0.029)
YDAG	0.18 (0.093)	0.19 (0.081)	ADAG	0.07 (0.102)	0.16 (0.090)
YSSTRC	0.25 (0.044)	0.31 (0.045)	ASSTRC	0.19 (0.040)	0.20 (0.032)
YWEATH	0.24 (0.081)	0.29 (0.079)	AWEATH	0.22 (0.098)	0.22 (0.102)
YCHAR	0.21 (0.024)	0.23 (0.019)	ACHAR	0.08 (0.039)	0.15 (0.046)
YFLROT	0.28 (0.055)	0.30 (0.054)	AFLROT	0.26 (0.036)	0.25 (0.041)
YDUST	0.19 (0.051)	0.23 (0.050)	ADUST	0.45 (0.111)	0.49 (0.117)
YGCOL	0.15 (0.011)	0.19 (0.015)	AGCOL	0.19 (0.031)	0.23 (0.040)
YCOLZ	0.11 (0.033)	0.13 (0.018)	ACOLZ	0.14 (0.040)	0.17 (0.037)
YCOLYZ	0.18 (0.030)	0.18 (0.020)	ACOLYZ	0.24 (0.033)	0.21 (0.028)
YCOLY	0.10 (0.038)	0.12 (0.026)	ACOLY	0.16 (0.042)	0.15 (0.050)
YCOLX	0.12 (0.037)	0.11 (0.029)	ACOLX	0.12 (0.069)	0.15 (0.059)
YSKINQ	0.11 (0.032)	0.15 (0.027)	ASKINQ	0.04 (0.091)	0.09 (0.110)
Mean[b]	0.21	0.21	Mean[b]	0.23	0.22

SE standard error of average accuracy of GEBV

[a] Trait names see Table 1

[b] Average accuracy across traits

(AFDCV, YBWR, YFD, YFDCV, and YSL) were those for which the number of significant SNPs ($P < 5 \times 10^{-7}$) was largest in the single-trait GWAS (Table 2).

Multi-PP

One hundred and two SNPs had an approximate multi-trait posterior probability (pp) higher than 0.3 (Fig. 3b), among which two SNPs had a $pp_{effect \neq 0}$ higher than 0.05 for four traits, 11 for three traits, 34 for two traits and the remaining 55 SNPs for one trait. Thus, although the multi-trait pp was calculated using all traits, the pp for the 102 identified SNPs was mainly influenced by between one and four traits. For example, Fig. 4b compares the multi-pp of SNPs on OAR3 around 59.0 Mb with the pp for the five traits YSL, YFD, YFDCV, AFDCV, and YBRWR. Two SNPs that were 39 kb apart had a high pp i.e. higher than 0.82 (Fig. 4b) and were in high LD ($r^2 = 0.44$), which indicates that they may tag the same QTL. There is another SNP with a multi-trait

$pp \approx 0.5$ that is located 0.5 Mb upstream of these two SNPs.

Multi-LGEBV

Local GEBV, using only the SNPs within a 250-kb window, were calculated for each animal using the BayesR estimates of SNP effects. The variance of local GEBV for each window and trait was calculated. A high variance indicates that within the 250-kb window there is a QTL for that trait. The highest percentage of variance of local GEBV was equal to 1.9% of the phenotypic variance for yearling staple length (YSL), which indicates that we did not detect QTL with very large effects for any of the traits. Figure 4c shows the variance for five traits for four windows around OAR3:59 Mb.

For each of the 9813 windows, a PC analysis yielded 44 eigenvalues and eigenvectors. A high first eigenvalue indicates a window that contains a QTL. We selected the 120 windows with the highest eigenvalues as windows that potentially contain a QTL (Fig. 3c). The distribution of the $\log_{10}$ of the first eigenvalues for 120 segments with the highest first eigenvalue is in Fig. 5a. If within the window, there is a single QTL that has an effect on multiple traits, we expect that the first eigenvector will explain most of the variance and that the other eigenvalues will be low. We selected 120 segments with the highest first eigenvalues across all 9813 segments and in 112 of these 120 selected segments, the PC1 explained more than 90% of the total variance. The distribution of the proportion of variance explained by PC1 eigenvalues across the 9813 windows is in Fig. 5b.

Figure 4c shows the first eigenvalues for four segments on OAR3. The second segment (58.75–59.25 Mb) had the highest eigenvalue, but the segment to its left had a high variance of local GEBV for YFD in spite of a lower first eigenvalue. This suggests that the first segment contains a more pleiotropic QTL while the second segment contains a QTL that affects YFD mainly.

In order to identify the SNP that is closest to the QTL (best SNP), for each of the 120 segments, we performed a new GWAS by using the eigenvector of PC1 as a new trait and fitting each SNP, one at a time, within each segment. Thus, the dependent variable was the linear combination of traits as defined by the first eigenvector. By this process, we selected the top SNPs (with the highest F value) in each of the top 120 segments. For example, Fig. 4d shows the eigenvalues ($\times 10^3$) of the four neighboring windows at around 59.0 Mb on OAR3 and the F values of the SNP effects ($\times 10^{-3}$) for PC1 in the corresponding windows. The second segment had a high eigenvalue that explains 98% of the total variance and three SNPs (circled in blue, orange and green) were most highly associated with the first eigenvector for this segment. The first segment had a comparably lower

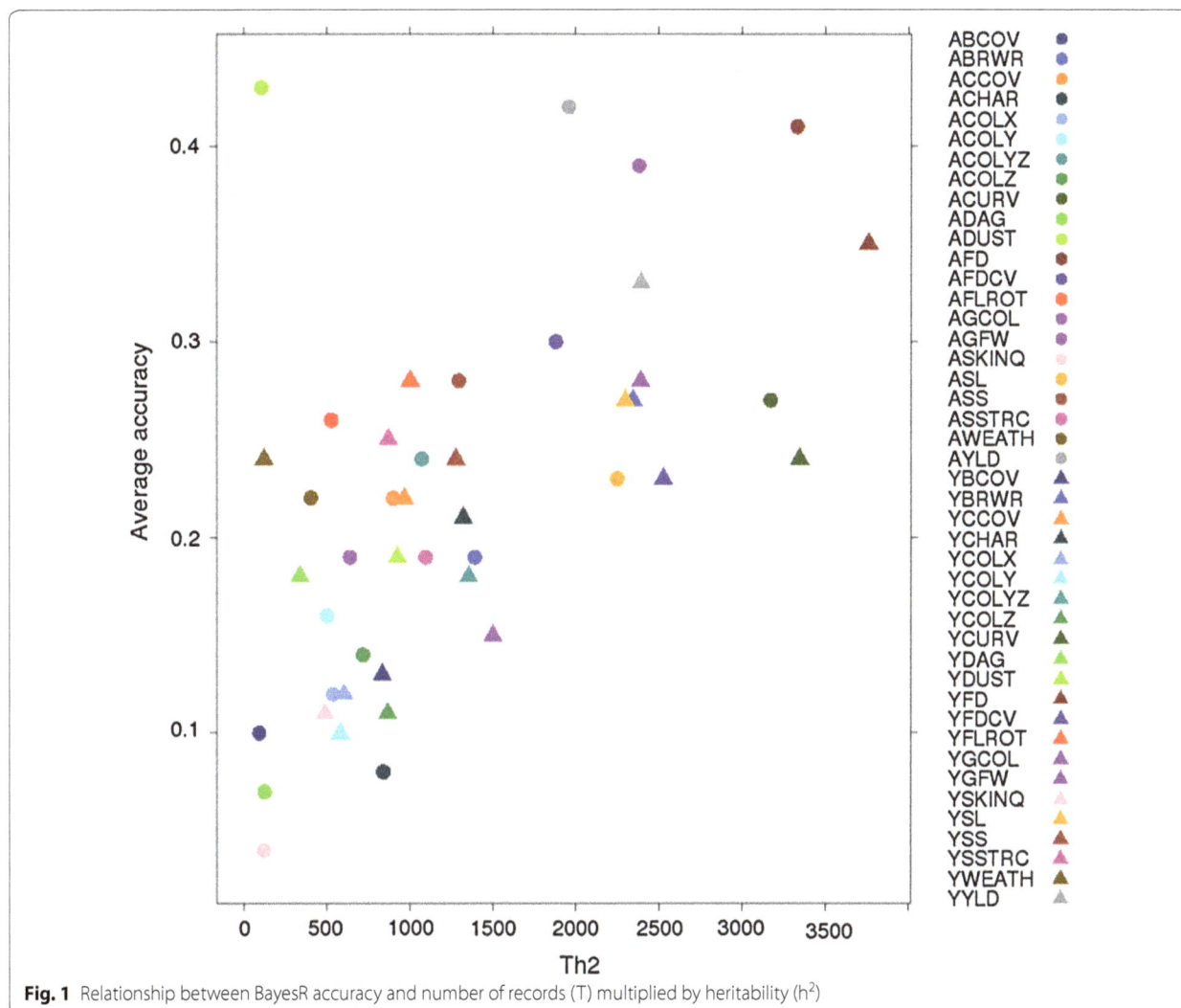

Fig. 1 Relationship between BayesR accuracy and number of records (T) multiplied by heritability (h^2)

first eigenvalue but this eigenvalue explained 96% of the total variance and there is only one SNP (circled in yellow) that was associated with the first eigenvector for that segment. The third segment had an even lower first eigenvalue (explaining 68% of the total segment variance), which probably indicates that there is no QTL in that segment. However, we did observe one SNP that was highly associated with the first eigenvector (Fig. 4d), which illustrates one of the drawbacks of this method. Since local GEBV are calculated from SNP genotypes, there will always be one or more SNPs associated with the first eigenvector but it should not be interpreted as evidence for a QTL unless the first eigenvalue is high.

Comparison and combination of results from the three multi-trait analyses

As described above, we selected 64 SNPs from the multi-GWAS (most significant SNP ($P < 5 \times 10^{-7}$) in each

1-Mb window), 102 SNPs with the highest multi-PP and 120 from the analysis of local GEBV. Among these, 75 SNPs overlapped across two or three analyses. Therefore, the total number of SNPs identified was 206. Of these 206 SNPs, seven were identified by all three methods. In addition to these seven SNPs, 64 were identified by both local GEBV and multi-PP, and two by multi-PP and multi-GWAS (Fig. 6 and see Additional file 1: Table S1). Of the 64 top multi-GWAS SNPs, 55 were not among the top SNPs selected by the other two multi-trait methods. In fact, 50 of these 55 SNPs were in segments that did not have a high eigenvalue based on local GEBV, and the first eigenvalue explained less than 90% of the total variance. Thus, it is possible that these SNPs are at some distance from the causal variant, which is located in another segment. Conversely, among the SNPs that were identified by the multi-LGEBV and multi-PP analyses, 38 (=9 + 13 + 16) were significant in the multi-GWAS

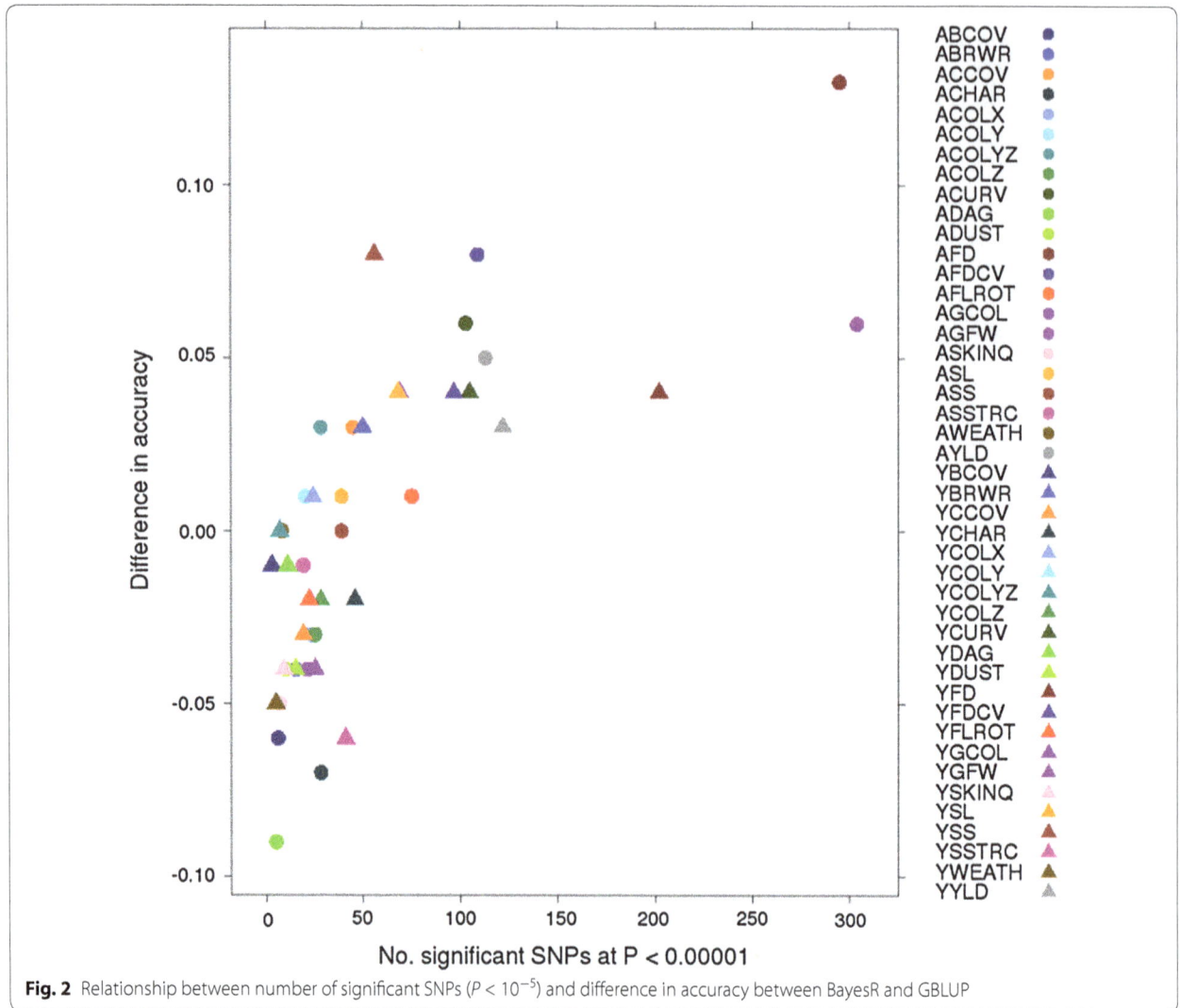

Fig. 2 Relationship between number of significant SNPs ($P < 10^{-5}$) and difference in accuracy between BayesR and GBLUP

($P < 10^{-5}$) although they were not the most significant SNPs in their 1-Mb window, or failed to reach a $P < 5 \times 10^{-7}$. By adding these 38 SNPs to the 64 SNPs from the multi-GWAS, 20 (=7 + 13) are common to all three multi-trait approaches, nine in the multi-LGEBV and multi-GWAS and 18 (=2 + 16) in the multi-PP and multi-GWAS (Fig. 6).

These results are illustrated in Fig. 4. The points that are circled in the same colour in Fig. 4a, b, d represent the same SNP. The second segment had the highest

eigenvalue and explained 98% of the total variance of local GEBV in that segment. The SNPs circled in blue, orange, and green are those that were most highly associated with the first eigenvector (Fig. 4d) and had the highest multi-PP (Fig. 4b). These were also highly significant in the multi-GWAS (Fig. 4a). The SNP circled in blue (at 59,019,274 bp on OAR3) was identified by all three multi-trait methods (see Additional file 1: Table S1). In all three methods, the same traits (YSL, YBRWR, YFDCV, AFDCV) contributed to the multi-trait statistics, which

(See figure on next page.)
Fig. 3 Manhattan plot of multi-GWAS (**a**), multi-PP (**b**), and multi-LGEBV (**c**). Y axes are $-\log_{10}$ (P values) of SNPs for multi-GWAS, multi-trait posterior probabilities for multi-PP, and eigenvalues ($\times 1000$) of the first principal component (PC1) of 9813 250 kb-windows for multi-LGEBV. Numbers on the x axes represent the number of ovine chromosomes (OAR) excluding OARX. SNPs in *red colour* represent the top selected SNPs from each of the three multi-trait analyses [the highest eigenvalues of 120 windows from multi-LGEBV, multi-trait posterior probabilities of 102 SNPs from multi-PP, and 102 multi-GWAS SNPs including 64 top SNPs ($P < 5 \times 10^{-7}$) in 1-Mb intervals and 38 SNPs ($P < 10^{-5}$)]

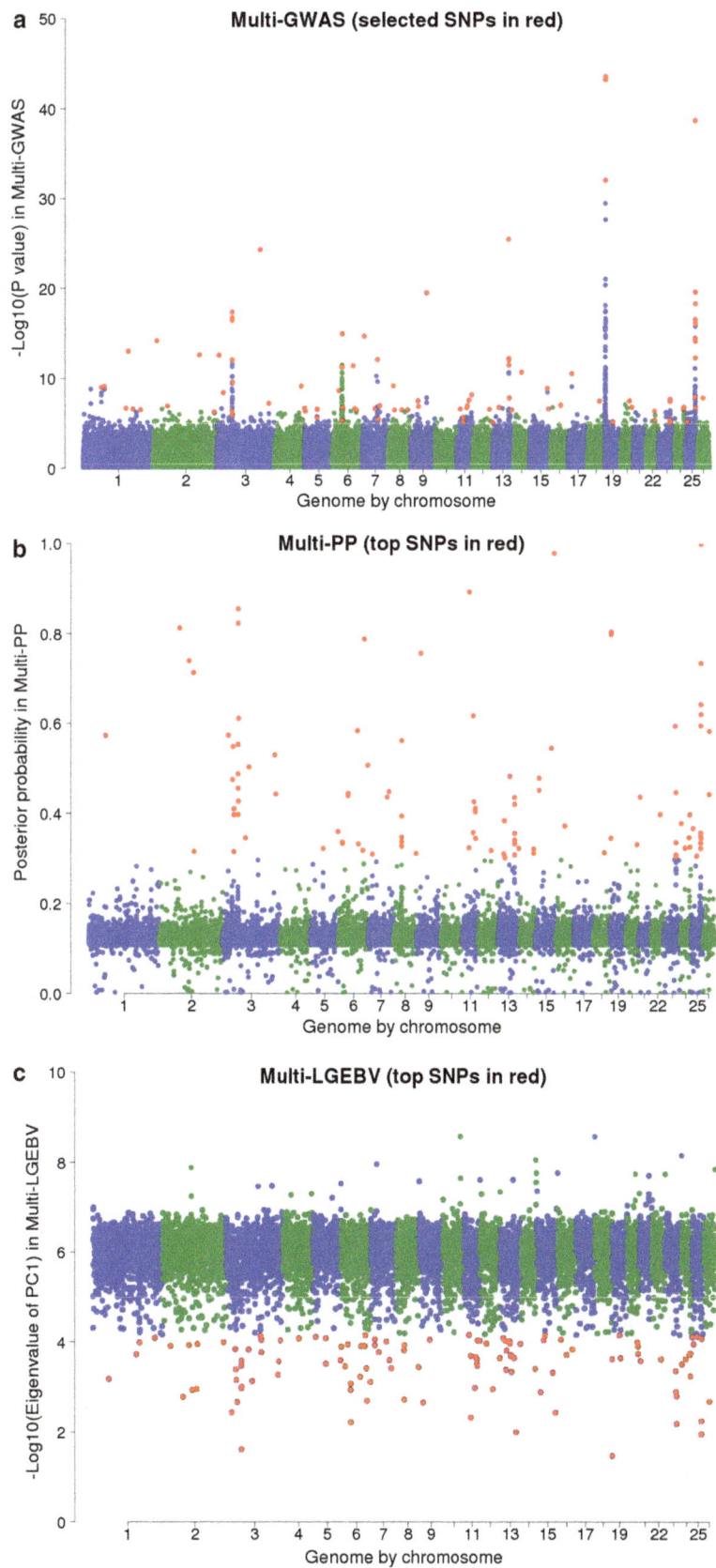

a Multi-GWAS (selected SNPs in red)

b Multi-PP (top SNPs in red)

c Multi-LGEBV (top SNPs in red)

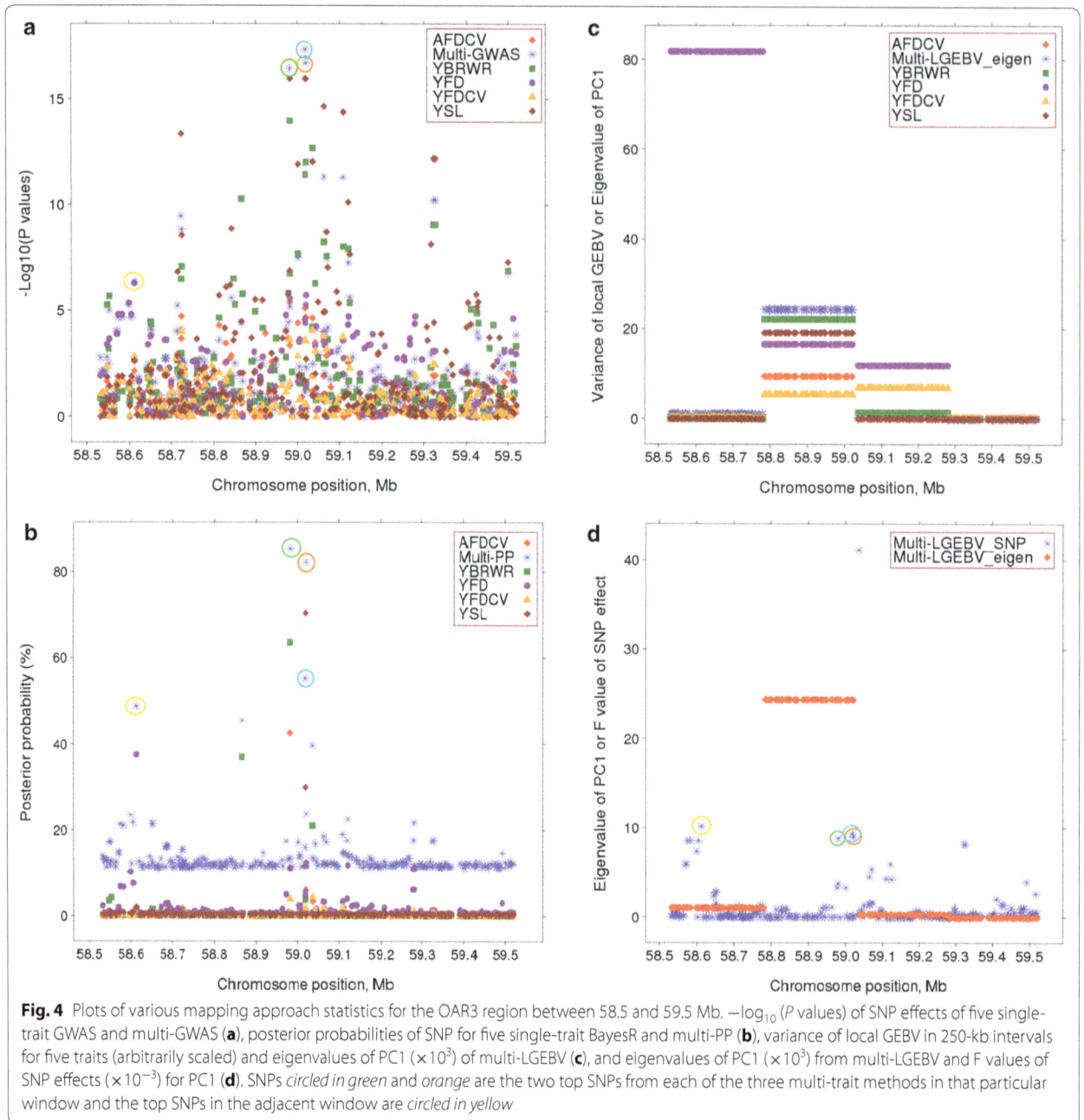

Fig. 4 Plots of various mapping approach statistics for the OAR3 region between 58.5 and 59.5 Mb. —$\log_{10}$ (*P* values) of SNP effects of five single-trait GWAS and multi-GWAS (**a**), posterior probabilities of SNP for five single-trait BayesR and multi-PP (**b**), variance of local GEBV in 250-kb intervals for five traits (arbitrarily scaled) and eigenvalues of PC1 ($\times 10^3$) of multi-LGEBV (**c**), and eigenvalues of PC1 ($\times 10^3$) from multi-LGEBV and F values of SNP effects ($\times 10^{-3}$) for PC1 (**d**). SNPs *circled in green* and *orange* are the two top SNPs from each of the three multi-trait methods in that particular window and the top SNPs in the adjacent window are *circled in yellow*

supports the conclusion that this is a pleiotropic QTL with effects on at least these four traits. The first segment in Fig. 4 had a high variance of local GEBV for YFD, but when this trait was combined with the other 43 traits, the first eigenvalue was lower than that of the second segment. However, the SNP that is circled in yellow was associated with the first eigenvector, had a high multi-PP and was significant ($P < 10^{-5}$) in the multi-GWAS. In all three analyses, this SNP was associated with YFD.

In Bayes R, which generates the statistics for multi-PP and local GEBV, all SNPs are fitted simultaneously in the model. Thus, it is possible that the QTL tracked by the yellow circled SNP differed from the QTL tracked by the blue, orange and green circled SNPs. This hypothesis is supported by the traits that are affected by the SNP. Using the effect of each SNP on the 44 traits as estimated by the GWAS, it is possible to calculate the correlation between any pair of SNPs. Figure 7 displays the correlation among

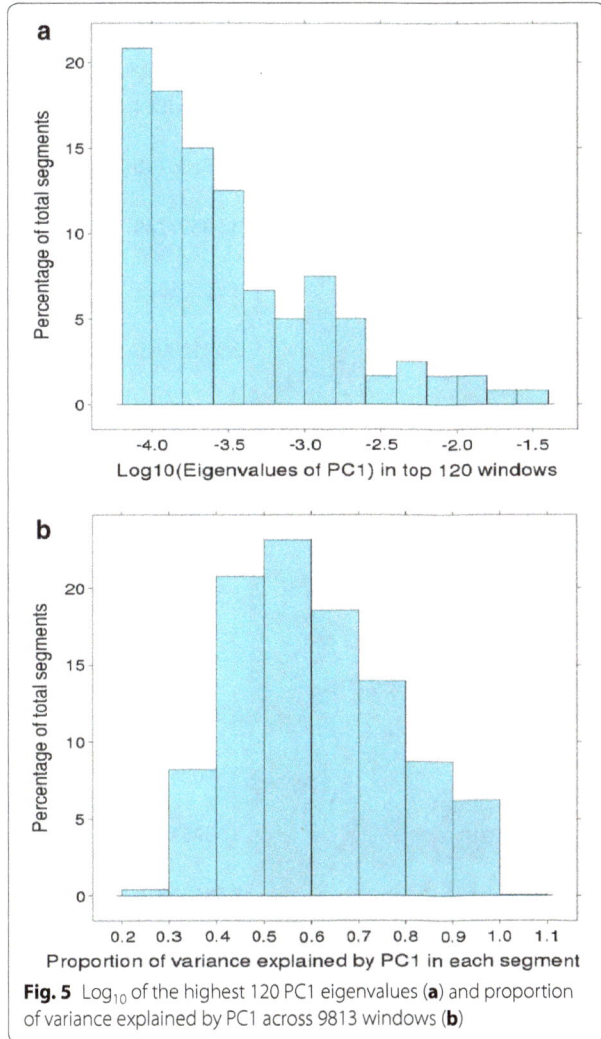

Fig. 5 Log_{10} of the highest 120 PC1 eigenvalues (**a**) and proportion of variance explained by PC1 across 9813 windows (**b**)

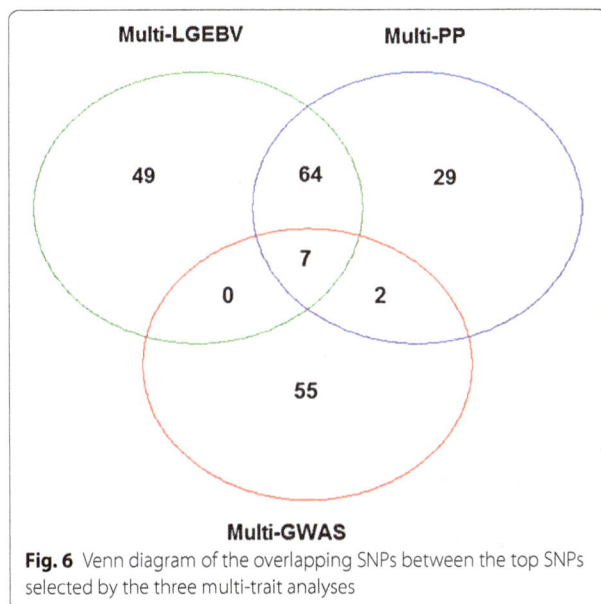

Fig. 6 Venn diagram of the overlapping SNPs between the top SNPs selected by the three multi-trait analyses

SNP effects in this region of OAR3 as a heat map. It shows that all the other SNPs except that circled in yellow have highly correlated effects and presumably track the same QTL. In the multi-GWAS (Fig. 4a), another significant SNP at 58.72 Mb was observed, which was not detected in either of the other analyses presumably because it tracked the same QTL as the SNP at 59.0 Mb circled in blue, orange and green colours. Figure 8 illustrates two more straightforward examples where the same SNP was identified by all three methods.

Validation of SNP effects
Validation of SNPs from single trait GWAS
The most significant SNPs ($P < 10^{-5}$) in each 1-Mb window were tested in the validation population. Table 4 shows the results for five traits, which were among those that had the largest number of significant associations (Table 2). The number of SNPs tested ranged from 14 (YYLD) to 101 (YFD). The proportion of these SNPs that were significant at $P < 0.05$ in the validation population varied from 0.07 (=1/14 for YYLD) to 0.30 (=7/23 for AYLD). The percentage of these SNPs that had an effect in the same direction in both the validation and training populations varied from 57 to 86% (Table 4). In addition, when a significance level was also imposed on SNPs discovered in the validation population, the percentage of SNP effects in the same direction across training and validation populations increased from 75 to 100% (small numbers of SNPs were discovered in the relatively small validation population partly due to lack of power).

Multi-trait validation using the linear index approach
Association between a SNP and its corresponding linear index was tested in the validation sample. The 105, 77, and 120 top SNPs were selected from the multi-GWAS ($P < 5 \times 10^{-7}$ in each 1-Mb window), multi-PP, and multi-LGEBV analyses, respectively, in the training population. These SNPs were tested in a GWAS in the validation population (see Table 5). Of the 105 multi-GWAS SNPs that were significant in the training population, 70% had an effect in the same direction in both the training and validation populations and 16 were also significant ($P < 0.05$) in the validation population of which 88% had an effect in the same direction in both the validation and training populations (Table 5). The results were slightly better for the 77 SNPs selected by the multi-PP analysis: 19 out of these 77 SNPs were significant and all had an effect in the same direction in both the validation and training populations. Thus, the number of validated SNPs (with a significant effect in the same direction) was largest with the multi-PP approach (19) followed by multi-LGEBV (15) and multi-GWAS ($14 = 16 \times 0.88$).

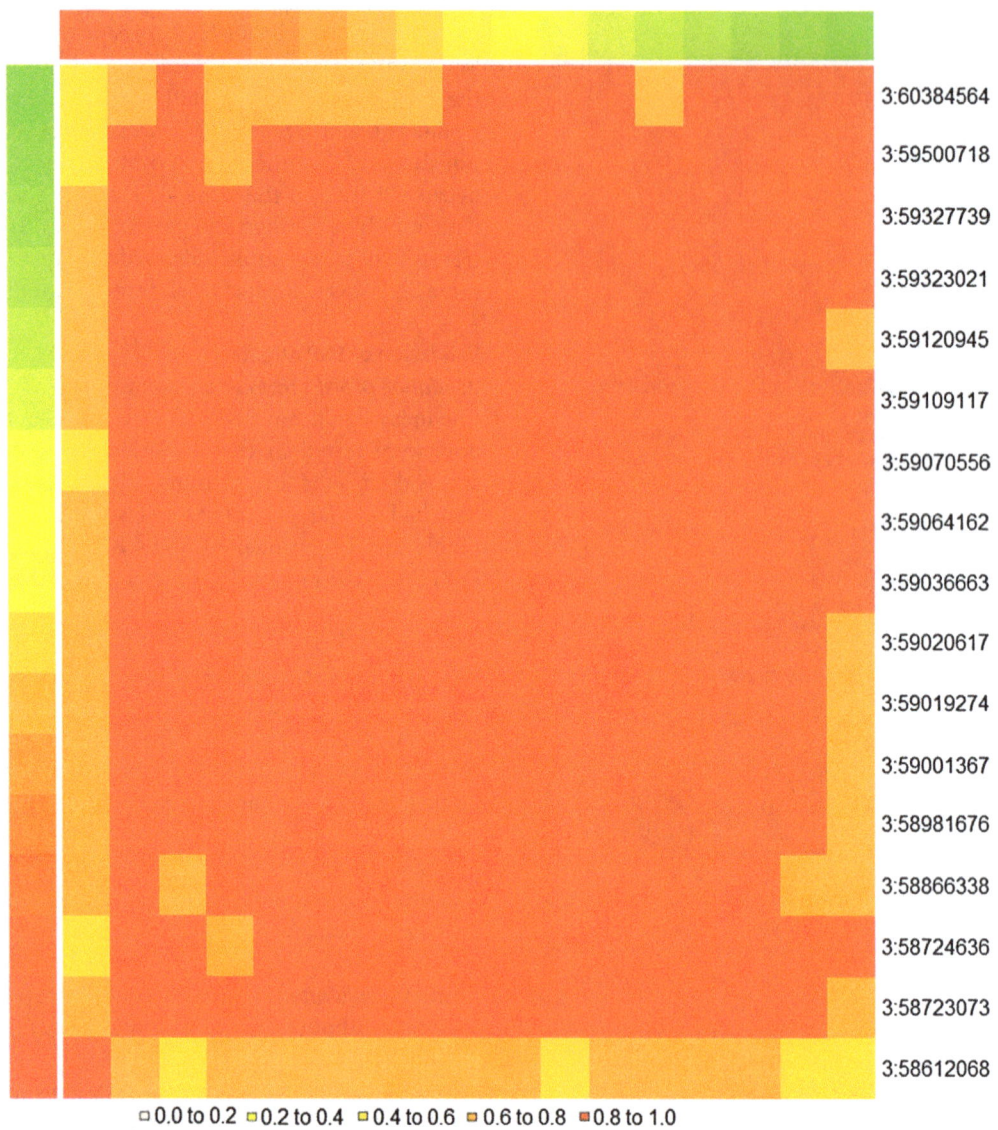

Fig. 7 Correlation matrix of the effects on 44 traits between the 16 top SNPs within the 58.5–59.5 Mb region on OAR3. *Numbers on the right* represent chromosome number and position in base pairs of these top SNPs

Examples of QTL with a similar pattern of effects across traits

If two QTL affect the same physiological pathway, one might expect that they have the same pattern of effects. We assessed the similarity of SNP effects by the correlation between pairs of SNPs across the 44 traits. Two SNPs might have a similar pattern because they affect the same pathway or because they tag the same QTL. Among the 206 SNPs, strong correlations (>0.8) were mainly found among the SNPs located on the same chromosome. For instance, six SNPs were on OAR3:59.0 Mb, three on OAR5:48.5 Mb, four on OAR6:37.5 Mb, five

on OAR8:31.2 Mb, six on OAR13:62.9 Mb, three on OAR19:0.6 Mb, five on OAR23:44.3 Mb, and 11 on OAR25:35.3 Mb. Some of these were in high LD (e.g. LD estimates (r^2) between three SNPs on OAR3 and OAR19 ranged from 0.50 to 0.95). It is likely that each of these clusters of highly correlated SNPs tag one major QTL. In a few cases, moderate correlations (0.6 to 0.8) existed between SNPs located on different chromosomes. For instance, the SNP at OAR3:59.0 Mb (near the *FOXI3* gene) has a similar pattern of effects as the SNP at OAR6:37.5 Mb (near *LCORL*) and OAR23:44.2 and OAR25:35.3 (near *MAT1A*). In each case, there is an

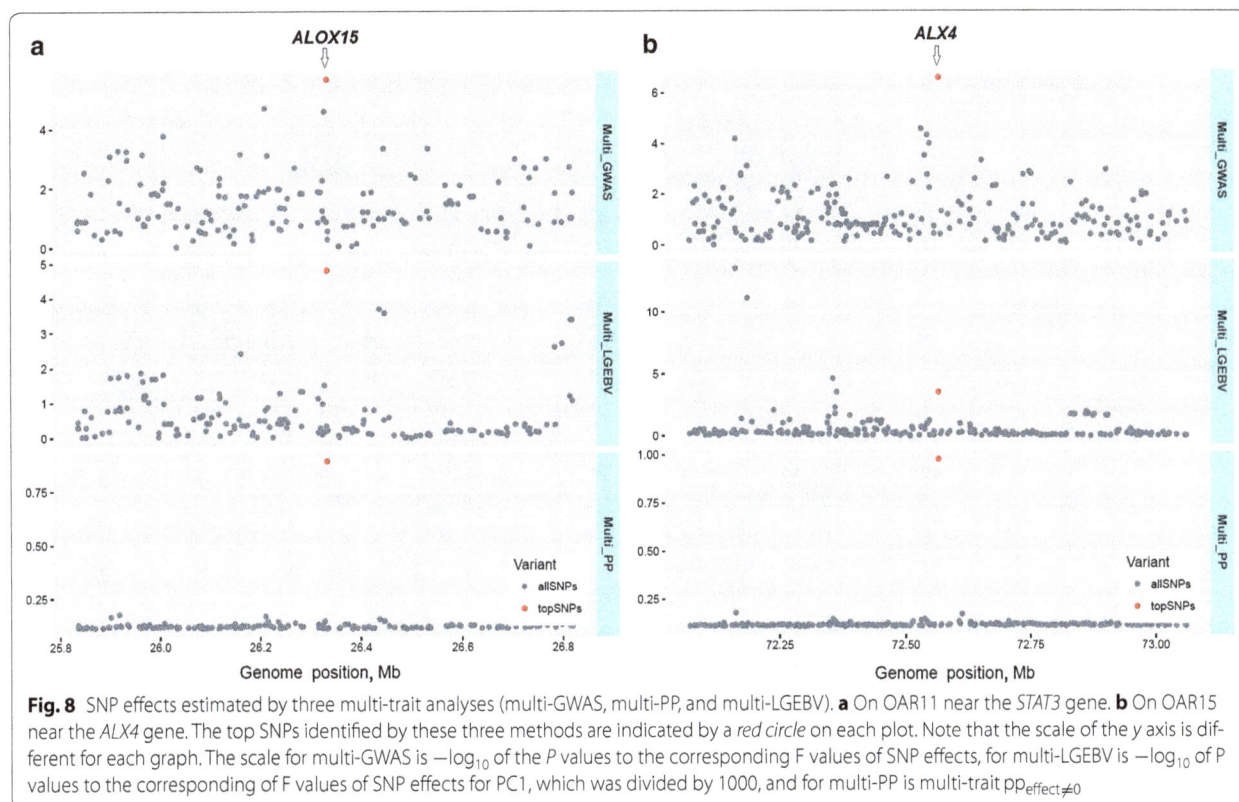

Fig. 8 SNP effects estimated by three multi-trait analyses (multi-GWAS, multi-PP, and multi-LGEBV). **a** On OAR11 near the *STAT3* gene. **b** On OAR15 near the *ALX4* gene. The top SNPs identified by these three methods are indicated by a *red circle* on each plot. Note that the scale of the y axis is different for each graph. The scale for multi-GWAS is $-\log_{10}$ of the P values to the corresponding F values of SNP effects, for multi-LGEBV is $-\log_{10}$ of P values to the corresponding of F values of SNP effects for PC1, which was divided by 1000, and for multi-PP is multi-trait $pp_{effect \neq 0}$

Table 4 Number of significant SNPs ($P < 10^{-5}$) in each 1-Mb region in the training population that were also significant in the validation population for the five individual single traits at two yearling (Y) and adult (A) ages

P value[a]	Number of SNPs	%-same	P value[a]	Number of SNPs	%-same
YGFW			AGFW		
0.05	2	100	0.05	7	100
All	20	75	All	41	80
YYLD			AYLD		
0.05	1	100	0.05	7	86
All	14	57	All	23	61
YSL			ASL		
0.05	3	100	0.05	4	75
All	24	67	All	27	74
YFD			AFD		
0.05	24	100	0.05	15	93
All	101	86	All	92	78
YCURV			ACURV		
0.05	6	100	0.05	2	100
All	30	83	All	28	82

%-same = percentage of SNPs, which have an effect in the same direction in both the training and validation populations

[a] P value in the validation population

Table 5 Validation of the top SNP effects from three multi-trait analyses

P value[a]	Number of SNPs	FDR %	%-same
SNPs from multi-GWAS			
0.05	16	29.3	88
All	105		70
SNPs from multi-PP			
0.05	19	16.1	100
All	77		75
SNPs from multi-LGEBV			
0.05	15	36.8	100
All	120		74

%-same = percentage of SNPs, which have an effect in the same direction in both the training and validation populations

[a] P value in the validation population

allele that increases SL, SS and FD, but decreases FDCV. However, the SNPs differ in their effects on other traits, such as GFW, thus it is less likely that they act through the same pathway. Similarly, the SNP at OAR13:62.9 (near *RALY*) and OAR15:72.6 (near *ALX4)* both have an allele that increases FD and FDCV.

Identification of candidate genes

We searched for genes within 50-kb genomic regions up and downstream from each of the 206 SNPs selected from the three multi-trait analyses (see Additional file 1: Table S1) and identified many genes with known effects on wool and hair growth (Table 6). We also found strong effects for SNPs near the genes *ALOX15*, *ANKS1B*, *ELOVL6*, *FASN*, *NCAPG*, *LCORL*, *FTL*, and *GK5*, which are associated with growth, fatness and body composition in sheep and humans [14, 31–33].

Discussion

Genomic prediction

Our results show that the estimated GEBV accuracies are affected by trait, size of the training population, and statistical method used (GBLUP vs. BayesR). As expected from theory [34] in which Th^2 is a critical parameter, traits with a high heritability and a large training population tended to result in higher accuracies than those with an average heritability across populations (Fig. 1).

On average, BayesR and GBLUP resulted in similar GEBV accuracies but BayesR resulted in higher accuracy for traits (GFW, YLD, SL, SS, FD, FDCV, and CURV) for which there was a large number of significant SNPs in the GWAS. In an earlier study on the genomic prediction of wool and carcass traits using the 50 k SNP chip based on the same population, Daetwyler et al. [22] found no clear differences in accuracy between GBLUP and BayesR. Kemper et al. [11] found that accuracy of across-breed

genomic predictions for selection candidates that are less related to the training animals was higher with BayesR than with GBLUP and the use of BayesR mapped QTL more accurately than GBLUP in dairy cattle. Many other studies in cattle and humans have reported that BayesR results in more accurate GEBV than GBLUP, in particular for traits for which mutations of moderate effect are segregating [29, 35–37].

There are few reports on the accuracy of genomic predictions in sheep [38–43]. Pickering et al. [43] reported the accuracy of genomic predictions for health traits including dagginess for several New Zealand breeds (Romney, Coopworth, Perendale, Texel and three breed crosses). For the dagginess traits, they found that accuracies of genomic predictions ranged from 0.11 to 0.56 for those breeds, while in our study an accuracy of 0.19 was estimated for YDAG using GBLUP. Daetwyler et al. [38] (on the same population as used here but with fewer records) reported that accuracies of GEBV with 50 k SNPs ranged from 0.15 to 0.79 for wool traits in Merino sheep and from −0.07 to 0.57 for meat traits in all breeds studied. These accuracies were higher than those that we obtained for wool traits. Several factors may explain the difference between our results and those in these previous studies: (1) only 107 animals were included in their validation set resulting in accuracies with larger standard errors; (2) some animals in their validation set were closely related to animals in the training dataset, whereas we deliberately limited relationships between training

Table 6 List of candidate genes with known effects on wool or hair growth

Gene code	Start and stop position of gene (bp)	Gene function	References
ITGA6	OAR2:136,225,512–136,275,286	Surface markers for epithelial stem cells within hair follicles	[56, 57]
ANTXR1	OAR3:38,977,681–39,235,539	Affects hair follicle growth and cycling; Alopecia	[58, 59]
FOXI3	OAR3:58,986,758–58,990,671	Regulates multiple aspects of hair follicle development and homeostasis	[47]
KRT86	OAR3:133,936,440–133,944,796	Maintains strength and elasticity of hair	[59, 60]
WNT1[a]	OAR3:137,053,186–137,056,187	Wnt/β-catenin signalling is necessary for hair follicle stem cell proliferation	[61]
BMPER	OAR4:62,662,041–62,917,582	BMP signalling controls the hair follicle cycle	[49]
FRAS1	OAR6:92,393,951–92,908,337	Basement membrane protein; dermal-epidermal adhesion	[62]
FGF5	OAR6:94,584,400–94,605,575	Induces regression of the human hair follicle; regulator of hair growth	[63]
FGF7	OAR7:57,779,972-57,841,735	Regulates cell proliferation and cell differentiation and is required for normal regulation of the hair growth cycle	[64]
STAT3	OAR11:41,903,051–41,934,839	Keratinocyte stem homeostasis; alters behaviour of hair follicle stem populations	[65]
OVOL2[a]	OAR13:37,399,961–37,424,537	Controls epithelial cell proliferation and differentiation in hair bulb and skin	[66]
EIF2S2	OAR13:62,907,171–62,923,869	Inhibition of eIF4E protects against cyclophosphamide-induced alopecia	[67]
ALX4	OAR15:72,556,058–72,606,253	Affects hair follicle growth and cycling; total alopecia (hair loss)	[59, 68]
MAML2	OAR15:13,755,425–13,774,117	Modifies Notch signalling that controls a cell fate switch in hair follicle stem cells	[69]
CUX1	OAR24:34,631,867–34,950,962	Essential for epithelial cell differentiation of the hair follicle in mice	[70]
ACHE[a]	OAR24:35,649,479–35,653,376	M4 muscarinic acetylcholine receptors play a key role in the control of murine hair follicle cycling and pigmentation	[71]

[a] Not the nearest gene to the particular SNP with a significant effect, but is a gene with a known effect on wool or hair

and validation animals; and (3) we used within-strain GEBV (i.e., adjusted for genetic differences between Merino strains) whereas they computed GEBV accuracy in the overall population.

Our validation approach, in which the relationship between training and validation populations is minimized, is relevant to the commercial use of genomic selection in which a sheep breeder relies on a training population that is not closely related to his own sheep. In our case, the average of the top 10% genomic relationships that a validation individual has with animals in the training population was equal to 0.02. When validation animals are more related to the training animals, it is likely that the estimated genomic prediction accuracies will be higher.

The level of genomic prediction accuracy may also depend on the accuracy of imputation [44]. However, based on the imputation tests from 50 k to HD SNPs (not reported here), the mean imputation accuracy between imputed and real non-50 k genotypes on our HD data (as the proportion of correctly imputed genotypes for non-HD SNPs) was higher than 0.98, which means that this is not a likely reason for the low prediction accuracies obtained with our data.

Overall, the accuracy of GEBV was low. The similar accuracies obtained with GBLUP and BayesR suggest that there are few QTL of moderate or large effect, which is supported by the single-trait GWAS results for many traits. Therefore, we investigated whether combining information from all traits could help to identify QTL for multiple traits.

Multi-trait analyses

It is known that genetic correlations exist between many of the traits studied here and thus there must be QTL with effects on multiple traits. We used three multi-trait analyses to map these QTL. All three methods involve a degree of approximation, thus it is difficult to apply precise significance tests. However, the fact that there is agreement among the results of the three multi-trait and with the single-trait results supports the conclusion that they detect pleiotropic QTL. This is also supported by the rates of validated individual SNPs.

The multi-trait GWAS method used here was previously shown to increase the power of QTL detection [14, 30]. In most 1-Mb intervals that we selected, the pattern of the SNP effects is similar across traits, thus the high correlation among traits, which implies that there is probably only one major QTL within a given interval. The most significant SNP in a region varies from one trait to another due to sampling error even if there is only one QTL. Unless the errors for different traits are highly correlated, the multi-trait analysis reduces the sampling

error, which results in a more precise localization of the QTL. However, because long range LD occurs in the ovine genome [45] a SNP that is located at a long distance from any QTL can still have a significant association with the trait. Combining this with the large number of QTL that affect most complex traits, a SNP associated with one QTL may merge with those associated with another nearby QTL, which decreases our ability to map the QTL. To overcome this difficulty, methods that fit all SNPs simultaneously, such as the BayesR method [29] used here, have been advocated for QTL mapping.

In dairy cattle, Kemper et al. [11] showed that BayesR maps QTL more precisely than GWAS. A multi-trait BayesR analysis with 44 traits would impose a very large computational burden, thus we used two approximate methods (multi-PP and multi-LGEBV) to combine the results from 44 single-trait BayesR analyses. In the single-trait BayesR analysis, high variance of local GEBV indicates the presence of a QTL in that 250-kb window. The equivalent multi-trait test is based on the first eigenvalue that is caused by one or more traits having a high variance of local GEBV. If there is more than one trait with a high variance and if the local GEBV are correlated between different traits (as expected if there is only one QTL), the first eigenvalue increases. Windows with more than one QTL can occur but for 112 of the 120 windows with the highest eigenvalues, the first eigenvalue explained more than 90% of the total variance, which indicates that windows with only one QTL predominated in our study. To identify the SNP that was located nearest to this QTL, we carried out a local GWAS using the local GEBV as a new trait.

Since multi-LGEBV and multi-PP both used the output from single-trait BayesR analyses, it is reassuring that they detected many common SNPs. In fact, of the 102 SNPs found by multi-PP, 64 were also detected among the 120 best SNPs from the multi-LGEBV (Fig. 6). However, there was less agreement between multi-GWAS and the BayesR based methods. In particular, when we considered only the top multi-GWAS SNPs ($P < 5 \times 10^{-7}$) in each 1-Mb window, the multi-GWAS detected 64 SNPs, but only seven were found by the other two methods. Fifty-five of the SNPs detected by the multi-GWAS were not identified by the other two methods but 50 of them are located in the windows, which had a low eigenvalue of PC1 with PC1 explaining less than 90% of the total variance. Decreasing the threshold for the multi-GWAS to $P < 10^{-5}$ (Fig. 6) improves the agreement between the multi-GWAS and the other two methods.

Validation of SNP effects

The proportion of SNPs that were confirmed in the validation set by the multi-trait methods was equal to that obtained for the best single trait (fibre diameter) and higher than that for most single traits. Among the three multi-trait methods, multi-PP had the highest validation rate. This was unexpected since the two other methods used GWAS in both training and validation, whereas multi-PP did not use GWAS methods in the discovery process. However, a multi-PP can miss some QTL since it relies on finding individual SNPs with a high posterior probability. In some cases, the evidence for a QTL is spread across many SNPs each with a low posterior probability, and thus the local GEBV variance is more likely to detect the QTL than the posterior probability.

Identification of QTL with similar patterns of pleiotropic effects

Stearns et al. [46] pointed out that the relative advantage of multivariate over univariate approaches varied with the level of genetic covariance between traits. In this study, some of the wool traits are genetically highly correlated. Previously, Bolormaa et al. [14] used a correlation matrix of pairs of SNP effects across 56 meat and body composition traits to perform a hierarchical clustering analysis. Using this approach, they identified at least four groups of QTL with similar patterns of pleiotropic effects on body composition (the population of sheep was similar to that used here). In our study on wool traits, the clustering analysis based on 206 SNPs was also done using GWAS SNP effects, but there were no clear-cut clusters of SNPs (results not shown).

Candidate genes

By exploiting pleiotropic effects for mapping QTL, we identified 206 putative QTL, which were close to 130 genes (within a distance of 50 kb on either side of each SNP). In some cases, the known function of the candidate gene fits the observed phenotype well. Table 6 provides a list of candidate genes with known effects on wool or hair growth. For instance, SNPs that are located around 59.0 Mb on OAR3 and near the *FOXI3* gene, are associated with multiple traits including wool quality and breech conformation traits (FDCV, SL, SS, FD, BRWR, BCOV, CCOV, and FLROT) at both ages. *FOXI3* regulates multiple aspects of hair follicle development and homeostasis and loss of *FOXI3* impedes hair follicle down-growth and progression of the hair cycle [47]. Shirokova et al. [47] showed that *FOXI3* displays a highly dynamic expression pattern during hair morphogenesis and cycling. In mice, absence of *FOXI3* results in a sparse fur phenotype and poor hair regeneration after hair plucking, and these effects are exacerbated with age due to impaired secondary hair germ activation leading to progressive depletion of stem cells. A SNP at 62.6 Mb on OAR4, which is located near (9.8 kb) the *BMPER* gene,

was detected through multi-LGEBV and multi-GWAS ($P = 2 \times 10^{-4}$). Bone morphogenetic protein (BMP) signalling regulates hair follicle cycle and development [48–51]. BMP signalling is a critical feature of the complex epithelial–mesenchymal cross-talk necessary to produce hair [50].

We also found significant SNPs (Table 7) close to genes (*ALOX15*, *ELOVL6*, *FASN*, *NCAPG*, and *GK5*) that are associated with variation in size, fatness and body composition in sheep, cattle and humans [7, 14, 32, 33]. This is not surprising since greasy fleece weight and fibre diameter have positive genetic correlations with yearling weight (0.23 and 0.17, respectively; [7]). The SNPs that are located within or near (<5 kb) *ALOX15*, *FASN*, *FTL*, and *NCAPG* were associated with GFW (up to $|t| = 8.9$), while the SNPs in *ELOVL6* and *GK5* had significant associations with FD (up to $|t| = 4.7$). Another SNP with an effect of $|t| = 4.3$ for GFW is located at 37,559,817 bp on OAR6 with the nearest gene being *LCORL* (at 107 kb). Not surprisingly, the effects of SNPs in *NCAPG* and near *LCORL* were highly correlated (r > 0.8) in our data, which indicates that there may be only one QTL in this region [52–54]. Furthermore, we found that the *LCORL* SNP identified in our study (at OAR6:37,559,817 bp) is located 29 kb from the *LCORL* SNP that was detected in a multi-trait GWAS across carcass and growth traits (at OAR6:37,530,647 bp) [14]. The effects of these two SNPs are strongly correlated (r > 0.8), which indicates that they may be in strong LD with the same underlying causal mutation with pleiotropic effects: i.e. simultaneously increasing carcass and skeletal weights and lean meat yield and decreasing dressing percentage, fatness, and muscling (i.e. CEMD), while increasing wool growth.

In a single-trait GWAS with the 50 K SNP chip in Chinese Merino sheep, Wang et al. [55] identified 28 SNPs that affect fibre diameter, fibre diameter coefficient of variance, and crimp and are located within 12 genes. However, we did not find any significant SNPs ($P < 10^{-5}$) within these genes.

Application to wool quality improvement

The patterns of the effects of the QTL that we studied here indicate that they have various degrees of usefulness for selection. Some QTL have an allele with desirable effects on more than one trait and appear to be good targets for selection. For instance, the QTL on OAR11 (located at the edge of the *FASN* gene) has an allele that increases greasy fleece weight, wool yield, and staple length and decreases fibre diameter. This pattern of quality (sheep with higher fleece, higher yield, longer staples, and finer wool) is desirable for sheep breeders and the wool industry because these traits affect the price paid for wool to the producer and the processing efficiency

Table 7 Examples of pleiotropic effects of SNPs selected from the multi-trait analyses on the individual traits (signed values with |t| > 1 are shown)

CHR:POS[a]	GFW	YLD	SL	SS	FD	FDC	CU	WR	BC	CC	DAG	SST	WE	CHA	FR	DU	COL	CLZ	CLYZ	CLY	CLX	SK	Gene code
1:244,308,221			2.0	1.3	-5.5	1.1	-1.3			-1.4					-1.3		-2.8		1.3				*TRPC1*
3:37,206,022		-1.5	-4.1	-1.1	-5.5	2.7	1.2		-1.1	1.0		-2.9			1.6		1.1	-1.9	-1.1				*LCLAT1*
3:39,080,949		2.1	3.2		-5.5		-3.7			1.0		2.7					1.2	-1.9		-2.0	-1.8		*ANTXR1*
3:58,612,068	-1.3	3.1	3.0		5.0	-3.2		-3.1	-1.6					-1.7	-2.6		1.3	-2.2		-2.4	-2.5	-1.5	*KRCC1*
3:59,019,274[6]			9.5	2.4	4.1	-3.7	1.1	-7.0	-4.3	-3.9	-1.4				-4.4	2.3			1.1			-4.9	*FOXI3*
3:60,384,564	3.5		-5.0			2.7	-2.5	5.6	1.8	1.2					3.3	-2.8	1.9		1.2			3.9	*TTL*
3:133,925,825	-1.4	1.6			-1.6	-1.3		-2.5	-1.0	-1.4		-2.9		-4.6	-1.5		-1.4					-1.7	*KRT86*
3:137,105,001	3.5		1.4					1.4				2.0	-1.3	1.4							1.2	1.2	*WNT1*
3:202,672,118	-2.0	-2.5			6.1	-2.6	3.0		-2.4	-3.2		1.9		2.3		1.3		-2.3	1.1	-2.2	-2.3		
4:62,652,234		-1.4	-1.8		-2.4	1.4	2.7	1.8				-1.7		-2.4				5.1	-3.0	4.5	4.5		*BMPER*
5:48,528,440[3]	2.8	2.2	3.3		2.0	1.3	-3.2	-1.3	-1.2		-1.6		2.0				2.7	2.9	-1.1	2.7	2.8		*FTL*
6:37,256,712[4]	4.4			-1.5	-5.5	3.6	-1.8					1.6		3.5	3.2		1.0	-1.1					*NCAPG/LCORL*
6:86,591,198	-1.1	1.3	1.2		-1.5	1.8	1.2	-2.4		2.7			2.0	3.5	3.0		-3.0	-1.3		-1.3	-1.4	-1.4	*GC*
6:94,602,390			-3.7		-3.7				1.5			-2.2							-1.1				*FGF5*
6:114,171,155	1.2				2.8			-1.5	-1.2		1.1		-1.4				-1.6	-5.3	2.5	-4.8	-4.8		*TRMT44*
7:57,834,812	3.4	1.5	1.8	-1.3	-2.2		-1.1	-1.5	-1.2	-1.3			1.4	1.2		1.0	1.7	-1.0	-1.1	-2.1	-1.5		*FGF7*
8:31,242,591[5]	-2.4	-3.1	-1.3	-1.3	-5.5		4.7	-1.3		-1.6	1.1	-1.7			-1.1		-1.5	1.3	-2.3				*ALOX15*
11:26,333,173	5.0	-1.6	1.0	-2.0	-1.9	2.7	-2.6		-1.1								1.5						*FASN*
11:49,956,916	4.5	2.1	2.1				-5.5	1.2	-1.0	1.3	2.3	-1.0		-1.3									*PIP4K2A*
13:22,846,602				-2.2	-1.3				1.3	5.3	1.9	-1.0					-2.5						
13:62,872,216[6]		-2.5	-3.8	1.8	-4.1	-5.5	1.7			1.1		-3.3	-2.7	-5.3				2.2		2.1	2.0		*RALY*
14:55,112,107[2]	-1.5	1.1	-1.1	1.1	-5.5	-4.6	2.3		-2.6			-3.3	-1.2	-1.6					-1.5				*BCL2L12*
15:13,764,013[2]		1.0	1.1	-3.3			-5.0			1.0			1.8				1.8	-2.0		-1.7	-1.6		*MAML2*
15:72,565,587			-1.5		-6.6	-1.3	-1.2		-1.0	-1.5		-2.8	-1.5	-2.7	1.3	-2.0	-1.3	2.7	-1.7	2.3	2.3		*ALX4*
19:658,455[3]	5.3	-11.3	3.3		2.3	-2.2	-2.1			-1.1	1.1	-3.9	-1.3	-3.6	-2.0	-2.0	-2.8	-1.2	1.7			-2.3	*LANCL2*
21:5,529,725	2.0	1.0			1.3	2.2	-4.9			1.6		-2.2	1.8					-1.2		-1.2	-1.2	-3.4	*NOX4*
22:34,770,924	1.4	1.2		-2.1	-4.3			-2.6		-1.6	1.4	1.2	1.3	1.4	2.4		2.1			-1.1	-1.0		*ATRNL1*
23:44,250,113[5]	-2.2		1.2	2.2	2.4	-5.3		-1.9	-2.4	-4.3	-3.1		1.6					-2.2		-2.2	-2.3		
24:35,659,971					-1.7		1.0	2.9		1.3	1.1						-1.5	4.1		5.1	5.1	1.1	*ACHE*
25:35,306,299[11]	-3.8	-4.0	-3.1	-8.7	-3.8	6.4	4.4	1.2	1.2	3.4	-1.8			2.9		1.3		2.1	-2.9				*MAT1A*
26:22,464,394			2.4		5.5					-1.1	1.1	3.7	1.3					-2.1	2.2	-1.6	2.0		*DLC1*

Traits at yearling age: FDC = FDCV; CU = CURV; WR = BRWR; BC = BCOV; CC = CCOV; SST = SSTRC; WE = WEATH; CHA = CHAR; FR = FLROT; DU = DUST; COL = GCOL; CLZ = COLZ; CLYZ = COLYZ; CLY = COLY; CLX = COLX; SK = SKINQ

[a] Ovine chromosome and position of SNPs, and the superscript numbers in brackets are the number of SNPs that have a similar pattern of pleiotropic effects within a distance of 1 to 1.5 Mb

and use of the wool in manufacturing. A SNP on OAR23 at about 44.2 Mb had an allele that increases staple strength and decreases fibre diameter coefficient of variance, breech wrinkle, breech cover, crutch cover and dag, which is a valuable pattern for resistance to breech strike and would reduce the need for flystrike prevention strategies such as mulesing and chemical treatment. SNPs on OAR3:137.1 Mb (*KRT86*), OAR3:133.9 Mb (*WNT1*), OAR8:89.0 Mb (*MPC*), and OAR12:49.9 Mb (*RSC1A1*) had an allele associated with better staple structure (staple with very fine bundles) and wool character (well-defined crimp with low variation along the staple), which is a desirable pattern for wool manufacturing. Generally, for a given trait, SNPs showed a similar association at both of the ages at which the trait was measured. However, none of the associations explained a large fraction of the variance for any trait. Therefore, although incorporation of the identified QTL may not increase the accuracy of single-trait EBV, they may be useful to manage unfavourable genetic correlations between traits.

Conclusions

For many wool traits, accuracy of genomic prediction was low (average over all traits = 0.22), especially for traits with a low heritability, few records and for which few QTL were identified. In an attempt to identify more QTL for these traits, we examined three approximate multi-trait methods. As well as a multi-trait GWAS, we describe two new multi-trait methods based on single-trait BayesR results. Collectively, these three methods mapped 206 putative QTL of which 20 were common to all methods. Sixteen genes that are located near a significant SNP have known effects on hair growth and a further five significant SNPs are near genes that were previously reported for QTL for growth and body composition. Future research should examine whether genomic prediction accuracy can be increased by using the QTL identified in this paper.

Additional file

Additional file 1: Table S1. Pleiotropic effects of 206 SNPs selected from multi-trait analyses on individual traits. The signed significant t-values are provided for 206 SNPs across 22 traits at two ages. Gene names are also provided if SNPs were located in or within a distance of 50 kb of the gene.

Authors' contributions

SB performed all analyses. SB, MEG, and HHD conceived the study and wrote the manuscript. SB, AAS and DJB were involved in preparing pedigree, phenotype and genotype data. SB, HDD, MEG, JHJvdW, AAS, DJB, SH, and NM contributed to the design of the study. All authors read and approved the final manuscript.

Author details

[1] Agriculture Victoria Research, AgriBio Centre, Bundoora, VIC 3083, Australia. [2] Animal Genetics and Breeding Unit, University of New England, Armidale, NSW 2351, Australia. [3] NSW Department of Primary Industries, Orange Agricultural Institute, Orange, NSW 2800, Australia. [4] School of Environmental and Rural Science, University of New England, Armidale, NSW 2351, Australia. [5] School of Land and Environment, University of Melbourne, Parkville, VIC 3010, Australia. [6] School of Applied Systems Biology, La Trobe University, Bundoora, VIC 3086, Australia. [7] Cooperative Research Centre for Sheep Industry Innovation, Armidale, NSW 2351, Australia.

Acknowledgements

The authors gratefully acknowledge funding or use of data from the Cooperative Research Centre for Sheep Industry Innovation, SheepGENOMICS project, Sheep Genetics, Australian Wool Innovation Ltd, and FarmIQ. The authors would like to extend their gratitude to Klint Gore (University of New England, Armidale, NSW, 2351, Australia) for preparing and cleaning genotype data and managing the CRC information nucleus database, and all staff involved at the Sheep CRC sites and SG sites across Australia. The ovine HD BeadChip was developed under the auspices of the International Sheep Genomics Consortium, in close collaboration between AgResearch New Zealand, CSIRO Livestock Industries, Baylor College of Medicine, and Agriculture Victoria, with work underwritten by FarmIQ (http://www.farmiq.co.nz/), a joint New Zealand government and industry Primary Growth Partnership programme.

Competing interests

The authors declare that they have no competing interests.

References

1. Nolan E. The economic value of wool attributes Phase 2. Report prepared for Australian Wool Innovation Limited. The University of Sydney School of Economics; 2014. p. 237.
2. Sackett D, Holmes P, Abbott K, Jephcott S, Barber M. Assessing the economic cost of endemic disease on the profitability of Australian beef cattle and sheep producers. Meat and Livestock Australia Ltd.; 2006.
3. Smith JL, Brewer HG, Dyall T. Heritability and phenotypic correlations for breech strike and breech strike resistance indicators in Merinos. Proc Assoc Adv Anim Breed Genet. 2009;18:334–7.
4. Brown DJ, Swan AA, Gill JS. Within- and across-flock genetic relationships for breech flystrike resistance indicator traits. Anim Prod Sci. 2010;50:1060–8.
5. Wuliji T, Dodds KG, Land JTJ, Andrews RN, Turner PR. Selection for ultrafine Merino sheep in New Zealand: heritability, phenotypic and genetic correlations of live weight, fleece weight and wool characteristics in yearlings. Anim Sci. 2010;72:241–50.
6. Huisman AE, Brown DJ. Genetic parameters for bodyweight, wool, and disease resistance and reproduction traits in Merino sheep 4 Genetic relationships between and within wool traits. Anim Prod Sci. 2009;49:289–96.
7. Safari E, Fogarty NM, Gilmour AR, Atkins KD, Mortimer SI, Swan AA, et al. Genetic correlations among and between wool, growth and reproduction traits in Merino sheep. J Anim Breed Genet. 2007;124:65–72.
8. Swan AA, Purvis IW, Piper LR. Genetic parameters for yearling wool production, wool quality and bodyweight traits in fine wool Merino sheep. Aust J Exp Agric. 2008;48:1168–76.
9. Brown DJ, Swan AA, Gill JS. Genetic correlations across ages for greasy fleece weight and fibre diameter in Merino sheep. Proc Assoc Adv Anim Breed Genet. 2013;20:110–3.
10. Meuwissen THE, Hayes BJ, Goddard ME. Prediction of total genetic value using genome-wide dense marker maps. Genetics. 2001;157:1819–29.
11. Kemper KE, Reich CM, Bowman PJ, Vander Jagt CV, Chamberlain AJ, Mason BA, et al. Improved precision of QTL mapping using a nonlinear Bayesian method in a multi-breed population leads to greater accuracy of across-breed genomic predictions. Genet Sel Evol. 2015;47:29.

12. Pryce JE, Hayes BJ, Bolormaa S, Goddard ME. Polymorphic regions affecting human height also control stature in cattle. Genetics. 2011;3:981–4.

13. Locke AE, Kahali B, Berndt SI, Justice AE, Pers TH, Day FR, et al. Genetic studies of body mass index yield new insights for obesity biology. Nature. 2015;518:197–206.

14. Bolormaa B, Hayes BJ, van der Werf JHJ, Pethick D, Goddard ME, Daetwyler HD. Detailed phenotyping identifies genes with pleiotropic effects on body composition. BMC Genomics. 2016;17:224.

15. van der Werf JHJ, Kinghorn BP, Banks RG. Design and role of an information nucleus in sheep breeding programs. Anim Prod Sci. 2010;50:998–1003.

16. White JD, Allingham PG, Gorman CM, Emery DL, Hynd P, Owens J, et al. Design and phenotyping procedures for recording wool, skin, parasite resistance, growth, carcass yield and quality traits of the SheepGENOMICS mapping flock. Anim Prod Sci. 2012;52:157–71.

17. Hatcher S, Hynd PI, Thornberry KJ, Gabb S. Can we breed Merino sheep with softer, whiter, more photostable wool? Anim Prod Sci. 2010;50:1089–97.

18. Australian Wool Innovation and Meat and Livestock Australia. Visual sheep scores. researcher ed. Sydney: Australian Wool Innovation and Meat and Livestock Australia; 2007.

19. Wood E. The basics of wool colour measurement. Wool Technol Sheep Breed. 2002;50:121–32.

20. Nicolazzi NL, Caprera A, Nazzicari N, Cozzi P, Strozzi F, Lawley C, et al. SNPchiMp v. 3: integrating and standardizing single nucleotide polymorphism data for livestock species. BMC Genomics. 2015;16:283.

21. Sargolzaei M, Chesnais J, Schenkel F. A new approach for efficient genotype imputation using information from relatives. BMC Genomics. 2014;15:478.

22. Daetwyler HD, Swan AA, van der Werf JHJ, Hayes BJ. Accuracy of pedigree and genomic predictions of carcass and novel meat quality traits in multi-breed sheep data assessed by cross-validation. Genet Sel Evol. 2012;44:33.

23. Gilmour AR, Gogel BJ, Cullis BR, Thompson R. ASReml user guide release 3.0 VSN. Hemel Hempstead: International Ltd; 2009.

24. Swan AA, Brown DJ, van der Werf JHJ. Genetic variation within and between sub-populations of the Australian Merino breed. Anim Prod Sci. 2014;56:87–94.

25. Bolormaa S, Hayes BJ, Savin K, Hawkin R, Barendse W, Arthur P, et al. Genome wide association studies for feedlot and growth traits in cattle. J Anim Sci. 2011;89:1684–97.

26. VanRaden PM. Efficient methods to compute genomic predictions. J Dairy Sci. 2008;91:4414–23.

27. Habier D, Fernando RL, Dekkers JCM. The impact of genetic relationship information on genome-assisted breeding values. Genetics. 2007;177:2389–97.

28. Yang J, Benyamin B, McEvoy NP, Gordon S, Henders AK, Nyholt DR, et al. Common SNPs explain a large proportion of the heritability for human height. Nat Genet. 2010;42:565–9.

29. Erbe M, Hayes BJ, Matukumalli LK, Goswami S, Bowman PJ, Reich M, et al. Improving accuracy of genomic predictions within and between dairy cattle breeds with imputed high-density single nucleotide polymorphism panels. J Dairy Sci. 2012;95:4114–29.

30. Bolormaa S, Pryce JE, Reverter A, Zhang Y, Barendse W, Kemper K, et al. A multi-trait, meta-analysis for detecting pleiotropic polymorphisms for stature, fatness and reproduction in beef cattle. PLoS Genet. 2014;10:e1004198.

31. Rahib L, MacLennan NK, Horvath S, Liao JC, Dipple KM. Glycerol kinase deficiency alters expression of genes involved in lipid metabolism, carbohydrate metabolism, and insulin signaling. Eur J Hum Genet. 2007;15:646–57.

32. Croteau-Chonka DC, Marvelle AF, Lange EM, Lee NR, Adair LS, Lange LA, et al. Genome-wide association study of anthropometric traits and evidence of interactions with age and study year in Filipino women. Obesity. 2011;19:1019–27.

33. Zhang L, Mousel MR, Wu X, Michal JJ, Zhou X, et al. Genome-wide genetic diversity and differentially selected regions among Suffolk, Rambouillet, Columbia, Polypay, and Targhee sheep. PLoS One. 2013;8:e65942.

34. Goddard ME. Genomic selection: prediction of accuracy and maximisation of long term response. Genetica. 2009;136:245–57.

35. Bolormaa S, Pryce JE, Kemper K, Savin K, Hayes BJ, Barendse W, et al. Accuracy of prediction of genomic breeding values for residual feed intake, carcass and meat quality traits in Bos taurus, Bos indicus and composite beef cattle. J Anim Sci. 2013;91:3088–104.

36. MacLeod IM, Hayes BJ, Goddard ME. The effects of demography and long-term selection on the accuracy of genomic prediction with sequence data. Genetics. 2014;198:1671–84.

37. Moser G, Lee SH, Hayes BJ, Goddard ME, Wray NR, Visscher PM. Simultaneous discovery, estimation and prediction analysis of complex traits using a Bayesian mixture model. PLoS Genet. 2015;11:e1004969.

38. Daetwyler HD, Hickey JM, Henshall JM, Dominik S, Gredler B, van der Werf JHJ, et al. Accuracy of estimated genomic breeding values for wool and meat traits in a multi-breed sheep population. Anim Prod Sci. 2010;50:1004–10.

39. Moghaddar N, Swan AA, van der Werf JH. Comparing genomic prediction accuracy from purebred, crossbred and combined purebred and crossbred reference populations in sheep. Genet Sel Evol. 2014;46:58.

40. Auvray B, McEwan JC, Newman SAN, Lee M, Dodds KG. Genomic prediction of breeding values in the New Zealand sheep industry using a 50 K SNP chip. J Anim Sci. 2014;92:4375–89.

41. Dodds KG, Auvray B, Newman SN, McEwan JC. Genomic breed prediction in New Zealand sheep. BMC Genet. 2014;15:92.

42. Lee MA, Cullen NG, Newman SAN, Dodds KG, McEwan JC, Shackell GH. Genetic analysis and genomic selection of stayability and productive life in New Zealand ewes. J Anim Sci. 2015;93:3268–77.

43. Pickering NK, Auvray B, Dodds KG, McEwan JC. Genomic prediction and genome-wide association study for dagginess and host internal parasite resistance in New Zealand sheep. BMC Genomics. 2015;16:958.

44. Bolormaa S, Gore K, van der Werf JHJ, Hayes B, Daetwyler HD. Design of a low-density SNP chip for the main Australian sheep breeds and its effect on imputation and genomic prediction accuracy. Anim Genet. 2015;46:544–56.

45. Kijas JW, Lenstra JA, Hayes B, Boitard S, Porto Neto L, San Cristobal M, et al. Genome-wide analysis of the world's sheep breeds reveals high levels of historic mixture and strong recent selection. PLoS Biol. 2012;10:e1001258.

46. Stearns TM, Beever JE, Southey BR, Ellis M, McKeith FK, Rodriguaz-Zas SL. Evaluation of approaches to detect quantitative traits loci for growth, carcass, and meat quality on swine chromosome 2, 6, 13 and 18. II. Multivariate and principal component analyses. J Anim Sci. 2005;83:2471–81.

47. Shirokova V, Biggs LC, Jussila M, Ohyama T, Groves AK, Mikkola ML. FOXI3 deficiency compromises hair follicle stem cell specification and activation. Stem Cells. 2016;34:1896–908.

48. Botchkarev VA, Sharov AA. BMP signaling in the control of skin development and hair follicle growth. Differentiation. 2004;72:512–26.

49. Zhang J, He XC, Tong WG, Johnson T, Wiedemann LM, Mishina AY, et al. Bone morphogenetic protein signaling inhibits hair follicle anagen induction by restricting epithelial stem/progenitor cell activation and expansion. Stem Cells. 2006;24:2826–39.

50. Rendl M, Polak L, Fuchs E. BMP signaling in dermal papilla cells is required for their hair follicle-inductive properties. Genes Dev. 2008;22:543–57.

51. Genander M, Cook PJ, Ramsköld D, Keyes BE, Mertz AF, Sandberg R, et al. BMP signaling and its pSMAD1/5 target genes differentially regulate hair follicle stem cell lineages. Cell Stem Cell. 2014;15:619–33.

52. Setoguchi K, Furuta M, Hirano T, Nagao T, Watanabe T, Sugimoto Y, et al. Cross-breed comparisons identified a critical 591-kb region for bovine carcass weight QTL (CW-2) on chromosome 6 and the Ile-442-Met substitution in NCAPG as a positional candidate. BMC Genet. 2009;10:43.

53. Tetens J, Widmann P, Kühn C, Thaller G. A genome-wide association study indicates LCORL/NCAPG as a candidate locus for withers height in German Warmblood horses. Anim Genet. 2013;44:467–71.

54. Saatchi M, Schnabel RD, Taylor JF, Garrick DJ. Large-effect pleiotropic or closely linked QTL segregate within and across ten US cattle breeds. BMC Genomics. 2014;15:442.

55. Wang Z, Zhang H, Yang H, Wang S, Rong E, Pei W, et al. Genome-wide association study for wool production traits in a Chinese Merino sheep population. PLoS One. 2014;9:e107101.

56. Ohyama M, Terunuma A, Tock CL, Radonovich MF, Pise-Masison CA, Hopping B, et al. Characterization and isolation of stem cell-enriched human hair follicle bulge cells. J Clin Invest. 2006;116:249–60.

57. Yang R, Zheng Y, Burrows M, Liu S, Wei Z, Nace A, et al. Generation of folliculogenic human epithelial stem cells from induced pluripotent stem cells. Nat Commun. 2014;5:3071.

58. Stranecky V, Hoischen A, Hartmannova H, Zaki MS, Chaudhary A, Zudaire E, et al. Mutations in *ANTXR1* cause GAPO syndrome. Am J Hum Genet. 2013;92:792–9.

59. Duverger O, Morasso MI. To grow or not to grow: hair morphogenesis and human genetic hair disorders. Semin Cell Dev Biol. 2014;25–26:22–33.

60. Yu Z, Wildermoth JE, Wallace OA, Gordon SW, Maqbool NJ, Maclean PH, et al. Annotation of sheep keratin intermediate filament genes and their patterns of expression. Exp Dermatol. 2011;7:582–8.

61. Choi YS, Zhang Y, Xu M, Yang Y, Ito M, Peng T, et al. Distinct functions for Wnt/β-catenin in hair follicle stem cell proliferation and survival and interfollicular epidermal homeostasis. Cell Stem Cell. 2013;13:720–33.

62. Clements SE, Techanukul T, Lai-Cheong JE, Mee JB, South AP, Pourreyron C, et al. Mutations in AEC syndrome skin reveal a role for p63 in basement membrane adhesion, skin barrier integrity and hair follicle biology. Br J Dermatol. 2012;167:134–44.

63. Higgins CA, Petukhova L, Harel S, Ho YY, Drill E, Shapiro L, et al. FGF5 is a crucial regulator of hair length in humans. Proc Natl Acad Sci USA. 2014;111:10648–53.

64. Rosenquist TA, Martin GR. Fibroblast growth factor signalling in the hair growth cycle: expression of the fibroblast growth factor receptor and ligand genes in the murine hair follicle. Dev Dyn. 1996;205:379–86.

65. Rao D, Macias E, Carbajal S, Kiguchi K, DiGiovanni J. Constitutive Stat3 activation alters behavior of hair follicle stem and progenitor cell populations. Mol Carcinog. 2015;54:121–33.

66. Ito T, Tsuji G, Ohno F, Uchi H, Nakahara T, Hashimoto-Hachiya A, et al. Activation of the OVOL1-OVOL2 axis in the hair bulb and in pilomatricoma. Am J Pathol. 2016;186:1036–43.

67. Nasr Z, Dow LE, Paquet M, Chu J, Ravindar K, Somaiah R, et al. Suppression of eukaryotic initiation factor 4E prevents chemotherapy-induced alopecia. BMC Pharmacol Toxicol. 2013;14:58.

68. Kayserili H, Uz E, Niessen C, Vargel I, Alanay Y, Tuncbilek G, et al. ALX4 dys-function disrupts craniofacial and epidermal development. Hum Mol Genet. 2009;18:4357–66.

69. Wu L, Sun T, Kobayashi K, Gao P, Griffin JD. Identification of a family of mastermind-like transcriptional coactivators for mammalian Notch receptors. Mol Cell Biol. 2002;22:7688–700.

70. Ellis T, Gambardella L, Horcher MSJC, Tschanz S, Capol J, Bertram P, et al. The transcriptional repressor CDP (Cutl1) is essential for epithelial cell differentiation of the lung and the hair follicle. Genes Dev. 2001;15:2307–19.

71. Hasse S, Chernyavsky AI, Grando SA, Paus R. The M4 muscarinic acetylcholine receptor plays a key role in the control of murine hair follicle cycling and pigmentation. Life Sci. 2007;80:2248–52.

6

An efficient exact method to obtain GBLUP and single-step GBLUP when the genomic relationship matrix is singular

Rohan L. Fernando[1]* ⓘ, Hao Cheng[1] and Dorian J. Garrick[1,2]

Abstract

Background: The mixed linear model employed for genomic best linear unbiased prediction (GBLUP) includes the breeding value for each animal as a random effect that has a mean of zero and a covariance matrix proportional to the genomic relationship matrix ($\mathbf{G}_{gg}$), where the inverse of $\mathbf{G}_{gg}$ is required to set up the usual mixed model equations (MME). When only some animals have genomic information, genomic predictions can be obtained by an extension known as single-step GBLUP, where the covariance matrix of breeding values is constructed by combining the pedigree-based additive relationship matrix with $\mathbf{G}_{gg}$. The inverse of the combined relationship matrix can be obtained efficiently, provided $\mathbf{G}_{gg}$ can be inverted. In some livestock species, however, the number N_g of animals with genomic information exceeds the number of marker covariates used to compute $\mathbf{G}_{gg}$, and this results in a singular $\mathbf{G}_{gg}$. For such a case, an efficient and exact method to obtain GBLUP and single-step GBLUP is presented here.

Results: Exact methods are already available to obtain GBLUP when $\mathbf{G}_{gg}$ is singular, but these require working with large dense matrices. Another approach is to modify $\mathbf{G}_{gg}$ to make it nonsingular by adding a small value to all its diagonals or regressing it towards the pedigree-based relationship matrix. This, however, results in the inverse of $\mathbf{G}_{gg}$ being dense and difficult to compute as N_g grows. The approach presented here recognizes that the number r of linearly independent genomic breeding values cannot exceed the number of marker covariates, and the mixed linear model used here for genomic prediction only fits these r linearly independent breeding values as random effects.

Conclusions: The exact method presented here was compared to Apy-GBLUP and to Apy single-step GBLUP, both of which are approximate methods that use a modified $\mathbf{G}_{gg}$ that has a sparse inverse which can be computed efficiently. In a small numerical example, predictions from the exact approach and Apy were almost identical, but the MME from Apy had a condition number about 1000 times larger than that from the exact approach, indicating ill-conditioning of the MME from Apy. The practical application of exact SSGBLUP is not more difficult than implementation of Apy.

Background

In animal breeding, two equivalent mixed linear models have been used for genomic prediction using phenotypes on genotyped individuals [1]. In the first, random effects of markers are explicitly included in the model [2, 3]. We will refer to this model as the marker effects model (MEM). In the second, the breeding value of each animal, which is a linear combination of the random marker effects, is included as a random effect [1, 2, 4, 5]. We will refer to this model as the breeding value model (BVM). The mixed model equations (MME) that corresponds to the MEM has order $p + k$, where p is the number of non-genetic effects and k is the number of marker covariates, and the MME that correspond to the BVM has order $p + N_g$, where N_g is the number of animals. When genomic data were first available, the number N_g of animals with genotypic and phenotypic records was much smaller than the number k of marker effects. Thus, genomic prediction with the BVM was more efficient than using the MEM [1, 5], and prediction using this approach is now known as GBLUP.

*Correspondence: rohan@iastate.edu
[1] Department of Animal Science, Iowa State University, Ames, IA 50011, USA
Full list of author information is available at the end of the article

However, at present, in some livestock species such as dairy cattle, N_g has increased to over 100,000 if not 1 million. When N_g exceeds k, the matrix $\mathbf{G}_{gg}$ of genomic relationships will have at least $n - k$ eigen values that are zero, and therefore $\mathbf{G}_{gg}$ is guaranteed to be singular. In practice, depending on the effective population size, some of the smallest of the k largest eigen values may be very near to zero if not zero. In either event, the MME that require the inverse of $\mathbf{G}_{gg}$ cannot be employed to obtain GBLUP. In that situation, an alternate form of the MME [4, 6–8] that can accommodate a singular $\mathbf{G}_{gg}$ can be employed, but this results in a completely dense set of MME of order $p + N_g$. Thus, when N_g is large, this formulation of the MME is not useful for computing GBLUP. An alternative is to use a modified matrix $\mathbf{G}^*$ obtained from $\mathbf{G}_{gg}$ by adding a small value to all its diagonals or by regressing it towards the pedigree-based relationship matrix, $\mathbf{A}$, so that it retains full rank, but this is no longer an exact representation of the model if the markers completely explain the breeding values. Furthermore, this modified relationship matrix still has a dense inverse, which may be impossible to compute when N_g is large.

Suppose the rank of $\mathbf{G}_{gg}$ is $r \leq k < N_g$. Then, we will show here how to obtain exact GBLUP without approximation from a set of MME that has order $p + r$, which can be much lower than $p + N_g$. We also show how this approach can be used to obtain exact single-step GBLUP without approximation when some animals have not been genotyped. These formulations are useful to better understand predictions that are obtained by using the recursive algorithm for "parents (core)" and "young (noncore)" animals i.e. Apy, which is gaining popularity [9–13] as an approach to approximate the inverse of $\mathbf{G}_{gg}$ [9] or $\mathbf{G}^*$ [13]. The exact inverse of the nonsingular matrix $\mathbf{G}^* = 0.95\mathbf{G}_{gg} + 0.05\mathbf{A}$ will be dense whereas Apy approximates this with a sparse inverse [9, 10]. We will show here that when a full-rank $\mathbf{G}^*$ is obtained by adding a small number to the diagonals of only noncore animals, the inverse calculated in Apy for a suitable choice of core animals will be sparse and an exact inverse of $\mathbf{G}^*$, but the inverse may be ill conditioned. The approximate inverse calculated in Apy cannot ever be that of $\mathbf{G}_{gg}$, which is singular when $r < N_g$. The Apy algorithm will never yield exact GBLUP predictions contrary to the claims in [9, 11], but it has been demonstrated to be a useful approximation for some choices of $\mathbf{G}^*$ [11–13].

Theory

Let $\mathbf{M}_g$ denote the centered marker genotype covariate matrix of order $N_g \times k$ with $N_g > k$, which is the case when the number N_g of genotyped animals is larger than the number k of marker covariates. Then, the row

rank r of $\mathbf{M}_g$ is $r \leq k < N_g$ [14]. Suppose $\mathbf{M}_g$ is ordered such that its first r rows are linearly independent and are denoted $\mathbf{M}_{g_i}$. It follows that the remaining $N_g - r$ dependent rows of $\mathbf{M}_g$, denoted $\mathbf{M}_{g_d}$ can be written as a linear combination:

$$\mathbf{M}_{g_d} = \mathbf{L}'\mathbf{M}_{g_i},$$

so that

$$\mathbf{M}_g = \begin{bmatrix} \mathbf{M}_{g_i} \\ \mathbf{M}_{g_d} \end{bmatrix} = \begin{bmatrix} \mathbf{M}_{g_i} \\ \mathbf{L}'\mathbf{M}_{g_i} \end{bmatrix}. \tag{1}$$

Now, a commonly-used form of the genomic relationship matrix [5] becomes

$$\mathbf{G}_{gg} = \frac{\mathbf{M}_g\mathbf{M}'_g}{k} = \frac{1}{k}\begin{bmatrix} \mathbf{M}_{g_i}\mathbf{M}'_{g_i} & \mathbf{M}_{g_i}\mathbf{M}'_{g_i}\mathbf{L} \\ \mathbf{L}'\mathbf{M}_{g_i}\mathbf{M}'_{g_i} & \mathbf{L}'\mathbf{M}_{g_i}\mathbf{M}'_{g_i}\mathbf{L} \end{bmatrix}, \tag{2}$$

where it can be seen that the last $N_g - r$ rows are a linear combination of the first r rows. The last $N_g - r$ columns of $\mathbf{G}$ are similarly a linear combination of the first r columns. Thus, in this case, $\mathbf{G}$ is singular and its inverse does not exist. It can be seen from (2) that $\mathbf{L}'$ can be written as:

$$\mathbf{L}' = \mathbf{G}_{g_d g_i}\mathbf{G}_{g_i g_i}^{-1}, \tag{3}$$

where $\mathbf{G}_{g_d g_i} = \frac{1}{k}\mathbf{M}_{g_d}\mathbf{M}'_{g_i}$ and $\mathbf{G}_{g_i g_i} = \frac{1}{k}\mathbf{M}_{g_i}\mathbf{M}'_{g_i}$.

GBLUP when G is singular

In the following, we will assume that the vector $\mathbf{u}_g$ of breeding values of animals can be adequately modeled as:

$$\mathbf{u}_g = \mathbf{M}_g\boldsymbol{\alpha}, \tag{4}$$

where the vector $\boldsymbol{\alpha}$ of marker effects is assumed to have zero mean and covariance matrix $\mathbf{I}\sigma_\alpha^2$. It follows that the covariance matrix of the breeding values is:

$$\begin{aligned} \mathrm{Var}(\mathbf{u}_g|\mathbf{M}_g) &= \mathbf{M}_g\mathbf{M}'_g\sigma_\alpha^2 \\ &= \mathbf{G}_{gg}k\sigma_\alpha^2 \\ &= \mathbf{G}_{gg}\sigma_u^2, \end{aligned} \tag{5}$$

where $\sigma_u^2 = k\sigma_\alpha^2$. To proceed, we further assume the following mixed linear model for the vector $\mathbf{y}$ of phenotypic values:

$$\mathbf{y} = \mathbf{X}\boldsymbol{\beta} + \mathbf{Z}\mathbf{u}_g + \mathbf{e}, \tag{6}$$

where $\boldsymbol{\beta}$ is a vector of non-genetic fixed effects, $\mathbf{X}$ and $\mathbf{Z}$ are incidence matrices relating $\boldsymbol{\beta}$ and $\mathbf{u}_g$ to $\mathbf{y}$, and $\mathbf{e}$ is a vector of residuals with zero mean and covariance matrix $\mathbf{I}\sigma_e^2$.

Here, we have assumed that the markers fully explain the breeding values. If this is not the case, a random polygenic residual effect with zero mean and covariance matrix that is proportional to $\mathbf{A}$ can be included in the model.

Strategy I
When $\mathbf{G}$ is singular, one strategy to get the BLUP of $\mathbf{u}$ is to use the formula:

$$
\begin{aligned}
\hat{\mathbf{u}}_g &= \mathrm{Cov}(\mathbf{u}_g, \mathbf{y}')\mathrm{Var}^{-1}(\mathbf{y})(\mathbf{y} - \mathbf{X}\hat{\boldsymbol{\beta}}) \\
&= \mathbf{G}_{gg}\mathbf{Z}'\mathbf{V}^{-1}(\mathbf{y} - \mathbf{X}\hat{\boldsymbol{\beta}}),
\end{aligned}
\tag{7}
$$

where $\hat{\boldsymbol{\beta}}$ is a solution to the system

$$
(\mathbf{X}'\mathbf{V}^{-1}\mathbf{X})\hat{\boldsymbol{\beta}} = \mathbf{X}'\mathbf{V}^{-1}\mathbf{y},
\tag{8}
$$

and $\mathbf{V} = (\mathbf{Z}\mathbf{G}_{gg}\mathbf{Z}'\sigma_u^2 + \mathbf{I}\sigma_e^2)$ [7]. When N_g is large, this strategy is not computationally feasible because the matrix $\mathbf{V}$ is dense, has order N_g, and its inverse is needed in (7) and (8).

Strategy II
Another strategy is to get the solution to the following MME as proposed by Harville [6]:

$$
\begin{bmatrix} \mathbf{X}'\mathbf{R}^{-1}\mathbf{X} & \mathbf{X}'\mathbf{R}^{-1}\mathbf{Z} \\ \mathbf{G}_{gg}\mathbf{Z}'\mathbf{R}^{-1}\mathbf{X} & \mathbf{G}_{gg}\mathbf{Z}'\mathbf{R}^{-1}\mathbf{Z} + \mathbf{I} \end{bmatrix} \begin{bmatrix} \hat{\boldsymbol{\beta}} \\ \hat{\mathbf{u}}_g \end{bmatrix} = \begin{bmatrix} \mathbf{X}'\mathbf{R}^{-1}\mathbf{y} \\ \mathbf{G}_{gg}\mathbf{Z}'\mathbf{R}^{-1}\mathbf{y} \end{bmatrix},
\tag{9}
$$

where $\mathbf{R}^{-1} = \mathbf{I}\frac{1}{\sigma_e^2}$. These MME are dense and have order $p + N_g$. Thus, although the above approaches do not require inverting $\mathbf{G}_{gg}$, explicitly using these MME do not provide a feasible approach as the number N_g of genotyped animals approaches or exceeds a million, because storing and solving such large and dense system of equations would exceed the capacity of the typical computer used for genetic evaluation. An implementation with iteration on data using the PCG algorithm may be feasible by computing matrix products like $\mathbf{G}_{gg}\mathbf{x}$ in parts as $\frac{1}{k}\mathbf{M_g}(\mathbf{M_g}'\mathbf{x})$ [15]. However, Aguilar et al. [16] reported these asymmetric equations do not scale up well and suffer convergence problems.

Strategy III
We show here that it is possible to obtain BLUP of $\mathbf{u}_g$ by solving a set of MME that has order $p + r$, which can be much lower than $p + N_g$. To do so, the breeding values of the r animals with genotypes $\mathbf{M}_{g_i}$ is denoted $\mathbf{u}_{g_i}$ and the breeding values of the $N_g - r$ animals with genotypes $\mathbf{M}_{g_d}$ is denoted $\mathbf{u}_{g_d}$. The model for the breeding values in (4) can be written as:

$$
\begin{aligned}
\mathbf{u}_g &= \mathbf{M}_g\boldsymbol{\alpha} \\
&= \begin{bmatrix} \mathbf{M}_{g_i} \\ \mathbf{M}_{g_d} \end{bmatrix} \boldsymbol{\alpha},
\end{aligned}
\tag{10}
$$

and writing $\mathbf{M}_{g_d} = \mathbf{L}'\mathbf{M}_{g_i}$ as in (1), this becomes:

$$
\begin{aligned}
\mathbf{u}_g &= \begin{bmatrix} \mathbf{M}_{g_i} \\ \mathbf{L}'\mathbf{M}_{g_i} \end{bmatrix} \boldsymbol{\alpha} \\
&= \begin{bmatrix} \mathbf{I} \\ \mathbf{L}' \end{bmatrix} \mathbf{M}_{g_i}\boldsymbol{\alpha} \\
&= \begin{bmatrix} \mathbf{I} \\ \mathbf{L}' \end{bmatrix} \mathbf{u}_{g_i}.
\end{aligned}
\tag{11}
$$

Note that the vector of breeding values given by (11) is identical to (4), and thus these two vectors have the same covariance matrix that is given by (5).

Now, using (11) for $\mathbf{u}_g$ in (6), the mixed linear model for the phenotypic values can be written in terms of $\mathbf{u}_{g_i}$ as:

$$
\mathbf{y} = \mathbf{X}\boldsymbol{\beta} + \mathbf{Z}\begin{bmatrix} \mathbf{I} \\ \mathbf{L}' \end{bmatrix} \mathbf{u}_{g_i} + \mathbf{e}.
\tag{12}
$$

The random effect $\mathbf{u}_{g_i}$ of this model has order r and can be much lower than N_g the order of $\mathbf{u}_g$. Furthermore, as $\mathbf{u}_g = \begin{bmatrix} \mathbf{I} \\ \mathbf{L}' \end{bmatrix} \mathbf{u}_{g_i}$, the models given by (6) and (12) have the same first and second moments, and thus they are equivalent models and yield the same BLUP for $\mathbf{u}_g$ [7]. The MME for the model (12) are

$$
\begin{bmatrix} \mathbf{X}'\mathbf{X} & \mathbf{X}'\mathbf{W} \\ \mathbf{W}'\mathbf{X} & \mathbf{W}'\mathbf{W} + \lambda\mathbf{G}_{g_i g_i}^{-1} \end{bmatrix} \begin{bmatrix} \hat{\boldsymbol{\beta}} \\ \hat{\mathbf{u}}_{g_i} \end{bmatrix} = \begin{bmatrix} \mathbf{X}'\mathbf{y} \\ \mathbf{W}'\mathbf{y} \end{bmatrix},
\tag{13}
$$

where $\mathbf{W} = \mathbf{Z}\begin{bmatrix} \mathbf{I} \\ \mathbf{L}' \end{bmatrix}$, $\mathbf{G}_{g_i g_i}^{-1}$ is the inverse of the $r \times r$ non-singular matrix $\mathbf{G}_{g_i g_i} = \frac{1}{k}(\mathbf{M}_{g_i}\mathbf{M}_{g_i}')$, and $\lambda = \frac{\sigma_e^2}{\sigma_u^2}$. The BLUP of $\mathbf{u}_{g_d}$ is obtained as $\hat{\mathbf{u}}_{g_d} = \mathbf{L}'\hat{\mathbf{u}}_{g_i}$.

Strategy IV
A key assumption in Strategy III is that the matrix $\mathbf{M}_g$ of marker covariates can be reordered such that the first r rows are linearly independent and the remaining dependent rows can be expressed as a linear combination of the first set of r linearly independent rows. Determining the precise rank of $\mathbf{M}_g$ may be inexact as the eigen values of $\mathbf{G}_{gg}$ decay slowly [17]. On the one hand, if the chosen $\mathbf{M}_{g_i}$ contains less rows than the rank of $\mathbf{G}_{gg}$, it would not be possible to express $\mathbf{M}_{g_d}$ as $\mathbf{M}_{g_d} = \mathbf{L}'\mathbf{M}_{g_i}$. On the other hand, if $\mathbf{M}_{g_d}$ contains more rows than the rank of $\mathbf{G}_{gg}$, $\mathbf{G}_{g_i g_i}$ will be singular. Even when the number of rows in $\mathbf{M}_{g_i}$ is equal to the rank of $\mathbf{G}_{gg}$, $\mathbf{G}_{g_i g_i}$ may be ill conditioned if the smallest eigen value of $\mathbf{G}_{g_i g_i}$ is close to zero. The condition number of a matrix is represented by the ratio of the largest to the smallest eigen value, and it is 1 for a perfectly conditioned matrix and a large number for an ill-conditioned matrix. There are many combinations of individuals that can be placed in $\mathbf{M}_{g_i}$, but the condition

number of the resultant $\mathbf{G}_{g_i g_i}$ may vary greatly according to the chosen combination. The condition number of $\mathbf{G}_{g_i g_i}$ will impact the condition number of the resultant MME, and poorly-conditioned equations take longer to solve iteratively than well-conditioned equations. In comparing the choice of core used in Apy in a pig evaluation, Ostersen et al. [18] reported similar numbers of PCG iterations for non-genomic analyses and 8 choices of core, but the correlation between the Apy-SSGBLUP and SSGBLUP ranged from 0.93 to more than 0.99 for genotyped animals. That paper did not report the criterion used to determine PCG convergence.

One way to improve the condition number of the MME is to fit an equivalent model obtained by orthonormalizing the rows of $\mathbf{M}_{g_i}$. Suppose $\mathbf{U} = \mathbf{TM}_{g_i}$, where $\mathbf{UU}' = \mathbf{I}$. Then, the transformed vector $\mathbf{v} = \mathbf{Tu}_{g_i}$ of breeding values will have a genomic covariance matrix:

$$
\begin{aligned}
\mathrm{Var}(\mathbf{v}) &= \mathbf{T}\mathrm{Var}(\mathbf{u}_{g_i})\mathbf{T}'\sigma_\alpha^2 \\
&= \mathbf{TM}_{g_i}\mathbf{M}'_{g_i}\mathbf{T}'\sigma_\alpha^2 \\
&= \mathbf{UU}'\sigma_\alpha^2 \\
&= \mathbf{I}\sigma_\alpha^2.
\end{aligned}
$$

Then, as in [17], formulating the model in terms of $\mathbf{v}$, which has a well-conditioned covariance matrix, will result in a well-conditioned MME.

Another way to improve the condition of the MME without explicitly reordering $\mathbf{M}_g$ is by using an RQ decomposition [19] that involves expressing $\mathbf{M}_g$ as $\mathbf{M}_g = \mathbf{RU}$, where $\mathbf{R}$ is a lower triangular $N_g \times k$ matrix and $\mathbf{U}$ is a $k \times k$ orthogonal matrix. The RQ decomposition applies to the rows of a matrix in the same manner that the QR decomposition is applied to the columns. Exploiting the decomposition, the model equation for the phenotypic values can be written in terms of $\mathbf{v} = \mathbf{U}\alpha$ as:

$$
\begin{aligned}
\mathbf{y} &= \mathbf{X}\boldsymbol{\beta} + \mathbf{ZM}_g\alpha + \mathbf{e} \\
&= \mathbf{X}\boldsymbol{\beta} + \mathbf{ZRU}\alpha + \mathbf{e} \\
&= \mathbf{X}\boldsymbol{\beta} + \mathbf{ZRv} + \mathbf{e} \\
&= \mathbf{X}\boldsymbol{\beta} + \mathbf{Wv} + \mathbf{e},
\end{aligned}
$$

where now $\mathbf{W} = \mathbf{ZR}$. The MME for this model are:

$$
\begin{bmatrix} \mathbf{X}'\mathbf{X} & \mathbf{X}'\mathbf{W} \\ \mathbf{W}'\mathbf{X} & \mathbf{W}'\mathbf{W} + \mathbf{I}\frac{\sigma_e^2}{\sigma_\alpha^2} \end{bmatrix} \begin{bmatrix} \hat{\boldsymbol{\beta}} \\ \hat{\mathbf{v}} \end{bmatrix} = \begin{bmatrix} \mathbf{X}'\mathbf{y} \\ \mathbf{W}'\mathbf{y} \end{bmatrix}, \quad (14)
$$

and predictions for all individuals on the original scale can be obtained as $\hat{\mathbf{u}}_g = \mathbf{R}\hat{\mathbf{v}}$. Also, the marker effects can be obtained as $\hat{\alpha} = \mathbf{U}'\hat{\mathbf{v}}$. Note that this factorization does not require us to know or determine the rank of $\mathbf{M}_g$. Furthermore, the orthogonal matrix $\mathbf{U}$ can be obtained by applying the RQ factorization to just the first k rows of $\mathbf{M}_g$, for which the number of operations is proportional to k^3 [19]. The matrix $\mathbf{R}$ can be obtained as $\mathbf{R} = \mathbf{M}_g\mathbf{U}'$.

Comparison to Apy-GBLUP

The efficient algorithm to obtain the inverse of the additive relationship matrix is based on the property that the additive relationships between an animal and any non-descendant (an individual that is not a descendant) can be written as a linear combination of the relationships between the non-descendant and the parents of the animal [20, 21]. This property of additive relationships also allows construction of the additive relationship matrix by the tabular method [22]. The so-called Apy algorithm [9, 10] attempts to extend this idea to genomic relationships by classifying animals into two groups: "core" and "noncore" animals. The Apy algorithm seems to imply that the relationship between a noncore animal and any other animal can be written as a linear combination of relationships between the other animal and the animals in the core group. We will refer to this property of the genomic relationships that is required for Apy as the Apy property. Provided this property holds, it is claimed that Apy results in an efficient inverse of $\mathbf{G}_{gg}$ that leads to exact calculations of GBLUP [9, 11]. However, when $N_g > k$, $\mathbf{G}_{gg}$ is singular and cannot have an inverse. Thus, Apy-GBLUP cannot be exact.

To better understand the matrix portrayed as an inverse by the Apy algorithm, the genomic relationship matrix is partitioned into the core and noncore animals as follows:

$$
\mathbf{G}_{gg} = \begin{bmatrix} \mathbf{G}_{cc} & \mathbf{G}_{cn} \\ \mathbf{G}_{nc} & \mathbf{G}_{nn} \end{bmatrix},
$$

where the subscripts c and n denote the core and noncore animals. The Apy algorithm implies that $\mathbf{G}_{cc}$ is nonsingular and that $\mathbf{G}_{nc}$ can be written as $\mathbf{G}_{nc} = \mathbf{PG}_{cc}$, where $\mathbf{P} = \mathbf{G}_{nc}\mathbf{G}_{cc}^{-1}$. Similarly, $\mathbf{G}_{cn} = \mathbf{G}_{cc}\mathbf{P}'$. Now, using these results, $\mathbf{G}_{gg}$ can be written as:

$$
\mathbf{G}_{gg} = \begin{bmatrix} \mathbf{G}_{cc} & \mathbf{G}_{cc}\mathbf{P}' \\ \mathbf{PG}_{cc} & \mathbf{PG}_{cc}\mathbf{P}' + \mathbf{D} \end{bmatrix},
$$

where

$$
\mathbf{D} = \mathbf{G}_{nn} - \mathbf{PG}_{cc}\mathbf{P}'. \quad (15)
$$

Assuming $\mathbf{D}$ is nonsingular, the inverse of $\mathbf{G}_{gg}$ can be obtained as follows. We start by expressing $\mathbf{G}_{gg}$ as:

$$
\mathbf{G}_{gg} = \begin{bmatrix} \mathbf{I} & \mathbf{0} \\ \mathbf{P} & \mathbf{I} \end{bmatrix} \begin{bmatrix} \mathbf{G}_{cc} & \mathbf{0} \\ \mathbf{0} & \mathbf{D} \end{bmatrix} \begin{bmatrix} \mathbf{I} & \mathbf{P}' \\ \mathbf{0} & \mathbf{I} \end{bmatrix}.
$$

Then, the inverse of $\mathbf{G}_{gg}$ can be written as:

$$
\begin{aligned}
\mathbf{G}_{gg}^{-1} &= \begin{bmatrix} \mathbf{I} & -\mathbf{P}' \\ \mathbf{0} & \mathbf{I} \end{bmatrix} \begin{bmatrix} \mathbf{G}_{cc}^{-1} & \mathbf{0} \\ \mathbf{0} & \mathbf{D}^{-1} \end{bmatrix} \begin{bmatrix} \mathbf{I} & \mathbf{0} \\ -\mathbf{P} & \mathbf{I} \end{bmatrix} \\
&= \begin{bmatrix} \mathbf{G}_{cc}^{-1} & \mathbf{0} \\ \mathbf{0} & \mathbf{0} \end{bmatrix} + \begin{bmatrix} -\mathbf{P}' \\ \mathbf{I} \end{bmatrix} \mathbf{D}^{-1} \begin{bmatrix} -\mathbf{P} & \mathbf{I} \end{bmatrix},
\end{aligned} \quad (16)
$$

Table 1 Pedigree for numerical example

Animal	Sire	Dam	PV	BV	EBV
1	0	0	99.25	−0.25	0.14
2	0	0	97.92	−0.94	−0.95
3	0	0	103.2	1.12	1.09
4	1	2	99.39	−1.01	−0.69
5	1	2	102.03	0.79	0.25
6	1	3	100.59	0.18	0.14
7	1	3	101.7	1.55	1.08

PV, BV and EBV are the phenotypic values, breeding values and the BLUPs of the BV

Table 2 Genotype covariates at four loci

Animal	Locus 1	Locus 2	Locus 3	Locus 4
1	0.0	0.0	−1.0	0.0
2	−1.0	1.0	0.0	0.0
3	1.0	0.0	−1.0	0.0
4	−1.0	0.0	0.0	1.0
5	0.0	1.0	0.0	1.0
6	0.0	1.0	−1.0	0.0
7	1.0	1.0	−1.0	0.0

which is identical to the formula given in Misztal et al. [9] provided $\mathbf{D}$ is diagonal.

To examine the situation when $N_g > k$ and $\mathbf{G}_{gg}$ is singular, suppose the animals with genotypes in $\mathbf{M}_{g_i}$ are considered as the core animals and those with genotypes in $\mathbf{M}_{g_d}$ are considered as the noncore animals. Then, $\mathbf{G}_{cc} = \mathbf{G}_{g_i g_i}, \mathbf{G}_{nc} = \mathbf{G}_{g_d g_i} = \frac{1}{k}\mathbf{L}'\mathbf{M}_{g_i}\mathbf{M}'_{g_{i'}}$ and $\mathbf{G}_{nn} = \mathbf{G}_{g_d g_d} = \frac{1}{k}\mathbf{L}'\mathbf{M}_{g_i}\mathbf{M}'_{g_i}\mathbf{L}$, and $\mathbf{P} = \mathbf{L}'$, where from (2) $\mathbf{L}'$ can also be written as $\mathbf{L}' = \mathbf{G}_{g_d g_i}\mathbf{G}_{g_i g_i}^{-1}$. Given this definition of the core and noncore animals, it can be seen from (2) that the Apy property holds, because the rows of $\mathbf{G}_{gg}$ for the noncore animals is a linear function of those in the core. Furthermore, it can be seen from (2) that:

$$\begin{aligned}
\mathbf{G}_{nn} &= \mathbf{G}_{g_d g_d} \\
&= \frac{1}{k}\mathbf{L}\mathbf{L}'\mathbf{M}_{g_i}\mathbf{M}'_{g_i} \\
&= \mathbf{P}\mathbf{G}_{cc}\mathbf{P}',
\end{aligned}$$

which shows that $\mathbf{D}$ is null, and thus (16) cannot be computed. In this situation, $\mathbf{D}$ can be replaced by $\mathbf{I}s$ for a positive scalar s. Then, (16) gives the exact inverse for a matrix $\mathbf{G}_{gg}^*$ of modified genomic relationships that is obtained by adding s to only the diagonals of the noncore group. If the scalar s is chosen to be small, $\mathbf{G}_{gg}^*$ will be close to $\mathbf{G}_{gg}$. Regardless of the size of s, the resulting inverse is sparse because the sub-matrix corresponding to $\mathbf{G}_{nn}$ in the inverse has non-zero elements only on the diagonal. If the core group is chosen such that $\mathbf{G}_{cc}$ has rank less than r the rank of $\mathbf{G}_{gg}$, the matrix $\mathbf{D}$ will not be null, but as can be seen by examining Eq. (2) and demonstrated in the numerical example, it is not likely to be diagonal as assumed in the Apy algorithm. In this case the inverse computed by the Apy algorithm is the inverse of:

$$\mathbf{G}^* = \begin{bmatrix} \mathbf{G}_{cc} & \mathbf{G}_{cc}\mathbf{P}' \\ \mathbf{P}\mathbf{G}_{cc} & \mathbf{P}\mathbf{G}_{cc}\,\mathbf{P}' + \mathrm{diag}(\mathbf{D}) \end{bmatrix}, \quad (17)$$

where diag($\mathbf{D}$) sets all the off-diagonal elements of $\mathbf{D}$ to zero, thus always leading to an approximation for that choice of core. Also, when $\mathbf{G}_{gg}$ is blended with $\mathbf{A}, \mathbf{D}$ will generally not be diagonal (see Additional file 3), and the inverse obtained in Apy is of $\mathbf{G}^*$ given by (17), where the off-diagonal elements of $\mathbf{D}$ have been set to zero.

Numerical example

A small example with seven animals is used to illustrate the calculation of GBLUP and Apy-GBLUP. The pedigree for the seven animals is in Table 1. Genotype covariates coded as −1, 0, 1 at four loci are in Table 2. Julia scripts and results for GBLUP by strategies I to IV and for Apy-GBLUP are in Additional file 1. Only the calculations by strategy III and by Apy-GBLUP are described below.

Strategy III

The first step in this approach is to reorder the rows of $\mathbf{M}_g$ such that the first r rows are linearly independent, where r is the rank of $\mathbf{M}_g$. As described below, this can be done using Gaussian elimination with pivoting on $\mathbf{M}_g$ to transform it to row echelon form, where all elements below the diagonal are zero. Starting in row $i = 1$, zeros are obtained below the diagonal by subtracting a multiple of row i from each subsequent row. Before doing these row operations to obtain zeros under the diagonal, the element with the largest absolute value is located in the sub-matrix comprising all rows below row $i − 1$ and all columns to the right of column $i − 1$. Then by swapping rows and columns, this element is moved to the ith diagonal. If the element with the largest absolute value is zero, Gaussian elimination is terminated. The rank of the matrix is the number of non-zero diagonals in the transformed matrix, and the rows used for Gaussian elimination provide a maximal set of linearly independent rows.

When Gaussian elimination was applied to genotype covariates in Table 2, the resulting matrix is in Table 3. All four diagonals of this matrix are non-zero, and so $\mathbf{M}_g$ has a rank equal to four. As a result of swapping rows, the rows were ordered as 2, 7, 1, 4, 5, 6, 3, where rows

Table 3 Genotype matrix transformed to row echelon form by Gaussian elimination with pivoting

−1.0	1.0	0.0	0.0
0.0	2.0	−1.0	0.0
0.0	0.0	−1.0	0.0
0.0	0.0	0.0	1.0
0.0	0.0	0.0	0.0
0.0	0.0	0.0	0.0
0.0	0.0	0.0	0.0

Table 5 The matrix $\mathbf{L}'$ that relates $\mathbf{M}_{g_i}$ to $\mathbf{M}_{g_d}$ as $\mathbf{M}_{g_d} = \mathbf{L}'\mathbf{M}_{g_i}$

0.0	1.0	−1.0	1.0
0.5	0.5	0.5	0.0
−0.5	0.5	0.5	0.0

2, 7, 1 and 4 were used for Gaussian elimination. Thus, these four rows are a linearly independent set, and they were taken to form $\mathbf{M}_{g_i}$ and rows 5, 6, and 3 were taken to form $\mathbf{M}_{g_d}$. The genomic relationship matrix that was constructed using the reordered genotype covariates is in Table 4. The upper-left, 4×4 sub matrix of this relationship matrix, denoted $\mathbf{G}_{g_i g_i}$, gives the relationships for individuals 2, 7, 1, and 4, and it is non-singular because the genotype covariates for these four individuals are linearly independent. Now, the matrix $\mathbf{L}'$ can be calculated using (3), and is in Table 5.

The last three rows of the matrix $\mathbf{G}_{gg}$ of genomic relationships in Table 4 can be written as a linear combination of the first four rows as shown in (2), by using the $\mathbf{L}'$ matrix in Table 5. Thus, breeding values for individuals 5, 6, and 3 can be written as:

$$\mathbf{u}_{g_d} = \mathbf{L}'\mathbf{u}_{g_i},$$

where $\mathbf{u}_{g_i}$ is the vector of breeding values for individuals 2, 7, 1, and 4. Now, the phenotypes for these seven individuals are modeled in terms of $\mathbf{u}_{g_i}$, which has a non-singular covariance matrix proportional to $\mathbf{G}_{g_i g_i}$. All seven individuals in this example have one phenotypic value, and so assuming that the vector $\boldsymbol{\beta}$ of fixed effects contains a single element for the overall mean, the matrix $\mathbf{X}$

Table 6 Mixed model equations for μ and u_{g_i}

	μ	u_2	u_7	u_1	u_4
μ	7.0	1.0	3.0	1.0	2.0
u_1	1.0	4.5	−1.0	1.0	−2.0
u_7	3.0	−1.0	5.5	−3.5	3.0
u_1	1.0	1.0	−3.5	9.5	−3.0
u_4	2.0	−2.0	3.0	−3.0	6.0
rhs	704.08	96.62	305.62	99.12	201.42
sol	100.43	−0.95	1.08	0.14	−0.69

The last two rows give the right-hand-side and the solutions of the equations

for this example is equal to a vector of seven 1s and $\mathbf{Z}$ is equal to an identity matrix of order seven. It follows that $\mathbf{W} = \begin{bmatrix} \mathbf{I} \\ \mathbf{L}' \end{bmatrix}$ for $\mathbf{L}'$ in Table 5. The MME to fit the overall mean (μ) and the breeding values $\mathbf{u}_{g_i}$ are in Table 6, where a value of 1.0 was used for $\lambda = \frac{\sigma_e^2}{\sigma_u^2}$. BLUP of $\mathbf{u}_g$ is obtained as $\hat{\mathbf{u}}_g = \mathbf{W}\hat{\mathbf{u}}_{g_i}$. Results for strategies I through IV are in Additional file 1, and they are all identical, as expected. The condition numbers of the left-hand-side of the MME for strategies II through IV were 10.9, 11.3, and 6.8, demonstrating the improved condition of the MME obtained by fitting a RQ transformed vector of breeding values.

Apy-GBLUP

Here, we can see that if animals 2, 7, 1, and 4 are used as the core group, the Apy property is met because the last three rows of $\mathbf{G}_{gg}$, which correspond to the animals in the noncore group, can be written as a linear combination of the first four rows, which correspond to the animals in the core group, using the $\mathbf{L}'$ matrix in Table 5 (see Additional file 1). Equation (2) shows that this property also holds for the columns of $\mathbf{G}_{gg}$, where the last three columns of $\mathbf{G}_{gg}$ can be written as a linear combination of the first four columns. In this case, the matrix $\mathbf{D}$, the inverse of which is needed in the Apy algorithm, is null (Additional file 1). In order to proceed with the Apy algorithm, we set $\mathbf{D} = \mathbf{I}s$ for a small value of s such as 0.0001. The inverse that is obtained from equation (16) will now be sparse because the sub-matrix corresponding to $\mathbf{G}_{nn}$ in

Table 4 Genomic relationship matrix

0.5	0.0	0.0	0.25	0.25	0.25	−0.25
0.0	0.75	0.25	−0.25	0.25	0.5	0.5
0.0	0.25	0.25	0.0	0.0	0.25	0.25
0.25	−0.25	0.0	0.5	0.25	0.0	−0.25
0.25	0.25	0.0	0.25	0.5	0.25	0.0
0.25	0.5	0.25	0.0	0.25	0.5	0.25
−0.25	0.5	0.25	−0.25	0.0	0.25	0.5

the inverse is diagonal. Inverting a modified $\mathbf{G}_{gg}$ matrix, $\mathbf{G}_{gg}^*$, by adding s to the diagonals of $\mathbf{G}_{gg}$ corresponding to the animals in the noncore group gives the same result (Additional file 1). Setting up and solving the MME for μ and $\mathbf{u}_g$ assuming $\mathrm{Var}(\mathbf{u}_g) = \mathbf{G}_{gg}^* \sigma_u^2$ give results that are approximate but very close in this instance to the exact BLUP results obtained by strategies I through IV (Additional file 1), but the condition number of these MME was 56,548, which indicates that they are ill-conditioned relative to those for strategies II through IV. However, if individuals 2, 7, and 1 are chosen as the core animals, the Apy property does not hold. In that case, the last four rows of $\mathbf{G}_{gg}$ cannot be written as a linear combination of the first three rows (Additional file 2). Furthermore, the matrix $\mathbf{D}$ computed by using equation (15) is not diagonal (Additional file 2). Now, the matrix $\mathbf{G}_{gg}^*$ that is inverted in the Apy algorithm deviates substantially from $\mathbf{G}_{gg}$, and as a result, solving the MME for μ and $\mathbf{u}_g$ assuming $\mathrm{Var}(\mathbf{u}_g) = \mathbf{G}_{gg}^* \sigma_u^2$ gives results that are substantially different from the exact BLUP (Additional file 2).

Recent publications [12, 13, 18] in which the Apy algorithm was applied to obtain a matrix portrayed as the inverse of the genomic relationship matrix use $0.95\mathbf{G}_{gg} + 0.05\mathbf{A}$ rather than the singular $\mathbf{G}_{gg}$. This approach applied to the example gives a solution that is neither the same as the exact solution obtained using any of the strategies I to IV (Additional file 1), nor the exact solution to the MME constructed with the blended genomic relationship matrix. However, the condition number of these equations was 62.1, which is much better than that obtained without blending but poorer than with strategies II through IV.

Exact single-step GBLUP when $\mathbf{G}_{gg}$ is singular

Single-step GBLUP (SS-GBLUP) was proposed [23, 24] to obtain genomic evaluations when genotypes are not available on all animals.

Strategy III

Let $\mathbf{u}_g$ denote the breeding values of animals with genotypes and $\mathbf{u}_m$ denote the breeding values of those without genotypes. Now, the mixed linear model for SSGBLUP can be written as:

$$\mathbf{y} = \mathbf{X}\boldsymbol{\beta} + \mathbf{Z}\begin{bmatrix} \mathbf{u}_m \\ \mathbf{u}_g \end{bmatrix} + \mathbf{e}.$$

It is convenient to similarly partition the vector of phenotypic values as $\mathbf{y} = \begin{bmatrix} \mathbf{y}_m \\ \mathbf{y}_g \end{bmatrix}$, where $\mathbf{y}_m$ are phenotypic values from animals that were not genotyped and $\mathbf{y}_g$ are from animals that were genotyped. However, because $\mathbf{G}_{gg}$ is singular, $\mathbf{u}_g$ is written as in Eq. (11) in terms of $\mathbf{u}_{g_i}$, and then the model becomes:

$$\begin{bmatrix} \mathbf{y}_m \\ \mathbf{y}_g \end{bmatrix} = \mathbf{X}\boldsymbol{\beta} + \mathbf{Z}\begin{bmatrix} \mathbf{I} & \mathbf{0} \\ \mathbf{0} & \mathbf{S} \end{bmatrix}\begin{bmatrix} \mathbf{u}_m \\ \mathbf{u}_{g_i} \end{bmatrix} + \mathbf{e}$$
$$= \mathbf{X}\boldsymbol{\beta} + \mathbf{W}_r\mathbf{u}_r + \mathbf{e}, \tag{18}$$

where $\mathbf{S} = \begin{bmatrix} \mathbf{I} \\ \mathbf{L}' \end{bmatrix}$, $\mathbf{W}_r = \mathbf{Z}\begin{bmatrix} \mathbf{I} & \mathbf{0} \\ \mathbf{0} & \mathbf{S} \end{bmatrix}$, and $\mathbf{u}_r = \begin{bmatrix} \mathbf{u}_m \\ \mathbf{u}_{g_i} \end{bmatrix}$. The MME that correspond to (18) are:

$$\begin{bmatrix} \mathbf{X}'\mathbf{X} & \mathbf{X}'\mathbf{W}_r \\ \mathbf{W}_r'\mathbf{X} & \mathbf{W}_r'\mathbf{W}_r + \mathbf{H}_r^{-1} \end{bmatrix}\begin{bmatrix} \hat{\boldsymbol{\beta}} \\ \hat{\mathbf{u}}_r \end{bmatrix} = \begin{bmatrix} \mathbf{X}'\mathbf{y} \\ \mathbf{W}_r'\mathbf{y} \end{bmatrix}, \tag{19}$$

where

$$\mathbf{H}_r = \mathrm{Var}(\mathbf{u}_r).$$

To obtain $\mathbf{H}_r^{-1}$, as in [23], $\mathbf{u}_m$ is written as:

$$\mathbf{u}_m = \mathbf{A}_{mg}\mathbf{A}_{gg}^{-1}\mathbf{u}_g + \mathbf{u}_m - \mathbf{A}_{mg}\mathbf{A}_{gg}^{-1}\mathbf{u}_g$$
$$= \mathbf{A}_{mg}\mathbf{A}_{gg}^{-1}\mathbf{u}_g + \boldsymbol{\epsilon}$$
$$= -(\mathbf{A}^{mm})^{-1}\mathbf{A}^{mg}\mathbf{u}_g + \boldsymbol{\epsilon},$$

where, in the last line, we have used the identity: $\mathbf{A}_{mg}\mathbf{A}_{gg}^{-1} = -(\mathbf{A}^{mm})^{-1}\mathbf{A}^{mg}$. Now, $\mathbf{u}_r$ is written in terms of $\mathbf{u}_{g_i}$ and $\boldsymbol{\epsilon}$ as:

$$\mathbf{u}_r = \begin{bmatrix} \mathbf{I} & -(\mathbf{A}^{mm})^{-1}\mathbf{A}^{mg}\mathbf{S} \\ \mathbf{0} & \mathbf{I} \end{bmatrix}\begin{bmatrix} \boldsymbol{\epsilon} \\ \mathbf{u}_{g_i} \end{bmatrix}$$
$$= \mathbf{T}\begin{bmatrix} \boldsymbol{\epsilon} \\ \mathbf{u}_{g_i} \end{bmatrix},$$

where $\mathbf{u}_g = \mathbf{S}\mathbf{u}_{g_i}$ and $\mathbf{T} = \begin{bmatrix} \mathbf{I} & -(\mathbf{A}^{mm})^{-1}\mathbf{A}^{mg}\mathbf{S} \\ \mathbf{0} & \mathbf{I} \end{bmatrix}$, and $\mathbf{H}_r$ is written as

$$\mathbf{H}_r = \mathbf{T}\mathrm{Var}\left(\begin{bmatrix} \boldsymbol{\epsilon} \\ \mathbf{u}_{g_i} \end{bmatrix}\right)\mathbf{T}'.$$

Following [23], $\mathrm{Var}(\boldsymbol{\epsilon}) = (\mathbf{A}^{mm})^{-1}\sigma_a^2$, $\mathrm{Var}(\mathbf{u}_{g_i}) = \mathbf{G}_{g_i g_i}\sigma_u^2$, and $\mathrm{Cov}(\boldsymbol{\epsilon}, \mathbf{u}_{g_i}) = 0$. Unlike in [23], two variance components σ_a^2 and σ_u^2 are used here, where σ_a^2 is the additive genetic variance and $\sigma_u^2 = k\sigma_\alpha^2$ stems from the prior used for the marker effects $(\boldsymbol{\alpha})$, and its relationship to the genetic variance may not be straightforward [25]. Finally, $\mathbf{H}_r^{-1}$ is as follows:

$$\mathbf{H}_r^{-1} = (\mathbf{T}')^{-1}\begin{bmatrix} \mathbf{A}^{mm}\frac{1}{\sigma_a^2} & \mathbf{0} \\ \mathbf{0} & \mathbf{G}_{g_i g_i}^{-1}\frac{1}{\sigma_u^2} \end{bmatrix}\mathbf{T}^{-1}$$

$$= \begin{bmatrix} \mathbf{I} & \mathbf{0} \\ \mathbf{S}'\mathbf{A}^{gm}(\mathbf{A}^{mm})^{-1} & \mathbf{I} \end{bmatrix}\begin{bmatrix} \mathbf{A}^{mm}\frac{1}{\sigma_a^2} & \mathbf{0} \\ \mathbf{0} & \mathbf{G}_{g_i g_i}^{-1}\frac{1}{\sigma_u^2} \end{bmatrix}$$

$$\begin{bmatrix} \mathbf{I} & (\mathbf{A}^{mm})^{-1}\mathbf{A}^{mg}\mathbf{S} \\ \mathbf{0} & \mathbf{I} \end{bmatrix}$$

$$= \begin{bmatrix} \mathbf{A}^{mm}\frac{1}{\sigma_a^2} & \mathbf{A}^{mg}\mathbf{S}\frac{1}{\sigma_a^2} \\ \mathbf{S}'\mathbf{A}^{gm}\frac{1}{\sigma_a^2} & \mathbf{Q}\frac{1}{\sigma_a^2} + \mathbf{G}_{g_i g_i}^{-1}\frac{1}{\sigma_u^2} \end{bmatrix},$$

where

$$\mathbf{Q} = \mathbf{S}'\mathbf{A}^{gm}(\mathbf{A}^{mm})^{-1}\mathbf{A}^{mg}\mathbf{S}.$$

Following [26], the MME given in equation (19) can be augmented to avoid the expression involving $(\mathbf{A}^{mm})^{-1}$ in $\mathbf{Q}$. To do so, equation (19) is first rewritten to show the partitions for $\mathbf{u}_m$ and $\mathbf{u}_{g_i}$. Note that the matrix $\mathbf{Z}$ has the following form:

$$\mathbf{Z} = \begin{bmatrix} \mathbf{Z}_m & 0 \\ 0 & \mathbf{Z}_g \end{bmatrix},$$

where $\mathbf{Z}_m$ relates $\mathbf{y}_m$ to $\mathbf{u}_m$ and $\mathbf{Z}_g$ relates $\mathbf{y}_g$ to $\mathbf{u}_g$. Then, $\mathbf{W}_r$ can be partitioned as:

$$\begin{aligned} \mathbf{W}_r &= \begin{bmatrix} \mathbf{Z}_m & 0 \\ 0 & \mathbf{Z}_g \end{bmatrix} \begin{bmatrix} \mathbf{I} & 0 \\ 0 & \mathbf{S} \end{bmatrix} \\ &= \begin{bmatrix} \mathbf{Z}_m & 0 \\ 0 & \mathbf{W}_{g_i} \end{bmatrix}, \end{aligned}$$

where $\mathbf{W}_{g_i} = \mathbf{Z}_g \mathbf{S}$. Now the MME that show the partitions for $\mathbf{u}_m$ and $\mathbf{u}_{g_i}$ are:

$$\begin{bmatrix} \mathbf{X}'\mathbf{X} & \mathbf{X}_m'\mathbf{Z}_m & \mathbf{X}_g'\mathbf{W}_{g_i} \\ \mathbf{Z}_m'\mathbf{X}_m & \mathbf{Z}_m'\mathbf{Z}_m + \mathbf{A}^{mm}\frac{\sigma_e^2}{\sigma_a^2} & \mathbf{A}^{mg}\mathbf{S}\frac{\sigma_e^2}{\sigma_a^2} \\ \mathbf{W}_{g_i}'\mathbf{X}_g & \mathbf{S}'\mathbf{A}^{gm}\frac{\sigma_e^2}{\sigma_a^2} & \mathbf{W}_{g_i}'\mathbf{W}_{g_i} + \mathbf{G}_{g_ig_i}^{-1}\frac{\sigma_e^2}{\sigma_u^2} + \mathbf{Q}\frac{\sigma_e^2}{\sigma_a^2} \end{bmatrix} \begin{bmatrix} \hat{\boldsymbol{\beta}} \\ \hat{\mathbf{u}}_m \\ \hat{\mathbf{u}}_{g_i} \end{bmatrix} = \begin{bmatrix} \mathbf{X}'\mathbf{y} \\ \mathbf{Z}_m'\mathbf{Y}_m \\ \mathbf{W}_{g_i}'\mathbf{Y}_g \end{bmatrix}, \tag{20}$$

where $\mathbf{X}_m$ and $\mathbf{X}_g$ are partitions of $\mathbf{X}$ corresponding to $\mathbf{y}_m$ and $\mathbf{y}_g$. Consider now the following augmented MME:

$$\begin{bmatrix} \mathbf{X}'\mathbf{X} & \mathbf{X}_m'\mathbf{Z}_m & \mathbf{X}_g'\mathbf{W}_{g_i} & 0 \\ \mathbf{Z}_m'\mathbf{X}_m & \mathbf{Z}_m'\mathbf{Z}_m + \mathbf{A}^{mm}\frac{\sigma_e^2}{\sigma_a^2} & \mathbf{A}^{mg}\mathbf{S}\frac{\sigma_e^2}{\sigma_a^2} & 0 \\ \mathbf{W}_{g_i}'\mathbf{X}_g & \mathbf{S}'\mathbf{A}^{gm}\frac{\sigma_e^2}{\sigma_a^2} & \mathbf{W}_{g_i}'\mathbf{W}_{g_i} + \mathbf{G}_{g_ig_i}^{-1}\frac{\sigma_e^2}{\sigma_u^2} & -\mathbf{S}'\mathbf{A}^{gm}\frac{\sigma_e^2}{\sigma_a^2} \\ 0 & 0 & -\mathbf{A}^{mg}\mathbf{S}\frac{\sigma_e^2}{\sigma_a^2} & -\mathbf{A}^{mm}\frac{\sigma_e^2}{\sigma_a^2} \end{bmatrix} \begin{bmatrix} \hat{\boldsymbol{\beta}} \\ \hat{\mathbf{u}}_m \\ \hat{\mathbf{u}}_{g_i} \\ \mathbf{c} \end{bmatrix} = \begin{bmatrix} \mathbf{X}'\mathbf{y} \\ \mathbf{Z}_m'\mathbf{Y}_m \\ \mathbf{W}_{g_i}'\mathbf{Y}_g \\ 0 \end{bmatrix}. \tag{21}$$

These equations do not have $\mathbf{Q}$ in them, and so they may be easier to construct. However, the left-hand-side is not positive definite and it has been reported that these equations are poorly conditioned [26]. Elimination of $\mathbf{c}$ from Eq. (21) results in equation (20), and thus, solutions for $\hat{\boldsymbol{\beta}}$, $\hat{\mathbf{u}}_m$ and for $\hat{\mathbf{u}}_{g_i}$ from Eq. (21) are identical to those from Eq. (20).

Strategy IV
The model for SSGBLUP can also be formulated in terms of $\mathbf{v}$ as:

$$\begin{bmatrix} \mathbf{y}_m \\ \mathbf{y}_g \end{bmatrix} = \mathbf{X}\boldsymbol{\beta} + \mathbf{Z}\begin{bmatrix} \mathbf{I} & 0 \\ 0 & \mathbf{R} \end{bmatrix}\begin{bmatrix} \mathbf{u}_m \\ \mathbf{v} \end{bmatrix} + \mathbf{e}, \tag{22}$$

and the MME corresponding to model (22) are:

$$\begin{bmatrix} \mathbf{X}'\mathbf{X} & \mathbf{X}_m'\mathbf{Z}_m & \mathbf{X}_g'\mathbf{W}_v \\ \mathbf{Z}_m'\mathbf{X}_m & \mathbf{Z}_m'\mathbf{Z}_m + \mathbf{A}^{mm}\frac{\sigma_e^2}{\sigma_a^2} & \mathbf{A}^{mg}\mathbf{R}\frac{\sigma_e^2}{\sigma_a^2} \\ \mathbf{W}_v'\mathbf{X}_g & \mathbf{R}'\mathbf{A}^{gm}\frac{\sigma_e^2}{\sigma_a^2} & \mathbf{W}_v'\mathbf{W}_v + \mathbf{I}\frac{\sigma_e^2}{\sigma_a^2} + \mathbf{Q}_v\frac{\sigma_e^2}{\sigma_a^2} \end{bmatrix} \begin{bmatrix} \hat{\boldsymbol{\beta}} \\ \hat{\mathbf{u}}_m \\ \hat{\mathbf{v}} \end{bmatrix} = \begin{bmatrix} \mathbf{X}'\mathbf{y} \\ \mathbf{Z}_m'\mathbf{Y}_m \\ \mathbf{W}_v'\mathbf{Y}_g \end{bmatrix}, \tag{23}$$

where $\mathbf{W}_v = \mathbf{Z}_g\mathbf{R}$ and $\mathbf{Q}_v = \mathbf{R}'\mathbf{A}^{gm}(\mathbf{A}^{mm})^{-1}\mathbf{A}^{mg}\mathbf{R}$.

Comparison to Apy-SSGBLUP
The SSGBLUP method given in [23] requires computing the inverse of the matrix $\mathbf{G}_{gg}$ of genomic relationships and of the matrix $\mathbf{A}_{gg}$ of additive relationships for the genotyped animals. At the time those papers were published, N_g was typically smaller than the number of markers so that $\mathbf{G}_{gg}$ was relatively small and of full rank. Since then N_g has greatly increased in most livestock applications. Computational effort in matrix manipulation is determined by the number of non-zero coefficients and these increase as N_g increases. To fully store a dense matrix of order one million in single precision requires about 4 TB. Therefore, it would be advantageous to have a sparse representation of all the large matrices involved in the MME.

Furthermore, the matrix $\mathbf{G}_{gg}$ is singular when $N_g > k$ and thus cannot be inverted when more animals than the number of SNPs have been genotyped. This suggests that there should be a sparse representation of $\mathbf{G}_{gg}$. Suppose $\mathbf{G}_{gg}$ has rank r and it is ordered such that the first r rows are linearly independent. Then, the sub-matrix of the first r rows and columns of $\mathbf{G}_{gg}$ denoted $\mathbf{G}_{cc}$ gives the genomic relationships among the r core animals of the Apy algorithm, and that sub-matrix is nonsingular. The remaining $n - r$ animals are referred to as noncore and their genomic relationship matrix is denoted $\mathbf{G}_{nn}$. When the genomic relationship matrix has not been blended, a nonsingular matrix $\mathbf{G}^*$ can be obtained by adding a small value to the diagonals of $\mathbf{G}_{gg}$ for the animals in the noncore group, and in the inverse of $\mathbf{G}^*$, the sub-matrix corresponding to $\mathbf{G}_{nn}$ will be diagonal. This exact inverse of that particular $\mathbf{G}^*$ can be obtained efficiently using the Apy algorithm given in [9]. When the genomic relationship matrix is blended with $\mathbf{A}$, the matrix resulting from the Apy algorithm is the exact inverse of the $\mathbf{G}^*$ where all the off-diagonals of $\mathbf{D}$ have been ignored. That matrix may or may not be a close approximation of the blended genomic relationship matrix depending on the size of the core, the particular animals chosen for the core, the relationship among noncore animals and the relationship between core and noncore animals. Regardless of the form of the matrix the Apy algorithm is applied to, the resultant inverse is sparse.

The MME for Apy-SSGBLUP, which includes the inverse of $\mathbf{G}^*$, are:

Both the MME for Apy-SSGBLUP and that for SSGBLUP using strategy IV (SIV-SSGBLUP), include equations for the same fixed effects and the random effects corresponding to the breeding values of animals that were not genotyped. In the MME for Apy-SSGBLUP, there is an additional vector of random effects corresponding to the breeding values for animals that were genotyped, which comprises sub-vectors representing core and noncore animals. In contrast, the MME for SSGBLUP using strategy IV contains a vector of random effects that is not larger than k regardless of the number of animals genotyped. If the core size in Apy-SSGBLUP was chosen to be k, Eq. (24) would contain an additional random effect of order equal to the number of noncore animals compared

$$
\begin{bmatrix}
\mathbf{X}'\mathbf{X} & \mathbf{X}'_m\mathbf{Z}_m & \mathbf{X}'_g\mathbf{Z}_g \\
\mathbf{Z}'_m\mathbf{X}_m & \mathbf{Z}'_m\mathbf{Z}_m + \mathbf{A}^{mm}\frac{\sigma^2_e}{\sigma^2_a} & \mathbf{A}^{mg}\frac{\sigma^2_e}{\sigma^2_a} \\
\mathbf{Z}'_g\mathbf{X}_g & \mathbf{A}^{gm}\frac{\sigma^2_e}{\sigma^2_a} & \mathbf{Z}'_g\mathbf{Z}_g + [(\mathbf{G}^*)^{-1} - \mathbf{A}^{-1}_{gg}]\frac{\sigma^2_e}{\sigma^2_u}
\end{bmatrix}
\begin{bmatrix}
\hat{\boldsymbol{\beta}} \\
\hat{\mathbf{u}}_m \\
\hat{\mathbf{u}}_g
\end{bmatrix}
=
\begin{bmatrix}
\mathbf{X}'\mathbf{y} \\
\mathbf{Z}'_m\mathbf{y}_m \\
\mathbf{Z}'_g\mathbf{y}_g
\end{bmatrix}.
\tag{24}
$$

In addition to the inverse $\mathbf{G}^*$, SSGBLUP requires the inverse of $\mathbf{A}_{gg}$. However, $\mathbf{A}^{-1}_{gg}$ is a dense matrix, and so subtracting it from the inverse of $\mathbf{G}^*$ will make the resultant matrix dense.

Part of the appeal of the Apy algorithm was to obtain a sparse representation of the MME for SSGBLUP. Accordingly, Misztal et al. [9] proposed that Apy could also be used to approximate the inverse of the nonsingular $\mathbf{A}_{gg}$. However, the nature of $\mathbf{A}_{gg}$ depends on the genotyping strategy such that genotyping unrelated individuals results in a diagonal $\mathbf{A}_{gg}$ whereas genotyping relatives results in non-zero off-diagonals between each related pair. If off-diagonal elements in the noncore sub matrix of $\mathbf{A}_{gg}$ are not well predicted by $\mathbf{P}\mathbf{G}_{cc}\mathbf{P}'$, the Apy inverse can significantly depart from its true inverse as easily demonstrated by using an example (see Additional file 1). This means the adequacy of Apy applied to $\mathbf{A}_{gg}$ will depend on the pedigree structure, the nature of the genotyping strategy, and the choice of core group. Presumably, this inadequacy of Apy for inverting $\mathbf{A}_{gg}$ has been recognized because recent implementations [13] have adopted an alternative approach that is computationally more demanding than applying Apy to approximate the inverse of $\mathbf{A}_{gg}$. Rather than forming $\mathbf{A}^{-1}_{gg}$ prior to solving the MME, a partitioned matrix inverse result is used to calculate products such as $\mathbf{A}^{-1}_{gg}\mathbf{x}$ as $\mathbf{A}^{gg}\mathbf{x} - \mathbf{A}^{gm}\mathbf{q}$, where $\mathbf{q}$ is the solution to $\mathbf{A}^{mm}\mathbf{q} = \mathbf{A}^{mg}\mathbf{x}$. This requires storing the sparse matrices $\mathbf{A}^{gg}, \mathbf{A}^{mg}$ and the sparse Cholesky factors of $\mathbf{A}^{mm}$. Each PCG iteration involves a matrix product $\mathbf{A}^{-1}_{gg}\mathbf{x}$ for a different vector $\mathbf{x}$, which requires one forward and one backward triangular solve to obtain $\mathbf{q}$, two sparse matrix vector multiplications, and one vector subtraction.

to Eq. (23), and this number increases with the number of animals genotyped.

Given a core of k animals, both MME contain a dense $k \times k$ matrix on the diagonal. Both MME contain the same sparse block on the diagonal for non-genotyped animals. Comparing the upper off-diagonals of the two sets of symmetric MME, that for Apy-SSGBLUP has the sparse $\mathbf{A}^{mg}$ matrix whereas SIV-SSGBLUP has the product of that $N_m \times N_g$ matrix with the mostly dense $N_g \times k$ matrix $\mathbf{R}$. Rather than forming the dense $N_m \times k$ product, matrix computations involving that matrix can be done more efficiently when $N_g < N_m$ in parts (e.g. $\mathbf{A}^{mg}\mathbf{R}\mathbf{x} = \mathbf{A}^{mg}(\mathbf{R}\mathbf{x})$) storing only $\mathbf{A}^{mg}$ and $\mathbf{R}$ in memory. The Apy-SSGBLUP MME contain on the upper diagonal a dense $N_c \times N_n$ block that does not appear in SIV-SSGBLUP and which increases in size as more animals are genotyped. The computation required to form the diagonal block of SIV-SSGBLUP involves computing $(\mathbf{A}^{mm})^{-1}\mathbf{A}^{mg}\mathbf{r}_i$, where $\mathbf{r}_i$ is column i of $\mathbf{R}$. This calculation is virtually identical to the computation of $\mathbf{A}^{-1}_{gg}\mathbf{x}$ in Apy-SSGBLUP, but the former needs to be done for each genotyped animal once whereas the latter needs to be done for each PCG iteration.

Discussion

When the number N_g of genotyped animals is larger than the number k of marker covariates, the matrix $\mathbf{G}_{gg}$ of genomic relationships becomes singular. In this situation, we have shown here how to obtain exact GBLUP without any approximation from either Eqs. (13) or (14) of order $p + r$ or $p + k$, where $r \leq k$ is the rank of $\mathbf{G}_{cc}$. The MME given by Eq. (9) can also be used to obtain GBLUP without approximation, but these asymmetric MME are of order $p + N_g$. When more individuals are genotyped and N_g grows, the order of those MME (9) also grows. In contrast, the order of the MME presented here (Eqs. 13 and 14) will remain constant even as N_g grows.

An alternative to these exact GBLUP calculations is used in Apy-GBLUP. Here, the pedigree is divided into two groups of animals: the core group and the noncore group. We have shown here that the inverse computed in the Apy algorithm is for a modified genomic relationship matrix, where the sub-matrix $\mathbf{G}_{nn}$ of genomic relationships among the noncore group of animals is replaced by $\mathbf{PG}_{cc}\mathbf{P}' + \mathrm{diag}(\mathbf{D})$. If the core group is chosen such that the rank of $\mathbf{G}_{cc}$ is equal to the rank of $\mathbf{G}_{gg}$, $\mathbf{D}$ will be null and the Apy algorithm will fail. In that case, the diagonals of $\mathbf{D}$ can be set to some small value, but this can result in ill-conditioned MME as shown by the example in Additional file 1. The MME can be ill-conditioned even when $\mathbf{D}$ is not null but contains very small values on the diagonal. Although the MME for Apy-GBLUP will also grow with N_g, it contains a $N_{g_d} \times N_{g_d}$ block that is diagonal, and thus is very sparse.

The approach presented here can also be used to obtain exact SSGBLUP when some animals are not genotyped. In contrast to the Apy algorithm, the method presented here is never an approximation. In agreement with [10], "BVs of core individuals can all be written as linear combinations of effective SNP effects" when SNP effects fully explain the BV. In contrast to the claim in [10] that "BVs of noncore individuals depend approximately only on the BVs of the core individuals" we have shown that the BVs of noncore individuals are an exact linear function of the BVs of the core individuals when the rank of $\mathbf{G}_{cc}$ is equal to the rank of $\mathbf{G}_{gg}$. This requires the core group to contain at least as many animals as the rank of $\mathbf{G}_{gg}$. When the number of genotyped animals exceeds the number of markers, $\mathbf{G}_{gg}$ will be singular and its rank cannot be greater than the number of markers. Only when the rank of $\mathbf{G}_{cc}$ is less than the rank of $\mathbf{G}_{gg}$, will the "BVs of noncore individuals depend approximately only on the BVs of the core individuals".

The Apy algorithm when applied to $\mathbf{A}_{gg}$ may or may not be a good approximation depending on the particular $\mathbf{A}_{gg}$. It will be exact for any core if genotyped animals are all unrelated as the matrix $\mathbf{D}$ is strictly diagonal. The quality of the approximation will erode with increases in the number of large-magnitude off-diagonal elements in $\mathbf{D}$. Demonstrating with real data that the Apy gives good results in one or more field data sets is no guarantee that it will perform well for all applications. This raises concerns that the same could be true for the application of Apy to the genomic relationship matrix. When the number of genotyped individuals increases and the number of core animals remains constant, there may be a large

increase in the number of off-diagonal coefficients in $\mathbf{D}$. Those coefficients are ignored in the Apy algorithm, and the predictions approximated by Apy are expected to deviate further from the exact predictions as more coefficients are ignored. Thus, inference that the Apy algorithm based on 100,000 or 500,000 genotyped animals is appropriate cannot be extrapolated to similar data structures with a million or more animals genotyped.

If SNP effects do not fully explain the BV, an additional polygenic effect for all animals can be readily fitted in addition to $\mathbf{u}_{g_i}$, the breeding values explained by the markers for a subset of genotyped animals. Lourenco et al. [12] used default options of BLUP90IOD2, which means they blended $\mathbf{G}_{gg}$ with $\mathbf{A}_{22}$, and included competitive results from Apy compared to exact predictions obtained by direct inversion for various analyses with N_g that were smaller than 52,000. Fragomeni et al. [11] limited their analyses to $N_g = 100{,}000$ in order to allow direct inversion of a $\mathbf{G}_{gg}$ matrix based on 42,503 SNPs but do not mention whether blending was used. In the absence of blending, the rank of $\mathbf{G}_{gg}$ matrix could not exceed 42,503 and a direct inverse of $\mathbf{G}_{gg}$ does not exist. Pocrnic et al. [27] simulated genotypes on 75,000 individuals and blended $\mathbf{G}_{gg}$ with $\mathbf{A}_{gg}$. They showed that Apy exceeded the accuracy of exact ssGBLUP by direct inversion. Their QTL effects were simulated from a Gamma distribution, which creates a few loci with large effects. In those circumstances, methods such as BayesB typically outperform GBLUP [3], and Apy with a small set of core individuals may similarly benefit from reducing the dimension of the model. Masuda et al. [13] blended $\mathbf{G}_{gg}$ with $\mathbf{A}_{gg}$ to guarantee nonsingularity of the blended matrix with N_g greater than 500,000. The exact inverse of that blended matrix will be a dense matrix of order N_g, which will make exact calculations computationally infeasible when N_g exceeds about 150,000. This makes it impossible to compare the accuracy of Apy approximations to exact predictions using that approach. They show high correlations between approximations for different core definitions but the correlations between their approximations and the exact predictions are not known.

There are no published results demonstrating the comparative accuracy of Apy and the exact approach when N_G is too large for direct inversion of $\mathbf{G}_{gg}$. However, using the exact SSGBLUP calculations presented in this paper such a comparison is feasible, requiring only special computation for $\mathbf{R}$ and $\mathbf{Q}$ in the MME (23). Computation of the matrix $\mathbf{Q}$ involves the same calculations as required to impute genotypes for non-genotyped animals as presented in Fernando et al. [28]. The computation of $\mathbf{R}$ is straightforward and analogous to matrix $\mathbf{P}$ that is fundamental to computations in the Apy algorithm. Accordingly, we do not consider that the practical application of exact SSGBLUP will be any more difficult than implementation of Apy.

Conclusions

When the number of genotyped animals exceeds the number of marker loci, the genomic relationship matrix cannot be full rank. We introduce an approach that partitions the genotyped animals into two sets, one of which can be referred to as core animals, and the other as non-core animals whose breeding values can be written as a linear function of the breeding values of core animals. The MME used for genomic prediction are then constructed with only the breeding values of the core animals, and with phenotpyes of the non-core animals contributing to the predictions for core animals through their linear relationships to the core animals. The estimated breeding values of the non-core animals are obtained as a linear function of the estimates of the breeding values of the core animals. This gives exact solutions for all animals. Another approach is to blend the genomic relationship matrix with a numerator relationship matrix or a scaled identity matrix to ensure the blended genomic relationship matrix is full rank. In that case, standard mixed model computing procedures can be used, but the increase in computing effort will be proportional to the cube of the number of animals genotyped. That effort can be reduced by approximating the inverse of the blended genomic relationship matrix using the Apy algorithm. That approximation also partitions the animals into core and non-core groups, but explicitly fits both sets of animals in the MME. In some cases, it has been reported that this approach gives useful approximations. However, the computing effort for that approximate approach is similar to that of the exact approach introduced here.

Additional files

Additional file 1. Julia scripts and results. A Jupyter notebook showing Julia scripts and results for GBLUP by strategies I to IV and for Apy-GBLUP.

Additional file 2. Results when Apy propery does not hold. Julia notebook showing results by Apy approach when Apy property does not hold.

Additional file 3. Magnitude of non-diagonal elements of D with blended G. Julia notebook showing that matrix **D** of (15) has large non-diagonal elements relative to the diagonals.

Authors' contributions
RLF conceived the initial idea for the exact approach, following discussions of the Apy approach with DJG. All the authors contributed to the subsequent development of the method. HC developed the Julia programs used in the numerical examples. RLF prepared the the manuscript with input from DJG. All authors read and approved the final manuscript.

Author details
[1] Department of Animal Science, Iowa State University, Ames, IA 50011, USA. [2] Institute of Veterinary, Animal and Biomedical Sciences, Massey University, Palmerston North, New Zealand.

Acknowledgements
The authors are grateful to an anonymous reviewer for suggesting the QR decomposition as an alternative to Gaussian elimination. This work was supported by the US Department of Agriculture, Agriculture and Food Research Initiative National Institute of Food and Agriculture Competitive Grant No. 2015-67015-22947.

Competing interests
The authors declare that they have no competing interests.

References
1. Strandén I, Garrick DJ. Technical note: derivation of equivalent computing algorithms for genomic predictions and reliabilities of animal merit. J Dairy Sci. 2009;92:2971–5.
2. Fernando RL. Genetic evaluation and selection using genotypic, phenotypic and pedigree information. In: Proceedings of the 6th World Congress on Genetics Applied to Livestock Production, Armidale, 11–16 January 1998, vol. 26. p. 329–36.
3. Meuwissen THE, Hayes BJ, Goddard ME. Prediction of total genetic value using genome-wide dense marker maps. Genetics. 2001;157:1819–29.
4. Nejati-Javaremi A, Smith C, Gibson JP. Effect of total allelic relationship on accuracy of evaluation and response to selection. J Anim Sci. 1997;75:1738–45.
5. VanRaden PM. Efficient methods to compute genomic predictions. J Dairy Sci. 2008;91:4414–23.
6. Harville DA. Maximum likelihood approaches to variance component estimation and to related problems. J Am Stat Assoc. 1976;72:320–40.
7. Henderson CR. Applications of linear models in animal breeding. Guelph: University of Guelph; 1984.
8. Legarra A, Ducrocq V. Computational strategies for national integration of phenotypic, genomic, and pedigree data in a single-step best linear unbiased prediction. J Dairy Sci. 2012;95:4629–45.
9. Misztal I, Legarra A, Aguilar I. Using recursion to compute the inverse of the genomic relationship matrix. J Dairy Sci. 2014;97:3943–52.
10. Misztal I. Inexpensive computation of the inverse of the genomic relationship matrix in populations with small effective population size. Genetics. 2015;202:401-9.
11. Fragomeni BO, Lourenco DAL, Tsuruta S, Masuda Y, Aguilar I, Legarra A, et al. Hot topic: use of genomic recursions in single-step genomic best linear unbiased predictor (BLUP) with a large number of genotypes. J Dairy Sci. 2015;98:4090–4.
12. Lourenco DAL, Tsuruta S, Fragomeni BO, Masuda Y, Aguilar I, Legarra A, et al. Genetic evaluation using single-step genomic best linear unbiased predictor in American Angus. J Anim Sci. 2015;93:2653–62.
13. Masuda Y, Misztal I, Tsuruta S, Legarra A, Aguilar I, Lourenco DAL, et al. Implementation of genomic recursions in single-step genomic best linear unbiased predictor for US Holsteins with a large number of genotyped animals. J Dairy Sci. 2016;99:1968–74.
14. Searle S. Matrix algebra for the biological sciences. New York: Wiley; 1966.
15. Misztal I, Legarra A, Aguilar I. Computing procedures for genetic evaluation including phenotypic, full pedigree, and genomic information. J Dairy Sci. 2009;92:4648–55.
16. Aguilar I. Genetic evaluation using unsymmetric single step genomic methodology with large number of genotypes. Interbull Bull. 2013;47:1–4.
17. Janss L, de los Campos G, Sheehan N, Sorensen D. Inferences from genomic models in stratified populations. Genetics. 2012;192:693–704.
18. Ostersen T, Christensen OF, Madsen P, Henryon M. Sparse single-step method for genomic evaluation in pigs. Genet Sel Evol. 2016;48:48.
19. Golub GH, Van Loan CF. Matrix computations, vol. 3. Baltimore: JHU Press; 2012.
20. Quaas RL. Computing the diagonal elements and inverse of a large numerator relationship matrix. Biometrics. 1976;32:949–53.
21. Chang HL, Fernando RL, Grossman M. On the principle underlying the tabular method to compute coancestry. Theor Appl Genet. 1991;81:233–8.

22. Emik LO, Terrill CE. Systematic procedures for calculating inbreeding coefficients. J Hered. 1949;40:51–5.

23. Legarra A, Aguilar I, Misztal I. A relationship matrix including full pedigree and genomic information. J Dairy Sci. 2009;92:4656–63.

24. Christensen OF, Lund MS. Genomic prediction when some animals are not genotyped. Genet Sel Evol. 2010;42:2.

25. Gianola D, de los Campos G, Hill WG, Manfredi E, Fernando R. Additive genetic variability and the Bayesian alphabet. Genetics. 2009;183:347–63.

26. Stranden I, Mantysaari EA. Comparison of some equivalent equations to solve single-step GBLUP. In: Proceedings of the 10th World Congress of Genetics Applied to Livestock Production, Vancouver, 17–22 August 2014.

27. Pocrnic I, Lourenco DAL, Masuda Y, Legarra A, Misztal I. The dimensionality of genomic information and its effect on genomic prediction. Genetics. 2016;203:573–81.

28. Fernando RL, Dekkers JC, Garrick DJ. A class of Bayesian methods to combine large numbers of genotyped and non-genotyped animals for whole-genome analyses. Genet Sel Evol. 2014;46:50.

7

Genetic tests for estimating dairy breed proportion and parentage assignment in East African crossbred cattle

Eva M. Strucken[1], Hawlader A. Al-Mamun[1], Cecilia Esquivelzeta-Rabell[2], Cedric Gondro[3], Okeyo A. Mwai[4] and John P. Gibson[1]* (ID)

Abstract

Background: Smallholder dairy farming in much of the developing world is based on the use of crossbred cows that combine local adaptation traits of indigenous breeds with high milk yield potential of exotic dairy breeds. Pedigree recording is rare in such systems which means that it is impossible to make informed breeding decisions. High-density single nucleotide polymorphism (SNP) assays allow accurate estimation of breed composition and parentage assignment but are too expensive for routine application. Our aim was to determine the level of accuracy achieved with low-density SNP assays.

Methods: We constructed subsets of 100 to 1500 SNPs from the 735k-SNP Illumina panel by selecting: (a) on high minor allele frequencies (MAF) in a crossbred population; (b) on large differences in allele frequency between ancestral breeds; (c) at random; or (d) with a differential evolution algorithm. These panels were tested on a dataset of 1933 crossbred dairy cattle from Kenya/Uganda and on crossbred populations from Ethiopia (N = 545) and Tanzania (N = 462). Dairy breed proportions were estimated by using the ADMIXTURE program, a regression approach, and SNP-best linear unbiased prediction, and tested against estimates obtained by ADMIXTURE based on the 735k-SNP panel. Performance for parentage assignment was based on opposing homozygotes which were used to calculate the separation value (*sv*) between true and false assignments.

Results: Panels of SNPs based on the largest differences in allele frequency between European dairy breeds and a combined Nelore/N'Dama population gave the best predictions of dairy breed proportion (r^2 = 0.962 to 0.994 for 100 to 1500 SNPs) with an average absolute bias of 0.026. Panels of SNPs based on the highest MAF in the crossbred population (Kenya/Uganda) gave the most accurate parentage assignments (*sv* = −1 to 15 for 100 to 1500 SNPs).

Conclusions: Due to the different required properties of SNPs, panels that did well for breed composition did poorly for parentage assignment and vice versa. A combined panel of 400 SNPs was not able to assign parentages correctly, thus we recommend the use of 200 SNPs either for breed proportion prediction or parentage assignment, independently.

Background

Based on bovine remains and terracotta figurines, it is assumed that the first domesticated cattle in Africa, around 5000 years ago, were humpless (*Bos taurus*) [1, 2]. Nowadays, the West African N'Dama cattle (*Bos taurus*) and closely related populations in West Africa are believed to be the only surviving population from the originally domesticated African cattle. Humped Zebu cattle (*Bos indicus*) were introduced to Africa with traders from Arabia 2000 to 3000 years ago [2, 3]. Crossbreeding of local African taurine with introduced indicine cattle created a variety of new populations that make up most of the native cattle of Africa today [4–6]. Based on

*Correspondence: jgibson5@une.edu.au
[1] School of Environmental and Rural Science, University of New England, Armidale 2350, Australia
Full list of author information is available at the end of the article

analyses of karyotypes and genetic markers, Frisch et al. [7] inferred that East African Zebu breeds are a mixture of *Bos indicus* and *Bos taurus*, and that Sanga breeds are *Bos taurus*. Subsequent studies using microsatellites and then single nucleotide polymorphisms (SNPs) confirmed the mixed ancestry of East African Zebu breeds but identified that the *Bos taurus* component is primarily African rather than European *Bos taurus* [8]. Hanotte et al. [8] also found that the ancestry of the tested Sanga breeds was also mixed but with substantially higher proportions of African *Bos taurus* than Zebu breeds.

During the second half of the twentieth century, globalization and an increasing demand for milk fostered a new wave of crossbreeding in some parts of Africa. Northern American and European *Bos taurus* dairy breeds, known for their high production levels, were imported and crossed to native breeds in an attempt to improve the level of milk production. For example, in Kenya, Ayrshire, Jersey, and Guernsey cattle were originally imported, then Friesian and later Holstein dominated bovine imports. In Uganda, imports of Friesian and later Holstein cattle dominated [9]. The rapid and large-scale expansion of the East African highland dairy smallholders indicates that, under appropriate conditions, crossbreeding and the use of crossbred cattle can yield significant increases in smallholder income.

Knowledge of breed composition is required to determine which crossbreeds perform best under the wide variety of smallholder dairy systems, and also, to make breeding decisions for producing progeny of the desired breed composition. Because of the lack of pedigree records, the breed composition of most animals is not known [10]. Furthermore, the lack of knowledge about breed proportions and about the relationships within and between populations may lead to the loss of native genetic resources and may build-up inbreeding depression [11, 12].

High-density (HD) SNPs can be used to assess the levels of genetic diversity between individuals [13], to determine coefficients of kinship between pairs of individuals allowing for parentage exclusion [14], to obtain accurate estimates of breed proportions in crossbred animals [15], and to trace animal products to their source [16]. The HD SNP panels are too expensive for routine use in smallholder systems. Genotyping a few hundred SNPs can be relatively inexpensive but how accurate are the estimates of breed composition or parentage assignment when using such small numbers of SNPs in crossbred dairy populations is not known.

The aim of this study was to determine the accuracy and bias when using small subsets of SNPs from a commercially available 735k-SNP panel to estimate breed proportion and parentage assignment in crossbred dairy

cattle populations in East Africa. We used a variety of methods to select the SNPs for reaching the highest possible accuracy (r^2) of estimated breed proportions and parentage assignment. Based on the history of cross-breeding in Africa, we included as baseline information the genotype frequencies in pure breeds such as the N'Dama (reference for African *Bos taurus*), Nelore (reference for pure *Bos indicus*), and several European and North American dairy breeds, which collectively represent the ancient and more recent ancestral gene pool of the crossbred dairy animals.

Methods

Animals

In total, 1933 crossbred dairy cows and local indigenous breeds of Ankole (n = 43), Nganda (n = 17), and Small East African Zebu (Zebu; n = 58) were sampled from 845 households that are distributed at five sites in Kenya and two sites in Uganda (Dairy Genetics East Africa, DGEA1, project). In addition, genotype datasets for N'Dama (as the reference African *Bos taurus* breed; n = 20), Nelore (as the reference *Bos indicus* breed; n = 20), Guernsey (n = 20), Holstein (n = 20), and Jersey (n = 20) were sourced from the International Bovine HapMap consortium. Furthermore, British Friesian (n = 25) from the SRUC in Scotland and Canadian Ayrshire (n = 20) from the Canadian Dairy Network (CDN) were used as reference breeds.

An independent population of 545 crossbred animals from Ethiopia (DGEA2 project) was sampled from 400 households at nine sites. Instead of the Kenyan and Ugandan indigenous breeds, we included the Ethiopian Begait Barka (n = 30), Danakil Harar (n = 30), Fogera (n = 29), and Boran (n = 30) in the analyses of breed composition. An independent Tanzanian dataset (DGEA2 project) consisted of 462 crossbred animals sampled from 326 households at three sites. Tanzanian indigenous breeds for the analysis of breed composition included Iringa Red (n = 13), Singida White (n = 22), and Tanzanian Boran (n = 20).

Genotype data

All animals were genotyped with the 777k-SNP BovineHD Beadchip (Illumina Inc., San Diego). In order to keep potentially interesting SNPs that could be excluded due to population stratification, criteria for genotype data filtering were applied per breed and focused on genotyping quality. Genotypes of the DGEA1 and 2 and SRUC data were filtered using 'SNPQC' an R pipeline for quality control of Illumina SNP genotyping array data described in [17] to eliminate SNPs that had a median GC score lower than 0.6 and a sample-wise call rate lower than 90%. Only the SNPs on the 29 autosomal bovine chromosomes were included in the analysis. Genotypes provided by the Bovine HapMap consortium and the Canadian Dairy Network were already quality-controlled. The cleaned population datasets were merged and included 735,293 SNPs. SNPs that were excluded after quality control in one breed but not in another breed were set to "not available" (NA) in the breed for which they were excluded.

We checked the relationships between animals based on the genomic relationship matrix [18], with missing genotypes being replaced by the average genotype (encoded as 0, 1, 2) across all animals:

$$\mathbf{GRM} = \mathbf{ZZ'}/2 * \sum p_l * (1 - p_l),$$

where $\mathbf{Z}$ is the centered genotype matrix and p is the allele frequency at locus l. Matrix $\mathbf{Z}$ was constructed by subtracting from the genotype matrix $\mathbf{M}$ the $\mathbf{P}$ matrix, which equaled 2*(p − 0.5). The centering of $\mathbf{Z}$ was achieved by subtracting −1 from $\mathbf{M}$.

Inbreeding coefficients (F_{IS}) were calculated per breed according to Weir and Cockerham [19].

Observed breed compositions

Breed proportions of crossbred animals from both crossbred populations were estimated by using the full quality-controlled data in the ADMIXTURE 1.23 program [20]. Analyses were performed by assuming that N'Dama, Nelore, Ayrshire, Friesian, Guernsey, Holstein, and Jersey represented ancestral populations. We used all 735k SNPs to estimate breed proportions to create a baseline for comparison with the estimates using subsets of SNPs. Dairy proportion was defined as the sum of breed proportions across all European dairy breeds that was estimated in the crossbred populations. The Kenyan/Ugandan dataset also included the local pure breeds of Ankole, Nganda, and Zebu whereas the Ethiopian dataset included Begait Barka, Danakil Harar, Fogera, and Ethiopian Boran, and the Tanzanian dataset included Iringa Red, Singida White, and Tanzanian Boran.

Observed pedigree

The pedigrees of the crossbred animals from Kenya/Uganda, Ethiopia, and Tanzania were reconstructed based on the presence or absence of opposing homozygotes [21, 22]. Opposing homozygotes (*opH*) occur if at the same SNP, two individuals carry opposite homozygous genotypes [21]. The more *opH* are found between two individuals, the less likely are these individuals related. Except for genotyping errors and mutations, a parent and offspring cannot display *opH*. The distribution of *opH* that are associated with parent–offspring or other relationships is specific to the allele frequencies of

the population and the number of SNPs used; however, with several tens of thousands SNPs or more, parent–offspring relationships can always be clearly separated from other relationships. By applying the approach of Strucken et al. [23] if there are less than 1000 *opH*, it is possible to unambiguously distinguish between parent–offspring and unrelated individual pairs in the DGEA1 and 2 crossbred populations.

The Kenyan/Ugandan crossbred population contained 171 cows with 189 offspring, of which 15 cows had two offspring and one cow had three offspring. The relationship between two parent individuals was similar to that between half-sibs. The Ethiopian dataset included 38 cows that each had one offspring, and the Tanzanian dataset included 31 cows and 34 offspring with three of these cows having two offspring.

Selection of subsets of SNPs

From the 735k SNPs in the Kenyan/Ugandan dataset, subsets of 100, 200, 300, 400, 500, 1000, and 1500 SNPs were chosen based on several selection criteria that are described below, resulting in SNPs located on all chromosomes except for the smaller panels with less than 200 SNPs; the number of SNPs was smaller on short than on long chromosomes. To minimize linkage disequilibrium, SNPs had to be at least one megabase (Mb) pair apart. Some of the methods to select SNPs were carried out within the crossbred population under investigation (e.g. with the highest minor allele frequency (MAF)), which implies that they should, ideally, be repeated when moving to a different population. However, the selected SNP panels were validated in independent crossbred populations to assess the potential for a wider application of our SNP panels. SNP panels were selected based on the criteria described in the following paragraphs.

Highest minor allele frequency

Allele frequencies were calculated for the crossbred animals. SNPs were sorted by MAF and subsets were selected based on the highest MAF in the crossbred animals. These subsets are not independent since larger subsets always included SNPs in the smaller subsets. The average distance between SNPs in the smallest and largest panels were 19.4 Mb [standard deviation (SD) = 16.2 Mb] and 1.7 Mb (SD = 0.7 Mb), respectively.

Differences in absolute frequency

Allele frequencies were calculated for the ancestral breeds. The weighted average allele frequency across breeds was calculated based on the number of animals in each breed sample. Weighted averages were calculated across the Nelore and N'Dama populations (NelNd) and across all European dairy breeds (EU). The differences in absolute frequencies were determined between Nelore and EU (NelEU), N'Dama and EU (NdEU), and NelNd and EU (NelNdEU). SNPs were sorted according to differences in absolute frequencies and subsets that had the largest differences were selected. As above, these subsets are not independent because larger subsets include all the SNPs in the smaller subsets. The average distance between SNPs per chromosome for the 100-SNP panel was 17.6 Mb (SD = 14.4 Mb), 17.4 Mb (SD = 16.3 Mb), and 15.6 Mb (SD = 14.2 Mb) for NelEU, NdEU, and NelNdEU, respectively. For the 1500-SNP panel, the average distance between SNPs per chromosome was 1.7 Mb (SD = 0.7 Mb), 1.7 Mb (SD 0.7 = Mb), and 1.7 Mb (SD = 0.6 Mb) for NelEU, NdEU, and NelNdEU, respectively.

Random selection

We selected 10 random samples for each subset and results were averaged across these random samples. These random panels were not restricted by SNP spacing (i.e. the 1 Mb pair restriction). The average distance between SNPs per chromosome ranged from 20.4 Mb (SD = 16 Mb) for the 100-SNP panel to 1.7 Mb (SD = 1.6 Mb) for the 1500-SNP panel.

ISAG panel and 50k-SNP chip

The official International Society for Animal Genetics (ISAG) panel for parentage assignment [24] consists of 100 core SNPs, which are mostly derived from European breeds, plus an additional 100 SNPs from *Bos indicus* animals. We also tested 47,810 SNPs from the Illumina 50k-SNP bovine chip v2 (San Diego, CA, USA).

Differential evolution (DE) algorithm

The differential evolution (DE) algorithm is based on Storn and Price [25] and ranks SNPs according to a random key (vector of real values; [26]). This key evolves to a higher rank as the SNP is more suited to solve a particular problem (e.g. estimation of breed proportion or parentage assignment [27, 28]).

In our study, the "all animals" set (including pure and crossbred animals) was split into a training and a test population. The DE algorithm was initiated in the training population with 100 random samples of SNPs ('parental sets') for each panel size (i.e. 100, 200, 300, 400, 500, 1000, 1500 SNPs). From these 100 parental sets, two sets were randomly selected to create an 'offspring set' consisting of 50% randomly sampled SNPs from each parental set. If this offspring set performed better than the initial 100 parental sets (according to a fitness function), then this offspring set was retained and the worst parental set was discarded. The dairy proportions were

estimated internally with a SNP-best linear unbiased prediction (BLUP) approach (see below), whereas the parentage test was based on number of *opH*.

The fitness function used to optimize prediction of dairy breed proportions was the coefficient of determination (r^2) between the subsets of SNPs and the dairy breed proportions predicted with the 735k SNPs in ADMIXTURE. To optimize parentage assignments, the fitness function was the percentage of correctly assigned parentages according to the reconstructed pedigree. This process was run for 2000 iterations/generations. No spacing restriction between SNPs was applied since the DE algorithm should select best SNPs by default.

The average distance between SNPs per chromosome for the panels to estimate breed proportions ranged from 23.6 Mb (SD = 12.8 Mb) for the 100-SNP panel to 1.6 Mb (SD = 1.6 Mb) for the 1500-SNP panel and for the panels to assign parentage, it ranged from 19.7 Mb (SD = 11.2 Mb) for the 100-SNP panel to 1.7 Mb (SD = 1.6 Mb) for the 1500-SNP panel.

Accuracy and bias of breed proportion prediction

Total dairy proportion for an animal was the sum of the estimated individual breed proportions for Ayrshire, Guernsey, Jersey, Holstein, and Friesian. Accuracy of the prediction of dairy proportions for all subsets of SNPs was assessed by the coefficient of determination (r^2) between observed (based on all 735k SNPs) and predicted (based on subsets of SNPs) dairy proportions of the 1933 crossbred animals. The linear bias of breed proportions estimated from subsets of SNPs was assessed as the average deviation or average absolute difference estimated minus the observed (735k SNPs) values.

Parentage assignment

The *opH* matrices were calculated for each subset of SNPs in the crossbred population. Subsequently, the separation value (*sv*) was used to quantify and visualize the performance of each SNP subset and was calculated as:

$$sv = \min(FR) - \max(TR), \tag{1}$$

where *FR* is the number of opposing homozygotes in false parent–offspring relationships according to the reconstructed pedigree information; and *TR* is the number of opposing homozygotes in true parent–offspring relationships [23, 29].

Regression and SNP-BLUP

Prediction of breed proportions was also made with a regression model and a SNP-BLUP approach to test the ability of the SNP panels to perform outside ADMIXTURE.

The regression method was based on Kuehn et al. [30] and described in Dodds et al. [31] for prediction of breed proportions:

$$\mathbf{y} = \mathbf{X}\hat{\mathbf{b}} + \mathbf{e}, \qquad (2)$$

where $\mathbf{y}$ are the proportions of the designated allele in the genotypes for each SNP of each animal (encoded as allele counts 0, 0.5, 1); $\mathbf{X}$ is a matrix of allele frequencies in each reference breed (ADMIXTURE P-file output); $\hat{\mathbf{b}}$ are the breed proportions of each animal for each reference breed (to be estimated); and $\mathbf{e}$ are the residual errors. Coefficients of determination were calculated between predictions of breed proportions in the ADMIXTURE analysis (735k SNPs) and predictions of the regression method. In addition, the ADMIXTURE P-file was replaced by observed allele frequencies in the ancestral populations.

The SNP-BLUP approach required the replacement of missing genotypes (NA) with the average allele count across all animals. Only SNPs with a call rate higher than 95% were used in this analysis to limit potential bias due to SNPs with only a few recorded genotypes. SNP-BLUP was performed as follows:

$$\left[\mathbf{Z}\mathbf{Z}' + \mathbf{I}\lambda\right] * \hat{\mathbf{g}} = [\mathbf{Z}\mathbf{y}], \qquad (3)$$

where $\hat{\mathbf{g}}$ is the effect of the SNPs to be estimated; $\mathbf{y}$ is a vector of dairy proportions (ADMIXTURE output for 735k SNPs) scaled with a mean = 0 and SD = 1; $\mathbf{Z}$ is a design matrix allocating SNP genotypes (multiplied by their allele frequencies) to records; $\mathbf{I}$ is an identity matrix and λ defines the contribution of genomic relationships. λ was set to $\lambda = \left(1 - \mathrm{h}^2\right)/(\mathrm{h}^2/\mathrm{d})$ with the heritability assumed to be $\mathrm{h}^2 = 0.99$ and d representing the average heterozygosity of the panel.

SNP effects ($\hat{\mathbf{g}}$) were subsequently multiplied by $\mathbf{Z}$ to obtain estimates of dairy proportions (i.e. GEBV) for each panel. The estimated dairy proportions had to be rescaled (reversing the scaling of $\mathbf{y}$) to be correctly interpreted. This approach was also used within the DE algorithm.

Validation

When SNPs are selected based on information that is independent of the test dataset, there is no ascertainment bias. Selection of SNPs based on MAF in the crossbred population is subject to trivial ascertainment bias due to binomial sampling variance of allele frequencies (approximately ± 0.01).

The linear regression and SNP-BLUP estimates of breed proportions are subject to ascertainment bias and thus require validation. Validation was achieved by using the SNP effects that were estimated in the Kenyan/Ugandan dataset to predict dairy proportions in the Ethiopian and Tanzanian dataset and vice versa. To determine whether population structure or random sampling caused bias in the estimates, we applied the SNP-BLUP approach to predict breed proportions for 50% of the animals in the Kenyan/Ugandan dataset by randomly selecting 50% of the animals in each breed (training dataset). Then, cross-validation of the estimates was performed on the other half of the Kenyan/Ugandan population as well as on the Ethiopian and Tanzanian crossbred animals (test datasets).

We further validated our sets of SNPs in independent crossbred populations. The subsets of SNPs that were selected from the Kenyan/Ugandan dataset were used to predict breed proportions and parentage assignment in the Ethiopian and Tanzanian datasets. The coefficient of determination and the absolute linear bias between the full dataset and the subsets within the Ethiopian and Tanzanian datasets were used to determine the performance of each subset of SNPs to accurately assign dairy proportions, and the sv was used for parentage assignment.

Results and discussion
Description of data

After merging the quality-controlled datasets for each breed, 4.8, 5.1, and 5.8% of genotypes were missing in the entire Kenyan/Ugandan, Ethiopian, and Tanzanian datasets, respectively. Within the crossbred animals, 4.9, 5.5, and 4.8% of genotypes were missing in the Kenyan/Ugandan, Ethiopian, and Tanzanian datasets, respectively. There was no general pattern of where the missing genotypes occurred along the genome.

The average inbreeding coefficient (F_{IS}) did not show any substantial average inbreeding in any of the breeds; however, the SD was very large (Table 1). The genomic relationship matrix (GRM) showed that the crossbred animals were mostly unrelated with no detectable inbreeding (Table 1). The assumed ancestral breeds included related individuals within the range of half-sib relations. Exceptions were the N'Dama and Nelore populations in which individuals appeared to be highly related and inbred, with Nelore showing an average diagonal element of 1.82 (Table 1). The high values of the GRM for the Nelore population can be explained by ascertainment bias [32] combined with how the GRM is calculated. Nelore is a pure *Bos indicus* breed and N'Dama represents a unique African *Bos taurus* breed. The largest proportion of SNPs on the 735k-Illumina chip was chosen based on high information content (high MAF) within non-African *Bos taurus* populations.

Figure 1 shows MAF and absolute allele frequencies for the various populations in our analyses and clearly illustrates the bias that is due to the criteria applied for selecting SNPs in the assay. The method of constructing the

Table 1 Average diagonal and off-diagonal elements of the GRM [18] and inbreeding coefficient (F_{IS}) ± SD (SE) in cattle

	Diag	Off-Diag	F_{IS}
Ayrshire	1.11 ± 0.044 (0.01)	0.34 ± 0.089 (0.02)	−0.024 ± 0.206 (0.05)
Friesian	1.01 ± 0.025 (0.005)	0.18 ± 0.05 (0.01)	−0.005 ± 0.191 (0.04)
Guernsey	1.16 ± 0.036 (0.008)	0.37 ± 0.089 (0.02)	0.018 ± 0.218 (0.05)
Holstein	1.11 ± 0.036 (0.008)	0.29 ± 0.089 (0.02)	−0.021 ± 0.210 (0.05)
Jersey	1.21 ± 0.040 (0.009)	0.48 ± 0.134 (0.03)	−0.0003 ± 0.218 (0.05)
N'Dama	1.28 ± 0.022 (0.005)	0.65 ± 0.027 (0.009)	0.013 ± 0.215 (0.05)
Nelore	1.82 ± 0.036 (0.008)	1.22 ± 0.045 (0.01)	0.003 ± 0.209 (0.05)
XBred[a]	0.98 ± 0.044 (0.001)	0.0004 ± 0.044 (0.001)	0.024 ± 0.037 (0.0008)
E XBred[a]	0.95 ± 0.070 (0.003)	0.01 ± 0.070 (0.003)	0.015 ± 0.050 (0.002)
T XBred[a]	0.954 ± 0.003 (0.0001)	0.004 ± 0.002 (0.0001)	0.032 ± 0.06 (0.003)

[a] XBred, Kenya/Uganda; E XBred, Ethiopia; T XBred, Tanzania

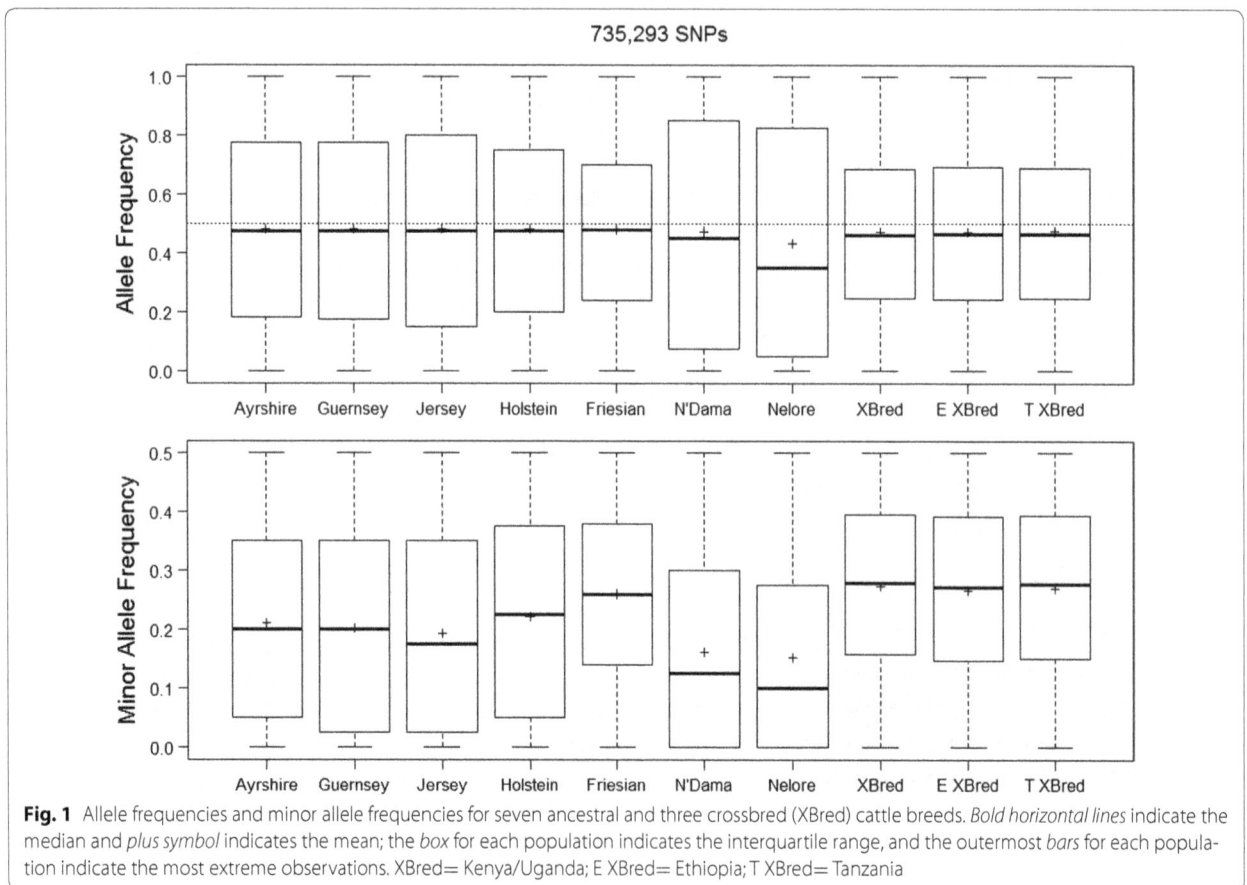

Fig. 1 Allele frequencies and minor allele frequencies for seven ancestral and three crossbred (XBred) cattle breeds. *Bold horizontal lines* indicate the median and *plus symbol* indicates the mean; the *box* for each population indicates the interquartile range, and the outermost *bars* for each population indicate the most extreme observations. XBred= Kenya/Uganda; E XBred= Ethiopia; T XBred= Tanzania

GRM across multiple breeds [18] centers the matrix by using most of the animals in the dataset. Thus, the level of inbreeding appears to be high in the N'Dama and Nelore populations, which represent a small number of animals and they have MAF that clearly differ from those of other groups. When the GRM was constructed by using only the Nelore animals (n = 20), the average of the diagonal elements was equal to 0.975, which is consistent with the

diagonal elements of the GRM for Nelore reported by Zavarez et al. [33].

The allele frequencies for the three crossbred populations showed a narrower inter-quartile range compared to the assumed ancestral populations (Fig. 1). Compared to the other breeds, more than twice the number of SNPs were not in Hardy–Weinberg equilibrium (the null-hypothesis was rejected) in the crossbred populations

(151,486 SNPs for the Kenyan/Ugandan, 58,493 for the Ethiopian, and 91,460 for the Tanzanian datasets), which is likely due to a proportion of the crossbred animals originating from the first generation progeny of crosses with pure dairy or indigenous breeds.

A principal component (PC) analysis based on the GRM for the combined Kenyan and Ugandan dataset separated the European dairy breeds from the Nelore breed and then from the African pure breeds and the first PC explained 86.91% of the genetic variation. The second PC clearly separated the Nelore and N'Dama breeds from the European breeds and crossbreds, with the East African indigenous breeds being intermediate; it explained 1.75% of the genetic variation (Fig. 2a). The second PC also separated Kenyan and Ugandan crossbred animals, which spread between their respective indigenous breeds (Ankole and Nganda in Uganda and Zebu in Kenya) and European ancestral breeds. Although the indigenous samples were collected from animals that phenotypically appeared as pure indigenous, they clearly included animals that were admixed with European *Bos taurus* genes. When considering only the clusters of apparently pure indigenous animals, the variation between these three indigenous breeds was substantially larger in both dimensions (PC1 and PC2) than the difference between the European *Bos taurus* dairy breeds.

The first two PC in the Ethiopian dataset explained 90.92 and 1.71% of the variation, respectively (Fig. 2b), and in the Tanzanian dataset, they explained 85.19 and 3.23% of the variation, respectively (Fig. 2c). Most of the Ethiopian crossbreds aligned with their respective indigenous breeds and with the Friesian and Holstein breeds; however, some animals were positioned between an unknown indigenous population and the Ayrshire population. The Tanzanian crossbred animals were positioned between their indigenous breeds and the Holstein and Ayrshire breeds. However, when all the data from DGEA1 and 2 were analyzed simultaneously and the results were plotted to show the third PC, the Tanzanian

crossbred animals were closer to the Friesian breed (see Additional file 1: Figure S1). Similar to the Ethiopian crossbreds, some Tanzanian crossbreds seemed to align with an unknown indigenous breed (Fig. 2c). The three-dimension PCA plot for the analysis that included indigenous breeds from all countries, showed that the Ethiopian and Tanzanian crossbred animals that were not aligned to a local indigenous breed, aligned with an unknown breed(s) between the East African Zebu and the Nganda breed (see Additional file 1: Figure S1). Crossbred animals from Tanzania that did not align with a local breed in the analysis were sampled from the Southern Highlands, whereas those from Ethiopia came from various locations across the country.

Description of the SNP panels

As expected given the sampling procedure applied, panels of SNPs that were selected on their highest MAF showed almost no variation in allele frequencies for the Kenya/Uganda crossbred animals with median and mean allele frequencies at 0.5. The panel that showed the next to lowest variation was the combined NdEU panel for which the interquartile range was between 0.35 and 0.6 (Fig. 3).

All other selection methods resulted in relatively large interquartile ranges. Deviation in mean and median allele frequency was largest for the NelEU panels (Fig. 3). Examination of these allele frequencies within the Nelore and EU breeds revealed that the shift in frequencies was very similar but in opposite directions for the European breeds versus the Nelore breed for all SNP panels (see Additional file 2: Figure S2). The observed frequency for the NelEU SNP panel in the crossbred animals (Fig. 3) most likely reflects that the average crossbred animal in this population was 69.7% (SD = 21.1%) European *Bos taurus*. A similar but less extreme effect was found for the NdEU SNP panel (see Additional file 2: Figure S2).

The ISAG SNP panel showed a narrow inter-quartile range with a mean and median at 0.5, and the

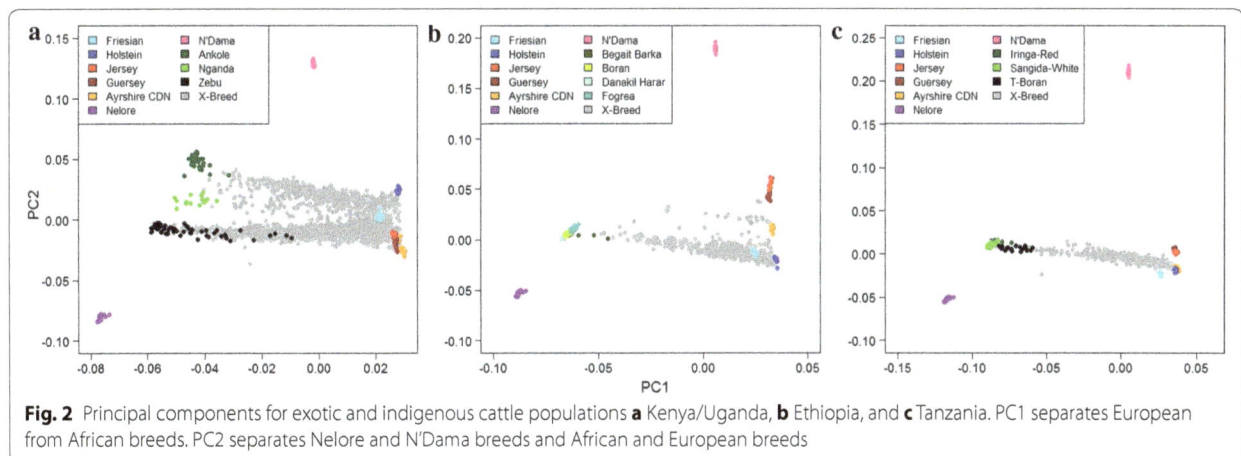

Fig. 2 Principal components for exotic and indigenous cattle populations **a** Kenya/Uganda, **b** Ethiopia, and **c** Tanzania. PC1 separates European from African breeds. PC2 separates Nelore and N'Dama breeds and African and European breeds

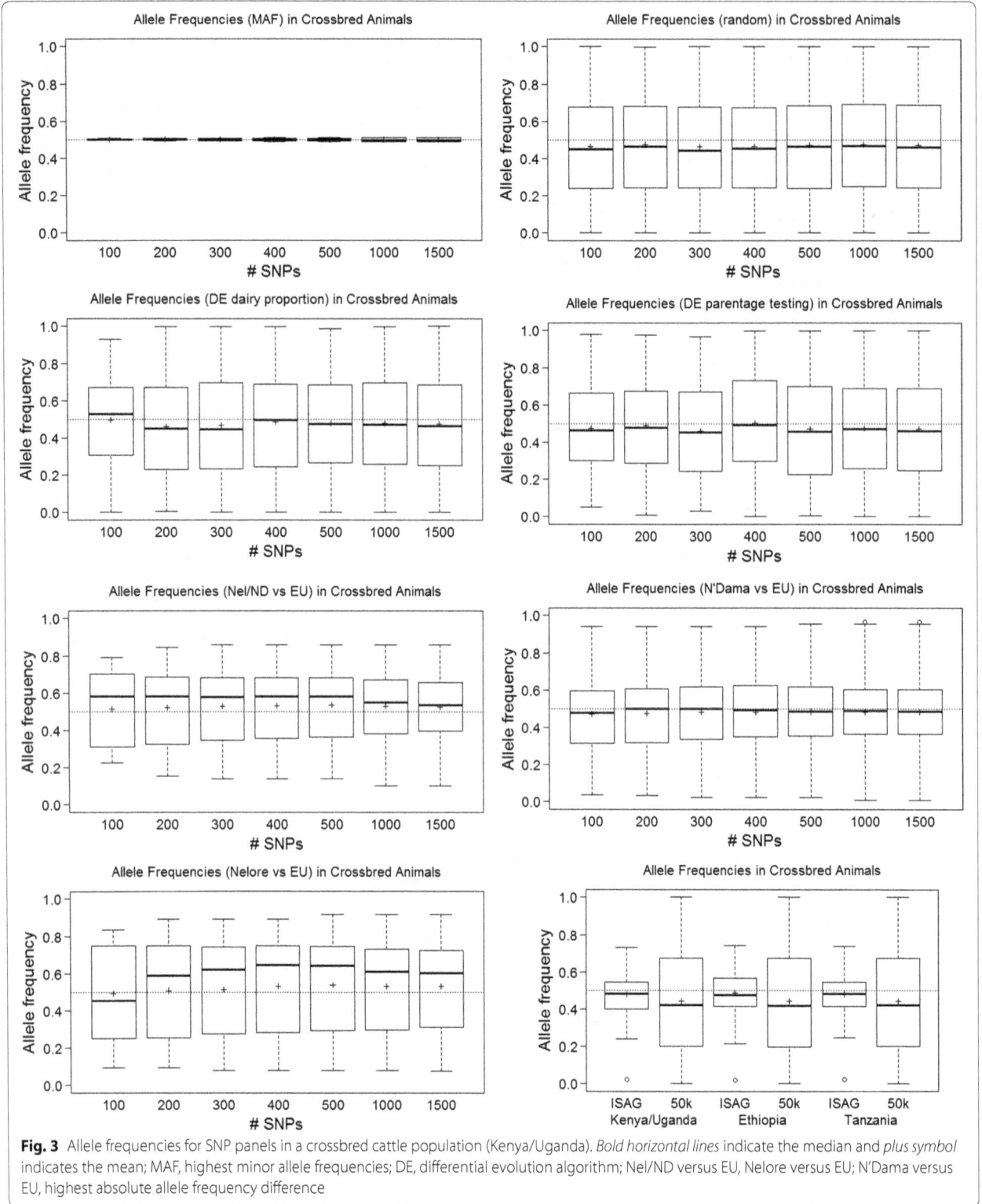

Fig. 3 Allele frequencies for SNP panels in a crossbred cattle population (Kenya/Uganda). *Bold horizontal lines* indicate the median and *plus symbol* indicates the mean; MAF, highest minor allele frequencies; DE, differential evolution algorithm; Nel/ND versus EU, Nelore versus EU; N'Dama versus EU, highest absolute allele frequency difference

inter-quartile range of the 50k-SNP panel was similar to that of the full 735k-SNP panel although the mean and median frequencies deviated more from 0.5 (Fig. 3).

The distributions of allele frequencies in the Ethiopian and Tanzanian crossbred animals were similar to that in the Kenya/Uganda crossbred animals, but with a wider range of frequencies for all SNP panels (see Additional file 3: Figure S3, Additional file 4: Figure S4).

Estimation of breed proportions

Proportions of dairy breed in the crossbred animals were on average equal to 0.70 (SD = 0.21), 0.78 (SD = 0.20), and 0.78 (SD = 0.18) for the Kenyan/Ugandan, Ethiopian, and Tanzanian datasets, respectively. This proportion was highest for Ayrshire in Kenyan crossbred animals and for Friesian in Ugandan crossbred animals, which was consistent with the PCA results (Fig. 4). Based on a study of smallholder cattle that were sampled from mostly peri-urban areas in Kenya, Gorbach et al. [12] reported that the crossbred cattle had very high dairy breed proportions,

which reflected the fact that their samples originated from a much smaller, more intensive and older dairy production area than in our study. They found that the main dairy breeds present in the crossbred individuals were Holstein and Jersey/Guernsey, the latter two being indistinguishable. However, their analysis did not include Ayrshire as a reference breed and Weerasinghe [15] showed that when Ayrshire was excluded from the ADMIXTURE analysis, most of the Ayrshire signal appeared as Guernsey or Jersey.

In Ethiopian crossbred animals, proportions of dairy breeds were highest for Holstein and Friesian (Fig. 4), which was consistent with the PCA analyses and the documented history of the country's specific cattle imports [15]. In Tanzanian crossbred animals, the Friesian breed

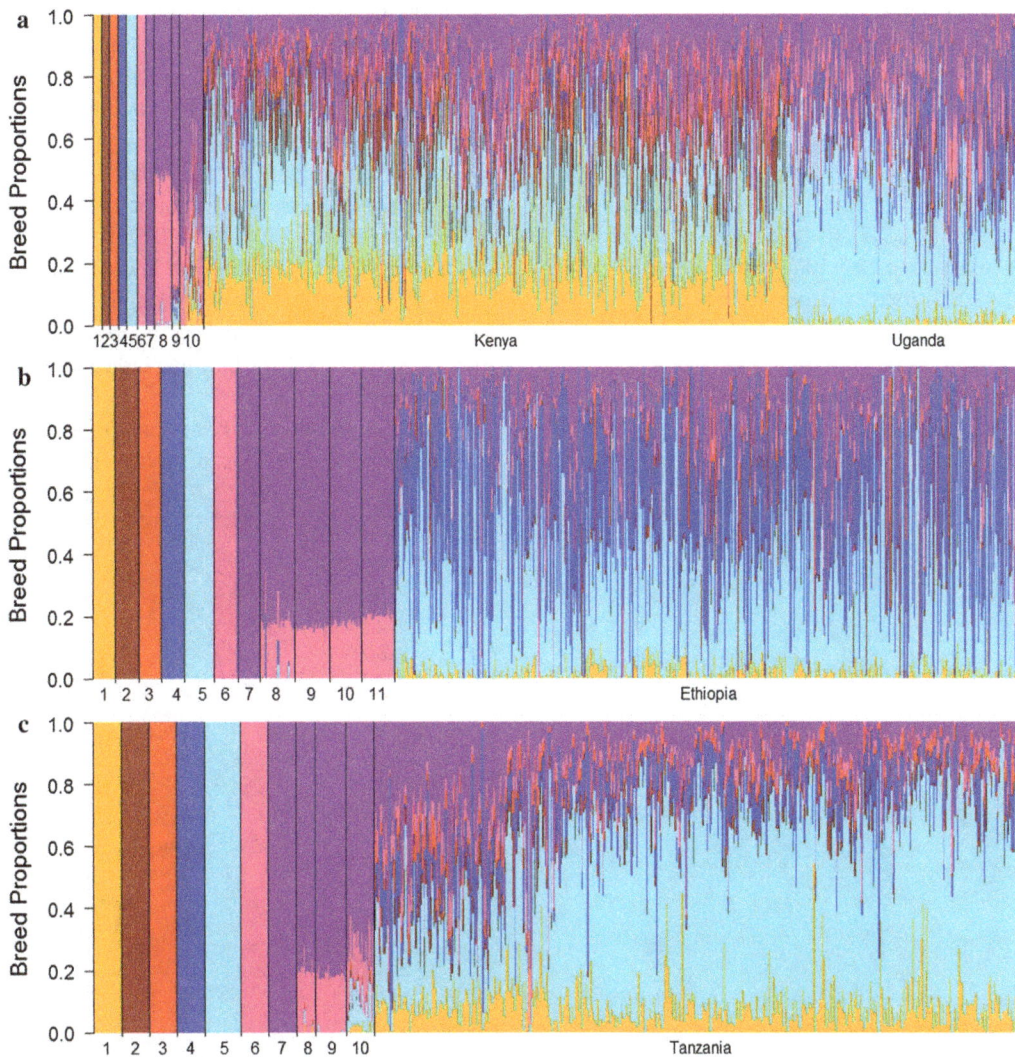

Fig. 4 Breed proportions of crossbred dairy cattle **a** Kenya/Uganda, **b** Ethiopia, and **c** Tanzania. Supervised ADMIXTURE analysis with seven fixed ancestral breeds: **a** *1* Ayrshire, *2* Guernsey, *3* Jersey, *4* Holstein, *5* Friesian, *6* N'Dama, *7* Nelore, *8* Ankole, *9* Nganda, *10* Zebu. **b** *1* Ayrshire, *2* Guernsey, *3* Jersey, *4* Holstein, *5* Friesian, *6* N'Dama, *7* Nelore, *8* Begait Barka, *9* Danakil Harar, *10* Ethiopian Boran, *11* Fogera. **c** *1* Ayrshire, *2* Guernsey, *3* Jersey, *4* Holstein, *5* Friesian, *6* N'Dama, *7* Nelore, *8* Iringa Red, *9* Singida White, *10* Tanzanian Boran

proportion was highest, which differed from the results of the PCA in which they aligned more closely with the Holstein and Ayrshire breeds (Fig. 4). Dairy breed proportions were highest in crossbred animals from the Southern Highland sampling site (0.84, SD = 0.12) compared to the other Tanzanian crossbreds, which was consistent with the PCA.

We selected three SNP panels (NelEU, NdEU, and NelNdEU) based on the largest differences in allele frequency between ancestral breeds. Our hypothesis was that SNPs that display the largest difference in allele frequency between the indigenous ancestral breeds and the dairy breeds will provide the most accurate estimates of total dairy breed proportion. Total dairy breed proportion was defined as the sum of breed proportions across Ayrshire, Guernsey, Jersey, Holstein, and Friesian breeds. Panels of SNPs that were selected by applying other methods were included to investigate the factors that determine accuracy of prediction and whether it was possible to develop SNP panels that could estimate both breed proportion and parentage assignment.

The various panels used in this study predicted dairy breed proportions in the Kenyan/Ugandan crossbreds with an r^2 of 0.725 to 0.963 (SE = 0.004–0.012) for the smallest subsets of 100 SNPs, and 0.977–0.994 (SE = 0.002–0.003) for the largest subsets of 1500 SNPs (Fig. 5a). As hypothesized, the NelNdEU SNP panel achieved the best results for all panel sizes, with an r^2 of 0.974 (SE = 0.004) with just 200 SNPs. The next best panel was the NdEU for all panel sizes except 100 SNPs, for which the DE algorithm performed slightly better

(Fig. 5a). Surprisingly, the NelEU SNP panel performed worse compared to the other panels selected for largest differences in allele frequency, with an r^2 of 0.852 (SE = 0.009) and 0.898 (SE = 0.007) for 100 and 200 SNPs, respectively, because as shown by Table 2 NelEU SNPs are efficient for distinguishing *Bos taurus* from *Bos indicus* but not for separating African from European *Bos taurus*.

The performance of the panels selected with the DE algorithm did not improve much as the number of SNPs increased, and hence were outperformed by the NelNdEU and NdEU panels for more than 100 SNPs. The DE algorithm was designed to optimize a panel for the prediction of the 735k ADMIXTURE estimates of dairy proportions. However, SNP-BLUP was used to estimate dairy proportions, rather than ADMIXTURE as for all other panels. In addition, when we predicted dairy proportions by using a SNP-BLUP approach independently of the DE algorithm (see next section), the DE-based panels continued to perform less well than the NelNdEU and NdEU panels, which indicated that the DE algorithm failed to find the optimal solution with the number of iterations performed. Esquivelzeta-Rabell et al. [28] used the DE algorithm to predict Korean Hanwoo proportions in a Chinese Yeonbyun population and reported r^2 of 0.69 and 0.88 for 100 and 1000 SNPs, respectively. These coefficients of determination are lower than those obtained by using the same number of SNPs and the DE algorithm but the genetic differences between Yeonbyun and Hanwoo are much smaller than those between indigenous and European dairy breeds in our study.

Figure 5d shows the relationship between bias and accuracy for all methods of SNP selection and size of

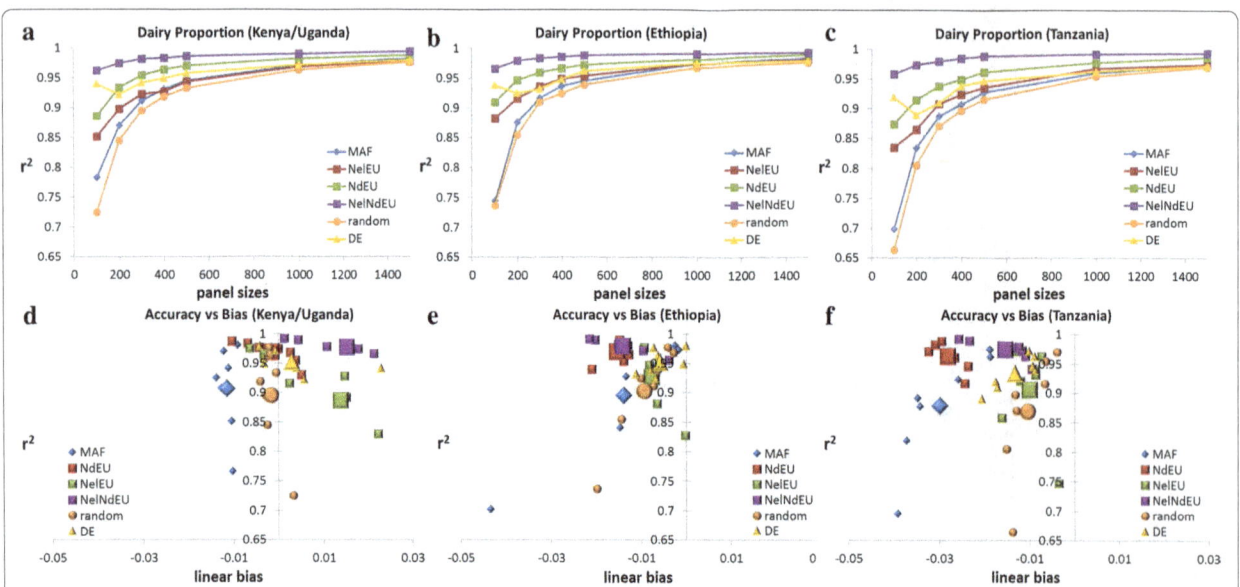

Fig. 5 Accuracy (r^2) of dairy proportion estimates (**a–c**) and accuracy versus bias (**d–f**) for different panel sizes. **d–f** large symbols show average linear bias across all panel sizes. Standard errors of accuracy ranged on average from 0.008 for 100 SNPs to 0.003 for 1500 SNPs (Kenya/Uganda), 0.015 to 0.005 (Ethiopia), and 0.02 to 0.008 (Tanzania)

Table 2 Accuracies (r^2) of individual breed proportions in crossbred dairy cattle (Kenya/Uganda) for 200 SNPs selected by different methods

	Dairy	Ayrshire	Friesian	Guernsey	Holstein	Jersey	N'Dama	Nelore	Mean ± SD[a]
MAF	0.870	0.486	0.312	0.375	0.483	0.165	0.397	0.874	0.442 ± 0.22
NelEU	0.898	0.120	0.140	0.056	0.198	0.034	0.484	0.944	0.282 ± 0.33
NdEU	0.934	0.373	0.263	0.265	0.437	0.141	0.538	0.867	0.412 ± 0.24
NelNdEU	0.974	0.075	0.132	0.013	0.007	0.004	0.511	0.190	0.133 ± 0.18
Random	0.845	0.405	0.243	0.276	0.414	0.133	0.373	0.874	0.388 ± 0.24
DE	0.922	0.421	0.049	0.281	0.402	0.141	0.427	0.925	0.378 ± 0.28
ISAG	0.831	0.452	0.197	0.270	0.306	0.127	0.482	0.759	0.378 ± 0.21
Mean ± SD	0.896 ± 0.05	0.333 ± 0.17	0.197 ± 0.09	0.219 ± 0.13	0.321 ± 0.17	0.107 ± 0.06	0.459 ± 0.06	0.776 ± 0.27	

Standard errors of breed-wise accuracies ranged from 0.017 to 0.02

[a] Excluding dairy

SNP panels in each of the three populations. The average absolute linear bias was smallest for the NelNdEU panel (0.026, SD = 0.009) followed by the NdEU (0.035, SD = 0.01), and the DE panel (0.036, SD = 0.009).

When regressing 735k SNP predictions on either the 200 or 400 SNP predictions (see Additional file 5: Figure S5, Additional file 6: Figure S6) for the best panel (NelNd:EU), the slope was greater than 1.0, with the highest bias obtained for low dairy breed proportions. ADMIXTURE forces estimates of breed proportions to be between 0 and 1, which might lead to an inherent bias at either end of the range of breed proportion estimates. To assess whether the linear bias stemmed intrinsically from this constraint on the ADMIXTURE estimates and whether a correction factor could be introduced, for the 200 and 400 NelNdEU SNP panels, we truncated the range of dairy proportion estimates from the 735k-SNP panel to between 0.1 and 0.9 or between 0.2 and 0.8. Absolute biases were not much affected by truncation of the data; they increased slightly at high dairy breed proportions and decreased slightly at low proportions (see Additional file 5: Figure S5, Additional file 6: Figure S6).

The estimates of dairy proportions were slightly more accurate and less biased for the Kenyan crossbred animals (r^2 = 0.972, SE = 0.005; average absolute bias = 0.028 SD = 0.022) than for the Ugandan crossbred animals (r^2 = 0.963, SE = 0.008; average absolute bias = 0.04, SD = 0.03). Kenyan crossbred animals are the result of crosses between European dairy breeds and Zebu whereas Ugandan crossbreds are crosses between European dairy breeds and Ankole and Nganda, which have much higher proportions of African *Bos taurus* ancestry than Zebu. This suggests that the bias observed for the crossbreds in these two countries is predominantly due to a tendency to over-predict the African *Bos taurus* proportion and under-predict the European *Bos taurus* proportion.

Validation of SNP panels in the Ethiopian and Tanzanian crossbred animals resulted in a similar ranking with the NelNdEU panel performing best for all panel sizes, and resulting in r^2 of 0.966, 0.980, and 0.993 in Ethiopian crossbreds, and 0.958, 0.974, and 0.994 in Tanzanian crossbreds for 100, 200, and 1500 SNPs, respectively (Fig. 5b, c). The worse performance was observed for the random panel closely followed by the MAF panel with r^2 of 0.745 and 0.699 (SE = 0.022 and 0.026) in Ethiopian and Tanzanian crossbreds for 100 SNPs, respectively, compared to an r^2 of 0.783 (SE = 0.012) in the Kenyan/Ugandan population. Average absolute bias was smallest for the NelNdEU panel in both datasets (0.024, SD = 0.003 for the Ethiopian dataset and 0.026, SD = 0.002 for the Tanzanian dataset), followed by the NdEU panel (0.033, SD = 0.009 for the Ethiopian dataset and 0.041, SD = 0.009 for the Tanzanian dataset). In both countries, panels under-predicted dairy proportions except for some NelEU panel sizes in Tanzania (Fig. 5e, f).

The full ISAG panel of 200 SNPs predicted dairy proportions with r^2 of 0.831 (SE = 0.009), 0.830 (SE = 0.018), and 0.768 (SE = 0.022) in the Kenyan/Ugandan, Ethiopian, and Tanzanian datasets. Average absolute bias of the ISAG panel was among the highest values for the 200-SNP panels, i.e. 0.069, 0.067, and 0.076 in the Kenyan/Ugandan, Ethiopian, and Tanzanian datasets, respectively. This poor performance is not unexpected since the ISAG panel was selected for parentage assignment, predominantly in *Bos taurus* populations. SNPs on the 50k-SNP v2 Illumina chip predicted dairy proportions with r^2 of 0.9987 (SE = 0.0008), 0.9989 (SE = 0.001), and 0.9985 (SE = 0.002) in the Kenyan/Ugandan, Ethiopian, and Tanzanian datasets, respectively. Absolute bias of predicted dairy proportions was equal to 0.006, 0.005, and 0.009 for the Kenyan/Ugandan, Ethiopian, and Tanzanian datasets, respectively.

When predicting the proportions of each of the seven ancestral breeds based on the various 200-SNP panels (Table 2), average accuracies across ancestral breeds were highest for selection of SNPs based on MAF ($r^2 = 0.442$, SE = 0.017) followed by the NdEU SNP selection ($r^2 = 0.412$, SE = 0.017). The panels based on maximizing indigenous versus dairy allele frequencies including the Nelore breed (NelNdEU, NelEU) gave very poor predictions of individual dairy breed proportions. This is due to the selection method that preferentially selects alleles at extreme frequencies between *Bos taurus* and *Bos indicus* breeds and hence results in low variance between dairy breeds.

With subsets of 200 SNPs, individual breed proportions were on average best predicted for the Nelore breed ($r^2 = 0.776$, SE = 0.009) followed by the N'Dama breed ($r^2 = 0.459$, SE = 0.017). Jersey breed proportions were poorly predicted with on average an r^2 of 0.107 (SE = 0.021, Table 2). The accuracy of individual breed proportion predictions is strongly influenced by two factors: (1) breeds that exhibit little variation in breed proportions in the crossbred animals (such as Jersey) will have their proportions predicted with lower r^2 since the residual errors account for a higher proportion of the total variation accuracy; and (2) breeds that are most genetically distant from the others (such as the Nelore breed) are more likely to display allele frequencies that differ from those of other breeds with most methods used to select SNP panels.

The NelEU panel performed well for the prediction of Nelore proportion, with an r^2 of 0.944 (SE = 0.005), but gave poorer predictions of total dairy proportion ($r^2 = 0.484$, SE = 0.016) because of its poor prediction of N'Dama (African *Bos taurus*) versus European *Bos taurus* proportions. Although this panel was not as good at predicting total dairy proportion in these African crossbred populations, it should perform better in populations in which the indigenous population is pure *Bos indicus*, as is the case in much of India.

Separating the crossbred animals according to their country of origin (Kenya vs. Uganda) improved the prediction of Nelore proportion in the Kenyan crossbred animals and of breed proportion for Holstein, Jersey, and N'Dama in the Ugandan crossbred animals with most panels. Ayrshire and Guernsey predictions were less accurate with most panels in both Kenyan and Ugandan crossbred animals.

Regression and SNP-BLUP

The regression method using all 735k SNPs predicted dairy proportions from a 735 k ADMIXTURE analysis with an r^2 of 0.9914 (SE = 0.002) and an absolute bias of 0.014 (SD = 0.018). However, the selected SNP panels gave poor predictions when using the regression method (Fig. 6a). Using 50k SNPs to predict breed proportions in sheep, Dodds et al. [31] reported accuracies of $r^2 = 0.941$. Frkonja et al. [34] used different prediction methods and compared the results to pedigree-based admixture estimates. All methods resulted in fairly low r^2 values (0.872–0.953) with 40,000 SNPs, with a partial least square regression approach performing best. Frkonja et al. [34] also reported no substantial loss in accuracy when the number of SNPs dropped to 4000 ($r^2 = 0.949$), but observed a significant loss in accuracy when it dropped to 400 ($r^2 = 0.912$). Our results show that the regression approach performs much worse than ADMIXTURE even with 1500 SNPs (Fig. 6a).

We tested the NelNdEU and NdEU panels to determine whether the use of true allele frequencies improved prediction of breed proportions compared to that of ADMIXTURE estimates, and found that prediction accuracy (r^2) decreased substantially (Fig. 6b).

When we applied the SNP-BLUP approach with each of the SNP panels and using all the data from each of the three populations, estimates of dairy proportions were more accurate than those obtained by ADMIXTURE

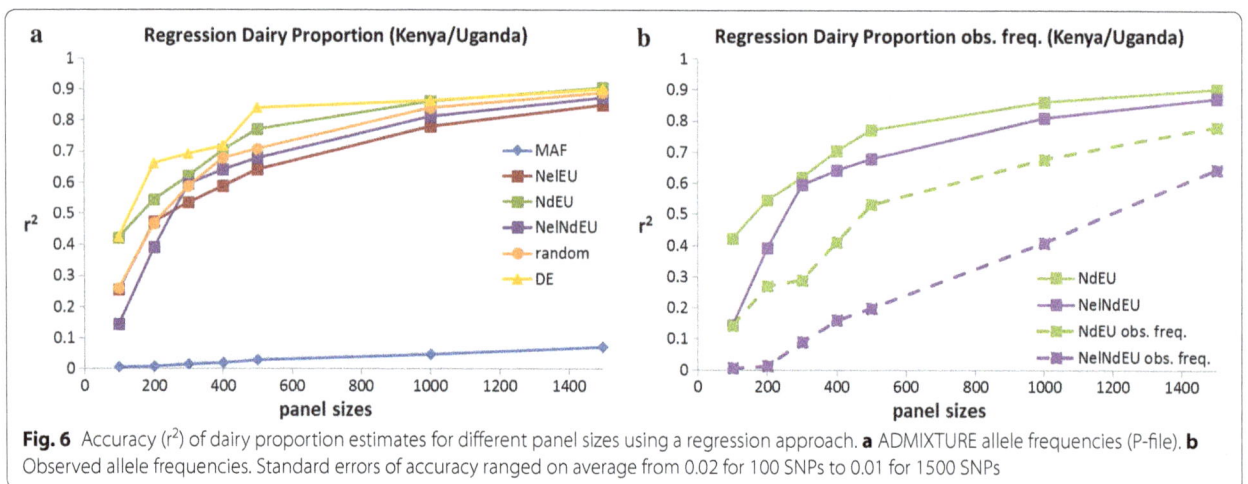

Fig. 6 Accuracy (r^2) of dairy proportion estimates for different panel sizes using a regression approach. **a** ADMIXTURE allele frequencies (P-file). **b** Observed allele frequencies. Standard errors of accuracy ranged on average from 0.02 for 100 SNPs to 0.01 for 1500 SNPs

except for the NdEU panel, and even much more than those obtained by the regression approach, as shown by the comparison of Fig. 7a–c with Fig. 5a–c. For example, in the Kenyan/Ugandan dataset (Fig. 7a), the NeINdEU panel achieved on average 0.008 higher r^2 values and the MAF panel 0.026 higher r^2 values with SNP-BLUP estimates compared to ADMIXTURE estimates. Estimates obtained by using ADMIXTURE are unbiased, because estimates for each SNP panel are obtained independently of the estimates using the 735k SNPs, against which they are tested. In contrast, the accuracies of SNP-BLUP estimates are subject to ascertainment bias that leads to overestimated accuracies. When the prediction equations obtained with SNP-BLUP for the Kenyan/Ugandan dataset were validated by applying them to the Ethiopian and Tanzanian datasets (Fig. 7d, e), accuracies of all panels were substantially lower and always higher with ADMIXTURE. A cross-validation within the Kenyan/Ugandan dataset resulted in similarly lower accuracies, which indicates that dairy proportions are generally overestimated due to ascertainment bias with the SNP-BLUP approach.

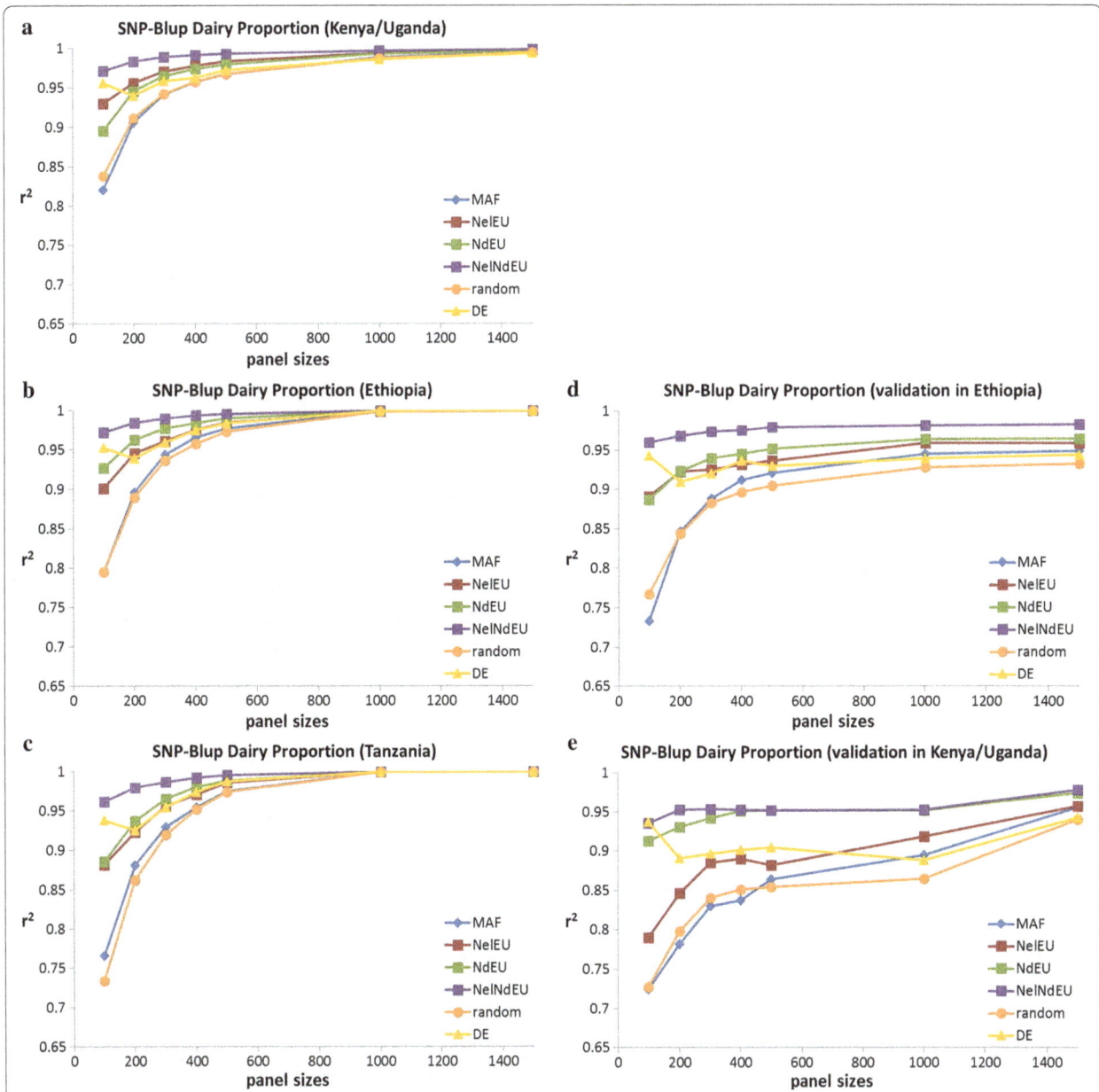

Fig. 7 Accuracy (r^2) and validation of dairy proportion estimates using a SNP-BLUP approach. **a–c** Discovery of SNP effects in three independent populations. **d, e** Validation of SNP effects estimated in the Kenyan/Ugandan dataset for two independent populations. Standard errors of accuracy ranged on average from 0.006 for 100 SNPs to 0.001 for 1500 SNPs in the Kenyan/Ugandan dataset

To assess whether this ascertainment bias results from random sampling or population structure, we used the NelNdEU panels and split the Kenyan/Ugandan dataset randomly into two equally-sized subsets. The numbers of animals from Kenya and Uganda were relatively evenly distributed between the two subsets. Accuracies of the predictions of dairy proportions obtained with SNP-BLUP on the first half of the dataset were very similar to those obtained using the full dataset. When SNP-BLUP equations were validated on the second half of the dataset, accuracies dropped and biases increased substantially. Using the effects of SNPs that were predicted with the first half of the Kenyan/Ugandan dataset to predict dairy proportions in the second half of the dataset (within-population validation) led to decreased accuracy [average reduction in $r^2 = -0.012$, SD = 0.004; (see Additional file 7: Figure S7a)] and increased bias. When these prediction equations were used on the Ethiopian and Tanzanian datasets, accuracies were lower (average $r^2 = 0.417$, SE = 0.02 for the Ethiopian dataset and average $r^2 = 0.485$, SE = 0.019 for the Tanzanian dataset) (see Additional file 7: Figure S7b). The reduction in accuracy in the within-population validation reflects ascertainment bias due to random sampling. However, reduction in accuracy was even larger in the cross-population validation, which indicates that population structure has a stronger impact on ascertainment bias. Since the SNP-BLUP approach incorporates the allele frequencies of the fitted dataset, predictions of dairy proportions will be less accurate and more biased if the validation dataset includes populations with different allele frequencies (population structure). On average, absolute differences in allele frequency were equal to 0.048 (SD = 0.039) and 0.031 (SD = 0.025) between Kenyan/Ugandan and Ethiopian, and Kenyan/Ugandan and Tanzanian datasets, respectively.

The poor performance of the estimates obtained with the panels that were optimized by using the DE algorithm indicates that either it did not properly search the parameter space to find the optimum panel, or that the number of iterations was insufficient to evolve to the optimum. Figure S8 (see Additional file 8: Figure S8) shows that accuracy continued to increase slowly after 10,000 iterations of the DE algorithm. The curve was too flat to make any prediction about what the asymptotic accuracy might be if a much larger number of iterations was run. However, given the substantial drops in accuracy seen in the validation datasets, there is no reason to believe that the DE algorithm would eventually produce more accurate estimates after validation than ADMIXTURE estimates.

Parentage assignment

The separation value (*sv*) provides a measure of the difference in *opH* in true parent–offspring relationships and other forms of relationships and in unrelated individuals. A *sv* lower than 0 indicates that the panel cannot reliably separate parent–offspring status from other relationships. SNPs with a high MAF have the highest probability of having *opH* between two unrelated individuals within a population under Hardy–Weinberg equilibrium. Thus, panels of SNPs that were selected for a high MAF in the crossbred population should perform best in assigning parentage using *opH* criteria. Our hypothesis was confirmed since SNP panels based on MAF achieved the highest *sv* in the Kenyan/Ugandan, Ethiopian, and Tanzanian datasets (Fig. 8a–c). However, none of the 100-SNP panels had a *sv* higher than 0. With 200 SNPs, only the panel of MAF-based SNPs had a positive *sv* in all three populations. As the number of SNPs in the panel increased, all methods used to select SNPs eventually achieved positive *sv*. Although the panel of MAF-based SNPs was based on allele frequencies in the Kenyan/Ugandan population, it performed well in all three crossbred populations.

The panel derived by the DE algorithm performed erratically and did not achieve positive *sv* in all three populations with less than 400 SNPs. Gondro et al. [27] reported positive *sv* for a 100-SNP panel derived by the DE algorithm in a Hanwoo cattle population. However, the Hanwoo cattle population was relatively small and all animals were part of parent–offspring pairs, which will tend to lead to higher *sv* than the much larger population of largely unrelated animals in which we tested the DE algorithm. For a larger crossbred sheep population, at least 400 SNPs were required to achieve a positive *sv* in both the discovery and validation dataset [27], which is consistent with our results.

Since *sv* is an integer variable it can be difficult for a DE algorithm to evolve to a higher *sv* once the panels of SNPs being evaluated by the algorithm all achieve the same *sv* and this may limit the ability of DE to find an optimum solution. The DE algorithm might perform better if it is initiated with prior knowledge on suitable SNP panels (e.g. panels of MAF-based SNPs, or SNP spacing restrictions) but it would still likely generate spuriously high *sv* values due to ascertainment bias.

The ISAG panel and the 50k-SNP chip yielded *sv* of 1 and 170, respectively, in the Kenyan/Ugandan dataset, 3 and 944, respectively, in the Ethiopian dataset, and 5 and 796, respectively, in the Tanzanian dataset.

The average *sv* of the randomly selected panels indicated that at least 300 SNPs are required to achieve a positive *sv*, which is in concordance with the findings of Strucken et al. [23], who reported that 340 randomly selected SNPs were needed for a positive *sv* in a composite cattle population.

We investigated whether the accuracy of the MAF-based panel was affected by breed composition. When using all

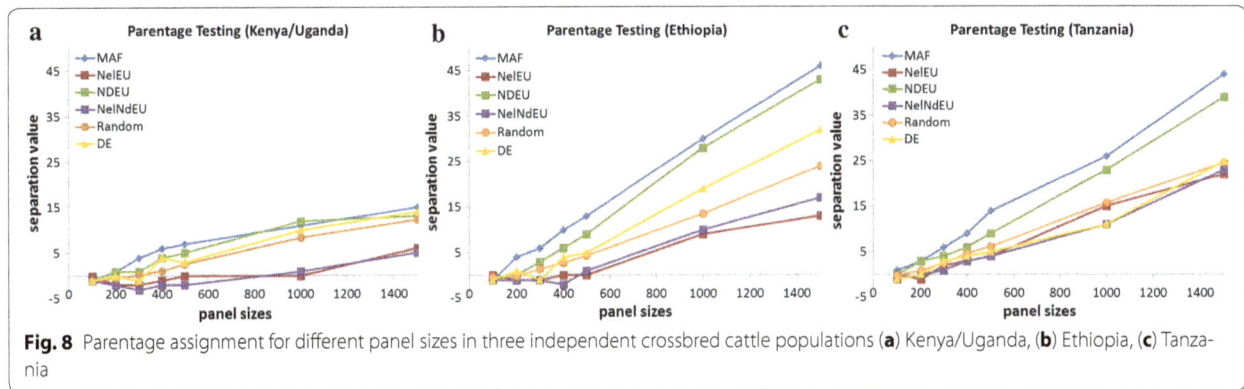

Fig. 8 Parentage assignment for different panel sizes in three independent crossbred cattle populations (**a**) Kenya/Uganda, (**b**) Ethiopia, (**c**) Tanzania

735k SNPs, dairy proportions of the animals in parent–offspring pairs ranged from 11 to 99%. Average opH counts for parent–offspring pairs with dairy proportions higher than 0.5 and those with dairy proportions lower than 0.5 were equal to 293 (SD = 93) and 420 (SD = 84), respectively. Although statistically highly significant (P < 0.0001), this difference is small since only a proportion of SNPs were tested with the SNP panels. MAF for animals with dairy proportions lower than 0.5 versus higher than 0.5 were virtually identical (0.2738 vs. 0.2734), and when using the MAF-based panel with either 200 or 400 SNPs, there was no significant correlation between opH and dairy

proportion. Therefore, parentage assignment of the MAF panel is not expected to be affected by dairy proportion.

Applications in the field

The NelNdEU panel was superior for the prediction of dairy proportions in all three populations: Kenyan/Ugandan, Ethiopian and Tanzanian. This panel was chosen based on allele frequencies in reference samples of Nelore, N'Dama, and *Bos taurus* dairy breeds, and no information from crossbred animals was used to select the SNPs. Thus, there is no ascertainment bias in the estimated accuracies with the NelNdEU panels using ADMIXTURE

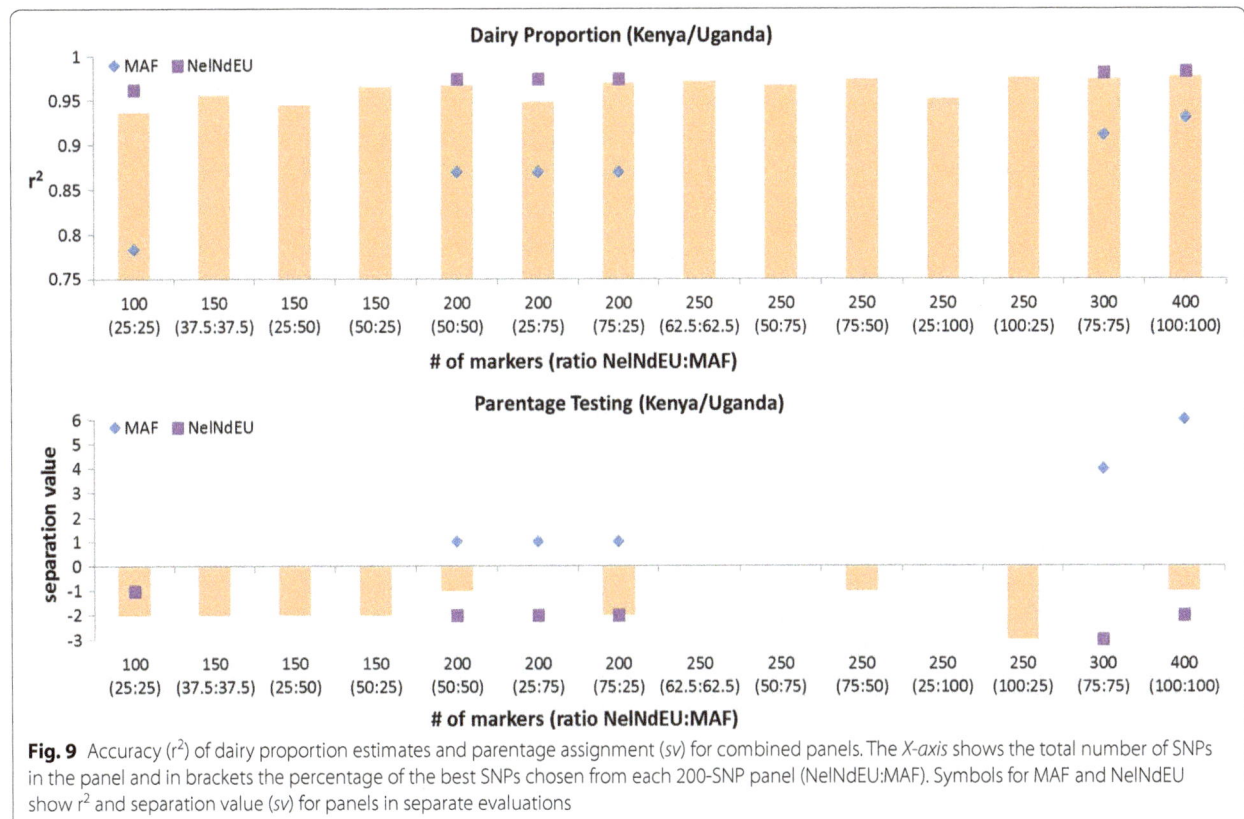

Fig. 9 Accuracy (r^2) of dairy proportion estimates and parentage assignment (sv) for combined panels. The *X*-axis shows the total number of SNPs in the panel and in brackets the percentage of the best SNPs chosen from each 200-SNP panel (NelNdEU:MAF). Symbols for MAF and NelNdEU show r^2 and separation value (sv) for panels in separate evaluations

in our datasets. The 200-SNP panel provides a good compromise between a small number of SNPs while achieving high accuracy and the lowest absolute bias ($r^2 = 0.974$ SE = 0.004; absolute bias = 0.031 using ADMIXTURE). In practice, not all of the 200 SNPs are available due to genotyping errors or to failure of some SNPs to work on a given assay platform. Therefore, we randomly selected 75 to 95% of the SNPs by simulating a genotyping failure of 5 to 25%. Accuracies of predicted dairy proportions remained high with an r^2 of 0.965 and 0.975 when 75 and 95% of the genotypes were available, respectively. Thus, the NelNdEU 200-SNP panel should perform well in the field for the prediction of dairy proportions.

The MAF-based panel performed best for parentage assignment and achieved a positive sv of 1 with 200 SNPs. Similarly, we calculated the sv assuming a random genotyping failure of 5 to 25%. The sv was positive with 1 opH between true and false parentages. Thus, the MAF-based 200-SNP panel should perform well in the field for parentage assignment.

The NelNdEU panel performed poorly for parentage assignment whereas the MAF-based panel performed poorly for the prediction of breed composition. We explored the possibility of having a single, combined panel that performed well for both prediction of breed composition and parentage assignment. We tested panel sizes from 100 to 400 SNPs, and different combinations of SNPs from the NelNdEU and MAF-based panels. In each case, the best SNPs for each selection criterion (i.e. large allele frequency difference between NelNd and EU or highest MAF) were chosen from each panel. Using all 200 SNPs from each panel (i.e. 400 SNPs) resulted in an r^2 of 0.978 (SE = 0.003; absolute bias = 0.027) for the prediction of breed proportion and a sv of -1. These results are slightly less good than those achieved by the NelNDEU and MAF-based 400-SNP panels (Fig. 9).

The combined panels performed relatively well for prediction of breed proportions, especially if more than 50% of SNPs were chosen from the NelNdEU panel. The combined panels did not achieve a positive sv, even if the majority of SNPs were chosen from the MAF panel. This is due to the fact that the 200-SNP panel chosen for high MAF is not performing well enough, i.e. the positive sv is not sufficiently high, to counteract the negative sv value of the NelNdEU panel. The 300 or 400 SNPs from the NelNdEU and MAF panels resulted in positive sv values. However, combining these to create a panel of 600 or 800 SNPs means doubling the number of SNPs. Therefore, we recommend using the 200 SNPs of the NelNdEU panel for prediction of breed proportions or the 200 SNPs of the MAF panel for parentage assignment.

The 1500 SNPs in the NelNdEU and MAF panels are provided in rank order of selection in Table S1 [see Additional file 9: Table S1], from which all the panels with smaller numbers of SNPs described in this paper can be reconstructed.

Conclusions

For East African crossbred dairy cattle populations, it is possible to create SNP panels with as few as 200 SNPs that will result in accurate estimates of dairy proportions and panels of similar size but with different SNPs to assign parentage accurately. A single combined panel of 400 SNPs achieved sufficient accuracies for breed proportion prediction but was not able to assign parentages correctly. Results of the 200-SNP panels chosen independently for breed proportion prediction and parentage assignment indicate that these panels should be reliable for animals that are crossbred to a wide range of African indigenous breeds. However, they are not expected to perform well outside of Africa where indigenous breeds do not originate from ancient crosses between African *Bos taurus* and *Bos indicus* populations. Alternative panels based on SNPs that differentiate *Bos indicus* from European *Bos taurus* should perform well in countries where the indigenous base population is *Bos indicus*, such as in south Asia, but this needs to be tested.

Additional files

Additional file 1: Figure S1. 3D-plots of principal components for reference, indigenous, and crossbred populations from Kenya/Uganda, Ethiopia, and Tanzania.

Additional file 2: Figure S2. Allele frequencies for SNP panels selected for the largest allele frequency differences of ancestral breeds, i.e. between Nelore or N'Dama and a weighted EU average in Nelore, N'Dama, or European populations.

Additional file 3: Figure S3. Allele frequencies for SNP panels in a crossbred cattle population (Ethiopia). Bold horizontal lines indicate the median and + indicates the mean.

Additional file 4: Figure S4. Allele frequencies for SNP panels in a crossbred cattle population (Tanzania). Bold horizontal lines indicate the median and + indicates the mean.

Additional file 5: Figure S5. Regression of estimates of dairy proportions (735k) on predictions of the NelNdEU 200-SNP panel.

Additional file 6: Figure S6. Regression of estimates of dairy proportions (735k) on predictions of the NelNdEU 400-SNP panel.

Additional file 7: Figure S7. Validation of dairy proportion estimates from one half of the Kenyan/Ugandan crossbreds (NelNdEU panel). (a) Validated in the other half of the Kenya/Uganda dataset. (b) Validated in independent populations from Ethiopia and Tanzania plus the second half of the Kenyan/Ugandan dataset.

Additional file 8: Figure S8. Accuracy (r^2) of dairy proportion estimates and parentage assignment with increasing number of iterations.

Additional file 9: Table S1. Best 1500 SNPs for breed proportion estimation (NelNdEU) and parentage assignment (MAF). They are presented in rank order, from best to worst SNP.

Authors' contributions
EMS performed panel selection, parentage assignment, ADMIXTURE and SNP-BLUP analyses, calculated breed diversity statistics, and prepared the manuscript including display items. HAM carried out the panel selection with the DE algorithm. CER carried out the random panel selection. CG conceived the DE algorithm analysis. OAM and JPG conceived the study and JPG critically revised the manuscript. All authors read and approved the final manuscript.

Author details
[1] School of Environmental and Rural Science, University of New England, Armidale 2350, Australia. [2] Pic Improvement Company (PIC), Genetic Services, Hendersonville, TN 37075, USA. [3] Michigan State University, Animal Science, East Lansing, Michigan 48824, USA. [4] International Livestock Research Institute, Nairobi, Kenya.

Acknowledgements
We would like to thank the smallholder farmers for participating in the DGEA project and providing hair samples of their animals for DNA testing. Julie Ojango, Denis Mujibi, James Rao and Absolomon Kiara, at ILRI, undertook farmer engagement, data and sample collection and storage, which were accessed for the present research.

Competing interests
The authors declare that they have no competing interests.

Funding
This study and the DGEA1 and 2 projects were funded by the Bill and Melinda Gates Foundation. HAM and CG were supported by a grant from the Next-Generation BioGreen 21 Program (No. PJ01134906), Rural Development Administration, Republic of Korea. Illumina Inc. and Genseek supported the project through reduced costs of the BovineHD genotyping Beadchip and laboratory assays, respectively.

References

1. Bradley DG, MacHugh DE, Loftus RT, Sow RS, Hoste CH, Cunningham EP. Zebu-taurine variation in Y chromosomal DNA: a sensitive assay for genetic introgression in west African trypanotolerant cattle populations. Anim Genet. 1994;25:7–12.
2. Mason IL. Evolution of domesticated animals. Upper Saddle River: Prentice Hall Press; 1984.
3. Epstein HE. The origin of the domestic animals of Africa, vol. 1. New York: Holmes & Meier Publishers; 1971.
4. Blench RM, MacDonald KC. The origins and development of African livestock: archaeology, genetics, linguistics and ethnography. London: UCL Press; 2000.
5. Zorloni A. Genus Bos: cattle breeds of the world. Kenilworth: MSO-AGVET Merck & Co., Inc.; 1985.
6. Mason IL. A world dictionary of livestock breeds, types and varieties. 4th ed. Wallingford: CABI Publishing; 1996.
7. Frisch JE, Drinkwater R, Harrison B, Johnson S. Classification of the southern African sanga and east African shorthorned zebu. Anim Genet. 1997;28:77–83.
8. Hanotte O, Tawah CL, Bradley DG, Okomo M, Verjee Y, Ochieng J, et al. Geographic distribution and frequency of a taurine *Bos taurus* and an indicine *Bos indicus* Y specific allele amongst sub-saharan African cattle breeds. Mol Ecol. 2000;9:387–96.
9. Rege JEO, Tawan CL. The state of African cattle genetic resources II. Geographical distribution, characteristics and uses of present-day breeds and strains. Anim Genet Resour Inf Bull. 1999;26:1–26.
10. Rege JEO, Kahi A, Okomo-Adhiambo M, Mwacharo J, Hanotte O. Zebu Cattle of Kenya: uses, performance, farmer preferences and measures of genetic diversity. Nairobi: ILRI (International Livestock Research Institute); 2001.
11. Amos W, Balmford A. When does conservation genetics matter? Heredity (Edinb). 2001;87:257–65.
12. Gorbach DM, Makgahlela ML, Reecy JM, Kemp SJ, Baltenweck I, Ouma R, et al. Use of SNP genotyping to determine pedigree and breed composition of dairy cattle in Kenya. J Anim Breed Genet. 2010;127:348–51.
13. Windig JJ, Engelsma KA. Perspectives of genomics for genetic conservation of livestock. Conserv Genet. 2010;11:635–41.
14. Werner FA, Durstewitz G, Habermann FA, Thaller G, Kramer W, Kollers S, et al. Detection and characterization of SNPs useful for identity control and parentage testing in major European dairy breeds. Anim Genet. 2004;35:44–9.
15. Weerasinghe WMSP. The accuracy and bias of estimates of breed composition and inference about genetic structure using high density SNP markers in Australian sheep breeds. Ph.D. Thesis, the University of New England. 2014.
16. Heaton MP, Leymaster KA, Kalbfleisch TS, Kijas JW, Clarke SM, McEwan J, et al. SNPs for parentage testing and traceability in globally diverse breeds of sheep. PLoS One. 2014;9:e94851.
17. Gondro C, Porto-Neto LR, Lee SH. SNPQC—an R pipeline for quality control of Illumina SNP genotyping array data. Anim Genet. 2014;45:758–61.
18. Van Raden PM. Efficient methods to compute genomic predictions. J Dairy Sci. 2008;91:4414–23.
19. Weir BS, Cockerham CC. Estimating F-statistics for the analysis of population structure. Evolution. 1984;38:1358–70.
20. Alexander DH, Novembre J, Lange K. Fast model-based estimation of ancestry in unrelated individuals. Genome Res. 2009;19:1655–64.
21. Hayes BJ. Efficient parentage assignment and pedigree reconstruction with dense single nucleotide polymorphism data. J Dairy Sci. 2011;94:2114–7.
22. Purcell S, Neale B, Todd-Brown K, Thomas L, Ferreira MA, Bender D, et al. PLINK: a tool set for whole-genome association and population-based linkage analyses. Am J Hum Genet. 2007;81:559–75.
23. Strucken EM, Lee SH, Lee HK, Song KD, Gibson JP, Gondro C. How many markers are enough? Factors influencing parentage testing in different livestock populations. J Anim Breed Genet. 2016;133:13–23.
24. Cattle molecular markers and parentage testing workshop. ISAG Conference (Chair: Romy Morrin O'Donnell, Marie-Yvonne Boscher). CMMPT, Cairns; 2012. p. 1–7.
25. Storn R, Price K. Differential evolution—a simple and efficient adaptive scheme for global optimization over continuous spaces, vol. 95. Berkeley, CA: International Computer Science Institute; 1995. p. 1–12.
26. Gondro C, Kwan P. Parallel evolutionary computation in R. In: Khosrow-Pour, M (ed) Bioinformatics: concepts, methodologies, tools, and applications, vol 1. Hershey, USA: IGI Global; 2013. p. 105–29.
27. Gondro C, Strucken EM, Lee HK, Song KD, Lee SW. Selection of SNP panels for parentage testing. In: Proceedings of the 10th world congress for genetics applied to livestock production, 17–22 August 2014; Vancouver. 2014.
28. Esquivelzeta-Rabell C, Al-Mamun HA, Lee SH, Lee HK, Song KD, Gondro C. Evolving to the best SNP panel for Hanwoo breed proportion estimates. Proc Assoc Advmt Anim Breed Genet. 2015;21:473–6.
29. Strucken EM, Gudex B, Ferdosi MH, Lee HK, Song KD, Gibson JP, et al. Performance of different SNP panels for parentage testing in two East Asian cattle breeds. Anim Genet. 2014;45:572–5.
30. Kuehn LA, Keele JW, Bennett GL, McDaneld TG, Smith TP, Snelling WM, et al. Predicting breed composition using breed frequencies of 50,000 markers from the US Meat Animal Research Center 2000 Bull Project. J Anim Sci. 2011;89:1742–50.
31. Dodds KG, Auvray B, Newman SA, McEwan JC. Genomic breed prediction in New Zealand sheep. BMC Genet. 2014;15:92.
32. Lachance J, Tishkoff SA. SNP ascertainment bias in population genetic analyses: why it is important, and how to correct it. BioEssays. 2013;35:780–6.
33. Zavarez LB, Utsunomiya YT, Carmo AS, Neves HH, Carvalheiro R, Ferencakovic M, et al. Assessment of autozygosity in Nellore cows (*Bos indicus*) through high-density SNP genotypes. Front Genet. 2015;6:5.
34. Frkonja A, Gredler B, Schnyder U, Curik I, Solkner J. Prediction of breed composition in an admixed cattle population. Anim Genet. 2012;43:696–703.

8

Incorporation of causative quantitative trait nucleotides in single-step GBLUP

Breno O. Fragomeni[1*], Daniela A. L. Lourenco[1], Yutaka Masuda[1], Andres Legarra[2] and Ignacy Misztal[1]

Abstract

Background: Much effort is put into identifying causative quantitative trait nucleotides (QTN) in animal breeding, empowered by the availability of dense single nucleotide polymorphism (SNP) information. Genomic selection using traditional SNP information is easily implemented for any number of genotyped individuals using single-step genomic best linear unbiased predictor (ssGBLUP) with the algorithm for proven and young (APY). Our aim was to investigate whether ssGBLUP is useful for genomic prediction when some or all QTN are known.

Methods: Simulations included 180,000 animals across 11 generations. Phenotypes were available for all animals in generations 6 to 10. Genotypes for 60,000 SNPs across 10 chromosomes were available for 29,000 individuals. The genetic variance was fully accounted for by 100 or 1000 biallelic QTN. Raw genomic relationship matrices (GRM) were computed from (a) unweighted SNPs, (b) unweighted SNPs and causative QTN, (c) SNPs and causative QTN weighted with results obtained with genome-wide association studies, (d) unweighted SNPs and causative QTN with simulated weights, (e) only unweighted causative QTN, (f–h) as in (b–d) but using only the top 10% causative QTN, and (i) using only causative QTN with simulated weight. Predictions were computed by pedigree-based BLUP (PBLUP) and ssGBLUP. Raw GRM were blended with 1 or 5% of the numerator relationship matrix, or 1% of the identity matrix. Inverses of GRM were obtained directly or with APY.

Results: Accuracy of breeding values for 5000 genotyped animals in the last generation with PBLUP was 0.32, and for ssGBLUP it increased to 0.49 with an unweighted GRM, 0.53 after adding unweighted QTN, 0.63 when QTN weights were estimated, and 0.89 when QTN weights were based on true effects known from the simulation. When the GRM was constructed from causative QTN only, accuracy was 0.95 and 0.99 with blending at 5 and 1%, respectively. Accuracies simulating 1000 QTN were generally lower, with a similar trend. Accuracies using the APY inverse were equal or higher than those with a regular inverse.

Conclusions: Single-step GBLUP can account for causative QTN via a weighted GRM. Accuracy gains are maximum when variances of causative QTN are known and blending is at 1%.

Background

Initially, genomic selection used a large set of single nucleotide polymorphisms (SNPs) for genetic evaluation without the explicit identification of quantitative trait loci (QTL) [1]. SNP estimation coupled with variable selection or weighting is a way to improve accuracy by emphasizing regions with major genes, which is generally called

Bayesian regression and we will use this term throughout the paper.

Those Bayesian methods could not be implemented directly for commercial populations, for which only a fraction of animals are genotyped. The methods were incorporated indirectly by using pseudo-observations and combining results with pedigree structure [2, 3]. Such a methodology called multistep is close to optimal only when pseudo-observations are very accurate (e.g., sires in dairy cattle or crop trials). When the structure of the genotyped dataset is more complex, problems such as double counting of contributions from pedigree

*Correspondence: fragomen@uga.edu
[1] Edgar L. Rhodes Center for Animal and Dairy Science, University of Georgia, Athens, GA, USA
Full list of author information is available at the end of the article

and phenotypes, and preselection bias [4] reduce accuracy. SNP best linear unbiased predictor (SNP BLUP) is equivalent to genomic BLUP (GBLUP) or BLUP with a genomic relationship matrix (GRM) [2]. Single-step GBLUP (ssGBLUP), which is an extension of GBLUP, can incorporate pedigree, genomic, and phenotypic information jointly by using a relationship matrix that combines pedigree and genomic relationships [5]; an equivalent ssGBLUP based on SNP effects only has also been implemented [6]. Due to its simplicity and accuracy, ssGBLUP is now a method of choice for genomic evaluation in many livestock species.

When the number of genotyped animals is small, the use of Bayesian regression was found to increase accuracy of genomic prediction for many traits [7, 8]. However, as the number of genotyped animals increases, the improvement in accuracy becomes smaller or is zero. For example, VanRaden [2] reported that the improvement from non-linear predictions for milk yield in US dairy cattle was 4% in 2008 but dropped to 1% in 2011 [9]. In other words, the influence of the prior vanishes with larger amounts of data, a well-known property of Bayesian inference. A small improvement could be an artifact due to the use of non-coding SNPs. If all causative SNPs are identified, only those markers need to be fit in the model and the accuracy could approach 100%.

When the number of genotyped animals is very large, the computing costs of ssGBLUP, especially for inverting the GRM, could be prohibitive. Such costs could be reduced if the dimensionality of the genomic information is limited and exploited to reduce computations. VanRaden [2] found that the GRM has limited dimensionality and that blending of GRM with pedigree relationships (numerator relationship matrix, NRM) was required for numerical stability of GBLUP. Dimensionality of the GRM can be understood as the number of linearly independent genotypes that are present in the GRM. This dimensionality of the genomic information can be equally assessed by the eigenvalues of the GRM, the eigenvalues of the design matrix of SNP-BLUP, and the squares of singular values from singular value decomposition of the matrix of SNP content (matrix containing genotyped animals in the rows and each SNP genotype in the columns), which are all identical. Indirectly assuming limited dimensionality, Misztal et al. [10] proposed a method for the inversion of GRM called algorithm for proven and young (APY) based on the inversion of a small matrix of "core" animals, followed by a sparse expression for the other individuals. APY has a cubic computational cost for the size of the core subset but cost is only linear for the remaining animals. If the size of the core subset is not too large, APY can successfully invert GRM for millions of animals at a small cost. When tested in Holsteins,

APY based on any core subset of more than 15,000 animals maximized the accuracy of genomic prediction [11]. APY was successfully used with several datasets that included up to 500,000 genotyped animals [12–14], which indicates that the dimensionality of the genomic information is indeed limited. Misztal [15] suggested that the dimensionality of the genomic information is proportional to effective population size (Ne). In simulations that involved populations with different Ne, accuracy was maximized when the number of animals in the core subset was equal to 4NeL, where L is genome length in Morgan [16]. However, accuracies decreased by less than 5% when the core subset size was equal to NeL. The number 4NeL (or NeL) is associated with the effective number of genomic segments, and was approximately 14,000 (3500) for Holsteins, 12,000 (3000) for Jerseys, 11,000 (2750) for Angus, and 4000 (1000) for pigs and broilers [17].

The concept of dimensionality of the genomic information, as described above, applies to generic GRM; however, it can also be applied to trait-specific or weighted GRM. If SNP selection for a specific trait results in only n SNPs being retained, the dimensionality cannot be greater than n. Subsequently, a trait-specific GRM that is created via SNP selection or GWAS is likely to have lower dimensionality than a generic GRM. Subsequently, the ratio of trait-specific to generic dimensionality could be an indicator of complexity of the trait. In particular, a low value of this ratio for a trait-specific GRM that results in the highest accuracy of GEBV would indicate that relatively few genes control this trait.

Recent advances in sequencing methodologies have renewed the interest in finding genes or QTN. If a trait is influenced by n QTN, the rank of the trait-specific genomic information (including GRM) is n, since only the QTN need to be used for the evaluation, and the accuracy of the genomic prediction reaches 100% if the dataset is large enough to estimate all QTN effects accurately. More realistically, if only a fraction of the causative QTN is identified, then both causative and non-causative SNPs must be used in the analyses. Some studies showed no improvement in accuracy of genetic evaluations when sequence data was included [18, 19], whereas other studies reported a small improvement [20–25]. Brøndum et al. [26] reported an important insight about the use of causative SNPs in genetic prediction i.e. they observed that including QTN with non-coding SNPs and using GBLUP or Bayesian regressions for the analyses did not result in any substantial increase in accuracy. However, accuracy increased when QTN were assigned more weight, in other words, higher a priori variance of their effects, to avoid these being heavily regressed towards zero like in SNP-BLUP. Thus, specific knowledge of those a priori variances is needed to correctly weight QTN.

If some causative QTN are identified, it would be useful to incorporate them in a simple analysis with increased gains in accuracy. The first goal of our study was to determine the properties of ssGBLUP when all or some QTN are identified and the second goal was to determine the dimensionality of genomic information when QTN are known and whether APY is applicable.

Methods

Heterogeneous SNP variances and weighted genomic relationship matrix

SNP-BLUP and GBLUP are equivalent models [2]. In particular, the breeding value is a linear function of SNP effects:

$$\mathbf{a} = \mathbf{Zs},$$

where $\mathbf{s}$ is a vector of SNP effects, $\mathbf{a}$ is a vector of breeding values, and $\mathbf{Z}$ is a matrix of gene content, centered on the allele frequencies that are obtained from the entire geno-typed population being evaluated. Assuming an equal distribution of SNP effects:

$$var(\mathbf{s}) = \mathbf{I}\sigma_s^2, var(\mathbf{a}) = \mathbf{G}\sigma_a^2 = \mathbf{ZZ}'\sigma_s^2,$$

where σ_s^2 is the SNP variance, $\mathbf{G}$ is a genomic relationship matrix (GRM), and σ_a^2 is the additive variance. GRM can be derived directly from the a priori SNP variance as:

$$\mathbf{G} = \mathbf{ZZ}'\frac{\sigma_s^2}{\sigma_a^2}.$$

Assuming that the additive variance and gene frequencies are known, and under certain assumptions including Hardy–Weinberg and linkage equilibrium, the SNP variance is estimated as follows:

$$\sigma_s^2 = \frac{\sigma_a^2}{\sum_i^m 2p_iq_i},$$

so that based on [2]:

$$\mathbf{G} = \mathbf{ZZ}'\frac{\sigma_s^2}{\sigma_a^2} = \frac{\mathbf{ZZ}'}{\sum_i^m 2p_iq_i},$$

where p_i is the allele frequency of the i-th SNP and $q_i = (1 - p_i)$. Allele frequencies were calculated using all genotypes in $\mathbf{G}$.

Assume a priori unequal SNP variances:

$$var(\mathbf{s}) = \begin{pmatrix} \sigma_{s,1}^2 & 0 & \dots & 0 \\ 0 & \sigma_{s,2}^2 & \dots & 0 \\ \dots & \dots & \dots & \dots \\ 0 & 0 & \dots & \sigma_{s,n}^2 \end{pmatrix},$$

where $\sigma_{s,i}^2$ is the variance of the i-th SNP effect and n is the number of SNPs. Then, it is possible to use a

SNP-BLUP with these variances [27] or, alternatively, GBLUP with a "weighted" genomic covariance matrix $Var(\mathbf{a}) = \mathbf{Z}var(\mathbf{s})\mathbf{Z}'$. Specifically, GRM can include a diagonal matrix $\mathbf{D}$ of "weights", such that:

$$Var(\mathbf{a}) = \mathbf{Z}var(\mathbf{s})\mathbf{Z}' = \frac{\mathbf{ZDZ}'}{\sum_{i=1}^m 2p_iq_i}\sigma_a^2 = \mathbf{G}\sigma_a^2,$$

where the factor $\sum_{i=1}^m 2p_iq_i$ is introduced for compatibility with the current software so that for the unweighted GRM $\mathbf{D} = \mathbf{I}$ and m is the number of SNPs. The contribution of locus i to the covariance matrix $\mathbf{G}$ must be equal to its contribution in $\mathbf{Z}var(\mathbf{s})\mathbf{Z}'$:

$$z_iz_i'd_i\frac{1}{\sum_{i=1}^m 2p_iq_i}\sigma_a^2 = z_iz_i'\sigma_{s,i}^2.$$

Thus,

$$\sigma_{s,i}^2 = d_i\frac{1}{\sum_{j=1}^m 2p_jq_j}\sigma_a^2, \quad \text{and} \quad d_i = \sigma_{s,i}^2\frac{\sum_{j=1}^m 2p_jq_j}{\sigma_a^2}.$$

In other words, d_i is proportional to $\sigma_{s,i}^2$. The genetic variance in the population is $\sigma_a^2 = \sum 2p_iq_i\sigma_{s,i}^2$, which means that all weights must average to 1. In practice, $\sigma_{s,i}^2$ are not available (or even estimated) and are often substituted by the squared effect of the SNP ($d_i \approx \hat{s}_i^2\frac{\sum_{j=1}^m 2p_jq_j}{\sigma_a^2}$). Because $\sum 2p_iq_i\hat{s}_i^2$ does not add up to the genetic variance of the population, σ_a^2, weights d_j are, after estimation, standardized to sum to 1. Thus, in practice d_i can be computed as equal to $\hat{s}_i^2$ and then scaled. Another approximation involves the squared effect of the SNP, weighted by the population heterozygosity ($d_i \approx 2p_iq_i\hat{s}_i^2\frac{\sum_{j=1}^m 2p_jq_j}{\sigma_a^2}$) [28], but this has no theoretical justification and gave poorer results in our study (not shown). Thus, here, the form $d_i \approx \hat{s}_i^2\frac{\sum_{j=1}^m 2p_jq_j}{\sigma_a^2}$ was used, by including either the estimated effect (for SNPs or QTN) or the true effect (of the QTN, in which case $\hat{s}_i^2 = s_i^2$).

Simulation

Using the software QMSim [29], we simulated a livestock population under selection for a single quantitative trait that has a heritability of 0.3. A historical population was generated by mutation and drift over 1000 generations, expanding from 1000 to 10,000 individuals, in order to create initial linkage disequilibrium (LD). For each replicate, 180,000 animals were simulated across 11 overlapping generations. Phenotypes were available for all animals in generations 6 to 10. For the first generation, 15,000 males and 15,000 females were simulated. A litter size of one individual was set resulting in 15,000 progeny in each generation, with a male to female ratio of 1:1. Sire and dam replacement rates of 20% were applied, animals were selected based on the highest estimated breeding values (EBV) estimated by BLUP at the end of each generation, and mating of selected animals was at random.

Genomic information was available only for animals in the last five generations. All animals with progenies were genotyped, i.e. 24,000 sires and dams. In addition, 5000 animals were randomly selected from the last generation to be genotyped. We simulated 10 chromosomes each 150 cM long and with evenly spaced 6000 SNPs, i.e. 60,000 SNPs in total. Each chromosome contained either 10 or 100 biallelic randomly located QTN (casual variants), i.e. 100 or 1000 QTN in total that are not included on the 60,000-SNP array. QTN effects were sampled from a gamma distribution with a shape parameter of 0.4 and scaled internally for a genetic variance of 0.3, and explained 100% of the genetic variance of the trait.

Analysis

We used two methods for genetic evaluation: PBLUP and ssGBLUP. Both included 75,000 phenotypes in generations 6 to 10 and all pedigree information. The linear model was the same for all analyses and scenarios:

$$\mathbf{y} = \mathbf{1}\mu + \mathbf{Wa} + \mathbf{e},$$

where $\mathbf{y}$ is the observation vector, μ is the mean, $\mathbf{a}$ is the vector of the animals' additive effects, $\mathbf{e}$ is the vector of residuals, and $\mathbf{W}$ is the incidence matrix. Assumptions for residual effects were the same in all methods:

$$\mathbf{e} \sim \mathrm{N}\left(\mathbf{0}, \mathbf{I}\sigma_e^2\right),$$

where σ_e^2 is the simulated residual variance, and $\mathbf{I}$ is an identity matrix with dimension equal to the number of animals.

The first method was PBLUP with $\mathbf{a} \sim \mathrm{N}\left(\mathbf{0}, \mathbf{A}\sigma_a^2\right)$, where σ_a^2 is the genetic additive variance and $\mathbf{A}$ is the numerator relationship matrix. The second method was ssGBLUP with $\mathbf{a} \sim \mathrm{N}\left(\mathbf{0}, \mathbf{H}\sigma_a^2\right)$, where $\mathbf{H}$ is defined as in Legarra et al. [30] and its inverse is the same as in BLUP is [4]:

$$\mathbf{H}^{-1} = \mathbf{A}^{-1} + \begin{pmatrix} \mathbf{0} & \mathbf{0} \\ \mathbf{0} & \mathbf{G}_b^{-1} - \mathbf{A}_{22}^{-1} \end{pmatrix},$$

where $\mathbf{A}_{22}^{-1}$ is the inverse of the numerator relationship matrix for genotyped animals, and $\mathbf{G}_b$ is a "blended" GRM as described next.

Matrix $\mathbf{G}$ was constructed using different combinations of SNPs and weights: (a) unweighted with 60,000 non-coding SNPs; (b) unweighted with non-coding SNPs and the 100 or 1000 causative QTN; (c) as in (b) but with weights in $\mathbf{D}$ calculated based on genome-wide association studies (GWAS) using iterative ssGBLUP as

in Wang et al. [31]; (d) as in (b) but unweighted for non-coding SNP ($d_i = c$, where c was a constant equal to the smaller simulated QTN variance) and with weights based on true QTN effects as $d_i = s_i^2 \frac{\sum_{j=1}^{2} p_i q_i}{\sigma_a^2}$; (e) unweighted using only 100 or 1000 causative QTN; (f–h) as (b–d) but using only 10% of the largest QTN; and (i) weighted by the true simulated variance using only 100 or 1000 causative QTN. Thus, QTN weights were proportional to s_i^2. Table 1 summarizes information about these scenarios. In an additional scenario, SNPs that are adjacent to causative variants received a weight equal to 0, while all other SNPs received the same constant for the polygenic effect, and causative SNPs received the simulated true effect as weight. The number of adjacent SNPs with weight equal to 0 started from 1 and increased until all non-coding SNPs had their weight set to 0.

Then, a scaled $\mathbf{G}_0$ was constructed as follows:

$$\mathbf{G}_0 = a\mathbf{I} + b\mathbf{G},$$

where constants a and b ensure equivalence of genomic and pedigree-based average relatedness and inbreeding [32], and $\mathbf{I}$ is an identity matrix with the same dimensions as $\mathbf{G}$. Because this $\mathbf{G}_0$ is not guaranteed to be positive definite [2], three alternative blended genomic matrices ($\mathbf{G}_b$) were constructed from $\mathbf{G}_0$ as $\mathbf{G}_b = (1 - \alpha)\mathbf{G}_0 + \alpha\mathbf{K}$, where α is a blending factor and $\mathbf{K}$ is a positive definite matrix. We considered three cases: blending with either $\alpha = 0.05$ or 0.01 of $\mathbf{A}_{22}$, or with $\alpha = 0.01$ of the identity matrix. The inverse of $\mathbf{G}_b$ was obtained either by direct inversion or by APY [15]. In the latter case, the number of core animals was either (a) the number of the largest eigenvalues explaining 98% of the variance of $\mathbf{G}_b$, or (b) twice the number of simulated QTN.

Table 1 Parameters for the analysis of scenarios

Scenario	60 k SNPs	Causative QTN	Weights GWAS	Causative variances
(a)	Yes			
(b)	Yes	Yes		
(c)	Yes	Yes	Yes	
(d)	Yes	Yes		Yes
(e)		Yes		
(f)	Yes	Top 10%		
(g)	Yes	Top 10%	Yes	
(h)	Yes	Top 10%		Yes
(i)		Yes		Yes

'60 k SNPs' defines scenarios that included the simulated SNPs

'Causative QTN' defines scenarios that included all or the top 10% simulated causative variants

'Weight GWAS' defines scenarios that used weights from the iterative GWAS approach

'Causative variance' defines scenarios that used true simulated variance for QTL

The quality of predictions was assessed for the 5000 genotyped animals in the last generation. The accuracy was measured as the Pearson correlation between the genomic EBV (GEBV) and the simulated true breeding value (TBV). All calculations were done by using the BLUPF90 program suite [33], preGSf90 [34] to calculate the genomic matrices and postGSf90 for the GWAS [34]. All analyses were replicated 10 times.

Results and discussion

We observed very little difference between the realized accuracies across the replicates (≤0.01), and standard errors were <0.005, thus only the results of one replicate are shown. Accuracies obtained with different options are in Figs. 1, 2, 3, 4 and 5. LD was measured by r^2 between adjacent SNPs with a mean (standard deviation) of 0.63 (0.06) across all chromosomes and generations.

Including only non-coding SNPs

The accuracies obtained with PBLUP and ssGBLUP using only non-coding SNPs are in Fig. 1 and, as expected, were higher for ssGBLUP than for PBLUP. Accuracies were much lower than the value of 0.8 found for dairy cattle [35] because the number of phenotypes was much smaller but accuracies were close to those found for the broiler population for which a similar number of phenotypes was available [36]. Using the APY inverse with 16,000 randomly selected core animals resulted in the same accuracies as using the regular inverse. When an

unweighted GRM was used to obtain the APY inverse, the optimum number of core animals was close to the number of the largest eigenvalues in the GRM that explained 98% of the variance [16], which in this case was close to 16,000.

Including causative QTN

Figure 2 presents the accuracies obtained when using non-coding SNPs and causative QTN together. Including causative QTN in the unweighted GRM increased accuracies by 0.04, which is similar to the 2.5% increase in reliability reported by VanRaden et al. [25]. Karaman et al. [37] found that, as in Bayesian regressions, GBLUP partially accounts for QTL regions, in particular for very large datasets because the variances of the SNP effects constitute prior information that vanishes as the amount of data increases. Using weighted GRM with weights obtained by GWAS as described by Wang et al. [31], the accuracy increased further, by 0.10 for the data with 100 QTN and by 0.05 with 1000 QTN. This increase was higher with 100 QTN because these have larger effects, and because there are fewer effects to be estimated by the model. Using GWAS for weighting SNP effects seems to have a limited success due to the structure of LD [17, 38]. GWAS as used in this study is relatively simple; in BayesR or BayesRC, several sets of prior variances are available, with the largest set being potentially useful for identifying causative QTN [19, 22]. When creating the GRM by using true effects for causative QTN with small

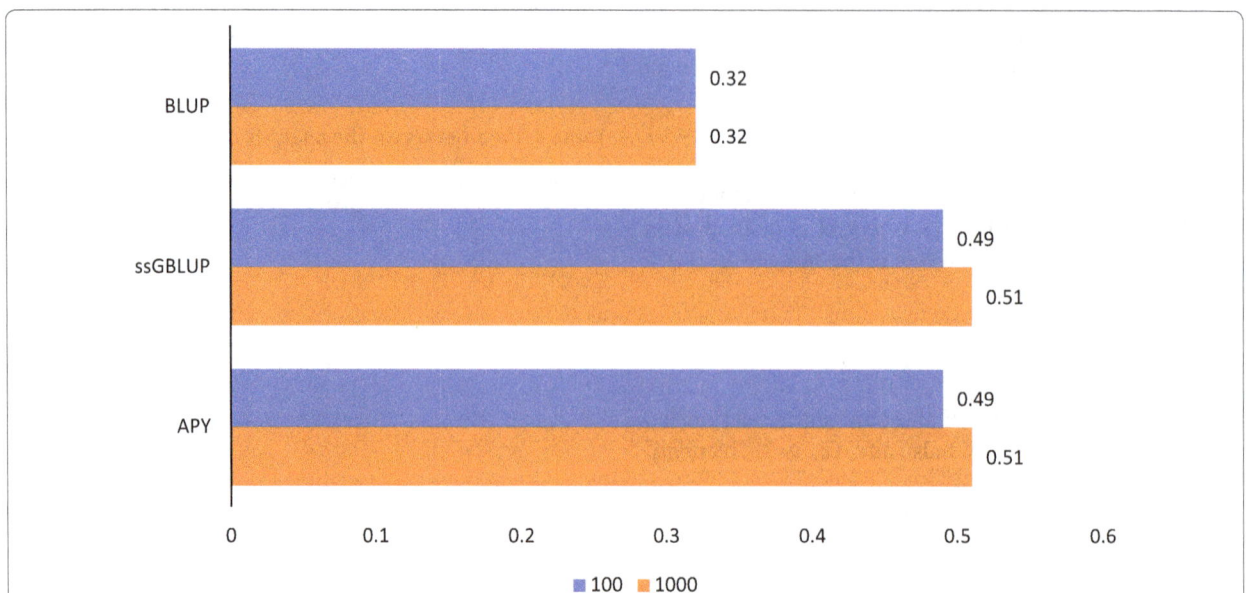

Fig. 1 Accuracies of predictions with BLUP and ssGBLUP. Predictions with only pedigree information (BLUP) or genomic information using unweighted GRM derived from 60 k SNPs and a regular inverse (ssGBLUP), and as ssGBLUP but with the GRM inverse derived using APY. The number of causative QTN is 100 or 1000

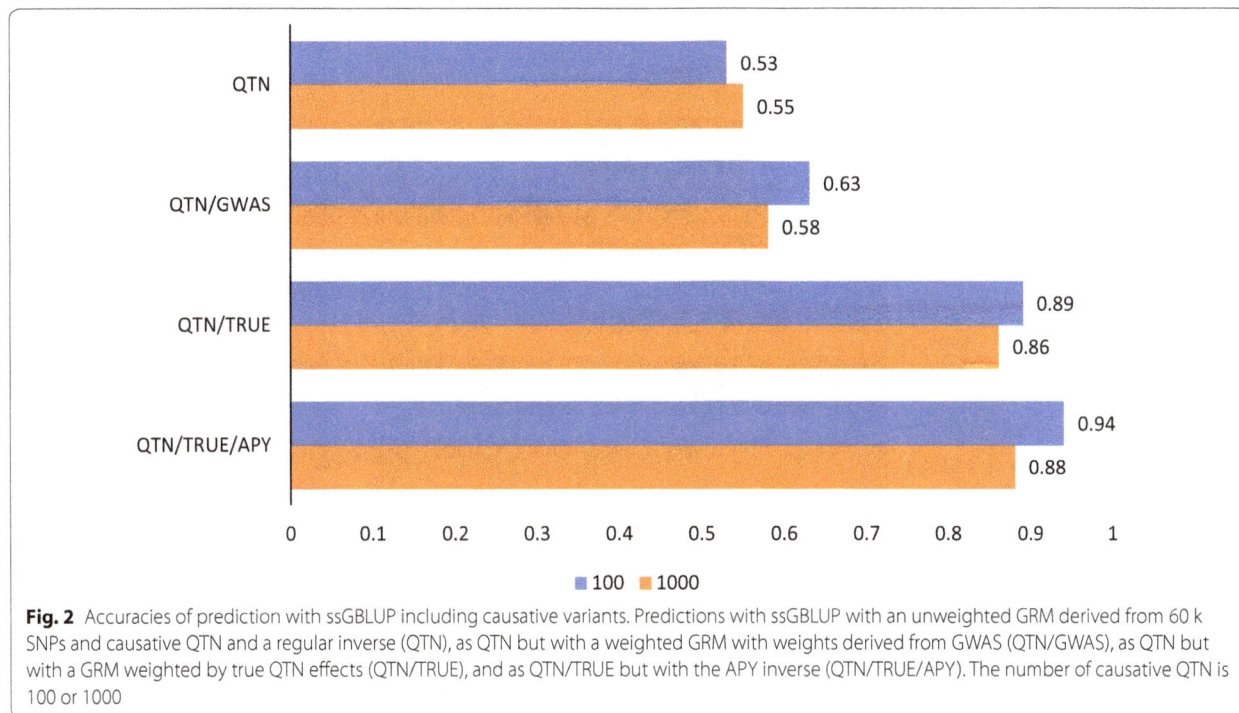

Fig. 2 Accuracies of prediction with ssGBLUP including causative variants. Predictions with ssGBLUP with an unweighted GRM derived from 60 k SNPs and causative QTN and a regular inverse (QTN), as QTN but with a weighted GRM with weights derived from GWAS (QTN/GWAS), as QTN but with a GRM weighted by true QTN effects (QTN/TRUE), and as QTN/TRUE but with the APY inverse (QTN/TRUE/APY). The number of causative QTN is 100 or 1000

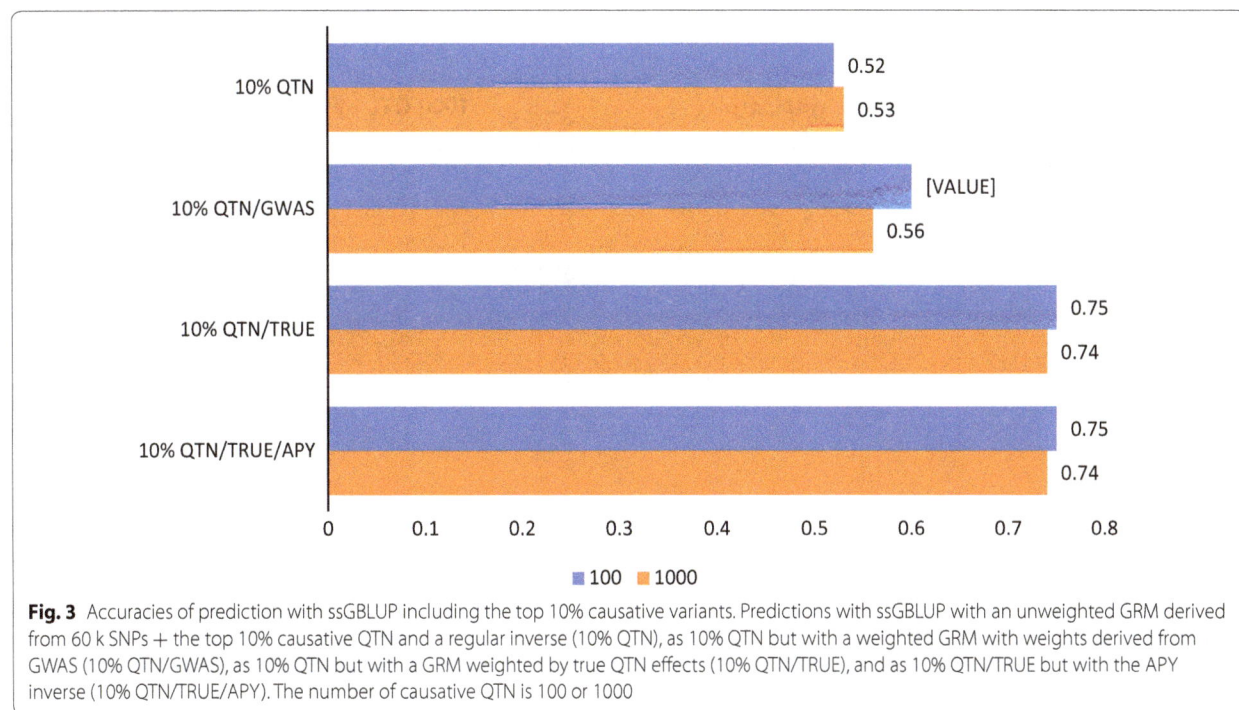

Fig. 3 Accuracies of prediction with ssGBLUP including the top 10% causative variants. Predictions with ssGBLUP with an unweighted GRM derived from 60 k SNPs + the top 10% causative QTN and a regular inverse (10% QTN), as 10% QTN but with a weighted GRM with weights derived from GWAS (10% QTN/GWAS), as 10% QTN but with a GRM weighted by true QTN effects (10% QTN/TRUE), and as 10% QTN/TRUE but with the APY inverse (10% QTN/TRUE/APY). The number of causative QTN is 100 or 1000

variances for the non-coding SNPs, accuracies increased substantially, i.e. by 0.36 with the 100 QTN data and 0.31 with the 1000 QTN data, as compared to the unweighted GRM including the causative variants. This confirms the assertion of Brøndum et al. [26] who reported that for

accuracy to increase substantially with causative QTN, it is necessary to weight them differently. When the previous analysis was repeated with the APY inverse, accuracies increased even further, to 0.94 and 0.88, respectively. As accuracies approach 1 in the analyses that fully exploit

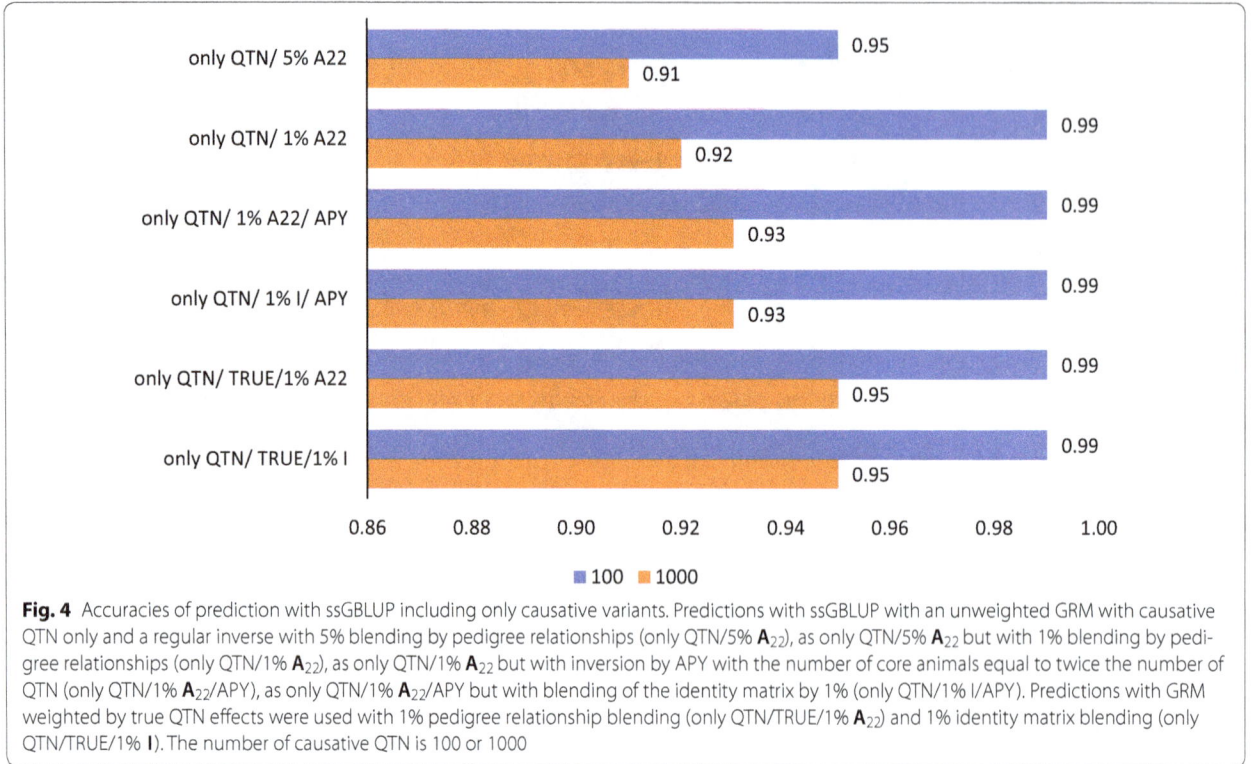

Fig. 4 Accuracies of prediction with ssGBLUP including only causative variants. Predictions with ssGBLUP with an unweighted GRM with causative QTN only and a regular inverse with 5% blending by pedigree relationships (only QTN/5% A_{22}), as only QTN/5% A_{22} but with 1% blending by pedigree relationships (only QTN/1% A_{22}), as only QTN/1% A_{22} but with inversion by APY with the number of core animals equal to twice the number of QTN (only QTN/1% A_{22}/APY), as only QTN/1% A_{22}/APY but with blending of the identity matrix by 1% (only QTN/1% I/APY). Predictions with GRM weighted by true QTN effects were used with 1% pedigree relationship blending (only QTN/TRUE/1% A_{22}) and 1% identity matrix blending (only QTN/TRUE/1% I). The number of causative QTN is 100 or 1000

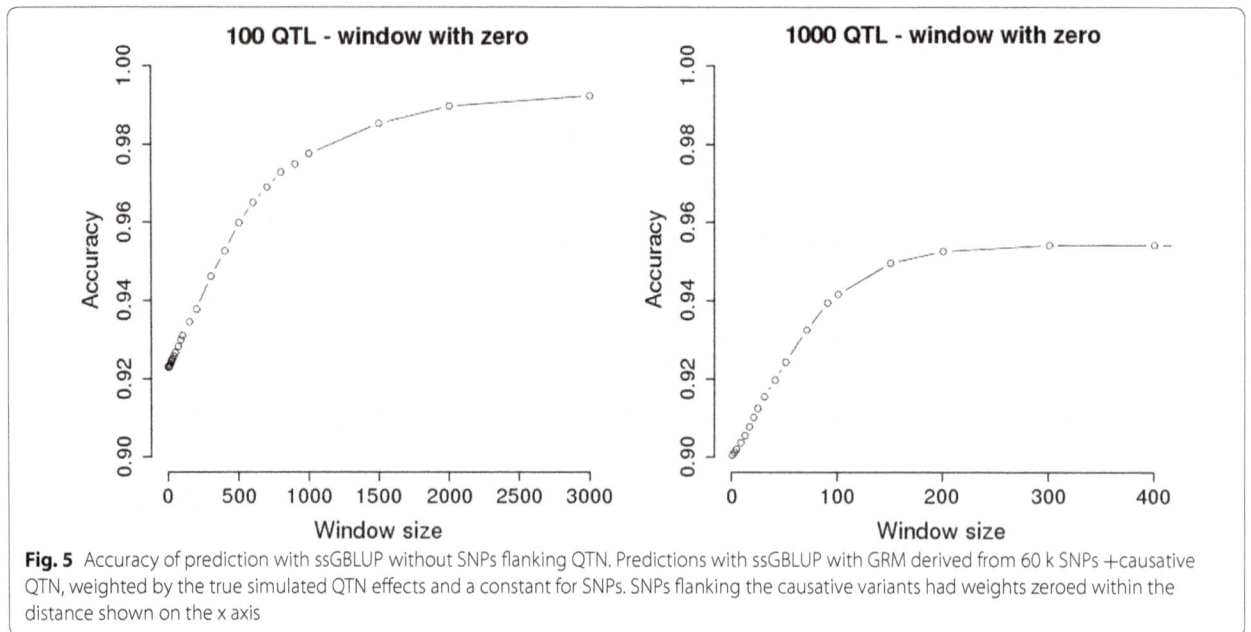

Fig. 5 Accuracy of prediction with ssGBLUP without SNPs flanking QTN. Predictions with ssGBLUP with GRM derived from 60 k SNPs +causative QTN, weighted by the true simulated QTN effects and a constant for SNPs. SNPs flanking the causative variants had weights zeroed within the distance shown on the x axis

all causative QTN, increases in accuracy with the APY inverse must be due to a decrease in noise from the non-coding SNPs. VanRaden et al. [25] obtained on average a 2.5% increase in reliability by incorporating potential causative SNPs while removing adjacent SNPs. Since one QTN generates a multi-SNP response [31, 39, 40], its incorporation in the analyses allows the removal of spurious effects of adjacent SNPs.

Analyses with the top 10% causative QTN

Identifying all causative QTN and their weights is unrealistic, and Fig. 3 presents accuracies for scenarios similar to those above but including only the top 10% causative QTN. Compared to the scenario including all causative QTN, considering only the top 10% resulted in a decreased accuracy, as expected. The reduction was small with unweighted GRM, larger with weights via GWAS, and largest with the true weights of causative SNPs. Using the APY inverse does not improve the accuracy as in scenarios that include all QTN, because the non-coding SNPs are not redundant anymore since they are proxies for the 90% missing causative QTN.

Analysis with causative QTN only

To investigate how blending of the GRM affects the accuracy with causative QTN, we conducted analyses using GRM calculated from QTN assuming equal weights and different blending factors (Fig. 4). While accuracies close to 1.00 were expected, the computed accuracies with blending factors of 5% and (1%) with the pedigree relationships (A_{22}) were equal to 0.95 and 0.91 and (0.99 and 0.92) with the 100 and 1000 QTN data, respectively. Using the APY inverse with the number of core animals equal to twice the number of QTN resulted in the same accuracy as with the 100 QTN data and increased by 0.01 with the 1000 QTN data. Accuracies obtained with a 1% blending factor with the identity matrix or A_{22} were identical.

When all causative QTN are known, blending with pedigree relationships only adds noise and is done for numerical stability. Blending at a 5% factor adds more noise than blending at 1%, and blending with the identity matrix may be slightly superior. The lower accuracy that is obtained with the 1000 QTN data can be explained by the use of an unweighted GRM. In SNP-BLUP, a large amount of data overwhelms the priors of variances when the number of SNPs is small (say 100) but less when it is larger (say 1000). Since SNP-BLUP and GBLUP are equivalent [2, 41], the same applies to GBLUP or ssGBLUP. When all causative SNPs are known, blending of GRM as used for the APY inverse is for numerical stability only. One way to eliminate blending is to estimate genomic breeding values by using a reduced model, which includes only the core animals in the equations and derives predictions for the remaining animals as linear functions of the core animals [42]. However, the optimal number of core animals is not an exact parameter, since varying the number of core animals by a factor of more than 2 (from 95 to 99% of the explained variance in GRM) changed the realized accuracy by 0.01 only [16].

Removing SNPs around causative QTN

Assigning zero as a weight for SNPs around causative variants increased the accuracy, until the weight of all non-causative SNPs was set to 0, which caused accuracies to reach the maximum of 0.99 for the 100-QTL scenario and 0.95 for the 1000-QTL scenario (Fig. 5). The shapes of the two curves were very similar, but scales differed i.e. in the 1000-QTL scenario, accuracy increased by a factor 10. This increase was observed because there were 10 times more SNPs with a zero weight in the scenario with more QTL. The shape of the curves showed that the difference in accuracy is bigger when the genomic segments with weights set to 0 are shorter. This can occur for two reasons. First, most of the non-causative SNPs had a weight set to 0 when the number of SNPs set to 0 was equal to 600 in the 100-QTL scenario or 60 in the 1000-QTL scenario; thus, random spacing of QTL could still allow a few SNPs to have a weight different from 0. Second, removing the SNPs that are located near causative variants is actually equivalent to removing SNPs that are "hitchhiking" because of LD. This is especially true for the SNPs that are located near QTL with a larger effect. Similar results were reported by VanRaden et al. [25] who found that removing SNPs around Manhattan plots peaks improved the resolution for potential causative variants in dairy cattle data. In drosophila, Ober et al. [43] showed that accuracy of phenotype prediction of phenotypes increased when non-causative SNPs were excluded from the analysis, but the pattern of accuracy fluctuated considerably, probably because of the small sample size.

Dimensionality of the genomic relationship matrix

Table 2 shows the number of eigenvalues required to explain a certain percentage of variance of GRM with various options. For unweighted and unblended GRM, the number of eigenvalues required to explain 90, 95 and 98% variance was about 8500, 12,000, and 17,000, respectively, with little difference between 100 and 1000 QTN datasets. According to Pocrnic et al. [16, 17], the optimal dimensionality of the genomic information—for prediction—corresponds to the number of eigenvalues associated with 98% of variance in GRM, and linked those values to the number of independent chromosome segment (ICS). While the GRM is not full rank, the NRM is full rank. In theory, the number of ICS depends on the effective population size and the length of genome but not on the number of QTN [44]. A blending factor of 5% with A_{22} increased the number of eigenvalues by 10 to 15%. Increasing the blending factor with A_{22} makes the blended $\mathbf{G}$ better conditioned numerically although the amount of information is not increased.

Table 2 Number of eigenvalues explaining 90, 95 or 98% of the variance for genomic relationship matrices

Option	Number of eigenvalues					
	100 QTN			1000 QTN		
	90% eigenvalue	95% eigenvalue	98% eigenvalue	90% eigenvalue	95% eigenvalue	98% eigenvalue
60 k	8496	12,185	16,978	8502	12,192	16,984
60 K-BL5	9553	13,787	19,111	9560	13,796	19,120
60 K-GWAS3	4571	7537	13,139	4757	7704	13,230
60 K-QTN-BL5	9553	13,788	19,112	9563	13,806	19,136
60 k-QTN-BL5-TRUE[d]	76	1803	5093	469	1942	5140
60 k-QTN10-BL5-TRUE[a,b,d]	4054	8972	15,886	7482	13,320	19,918
60 K-QTN-BL5-GWAS3	4082	7084	12,880	4627	7594	13,186
QTN	88	94	98	793	872	930
QTN-BL5[c]	94	122	7639	863	980	7925
QTN-BL1[c]	89	95	127	806	888	995

Options used to construct the genomic relation matrix: 60 k non-coding SNPs (60 k), all causative QTN (QTN), the top 10% causative SNPs (QTN10), blending at 5% (BL5) or 1% (BL1), weighted by the 3rd iteration of the single-step GWAS (GWAS3), and weighted by true QTN effects (TRUE) for datasets with 100 or 1000 causative QTN

[a] 10 eigenvalues explained 76% of the variance of **G** for the 100-QTN scenario

[b] 100 eigenvalues explained 71% of the variance of **G**

[c] Eigenvalues after number of QTN (100 or 1000) had values approaching 0 (below 10E−4)

[d] Simulated true weights for QTN and a constant equal to the minimum QTN value for SNPs

With GRM weighted by GWAS, the dimensionality was reduced especially at the 90% level. The reduction was bigger with fewer QTN, which indicated lower complexity of the trait as expected, but this difference was small. This could be due to limited efficiency of the method used for GWAS in this study. This method [31] estimates variances of SNP effects jointly, as opposed to sequentially in Bayesian methods, as squares of the SNP effects. Subsequently, the method is inefficient for QTL with small effects. Possible solutions include limiting the changes of variances from round to round as in NonlinearA [2], or setting the lower bound on the variance as in FastBayesA [45].

When the GRM was constructed using the QTN information only, the number of eigenvalues required to explain 90, 95 and 98% variance was close to the number of simulated QTN, especially for the scenario with 100 QTN. QTN were distributed randomly, and likely, QTN in large LD to adjacent QTN contributed little information, with more such QTN for the 1000-QTN scenario.

In a population with a different structure, QTN may be in LD with each other, and thus this number is expected to be smaller. Blending increased the dimensionality, especially at the 98% level. While this increase was at most 30% with a 1% blending factor, the increase was up to 8 (1000 QTN) and 77 times (100 QTN) with the 5% blending factor. While the extra dimensionality added noise, it made the matrix more stable to explicit inversion.

The numbers of eigenvalues obtained with the 10% top QTN are in between those obtained with no causative SNPs and with only causative SNPs. In general, the dimensionality of unweighted GRM could be equal to the number of ICS or close to 4NeL and the dimensionality of GRM constructed with causative QTN only would be equal to the number of those QTN or smaller (if some causative QTN have very little effect or are in LD). With GRM uniformly weighted for SNPs (with SNP weights accounting for a small proportion of the total genetic variance) and with true variances for all or the top 10% causative QTN, intermediate numbers of eigenvalues will be obtained.

Conclusions

Information on causative QTN can be included in single-step GBLUP via a weighted GRM. To obtain a high accuracy of prediction, the matrix has to be constructed using realistic weights for the causative QTN, by possibly eliminating non-coding SNPs that are located close to causative QTN, and with very little blending with pedigree information, i.e. the minimum required for stability. Use of the APY algorithm for inversion of GRM results in increased or similar accuracy as with the regular inverse but at much reduced cost, regardless of the inclusion of SNPs, QTN, or both. Finally, the dimensionality of the genomic information is roughly the number of independent chromosome segments for unweighted GRM, the number of causative QTN for

GRM weighted with their exact weights, and in between with a fraction of causative QTN or with GRM using weights from GWAS.

Authors' contributions
BOF, DALL and IM developed the experimental designs. BOF performed the analysis. BOF, IM and AL drafted the paper. YM developed and modified software for analysis. All authors read and approved the final manuscript.

Author details
[1] Edgar L. Rhodes Center for Animal and Dairy Science, University of Georgia, Athens, GA, USA. [2] GenPhySE, INRA, INPT, INP-ENVT, Université de Toulouse, 31326 Castanet-Tolosan, France.

Acknowledgements
The authors thank Paul M. VanRaden and Melvin E. Tooker for their very helpful comments and suggestions in the experimental design.

Competing interests
The authors declare that they have no competing interests.

Funding
BOF, DALL, YM and IM were supported by grants from Zoetis (Florham Park, NJ), Cobb-Vantress Inc. (Siloam Springs, AR), Smithfield Premium Genetics (Rose Hill, NC), American Angus Association (St. Joseph, MO), Holstein Association USA (Brattleboro, VT), Pig Improvement Company (Hendersonville, TN), and by Agriculture and Food Research Initiative Competitive Grants No. 2015-67015-22936 from the US Department of Agriculture's National Institute of Food and Agriculture (Washington, DC). AL was supported by INRA metaprogram SelGen and projects X-Gen and GenSSeq.

References
1. Meuwissen TH, Hayes BJ, Goddard ME. Prediction of total genetic value using genome-wide dense marker maps. Genetics. 2001;157:1819–29.
2. VanRaden PM. Efficient methods to compute genomic predictions. J Dairy Sci. 2008;91:4414–23.
3. Garrick DJ, Taylor JF, Fernando RL. Deregressing estimated breeding values and weighting information for genomic regression analyses. Genet Sel Evol. 2009;41:55.
4. Legarra A, Christensen OF, Aguilar I, Misztal I. Single step, a general approach for genomic selection. Livest Sci. 2014;166:54–65.
5. Aguilar I, Misztal I, Johnson DL, Legarra A, Tsuruta S, Lawlor TJ. Hot topic: a unified approach to utilize phenotypic, full pedigree, and genomic information for genetic evaluation of Holstein final score. J Dairy Sci. 2010;93:743–52.
6. Fernando RL, Dekkers JCM, Garrick DJ. A class of Bayesian methods to combine large numbers of genotyped and non-genotyped animals for whole-genome analyses. Genet Sel Evol. 2014;46:50.
7. VanRaden PM, Van Tassell CP, Wiggans GR, Sonstegard TS, Schnabel RD, Taylor JF, et al. Invited review: reliability of genomic predictions for North American Holstein bulls. J Dairy Sci. 2009;92:16–24.
8. Meuwissen T, Hayes B, Goddard M. Genomic selection: a paradigm shift in animal breeding. Anim Front. 2016;6:6–14.
9. VanRaden PM, O'Connell JR, Wiggans GR, Weigel KA. Genomic evaluations with many more genotypes. Genet Sel Evol. 2011;43:10.
10. Misztal I, Legarra A, Aguilar I. Using recursion to compute the inverse of the genomic relationship matrix. J Dairy Sci. 2014;97:3643–52.
11. Fragomeni BO, Lourenco DAL, Tsuruta S, Masuda Y, Aguilar I, Legarra A, et al. Hot topic: use of genomic recursions in single-step genomic best linear unbiased predictor (BLUP) with a large number of genotypes. J Dairy Sci. 2015;98:4090–4.
12. Masuda Y, Misztal I, Tsuruta S, Legarra A, Aguilar I, Lourenco DAL, et al. Implementation of genomic recursions in single-step genomic best linear unbiased predictor for US Holsteins with a large number of genotyped animals. J Dairy Sci. 2016;99:1968–74.
13. Lourenco DAL, Tsuruta S, Fragomeni BO, Masuda Y, Aguilar I, Legarra A, et al. Genetic evaluation using single-step genomic BLUP in American Angus. J Anim Sci. 2015;93:2653–62.
14. Ostersen T, Christensen OF, Madsen P, Henryon M. Sparse single-step method for genomic evaluation in pigs. Genet Sel Evol. 2016;48:48.
15. Misztal I. Inexpensive computation of the inverse of the genomic relationship matrix in populations with small effective population size. Genetics. 2016;202:401–9.
16. Pocrnic I, Lourenco DA, Masuda Y, Legarra A, Misztal I. The dimensionality of genomic information and its effect on genomic prediction. Genetics. 2016;203:573–81.
17. Pocrnic I, Lourenco DA, Masuda Y, Misztal I. Dimensionality of genomic information and performance of the algorithm for proven and young for different livestock species. Genet Sel Evol. 2016;48:82.
18. Veerkamp RF, Bouwman AC, Schrooten C, Calus MP. Genomic prediction using preselected DNA variants from a GWAS with whole-genome sequence data in Holstein-Friesian cattle. Genet Sel Evol. 2016;48:95.
19. Erbe M, Frischknecht M, Pausch H, Emmerling R, Meuwissen TH, Gredler B, et al. Genomic prediction using imputed sequence data in dairy and dual purpose breeds. J Anim Sci. 2016;94:198–9.
20. MacLeod IM, Hayes BJ, Goddard ME. The effects of demography and long term selection on the accuracy of genomic prediction with sequence data. Genetics. 2014;198:1671–84.
21. Pérez-Enciso M, Rincón JC, Legarra A. Sequence-vs. chip-assisted genomic selection. Genet Sel Evol. 2015;47:43.
22. MacLeod IM, Bowman PJ, Vander Jagt CJ, Haile-Mariam M, Kemper KE, Chamberlain AJ, et al. Exploiting biological priors and sequence variants enhances QTL discovery and genomic prediction of complex traits. BMC Genomics. 2016;17:144.
23. Pérez-Enciso M, Forneris N, de los Campos G, Legarra A. Evaluating sequence-based genomic prediction with an efficient new simulator. Genetics. 2016;205:939–53.
24. Wiggans GR, Cooper TA, VanRaden PM, Van Tassell CP, Bickhart DM, Sonstegard TS. Increasing the number of single nucleotide polymorphisms used in genomic evaluation of dairy cattle. J Dairy Sci. 2016;99:4504–11.
25. VanRaden PM, Tooker ME, O'Connell JR, Cole JB, Bickhart DM. Selecting sequence variants to improve genomic predictions for dairy cattle. Genet Sel Evol. 2017;49:32.
26. Brøndum RF, Su G, Janss L, Sahana G, Guldbrandtsen B, Boichard D, Lund MS. Quantitative trait loci markers derived from whole genome sequence data increases the reliability of genomic prediction. J Dairy Sci. 2015;98:4107–16.
27. Legarra A, Robert-Granié C, Croiseau P, Guillaume F, Fritz S. Improved Lasso for genomic selection. Genet Res (Camb). 2011;93:77–87.
28. Zhang Z, Liu J, Ding X, Bijma P, de Koning DJ, Zhang Q. Best linear unbiased prediction of genomic breeding values using a trait-specific marker-derived relationship matrix. PLoS One. 2010;5:e12648.
29. Sargolzaei M, Schenkel FS. QMSim: a large-scale genome simulator for livestock. Bioinformatics. 2009;25:680–1.
30. Legarra A, Aguilar I, Misztal I. A relationship matrix including full pedigree and genomic information. J Dairy Sci. 2009;92:4656–63.
31. Wang H, Misztal I, Aguilar I, Legarra A, Muir WM. Genome-wide association mapping including phenotypes from relatives without genotypes. Genet Res (Camb). 2012;94:73–83.
32. Vitezica ZG, Aguilar I, Misztal I, Legarra A. Bias in genomic predictions for populations under selection. Genet Res (Camb). 2011;93:357–66.

33. Misztal I, Tsuruta S, Lourenco D, Aguilar I, Legarra A, Vitezica Z. Manual for BLUPF90 family of programs. Athens: University of Georgia; 2014.

34. Aguilar I, Misztal I, Tsuruta S, Legarra A, Wang H. PREGSF90–POSTGSF90: computational tools for the implementation of single-step genomic selection and genome-wide association with ungenotyped individuals in BLUPF90 programs. In: Proceedings of the 10th world congress on genetics applied to livestock production, 18–22 Aug 2014. Vancouver; 2014.

35. Wiggans GR, VanRaden PM, Cooper TA. The genomic evaluation system in the United States: past, present, future. J Dairy Sci. 2011;94:3202–11.

36. Lourenco DA, Fragomeni BO, Tsuruta S, Aguilar I, Zumbach B, Hawken RJ, et al. Accuracy of estimated breeding values with genomic information on males, females, or both: an example on broiler chicken. Genet Sel Evol. 2015;47:56.

37. Karaman E, Cheng H, Firat MZ, Garrick DJ, Fernando RL. An upper bound for accuracy of prediction using GBLUP. PLoS One. 2016;11:e0161054.

38. Erbe M, Hayes BJ, Matukumalli LK, Goswami S, Bowman PJ, Reich CM, et al. Improving accuracy of genomic predictions within and between dairy cattle breeds with imputed high-density single nucleotide polymorphism panels. J Dairy Sci. 2012;95:4114–29.

39. Su G, Christensen OF, Janss L, Lund MS. Comparison of genomic predictions using genomic relationship matrices built with different weighting factors to account for locus-specific variances. J Dairy Sci. 2014;97:6547–59.

40. Hassani S, Saatchi M, Fernando RL, Garrick DJ. Accuracy of prediction of simulated polygenic phenotypes and their underlying quantitative trait loci genotypes using real or imputed whole-genome markers in cattle. Genet Sel Evol. 2015;47:99.

41. Strandén I, Christensen OF. Allele coding in genomic evaluation. Genet Sel Evol. 2011;43:25.

42. Fernando RL, Cheng H, Garrick DJ. An efficient exact method to obtain GBLUP and single-step GBLUP when the genomic relationship matrix is singular. Genet Sel Evol. 2016;48:80.

43. Ober U, Huang W, Magwire M, Schlather M, Simianer H, Mackay TF. Accounting for genetic architecture improves sequence based genomic prediction for a drosophila fitness trait. PLoS One. 2015;10:e0126880.

44. Stam P. The distribution of the fraction of the genome identical by descent in finite random mating populations. Genet Res (Camb). 1980;35:131–55.

45. Sun X, Qu L, Garrick DJ, Dekkers JC, Fernando RL. A fast EM algorithm for BayesA-like prediction of genomic breeding values. PLoS One. 2012;7:e49157.

9

Within-breed and multi-breed GWAS on imputed whole-genome sequence variants reveal candidate mutations affecting milk protein composition in dairy cattle

Marie-Pierre Sanchez[1*], Armelle Govignon-Gion[1,2], Pascal Croiseau[1], Sébastien Fritz[1,3], Chris Hozé[1,3], Guy Miranda[1], Patrice Martin[1], Anne Barbat-Leterrier[1], Rabia Letaïef[1], Dominique Rocha[1], Mickaël Brochard[2], Mekki Boussaha[1] and Didier Boichard[1]

Abstract

Background: Genome-wide association studies (GWAS) were performed at the sequence level to identify candidate mutations that affect the expression of six major milk proteins in Montbéliarde (MON), Normande (NOR), and Holstein (HOL) dairy cattle. Whey protein (α-lactalbumin and β-lactoglobulin) and casein (αs1, αs2, β, and κ) contents were estimated by mid-infrared (MIR) spectrometry, with medium to high accuracy ($0.59 \leq R^2 \leq 0.92$), for 848,068 test-day milk samples from 156,660 cows in the first three lactations. Milk composition was evaluated as average test-day measurements adjusted for environmental effects. Next, we genotyped a subset of 8080 cows (2967 MON, 2737 NOR, and 2306 HOL) with the BovineSNP50 Beadchip. For each breed, genotypes were first imputed to high-density (HD) using HD single nucleotide polymorphisms (SNPs) genotypes of 522 MON, 546 NOR, and 776 HOL bulls. The resulting HD SNP genotypes were subsequently imputed to the sequence level using 27 million high-quality sequence variants selected from Run4 of the 1000 Bull Genomes consortium (1147 bulls). Within-breed, multi-breed, and conditional GWAS were performed.

Results: Thirty-four distinct genomic regions were identified. Three regions on chromosomes 6, 11, and 20 had very significant effects on milk composition and were shared across the three breeds. Other significant effects, which partially overlapped across breeds, were found on almost all the autosomes. Multi-breed analyses provided a larger number of significant genomic regions with smaller confidence intervals than within-breed analyses. Combinations of within-breed, multi-breed, and conditional analyses led to the identification of putative causative variants in several candidate genes that presented significant protein–protein interactions enrichment, including those with previously described effects on milk composition (*SLC37A1*, *MGST1*, *ABCG2*, *CSN1S1*, *CSN2*, *CSN1S2*, *CSN3*, *PAEP*, *DGAT1*, *AGPAT6*) and those with effects reported for the first time here (*ALPL*, *ANKH*, *PICALM*).

Conclusions: GWAS applied to fine-scale phenotypes, multiple breeds, and whole-genome sequences seems to be effective to identify candidate gene variants. However, although we identified functional links between some candidate genes and milk phenotypes, the causality between candidate variants and milk protein composition remains to be demonstrated. Nevertheless, the identification of potential causative mutations that underlie milk protein composition may have immediate applications for improvements in cheese-making.

*Correspondence: marie-pierre.sanchez@inra.fr
[1] GABI, INRA, AgroParisTech, Université Paris Saclay, 78350 Jouy-en-Josas, France
Full list of author information is available at the end of the article

Background

In cattle, milk protein composition is mostly influenced by genetic factors [1–4] and is of interest because it determines cheese-making properties [5]. Bovine milk protein composition can be predicted at a large scale by analyzing mid-infrared (MIR) spectra, which is routinely performed [6, 7]. Combined with cow genotyping, this technique may open avenues to investigate the genomic regions that influence milk protein composition. In a previous genome-wide association study (GWAS) based on the bovine 50 K single nucleotide polymorphism (SNP) array, we highlighted numerous genomic regions with very significant effects on milk protein composition in the three main breeds of French dairy cattle: Holstein (HOL), Montbéliarde (MON), and Normande (NOR) [8]. However, because the 50 K SNP array contains only a small fraction of the total number of genomic variants, we were not able to directly pinpoint candidate mutations.

In Run4 of the 1000 bull genome reference population, a database containing more than 56 million SNPs and small insertions/deletions (InDel) was constructed by analyzing whole-genome sequences (WGS) from 1147 bulls representing 27 different breeds, including 288 HOL, 28 MON and 24 NOR bulls. These data can then be used to impute WGS from experimentally or routinely obtained 50 K SNP genotypes [9]. In this way, imputed WGS can be obtained for a large number of animals and in particular, those with phenotypes.

Since WGS contain almost all the genomic variants, they should contain the causal mutations for a given trait and, thus they provide a much higher GWAS resolution. However, due to the long-range linkage disequilibrium that exists within dairy cattle breeds, the resolution of within-breed GWAS is often limited. For causal mutations that are shared among breeds, a multi-breed model can be used to refine regions that harbour quantitative trait loci (QTL). This approach takes advantage of the historical recombination events that have occurred in each breed, resulting in linkage disequilibrium over shorter distances and better resolution [10].

Here, we report the results of a GWAS at the sequence level for six major milk protein contents, namely α-lactalbumin and β-lactoglobulin and αs1, αs2, β, and κ caseins from HOL, MON, and NOR cows. The results of within-breed, multi-breed, and conditional analyses, that fit the most significant variant in addition to other tested variants, are examined together in order to pinpoint potential candidate variants in each genomic region.

Methods
Animals, phenotypes, and genotypes

For this study, we did not perform any animal experiment, thus no ethical approval was required. Details on the animals and milk analyses are in Sanchez et al. [8]. Briefly, MIR spectra were obtained for 848,068 milk samples from 156,660 cows of the three main French dairy breeds: Montbéliarde (MON), Normande (NOR), and Holstein (HOL). These spectra were used to predict milk protein content (PC) and milk protein composition with the equations derived as described by Ferrand et al. [7]. More details about the method and the calibration population used are in Sanchez et al. [4]. The contents of the six main milk proteins (α_{s1}-CN, α_{s2}-CN, β-CN, κ-CN, α-LA, and β-LG) were predicted in g/100 g protein. Total casein content and total whey protein content were also analyzed (Σ-CN and Σ-WP, respectively). In order to adjust phenotypes for non-genetic effects, a within-breed mixed model was applied to test-day data using the GENEKIT software [11]. This single-trait repeatability model included genetic, permanent environmental, and residual random effects, as well as herd × test-day, parity × stage of lactation, year × month of calving, and spectrometer × test month fixed effects. We applied this model to data from the first two lactations that included at least three test-day records across lactations during the study period. Then, test-day data were corrected for all non-genetic effects included in the model and averaged per cow. Thus, for each trait and each cow, a single phenotype was defined and subsequently used in GWAS analyses. In total, 293,780, 58,594, and 72,973 test-day records were analyzed, which corresponded to 44,959 MON, 12,428 NOR, and 14,530 HOL cows, i.e. an average of 6.5, 4.7, and 5.0 test-day records per cow, respectively.

Among these cows, 8010 were genotyped with the Illumina BovineSNP50 BeadChip (Illumina Inc., San Diego). We applied the following quality control filters: the individual call rate had to be higher than 95%, the SNP call rate higher than 90%, the minor allele frequency (MAF) higher than 5%, and genotype frequencies had to be in Hardy–Weinberg equilibrium with $P > 10^{-4}$. The final dataset included between 37,332 and 41,028 SNPs (Table 1), depending on the within-breed or multi-breed population considered, for 7907 cows (3032 MON, 2659 NOR, and 2216 HOL) with phenotypes.

Imputation to whole-genome sequences

The 50 K SNP genotypes of the 7907 cows were imputed to whole-genome sequence (WGS) using FImpute software, which accurately and quickly analyzes large datasets [12]. A two-step approach was applied in order to improve the accuracy of results: from 50 to 777 K high-density (HD) SNPs, and then, from imputed HD SNPs to WGS [13]. All imputations were performed separately for each breed using either a breed-specific (from 50 K to HD SNPs) or a multi-breed (from HD SNPs to WGS)

Table 1 Features of the Montbéliarde (MON), Normande (NOR), Holstein (HOL), and multi-breed populations

Number of	MON	NOR	HOL	Multi-breed
Phenotyped cows	44,959	12,428	14,530	71,917
Total test-day records	293,780	58,594	72,973	425,347
Test-day records per cow	6.5	4.7	5	5.9
Genotyped cows	3032	2659	2216	7907
Polymorphic 50 K SNPs	37,332	37,690	39,158	41,028
Polymorphic HD SNPs	548,185	549,359	553,712	586,749
Polymorphic sequence variants	15,957,336	14,809,860	15,116,501	18,366,748
Sequence variants (MAF ≥ 2%)	11,755,172	11,445,432	11,592,432	13,534,013

reference panel depending on the targeted density [14]. In each MON, NOR and HOL breed, imputations to the HD SNP level were performed using a within-breed reference population that included respectively 522 MON, 546 NOR, and 776 HOL bulls that had been genotyped with the Illumina BovineHD BeadChip (Illumina Inc., San Diego, CA). Around 550,000 SNPs were retained in each breed after removing SNPs that failed in the quality control filters, as described above for the 50 K (Table 1). WGS variants were imputed from HD SNP genotypes using WGS variants of the 1147 bulls from Run4 of the 1000 Bull Genomes consortium; these bulls represent 27 cattle breeds (see Additional file 1: Table S1), with 288 HOL, 28 MON, and 24 NOR individuals [9]. The protocol used was defined in the "1000 bull genomes" consortium [9]. Whole-genomes of all individuals were used for 2×100 bp paired-end sequencing using Illumina sequencing-by-synthesis technology and sequence reads were further filtered for quality and subsequently aligned to the UMD3.1 reference sequence, as previously described [9, 15]. Small genomic variations (SNPs and InDel) were detected using SAMtools 0.0.18 [16]. Raw variants were further filtered to produce 27,754,235 autosomal variants [15]. Filtered variants were subsequently annotated with the Ensembl variant effect predictor (VEP) pipeline v81 [17] and effect of the amino acid changes was predicted using the SIFT tool [18].

Precision of imputation from HD to sequence was assessed by comparing imputed genotypes with those obtained by re-genotyping a subset of the same cows with a custom chip. This additional information was not used in the imputation process. Two datasets were available: (1) a group of 168 Holstein cows that were genotyped with the first version (V1) of the EuroG10k Illumina chip,

with 721 additional markers; and (2) a group of 2142 Montbéliarde cows that were genotyped with the fourth version (V4) of the same EuroG10k chip containing 3082 additional SNPs. Only SNPs with good technical quality (call rate > 95%, validation of the clusters by visual inspection, within-breed allelic frequency not significantly different across chip versions) were used. Imputation accuracy was measured by the squared correlation between true and imputed genotypes and by the genotypic and allelic concordance rate.

In order to remove SNPs with the lowest accuracies of imputation, only variants with a MAF higher than 0.02 were retained for further association analyses. Thus, about 11 million variants were included in each within-breed analysis and around 13 million were included in multi-breed analyses (Table 1).

Whole-genome sequence association analyses

We performed single-trait association analyses between all the polymorphic variants and the nine measured milk protein composition traits: PC, α-LA, β-LG, α_{s1}-CN, α_{s2}-CN, β-CN, κ-CN, Σ-CN, and Σ-WP (Table 2).

All association analyses were performed using the *mlma* option of the GCTA software, which applies a mixed linear model that includes the candidate variant [19]:

$$\mathbf{y} = \mathbf{1}\mu + \mathbf{x}b + \mathbf{u} + \mathbf{e}, \tag{1}$$

where $\mathbf{y}$ is the vector of pre-adjusted phenotypes, averaged per cow; μ is the overall mean; b is the additive fixed effect of the candidate variant to be tested for association; $\mathbf{x}$ is the vector of imputed genotypes coded as 0, 1, or 2 (number of copies of the second allele); $\mathbf{u} \sim N(\mathbf{0}, \mathbf{G}\sigma_u^2)$ is the vector of random polygenic effects, with $\mathbf{G}$ the genomic relationship matrix (GRM), calculated by using the HD SNP genotypes [20], and σ_u^2 the polygenic variance, estimated based on the null model ($\mathbf{y} = \mathbf{1}\mu + \mathbf{u} + \mathbf{e}$) and then fixed while testing for the association between each variant and the trait; and $\mathbf{e} \sim N(\mathbf{0}, \mathbf{I}\sigma_e^2)$ is the vector of random residual effects, with $\mathbf{I}$ the identity matrix and σ_e^2 the residual variance. Within-breed, the number of test-day records did not differ very much across cows, thus, the residual variance was assumed to be constant across cows.

For multi-breed association analyses, Model (2) was applied by adding a fixed breed effect $\mathbf{v}$ to Model (1), with $\mathbf{W}$ as the incidence matrix relating phenotypes to breed effect (three levels), and $\mathbf{x}$, $\mathbf{b}$, $\mathbf{u}$, and $\mathbf{e}$ as defined previously:

$$\mathbf{y} = \mathbf{W}\mathbf{v} + \mathbf{x}b + \mathbf{u} + \mathbf{e}. \tag{2}$$

The Bonferroni correction was applied to the thresholds in order to account for multiple testing. A very stringent

Table 2 MIR predictions for milk protein composition in Montbéliarde (MON), Normande (NOR), and Holstein (HOL) cows

Trait		Accuracy[a]		Means ± standard deviations[b]		
		R^2	RE	MON	NOR	HOL
PC	Protein content	1.00	0.73	3.4 ± 0.4	3.6 ± 0.4	3.3 ± 0.4
α-LA	α-lactalbumin	0.59	14.4	4.07 ± 0.28	4.16 ± 0.36	4.27 ± 0.42
β-LG	β-lactoglobulin	0.74	11.7	8.25 ± 1.12	7.94 ± 1.03	8.46 ± 1.17
α_{s1}-CN	α_{s1}-casein	0.88	4.7	27.8 ± 0.55	27.8 ± 0.68	27.9 ± 0.69
α_{s2}-CN	α_{s2}-casein	0.82	7.5	9.53 ± 0.30	9.89 ± 0.33	9.69 ± 0.39
β-CN	β-casein	0.92	3.7	36.6 ± 0.88	36.2 ± 1.2	36.2 ± 1.2
κ-CN	κ-casein	0.80	8.4	9.75 ± 0.60	9.87 ± 0.48	9.43 ± 0.58
Σ-CN	Sum of caseins	0.97	2.7	83.7 ± 0.94	83.7 ± 1.5	83.1 ± 1.4
Σ-WP	Sum of whey proteins	0.73	8.9	12.6 ± 1.1	11.9 ± 1.2	12.6 ± 1.3

[a] Accuracy of MIR predictions (R^2 = coefficient of determination and RE = relative error) estimated by Ferrand et al. [7] for protein composition expressed as g/100 g milk

[b] g/100 g milk for protein content (PC) and g/100 g protein for other traits

correction was used, which considered all 13 million tests as independent. Therefore, the 5% genome-wide threshold of significance corresponded to a nominal P value of 3.7×10^{-9} ($-\log_{10}(P) = 8.4$). QTL regions were identified by grouping significant results that were located within the same 2 million base-pair (Mbp) interval in a single genomic region, regardless of the breeds or traits under study. QTL regions were determined by considering positions of variants included in the upper third of the peak. For a given trait in a given breed, when two consecutive QTL regions had overlapping confidence intervals, or when the distance between the limits of the confidence intervals was less than 1 Mbp, only the confidence interval that presented the most significant results was retained.

Conditional association analyses

In the most significant QTL regions, conditional analyses were carried out using the *cojo* option of GCTA [21] in order to conclude if multiple significant variants in a genomic region were due to LD with the same causal mutation or to the presence of multiple causal mutations. Association analyses were performed by including in the model the most significant variant or the putative causal mutation as a fixed effect and by testing all variants that were not in strong LD with the conditional variant ($r^2 < 0.9$).

Annotation and protein interactions

Sequence-derived polymorphisms were extracted for candidate mutation regions from the corresponding VCF files [22]. All variants with a $-\log_{10}(P)$ higher than 8.4 and located within confidence intervals were annotated. To avoid missing important genes, confidence intervals were extended by 100 kb on each side.

In addition, functional protein–protein interactions (PPI) encoded by candidate genes were investigated, as well as gene ontology (GO) enrichment, using the STRING Genomics 10.0 database of protein–protein interaction (PPI) networks [23]. This database provides (1) known PPI from curated databases or experiments and (2) PPI predicted on the basis of gene neighborhood, gene fusions, gene co-occurrence, text mining in literature, co-expression, or protein homology. A global PPI network was constructed which retained only interactions with a high level of confidence (score > 0.4).

Results

The results of imputation accuracy at the sequence level for SNPs used in the GWAS analyses (MAF ≥ 2%) are in Table 3. Squared correlations between imputed and true genotypes in the validation set reached 76 and 84%, in Montbeliarde and Holstein breeds, respectively. This table also presents the overall results of concordance rate. Figure 1 shows the imputation precision according to MAF in the two breeds.

Among the 13 million tested variants, 71,755 had genome-wide significant effects ($-\log_{10}(P) \geq 8.4$) in at least one within-or multi-breed analysis and for at least one milk protein composition trait.

Among these, 29,722, 27,787, and 30,988 were found in within-breed MON, NOR, and HOL analyses, respectively. Some of these variants had significant effects in multiple breeds: 7343 in both MON and NOR, 8055 in NOR and HOL, 8068 in HOL and MON, and 3080 in all three breeds (Table 4; Fig. 2a).

For each trait, the number of significantly associated variants was relatively consistent between breeds. It was lower (from 193 to 2394) for α_{s2}-CN, β-CN, α_{s1}-CN, and PC; higher (from 8716 to 19,952) for β-LG, κ-WP, and

Table 3 Accuracies of imputations on whole-genome sequences in Holstein (HOL) and Montbéliarde (MON) breeds

Breed	HOL	MON
Number of cows	168	2142
EuroG10k chip version	V1	V4
Number of markers in the custom part	721	3082
Number of markers after quality control and MAF ≥ 0.02	221	1108
R^2 (%)	83.7	76.1
Genotypic concordance rate (%)	93.7	89.7
Allelic concordance rate (%)	96.5	94.0

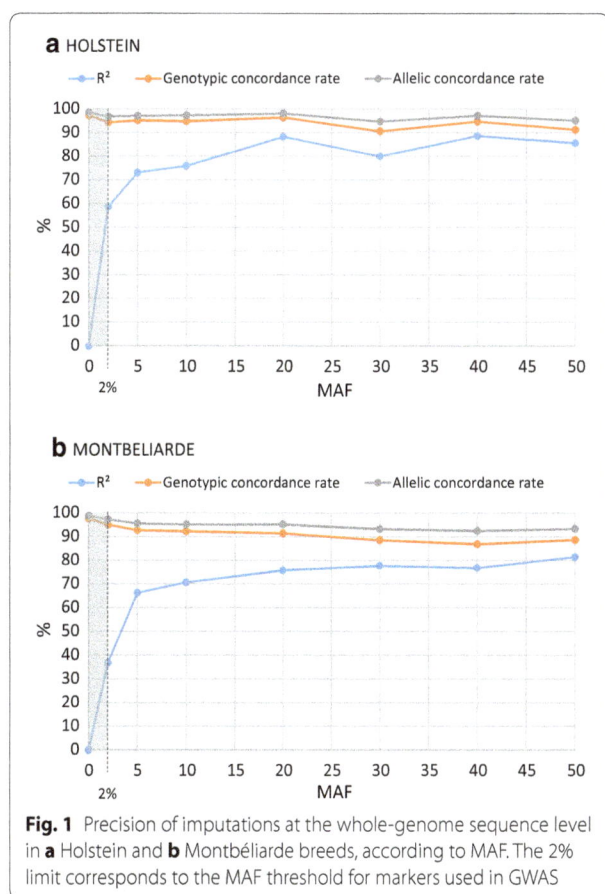

Fig. 1 Precision of imputations at the whole-genome sequence level in **a** Holstein and **b** Montbéliarde breeds, according to MAF. The 2% limit corresponds to the MAF threshold for markers used in GWAS

QTL regions were defined by merging the overlapping QTL regions obtained for the different traits and breeds and by grouping the corresponding significant results. Confidence intervals of these regions were defined as described in the Methods section. Thus, 34 QTL regions with significant effects on one or several milk protein composition traits were identified in within-breed and/or multi-breed analyses (see Additional file 2: Table S2). Three of these, located on chromosomes 6, 11, and 14, had significant pleiotropic effects on almost all protein composition traits analyzed (see Additional file 3: Table S3), while most (31 QTL) generally affected only one trait (see Additional file 4: Table S4).

The 34 QTL were distributed on 17 of the 29 bovine autosomes, with one to seven QTL per chromosome. Almost all of them (31) were detected in multi-breed analyses while 11, 8, and 11 QTL regions were found in MON, NOR, and HOL within-breed analyses, respectively. Four QTL regions, located on *Bos taurus* chromosome BTA6 (two regions at 45.8–46.9 Mbp and 85.2–87.4 Mbp), BTA11 (103.3 Mbp), and BTA20 (58.3–58.4 Mbp), were detected in three breeds. One additional region on BTA29 at about 9.6 Mbp was common to MON and HOL, and another region on BTA14 at 1.7–1.8 Mbp was common to HOL and NOR (Fig. 2b). The six QTL shared between two or three breeds had the most significant effects, along with one QTL detected only in the NOR breed on BTA2, at 131.8 Mbp ($-\log_{10}(P) \geq 20$; P value $< 10^{-13}$ after Bonferroni correction).

Multi-breed analyses led to the detection of a larger number of QTL regions than within-breed analyses: 14 of the 31 QTL detected in multi-breed analyses were not found in within-breed analyses. For the 17 QTL regions found in both within-and multi-breed analyses, the $-\log_{10}(P)$ value of the most significant (top) variant was almost always higher in multi-than in within-breed analyses; this was true even for most of the regions that had significant effects in only one within-breed analysis. For these QTL, the mean $-\log_{10}(P)$ value of the most significant (top) variant was 64 in multi-breed analyses versus 49, 46, and 42 in MON, NOR and HOL within-breed analyses, respectively. In addition, the QTL confidence intervals generated by multi-breed analyses contained a smaller number of variants than those produced by within-breed analyses. For the 17 QTL regions, an average of 134 variants (2–374) were found in multi-breed analyses versus 189 (39–335), 287 (61–872), and 308 (9–1236) in MON, NOR, and HOL within-breed analyses, respectively. However, in some QTL regions, specifically those located on BTA2 (131.8 Mbp), 6 (38 Mbp), and 19 (61 Mbp), the number of significant variants was smaller in within-breed analyses than in the multi-breed analysis.

Σ-CN; and intermediate (from 4110 to 8248) for α-LA and κ-CN. Among these variants, 0 (PC) to 2266 (β-LG) were shared among the three breeds. Multi-breed analyses were more powerful, and detected a larger number of distinct variants with significant effects (34,248) than any of the within-breed analyses. However, the number of variants detected per trait was larger in one of the within-breed analyses than in the multi-breed analysis for PC, α-LA, β-LG, Σ-CN, and Σ-WP (Table 4), probably because of the long-range within-breed LD.

Table 4 Number of variants with genome-wide significant effects ($-\log_{10}(P) > 8.4$) for milk composition traits in within- and multi-breed analyses

Trait	Within-breed analyses				Multi-breed analyses
	MON[a]	NOR[a]	HOL[a]	Shared among three breeds	
PC	1905	1201	2394	0	2350
α-LA	4590	6490	8248	213	7224
β-LG	19,952	16,048	15,517	2266	18,612
$α_{s1}$-CN	2232	708	629	182	2280
$α_{s2}$-CN	866	193	636	1	1947
β-CN	665	734	524	96	1652
κ-CN	4110	5878	6532	553	7012
Σ-CN	13,920	8716	11,833	961	12,698
Σ-WP	16,583	13,126	15,327	1916	16,546
Total number of distinct variants	29,722	27,787	30,988	3080	34,248

[a] Montbéliarde (MON), Normande (NOR), and Holstein (HOL) cows

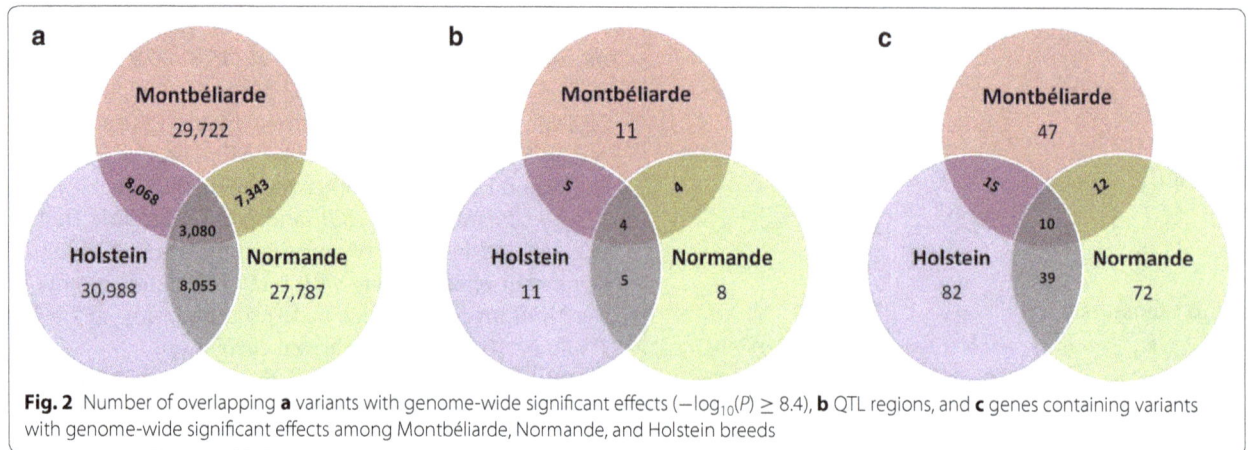

Fig. 2 Number of overlapping **a** variants with genome-wide significant effects ($-\log_{10}(P) \geq 8.4$), **b** QTL regions, and **c** genes containing variants with genome-wide significant effects among Montbéliarde, Normande, and Holstein breeds

Manhattan plots of three of the most significant QTL regions are in Fig. 3 for the three densities of markers (50 K SNP, HD SNP, or sequence). In each of these regions, several peaks are detected with the WGS data, whereas with the 50 K SNP density and in some cases with the HD SNP density, only one peak was observed.

All variants included within confidence intervals (+100 kb on each side) were functionally annotated (Table 5) and (see Additional file 5: Table S5). The percentage of variants that were located within genes ranged from 60.5% in HOL to 73.4% in NOR within-breed analyses, and it was intermediate in multi-breed analyses (65.8%). The vast majority of the genic variants were located within introns and in upstream or downstream regions. A total of 25, 82, 72, and 56 missense variants were found in MON, NOR, HOL, and multi-breed analyses, respectively; among these, we detected the previously reported missense mutations in the *PAEP* (103,303,475 bp) and *DGAT1* (1,802,266 bp) genes.

In 29 QTL regions, annotation led to the identification of candidate genes for milk protein composition. In total, 47, 72, and 82 candidate genes were identified in MON, NOR, and HOL within-breed analyses (109 in multi-breed analyses). Some of these were shared across breeds: 12 were found in both MON and NOR, 15 in MON and HOL, 39 in HOL and NOR, among which 10 were common to the three breeds (Fig. 2c). However, within a given region, the top variant was always different among the different breeds. The top variant was located in a gene in 21 of these regions, while in the remaining eight regions, the top variant was intergenic. However, these eight regions contained other variants located within confidence intervals that were annotated in genes, and of these, the most significant one was denoted the top genic variant. Genic variants with the most significant results were located within intron regions for 15 QTL and mainly upstream or downstream regulatory regions for 14 QTL. In total, 22 genes were identified as

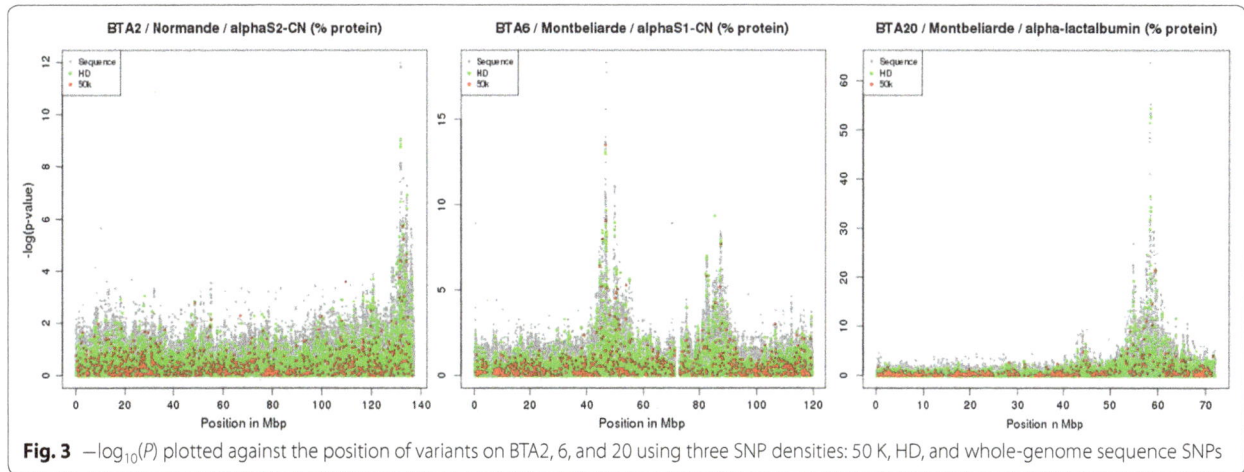

Fig. 3 $-\log_{10}(P)$ plotted against the position of variants on BTA2, 6, and 20 using three SNP densities: 50 K, HD, and whole-genome sequence SNPs

Table 5 Functional annotations of variants included within confidence intervals (±100 kb) of the 34 QTL in the three within-and multi-breed analyses

Functional annotation	Within-breed analyses			Multi-breed analyses
	MON[a]	NOR[a]	HOL[a]	
Intergenic	1514	1465	2676	1971
Intronic	1079	1804	1937	1737
3′ UTR	11	14	69	35
5′ UTR	14	27	16	18
Downstream	710	988	1276	1159
Inframe insertion	0	0	1	0
Missense	25	82	72	56
Splice acceptor	0	0	3	0
Synonymous	30	114	118	91
Upstream	509	1009	612	685
% genic	61.1	73.4	60.5	65.8
% genic non intronic	33.4	40.6	32.0	35.5

[a] Montbéliarde (MON), Normande (NOR), and Holstein (HOL) cows

the best candidates to explain the majority of the variability of milk protein composition in MON, NOR, and HOL cows. They were located on BTA1 (*SLC37A1*), BTA2 (*ALPL*), BTA5 (*MGST1*), BTA6 (*ABCG2*, *MEPE*, *PKD2*, *HERC3*, *SEPSECS*, *SEL1L3*, *DHX15*, *CSN1S1*, *CSN2*, *CSN1S2*, and *CSN3*), BTA11 (*PAEP*), BTA14 (*DGAT1*, *RECQL4*, *MROH1*, and *BOP1*), BTA20 (*ANKH*), BTA27 (*AGPAT6*), and BTA29 (*PICALM*).

Protein–protein interactions (PPI), as well as GO enrichment, were investigated for the 22 most plausible candidate genes of our study. Network proteins encoded by these genes had significantly more interactions than expected (10 edges identified; PPI enrichment P value $= 3.4 \times 10^{-9}$; Fig. 4), while GO terms for

12 biological processes, seven cellular components, and one molecular function were significantly (FDR < 0.05) enriched with two to nine of these genes for milk protein composition (Table 6).

Discussion

In this paper, we report the results of a whole-genome sequence scan for milk protein composition predicted from MIR spectra. We conducted within-and multi-breed analyses using imputed WGS of 7907 cows from three French dairy breeds. This approach led to the detection of 34 distinct regions that affect the protein composition of milk. The use of imputed WGS enabled us to confirm 22 of the 39 QTL that were previously detected from 50 K SNP genotypes [8] and to identify 12 novel QTL. In addition to genetic parameter results [4] and QTL detection results with the 50 K chip [8], these results confirm that MIR predictions are sufficiently accurate for genetic investigations. Repeated test-day records compensated for the moderate MIR prediction accuracy of some proteins.

Seventeen QTL that had been detected with 50 K SNP genotype data were not found with imputed WGS, possibly because different methods were used in the two studies (linkage disequilibrium and linkage analysis in the 50 K SNP study versus GWAS in the current imputed-WGS study) and also possibly because of the more stringent significance thresholds applied here. For GWAS on WGS data, the very stringent threshold that we used (with Bonferroni correction considering all variants as independent) probably reduced the detection power but minimized the number of false positive QTL.

Instead, the better resolution of the WGS data, combined with the power of the multi-breed GWAS approach, led to the detection of 12 QTL that were not previously found in the 50 K SNP study. To evaluate the

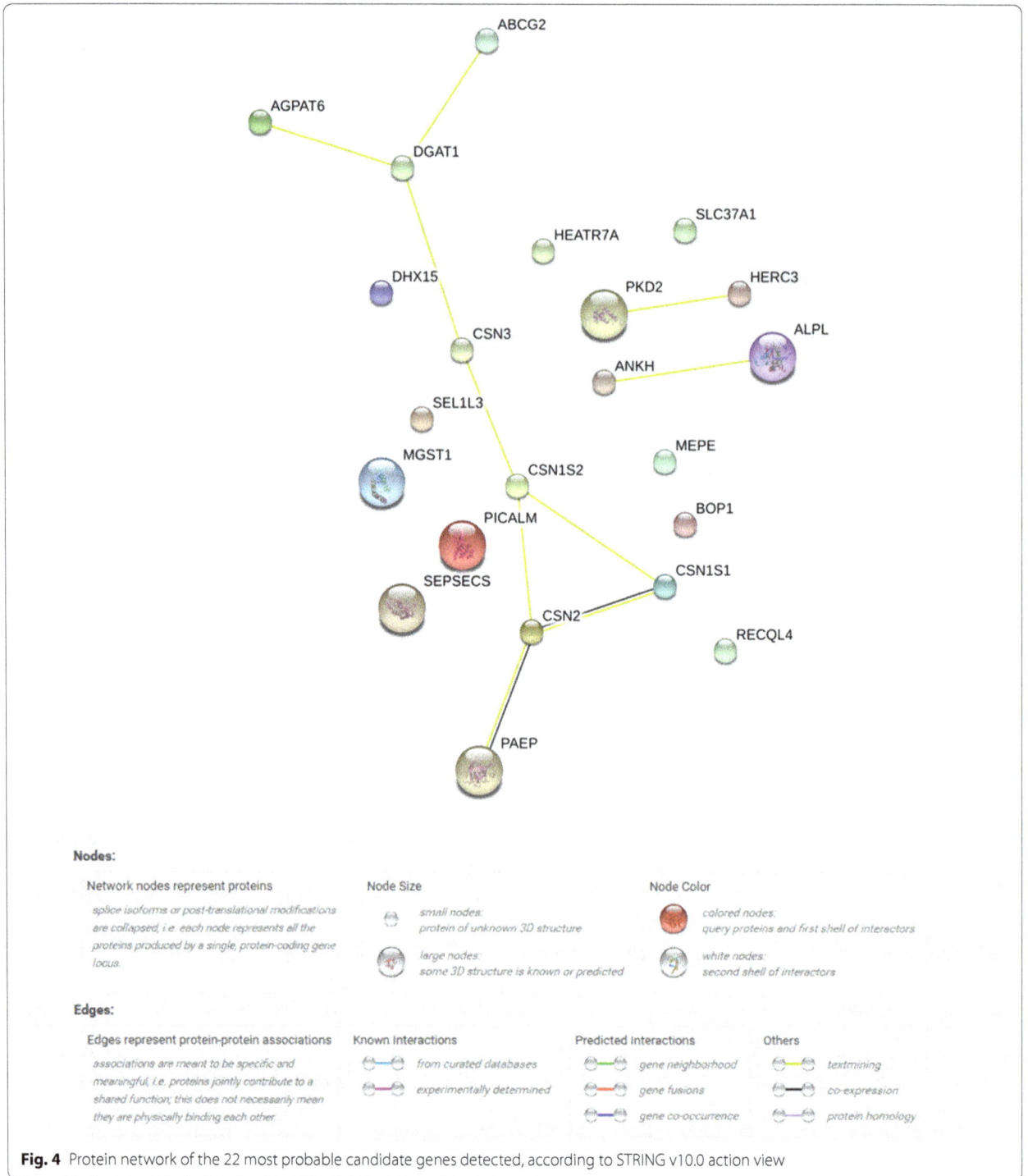

Fig. 4 Protein network of the 22 most probable candidate genes detected, according to STRING v10.0 action view

impact of marker density on GWAS results, we extracted 50 K and HD GWAS results from the WGS results. In several genomic regions, for example the regions on BTA2, 6, and 20 (Fig. 3), the increased resolution of the WGS data clearly makes it possible to identify two or more peaks whereas analysis of the 50 K SNP data detected only one peak.

Furthermore, the WGS resolution enables the use of a multi-breed approach, which is expected to better estimate the effects of rare variants and to reduce

Table 6 Gene Ontology (GO) functional enrichment with false discovery rate (FDR) < 0.05

	Pathway ID	Pathway description	Gene count	FDR	Genes
Biological process	GO.1903494	Response to dehydroepiandrosterone	4	1.73e−08	CSN1S1, CSN1S2, CSN2, CSN3
	GO.1903496	Response to 11-deoxycorticosterone	4	1.73e−08	CSN1S1, CSN1S2, CSN2, CSN3
	GO.0032570	Response to progesterone	4	1.81e−07	CSN1S1, CSN1S2, CSN2, CSN3
	GO.0097305	Response to alcohol	5	3.69e−07	ALPL, CSN1S1, CSN1S2, CSN2, CSN3
	GO.0032355	Response to estradiol	4	2.34e−06	CSN1S1, CSN1S2, CSN2, CSN3
	GO.1901700	Response to oxygen-containing compound	6	9.04e−05	ALPL, CSN1S1, CSN1S2, CSN2, CSN3, PKD2
	GO.0014070	Response to organic cyclic compound	5	0.000176	ALPL, CSN1S1, CSN1S2, CSN2, CSN3
	GO.0033993	Response to lipid	5	0.000181	ALPL, CSN1S1, CSN1S2, CSN2, CSN3
	GO.0009719	Response to endogenous stimulus	5	0.00205	CSN1S1, CSN1S2, CSN2, CSN3, PKD2
	GO.0048732	Gland development	3	0.0281	CSN2, CSN3, PKD2
	GO.0060416	Response to growth hormone	2	0.0281	CSN1S1, CSN1S2
	GO.0007595	Lactation	2	0.0298	CSN2, CSN3
Cellular component	GO.0005796	Golgi lumen	4	1.97e−08	CSN1S1, CSN1S2, CSN2, CSN3
	GO.0012505	Endomembrane system	8	0.00253	AGPAT6, CSN1S1, CSN1S2, CSN2, CSN3, DGAT1, MGST1, PKD2
	GO.0005576	Extracellular region	7	0.0372	ALPL, CSN1S1, CSN1S2, CSN2, CSN3, PAEP, PKD2
	GO.0005789	Endoplasmic reticulum membrane	4	0.0372	AGPAT6, DGAT1, MGST1, PKD2
	GO.0042175	Nuclear outer membrane-endoplasmic reticulum membrane network	4	0.0372	AGPAT6, DGAT1, MGST1, PKD2
	GO.0044444	Cytoplasmic part	9	0.0372	ABCG2, AGPAT6, CSN1S1, CSN1S2, CSN2, CSN3, DGAT1, MGST1, PKD2
	GO.0044446	Intracellular organelle part	9	0.0372	ABCG2, AGPAT6, CSN1S1, CSN1S2, CSN2, CSN3, DGAT1, MGST1, PKD2
Molecular function	GO.0035375	Zymogen binding	2	0.0177	CSN1S2, CSN3

LD between neighboring variants. With the multi-breed analysis, we detected 14 QTL that were not detected in any of our within-breed analyses (see, for example, regions of the *MGST1* and *AGPAT6* genes described below). For QTL that were detected in both within-and multi-breed analyses, the multi-breed approach provided smaller confidence intervals of the QTL than within-breed analyses. The three French breeds used in our study are not strongly related. Based on 50 K SNP data, Gautier et al. [24] reported a partitioning of the genetic diversity of cattle into distinct groups of breeds with high geographical consistency. The three breeds were classified into three distinct groups: from Eastern France and Alps for MON, from Northern European for HOL and from the Channel Islands and Northwestern France for NOR. Thus, our results illustrate the extent to which a multi-breed approach can complement and enhance the information gained from within-breed analyses even if breeds pooled in multi-breed analyses have different genetic origins.

In a previous study [25], the imputation from 50 K to HD SNP densities was found to be very accurate in all three breeds with the number of HD genotypes used here

(>500) in calibration. For the second imputation step, from HD SNPs to WGS, we used the Run 4 reference population of the 1000 Bull Genomes consortium, which contained 1147 bulls, of which 288 were HOL, 28 were MON, and 24 were NOR. Due to the larger number of sequenced HOL bulls compared to the two other breeds, imputation is more accurate with HOL data than with MON data. In NOR, we anticipate that imputation accuracy is close to that obtained in MON due to similar population structures and similar numbers of whole-genome sequences for major ancestors in both breeds. These results are in agreement with or are better than those already published in cattle. Daetwyler et al. [9] showed that the use of the 1000 Bull Genome multi-breed population (Run 2, 234 bulls) led to a similar imputation accuracy among data obtained from Holstein–Friesian, Fleckvieh, and Jersey cattle (near 80% of correlation) in spite of differences in the number of bulls in the reference population (129 Holstein–Friesian, 43 Fleckvieh, and 15 Jersey). Among the *PhénoFinlait* cows genotyped with the 50 K SNP Beadchip and then imputed to WGS, 1077 MON, 238 NOR, and 498 HOL originated from nine MON, five NOR and eight HOL bulls with WGS available in the Run4 reference population, i.e. 36, 9 and 22% of the PhénoFinlait cows, respectively. As expected, imputation accuracy dropped for variants with a low MAF. In order to limit the impact of imputation errors on the GWAS results, variants with a MAF lower than 2% were discarded from the analyses and almost all the genetic variants proposed as candidate variants in this study have moderate to high MAF.

Combining within-breed, multi-breed, and conditional GWAS analyses with functional annotations appears to be a good strategy for the differentiation of shared and breed-specific QTL. This approach also enables the direct identification of candidate genes with a very small number of candidate variants, or even in some cases, one unique variant which appears to be the best candidate to explain the observed effects.

On average, depending on the breed, between 60 and 73% of the QTL variants that we detected in the GWAS were located in genes; this is about twice as high as the percentage of genic variants at the whole-genome scale (35%; [15]). The most significant variants were located in 49 distinct genes, of which 22 were of particular interest, either because they were found in more than one breed or associated with several traits, or because they were previously described as influencing milk composition. These 22 genes, which are located in 11 distinct genomic regions and present significant protein–protein enrichment, are the most plausible candidates to explain a large part of the variation in milk protein composition among MON, NOR, and HOL cows. In four genomic regions (on BTA1, 2, 11, and 27), we identified one unique candidate variant (or a few candidate variants in LD) shared by all three breeds (in the *SLC37A1*, *ALPL*, *PAEP*, and *AGPAT6* genes, respectively). In three other genes, we suggest the presence of a breed-specific candidate variant (*MGST1* on BTA5 and *PICALM* on BTA29) or several candidate causative variants (*ANKH* on BTA20). Finally, four regions, including the *DGAT1* region on BTA14 and three regions on BTA6 (*ABCG2* region, a region at about 46 Mbp, and the casein gene cluster), were more complex, because they contained several candidate genes, each with several candidate variants. Eight of these candidate genes (*SLC37A1*, *MGST1*, *CSN1S1*, *CSN2*, *CSN1S2*, *CSN3*, *PAEP*, and *ANKH*) are known to be overexpressed in the mammary gland compared to other 17 tissues [26] and between two and nine of them are associated with one of the 20 GO terms in our study. The next sections describe these regions in more detail.

SLC37A1 (BTA1) and αs1-CN/ α-LA

The *SLC37A1* (*solute carrier family 37, member A1*) gene, which encodes a glucose-6-phosphate transporter that is involved in the homeostasis of blood glucose, is highly expressed in the mammary gland [27]. It could be a good candidate gene to explain the effects of the QTL identified on BTA1 on αs1-CN in both MON and multi-breed analyses and on the α-LA phenotype in the multi-breed analysis. In total, 138 distinct variants of this gene were located within the confidence intervals of the QTL, of which 133 were intronic, two were synonymous, and three were downstream (see Additional file 6: Figure S1a). For the αs1-CN/MON, αs1-CN/multi, and α-LA/multi results, the 80, 81, and 74 most significant variants in the peaks, respectively, were in intronic regions of *SLC37A1*. One downstream variant was detected for αs1-CN in the MON analysis, which ranked 104th among the significant variants, while multi-breed analyses revealed three downstream variants that ranked 81st, 87th, and 103rd. All intronic variants that are located at the top of the peaks are in strong LD but only one variant (indel), located at 144,397,274 bp, was common to all three TOP10 lists; it was 1st in the αs1-CN/MON ranking, 9th in the αs1-CN/multi-breed ranking, and 4th in the α-LA/multi-breed ranking. The top1 intronic variant detected in the αs1-CN/multi-breeds analysis, at 144,398,814 bp, ranked 75th in the αs1-CN/MON peak and 76th in the α-LA/multi-breed peak.

Two previous studies described the effects of *SLC37A1* gene variants on milk production traits. In an analysis of HD SNP genotypes, Kemper et al. [27] described six variants that are located between 144.325 and 144.525 Mbp in this region; the variant with the most significant effect was located in an intronic region of the gene (144,414,936 bp). In our study, this variant was included within the confidence interval of the QTL detected by the multi-breed analysis ($-\log_{10}(P) = 10.2$), but it ranked 101st. Two other intronic variants in strong LD in the *SLC37A1* gene, at 144,367,474 and 144,377,960 bp, were previously proposed as the best candidate mutations for changes in phosphorus concentration and milk production traits [28]. However, in our study in spite of relatively high MAF values (from 0.30 to 0.41 depending on the breed), these variants had a $-\log_{10}(P)$ value lower than 6 for all analyzed traits. In another study of targeted QTL regions after imputation to WGS level, the variant with the most significant effects was located at 144,381,564 bp [29]. This variant is close to the candidate variant identified in our analysis, but it can be excluded as the causal variant in our populations since it is monomorphic in the MON, NOR, and HOL individuals analyzed here.

The conditional analyses that we performed included the two best candidate variants as well as the candidate variant described by Kemper et al. [27]. These revealed that including the variant located at 144,398,814 bp in the model completely removed the original signal while with each of the two other variants, a less significant peak persisted (see Additional file 6: Figure S1a). This variant, which has contrasting effects on αs1-CN and α-LA phenotypes, but with a more marked effect on the former, therefore constitutes the most probable candidate variant for the effects detected in our study.

ALPL (BTA2) and αs2-CN

The QTL identified on BTA2 at 131.8 Mbp had significant effects on several traits (αs2-CN, β-CN, and κ-CN). In particular, although the αs2-CN-associated peaks were detected in all within-and multi-breed analyses, even if in the MON and HOL analyses, the maximal $-\log_{10}(P)$ values did not reach the stringent threshold of 8.4 that we applied in this study (7 and 6.9, respectively; see Additional file 6: Figure S1b). In all analyses, the most significant variants were located in intronic regions of the *ALPL* (*alkaline phosphatase*) gene, which encodes a member of the alkaline phosphatase family of proteins. The most significant variant differed among the three within-breed analyses: it was located at 131,806,882 bp in NOR, 131,850,456 bp in MON, and 131,808,301 bp in HOL sequences. Instead, the top-ranked variant in the peak detected in the multi-breed GWAS was located at 131,806,882 bp. All three single-breed conditional analyses that included each of these variants as fixed effects lacked peaks (see Additional file 6: Figure S1b). These results suggest that all three intronic variants are in strong LD in the three breeds and that the causal mutation could be shared among breeds. Among all the variants at the top of the peaks, the intronic variant at 131,806,882 bp appears to be the most probable candidate variant in the *ALPL* gene for the observed effects on αs2-CN content; it ranked 1st, 6th, 26th, and 1st in the NOR, MON, HOL, and multi-breed peaks, respectively.

MGST1 (BTA5) and milk protein content (PC)

One region on BTA5 that contains 63 variants affected PC in the multi-breed analysis. The MON and NOR within-breed analyses revealed no peaks ($-\log_{10}(P) < 6$), whereas the HOL analysis detected a single peak with a $-\log_{10}(P) = 8$, which was close to the significance threshold of 8.4. Only one gene, *MGST1* (*microsomal glutathione S-transferase 1*), was present within the confidence interval obtained in the multi-breed analysis. Fifty-one variants were located in intronic (29), exonic (1 synonymous), 5'-UTR (2), or regulatory (19 in the upstream region) regions of the gene. The variant with the most significant effects was located at 93,950,211 bp in the upstream region and its $-\log_{10}(P)$ value was 9.3, versus a value of 8.0 for the variants that ranked 2nd (93,950,116 bp and 93,950,288 bp), which were located, respectively, in the 5'-UTR and upstream regions of the gene. The MAF value for these variants was low in the

MON population (0.006; <MAF threshold of 0.02) and ranged from 0.08 to 0.12 in NOR, from 0.37 to 0.42 in HOL, and from 0.19 to 0.22 in the multi-breed population. Thus, the fact that peaks were detected only in HOL (close to significance) and multi-breed (significant) analyses could be due to the relatively low MAF for these variants in MON and NOR. The most significant variants in our study are located near a variant that was reported by Raven et al. [29] to be responsible for changes in fat percentage in Holstein cows (at 93,951,731 bp (upstream) and ranked 23rd in our study) and also near variants previously linked to fat yield by Iso-Touru et al. [30] and Van den Berg et al. [31] (93,945,694 and 93,945,738 bp, respectively; both were intronic variants and were not significant here). Conditional analyses including each of the six variants as a fixed effect showed that all variants except those reported by Iso-Touru et al. [30] and Van den Berg et al. [31] explained the effects observed in our study (see Additional file 6: Figure S1c). Thus, the effects observed on fat content by Raven et al. [29] and on protein content in our study could be explained by the same causative variant. Recently, Littlejohn et al. [32] confirmed that *MGST1* has causative pleiotropic effects on milk composition (percentage and yield of fat, protein, and lactose). These authors failed to identify causative variants in the gene but they pointed to a cluster of 17 variants that were grouped in a 10-kbp segment of the *MGST1* gene (93,944,937–93,954,751). Only one of these 17 variants is located in the confidence interval of the QTL that we detected and this is an intronic variant (93,949,810 bp) that ranked 7th in the peak in spite of having a higher MAF (0.32) than the most significant variants (MAF = 0.19–0.22). Thus, our study highlights three new candidate mutations in the *MGST1* gene, which are located very close to each other, in the 5'-UTR region (93,950,116 bp) or in the upstream region (93,950,211 and 93,950,288 bp) of the gene.

ABCG2, MEPE, PKD2, and HERC3 (BTA6) and αs1-CN

Several QTL were found on BTA6. The first one, detected in HOL and multi-breed analyses, was located in the 37.6–38.4 Mbp region, which contains the Y581S polymorphism of the *ABCG2* gene (38,027,010 bp) that was described by Cohen-Zinder et al. [33] as a causative mutation for changes in milk yield and composition. This missense variant had MAF values of 0.0029 and 0.0018 in HOL and multi-breed populations, respectively, and therefore did not pass the MAF filter in both analyses. In spite of a low MAF, the Y581S polymorphism had a highly significant effect on the αs1-CN phenotype in both HOL and multi-breed analyses, with $-\log_{10}(P)$ values of 31 and 21, respectively; these values were higher than those of the top variant in the peaks after filtering for MAF (20 and 15, respectively). However, among the sires of the HOL cows, six bulls were previously found to be heterozygous for the QTL detected in this region, but homozygous for this mutation [8]. Thus, we suggest that other mutations could be responsible for the QTL that affects milk protein composition.

In the HOL analysis, nine variants with MAF ranging from 0.022 to 0.041 were located within the confidence interval of the QTL. The most significant variants were located in intronic regions of the *ABCG2* gene, at 38,015,146 and 38,020,110 bp. Other variants, which are located in three other genes, i.e. *MEPE* (one downstream), *PKD2* (one intronic), and *HERC3* (two intronic), also had highly significant effects on αs1-CN. Due to the relatively low MAF of the candidate variants located in this region, these results require further analyses, including a larger number of animals and more accurate imputation or direct genotyping.

SEL1L3, SEPSECS, and DHX15 (BTA6) and αs1-CN
In all within-breed and multi-breed analyses, the αs1-CN phenotype was affected by another region of BTA6 at 45.8–46.9 Mbp. However, the most likely candidate genes differed among breeds. In MON, the nine variants with the most significant effects were located in intronic regions of the *SEL1L3* gene (max. at 46,874,151 bp). In NOR, the top 116 variants in the peak were intergenic, while the genic variant with the most significant effects was located in an intron of the *SEPSECS* gene (46,277,697 bp). In HOL, the most significant genic variants (*DHX15*) ranked 16th in the peak (45,639,181 and 45,640,564 bp). Finally, among the top 80 variants detected by the multi-breed analysis, only one was genic, which was located in an intron of the *SEL1L3* gene (46,874,514 bp, ranked 3rd in the MON analysis). There is insufficient concordance among these results to propose a single set of candidate variants.

Pleiotropic effects of the casein gene region (BTA6)
On BTA6, we found a QTL that affected both the overall protein content of milk and the content of all four individual caseins in all three breeds. Variants with the most significant effects were located in an 840-kb interval that contains the 250-kb casein gene cluster (87,062,878–87,903,002 bp); other variants with effects on αs1-CN and β-CN in MON were located at 85.2 Mbp. In all within- and multi-breed analyses, the most significant effects were detected for the κ-CN phenotype, followed by αs1, αs2, or β-CN depending on the breed. In each analysis, the variant with the most significant effects on κ-CN was located within or in the immediate vicinity of the *CSN3* gene, which encodes the κ casein: at 87,376,747 bp (upstream) in NOR, 87,392,592 bp (5'-UTR) in MON and

multi-breed, and 87,394,293 bp (downstream) in HOL. Each of these variants, as well as the κ casein A/B variant (87,390,576 bp, missense), was therefore included as a fixed effect in the conditional analyses. The results were breed-specific: in MON, the κ-CN-associated peak disappeared after fixing the upstream, missense, or 5'-UTR variant; in HOL, the peak disappeared after fixing the upstream, 3'-UTR, or downstream variant; but in NOR, the peak remained with the inclusion in the model of any of the four variants. Thus, none of the four candidate variants succeeded in explaining all the effects observed on κ-CN in the three breeds.

Instead, the peaks associated with the αs2-CN and β-CN phenotypes in NOR and the PC and αs2-CN phenotypes in MON could be explained by two distinct groups of six SNPs in complete LD, which were respectively located in the *CSN2* gene (three downstream and three intronic) and in the upstream region of the *ODAM* (odontogenic ameloblast-associated) gene (between the *CSN1S2* and *CSN3* genes).

Finally, the A1/B and A2 variants of *CSN2*, which ranked 147th and 86th, respectively, for their effects on PC and αs2-CN in NOR, were responsible for the αs2-CN phenotype in NOR but not for any other effect on the other traits or in the other breeds.

These results illustrate the complexity that is inherent with the analysis of the casein gene cluster, which contains the four genes *CSN1S1-CSN2-CSN1S2-CSN3* (encoding, respectively, αs1, β, αs2, and κ caseins). The polymorphisms of the amino-acid sequences of caseins are well known, and the effects on milk composition and cheese-making abilities have been well described (reviewed in Grosclaude et al. [34] and Caroli et al. [35]). Nevertheless, the effects of known polymorphisms are not always consistent between studies, likely because variations in the content of individual caseins are caused by several linked polymorphisms in the casein genes. Thus, it is likely that the most significant variants highlighted in our study are those that better explain haplotype effects. A multi-marker approach could facilitate efforts to distinguish the effects of all the causal polymorphisms located in this region.

Pleiotropic effects of the PAEP gene region (BTA11)
The most significant effects on protein composition were found for variants that are located on BTA11. Contents of each individual protein in milk, with the exception of αs2-CN, were affected by this region in all three breeds. Effects were most significant for β-LG and, to a lesser extent, for κ-CN in all within-and multi-breed analyses. All of the most significant variants were located in or close to the *PAEP* (*progestagen-associated endometrial protein*) gene, also named *LGB* gene, which encodes the

β-LG protein. The β-LG protein variants A and B, which are common in most cattle breeds, are associated with different β-LG levels in milk [34]. They differ by two amino-acid substitutions, caused by two missense mutations at 103,303,475 and 103,304,757 bp [36]. Interestingly, although these two variants had highly significant effects on β-LG in our study, they did not rank high in the peaks. In the MON and NOR analyses, both mutations were in complete LD and ranked 85th and 213rd, respectively, while in HOL, the two mutations ranked 48th and 109th, respectively (116th and 120th in multi-breed analysis). As suggested by Ganai et al. [36], differences in β-LG content may be caused by different levels of expression of the A and B alleles rather than by the direct effect of amino-acid substitutions. Among the top 30 variants in the within-and multi-breed analyses, only one, located at 103,298,431 bp in the upstream region of the *PAEP* gene, was shared by the four analyses. Moreover, this variant is one of the most significant in each analysis, ranking 6th, 4th, 1st, and 3rd, respectively, in the MON, NOR, HOL, and multi-breed analyses. The inclusion in conditional analyses of one of the causal missense variants or the most probable upstream variant identified in our study led to similar results in MON and HOL but not in NOR (see Additional file 6: Figure S1d). A peak remained in the conditional NOR analysis when missense mutations were fixed, but disappeared with the inclusion of the upstream variant at 103,298,431 bp. Thus, these results indicate that the missense mutations that cause the A and B variant protein polymorphisms do not explain all the variation associated with this region. Another variant, which is located in a regulatory region of *PAEP*, is more or less linked to the missense variants depending on the breed and appears to be a good candidate to explain different levels of expression of β-LG protein variants.

Pleiotropic effects of the *DGAT1* gene region (BTA14)

Very significant effects on different protein composition traits were associated with the region of the *DGAT1* gene in NOR and HOL but not in MON. This region affected PC and κ-CN in both NOR and HOL; αs1-CN, β-CN, and α-LA only in NOR, and αs2-CN only in HOL. Moreover, individual proteins with the lowest *P* value were κ-CN in NOR and αs2-CN in HOL. The A allele of the *DGAT1* K232A polymorphism, which decreases fat and protein percentages as well as fat yield, and increases milk and protein yields [37], was present at a frequency of 9.4% in NOR, 15.8% in HOL, and only 0.6% in MON. However, our study confirmed that this causative variant was not the most significant for all traits analyzed. It ranked 18th to 72th among variants in the NOR analysis, depending on the trait, and outside the confidence interval for

all traits in HOL. These results suggest, first, that not all variations observed in this region are associated with the K232A polymorphism and, second, that other specific causative mutations could explain the effects detected in NOR and HOL.

A large number of genes are annotated in the 1-Mbp region between 1.5 and 2.5 Mbp on BTA14 and, depending on the trait and the breed, between 66 and 494 variants located within the confidence intervals of this QTL are located in 17 to 30 of those genes. Among the top 50 variants for all traits, six were missense variants, of which two were found in NOR (*DGAT1* and *BOP1*) and four in HOL (three in *RECQL4* and one in *MROH1*). In this region, no variant remained significant in the conditional analyses for NOR when the *DGAT1* (K232A) or *BOP1* (1,842,678 bp) variants were included, and for HOL when the variants in *RECQL4* (one of the three variants in complete LD: 1,617,841, 1,618,978 and 1,619,555 bp) or *MROH1* (1,878,165 bp) were included. In contrast, a less significant peak persisted when the *DGAT1* or *BOP1* variant was included in the HOL analyses and when the *RECQL4* or *MROH1* variant was included in the NOR analyses (see Additional file 6: Figure S1e). Among the six missense variants, only the *RECQL4* variant at 1,617,841 bp has a predicted deleterious effect, with a SIFT score of zero. Therefore, in addition to the *DGAT1* K232A polymorphism previously identified as having effects on milk composition, we report additional candidate missense mutations in *BOP1*, *MROH1*, and *RECQL4* genes, which could be partly responsible for the effects associated with the centromeric end of BTA14.

ANKH (BTA20) and α-LA

The GWAS on WGS data detected a QTL with very significant effects on the α-LA phenotype in all three within-breed analyses and in the multi-breed analysis; this confirmed our previous report based on a GWAS using 50 K SNP data [8]. Confidence intervals of the QTL included between one and four genes depending on the within- or multi-breed analysis, and *ANKH* was the only gene to be highlighted in all four analyses. *ANKH* encodes an inorganic pyrophosphate transport regulator that helps to prevent the deposition of minerals (calcium and phosphorous) in bones and α-LA exhibits a high affinity to metal ions, calcium in particular. In addition, *ANKH* is highly expressed in mammary tissue in Holstein and Jersey cows [27] and we observed a significant interaction between *ANKH* and *ALPL* (candidate gene on BTA2 for effects on αs2-CN), which suggests a functional link between these two genes (Fig. 4). Thus, *ANKH* constitutes a good functional candidate for effects on α-LA in HOL, MON, and NOR. However, none of the top 50 variants in this QTL were shared among the three breeds. In each breed, the most significant variant was located either in intronic regions of the *ANKH* gene (at 58,422,697 bp in NOR and at 58,450,656 bp in multi-breed analyses) or in an intergenic region. In MON and HOL, for which the most significant variants were intergenic, *ANKH* intronic variants ranked 2nd (at 58,446,560 bp) and 13th (at 58,491,204 bp), respectively. After fixing the most significant variant from each within-breed analysis, a peak remained in all conditional analyses (see Additional file 6: Figure S1f), which suggests that several causative mutations in the *ANKH* gene could be responsible for the variation of the amount of α-LA in milk. The most significant variants could be those that are most tightly linked to the causative mutations in each breed, which could explain why they were breed-specific.

AGPAT6 (BTA27) and κ-CN

The multi-breed analysis detected a QTL for κ-CN content located at about 36.2 Mbp on BTA27, while in within-breed analyses, peaks were present in MON and NOR but they did not reach significance ($-\log_{10}(P) < 8.4$), and no peak was observed in HOL (see Additional file 6: Figure S1g). In the multi-breed analysis, the four most significant variants were located in an intergenic region but the variants that ranked 5th to 17th were located in the *AGPAT6* gene, which was previously described as a functional gene for milk fat content with pleiotropic effects on other milk components, in particular protein content [38]. The five most significant variants in the gene were in complete LD and located in the upstream region (at 36,209,319, 36,211,252, 36,211,258, and 36,211,708 bp) or in the 5′-UTR region (at 36,212,352 bp) of the *AGPAT6* gene. For the five linked variants, MAF were equal to 0.46 in MON, 0.47 in NOR, and 0.39 in HOL (0.44 in multi-breed population). When the κ-CN phenotype was conditioned on the effect of any of these mutations, the association signals completely disappeared in the MON, NOR, and multi-breed analyses (see Additional file 6: Figure S1g). The four variants located in the upstream region were previously identified as candidate causal polymorphisms in both Holstein and Fleckvieh cows by Daetwyler et al. [9]. These authors pointed to the polymorphism at 36,211,252 bp as the most plausible causative mutation because it presented a high probability of being within a transcription binding site. In addition, Littlejohn et al. [38] described strong associations between milk composition traits (fat, protein, and lactose) and 10 variants in the *AGPAT6* gene. Three of these 10 variants were among the most significant variants in our study, located at 36,209,319, 36,211,708, and 36,212,352 bp. Thus, we identified five putative causative variants in the *AGPAT6* gene for milk protein composition; of these, the

variant at 36,212,352 bp appears to be the most plausible causative mutation because it is located in the 5′-UTR region of the *AGPAT6* gene. However, the lack of a significant effect in the HOL analyses, in spite of the high MAF of the candidate variants, probably reflects additional effects yet to be explained.

PICALM (BTA29) and αs1-CN

The αs1-CN phenotype was influenced by a genomic region that is located at about 9.5 Mbp on BTA29. Significant associations were found in MON, HOL, and multi-breed analyses, and a peak close to significance was found in NOR ($-\log_{10}(P) = 7.9$) (see Additional file 6: Figure S1h). In the MON and HOL analyses, the most significant variants were intergenic and, likewise, in the multi-breed analysis, all nine variants located within the confidence interval were intergenic. The most significant non-intergenic variants were located in the *PICALM* gene in MON and HOL. Two intronic variants ranked 11th in the peak detected in MON (9,651,065 and 9,656,439 bp) and one variant that ranked 10th in the HOL analysis, is located in the upstream region of the gene (9,611,304 bp). When conditional GWAS analyses were performed, the inclusion of the intronic variants removed the peak in MON but not in HOL analyses, and conversely, inclusion of the upstream variant removed the peak in HOL but not in MON analyses. In NOR, the peak in question persisted when either intronic or upstream variants were fixed (see Additional file 6: Figure S1h). These results suggest that either the causative variant is different between breeds or that several linked causative variants explain the significant effects observed in this region. The *PICALM* gene encodes a phosphatidylinositol-binding clathrin assembly protein, and polymorphisms in this gene are associated with the risk of Alzheimer's disease [39] in humans. However, to date, no link was reported between polymorphisms in this gene and bovine milk composition

Conclusions

Our study provides evidence that a GWAS-based approach applied to fine-scale phenotypes, whole-genome sequences, and multiple breeds provides enough resolution to identify candidate genes and directly pinpoint a limited number of candidate variants in most of these genes. Several variants, some shared among breeds, were identified as plausible candidate mutations for changes in milk protein composition in the three main French dairy cattle breeds. They were located both in genes that had previously been found to affect milk composition (*SLC37A1, MGST1, ABCG2, CSN1S1, CSN2, CSN1S2, CSN3, PAEP, DGAT1, AGPAT6*) and in genes for which no such relationship was known (*ALPL,*

ANKH, PICALM). In the future, functional analyses will enable the establishment of causative links between these candidate variants and milk protein phenotypes. However, even before such studies are completed, our results offer the opportunity to improve cheese-making properties through the identification of genetic variants associated with changes in milk composition. Direct consequences of these results on practical selection are not obvious and depend on potential premiums on protein composition and on incentives proposed by the milk processing industry. Nevertheless, it would be desirable to favour caseins against whey proteins at least for milk collected for cheese production. Such an option could be implemented by including variants that affect individual proteins in genomic evaluation models.

Additional files

Additional file 1: Table S1. The 1000 bull genome population (RUN4). (Daetwyler HD, personal communication).

Additional file 2: Table S2. Number of variants included within confidence intervals for each QTL region and trait, regardless of breed.

Additional file 3: Table S3. Description of the pleiotropic QTL regions detected in within-breed (MON, NOR, or HOL) or multi-breed (Multi) analyses.

Additional file 4: Table S4. Description of other significant QTL regions detected in within-breed (MON, NOR, or HOL) or multi-breed (Multi) analyses.

Additional file 5: Table S5. Functional annotations of variants included within confidence intervals (± 100 kb) of the 34 QTL for each trait in the three within-breed Montbéliarde (MON), Normande (NOR), and Holstein (HOL) or in multi-breed analyses.

Additional file 6: Figure S1. $-\log_{10}(P)$ plotted against the position of variants detected by GWAS (in *grey*) and conditional GWAS (GWAS_COJO; in *blue*) **a** On BTA1, **b** BTA2, **c** BTA5, **d** BTA11, **e** BTA14, **f** BTA20, **g** BTA27 and **h** BTA29

Authors' contributions
MPS analyzed the data and wrote the manuscript. DB, SF, and MBr designed the *PhénoFinlait* project. GM and PM provided reference analyses for milk protein composition. AG, PC, CH, AB, and MBo provided support in computing. RL and DR contributed to the estimation of imputation accuracy. All authors read and approved the final manuscript.

Author details
[1] GABI, INRA, AgroParisTech, Université Paris Saclay, 78350 Jouy-en-Josas, France. [2] Institut de l'Elevage, 75012 Paris, France. [3] Allice, 75012 Paris, France.

Acknowledgements
The authors gratefully acknowledge the breeders who participated in the *PhénoFinlait* program; colleagues from the Institut de l'Elevage and INRA who designed and coordinated the farm sampling program and data collection; the partners of the program, laboratories, manufacturers, and DHI organizations who provided data; Marion Ferrand who developed the MIR prediction equations; and the members of the *PhénoFinlait* scientific committee who advised and managed this work. The authors would also like to thank the contribution of the 1000 Bull Genomes consortium.

Competing interests
The authors declare that they have no competing interests.

Funding
The *PhénoFinlait* program received financial support from ANR (ANR-08-GANI-034 Lactoscan), APIS-GENE, CASDAR, CNIEL, FranceAgriMer, France Génétique Elevage, and the French Ministry of Agriculture. The *Cartoseq* project was funded by ANR (ANR10-GENM-0018) and APIS-GENE.

References
1. Schopen GC, Heck JM, Bovenhuis H, Visker MH, van Valenberg HJ, van Arendonk JA. Genetic parameters for major milk proteins in Dutch Holstein-Friesians. J Dairy Sci. 2009;92:1182–91.
2. Bonfatti V, Cecchinato A, Gallo L, Blasco A, Carnier P. Genetic analysis of detailed milk protein composition and coagulation properties in Simmental cattle. J Dairy Sci. 2011;94:5183–93.
3. Gebreyesus G, Lund MS, Janss L, Poulsen NA, Larsen LB, Bovenhuis H, et al. Short communication: multi-trait estimation of genetic parameters for milk protein composition in the Danish Holstein. J Dairy Sci. 2016;99:2863–6.
4. Sanchez MP, Ferrand M, Gelé M, Pourchet D, Miranda G, Martin P, et al. Short communication: genetic parameters for milk protein composition predicted using mid-infrared spectroscopy in the French Montbéliarde, Normande, and Holstein dairy cattle breeds. J Dairy Sci. 2017;100:6371–5.
5. Wedholm A, Larsen LB, Lindmark-Månsson H, Karlsson AH, Andrén A. Effect of protein composition on the cheese-making properties of milk from individual dairy cows. J Dairy Sci. 2006;89:3296–305.
6. Bonfatti V, Di Martino G, Carnier P. Effectiveness of mid-infrared spectroscopy for the prediction of detailed protein composition and contents of protein genetic variants of individual milk of Simmental cows. J Dairy Sci. 2011;94:5776–85.
7. Ferrand M, Miranda G, Guisnel S, Larroque H, Leray O, Lahalle F, et al. Determination of protein composition in milk by mid-infrared spectrometry. In Proceedings of the international strategies and new developments in milk analysis VI ICAR Reference Laboratory Network Meeting: 28 May 2012; Cork. 2013;16:41–5.
8. Sanchez MP, Govignon-Gion A, Ferrand M, Gele M, Pourchet D, Amigues Y, et al. Whole-genome scan to detect quantitative trait loci associated with milk protein composition in 3 French dairy cattle breeds. J Dairy Sci. 2016;99:8203–15.
9. Daetwyler HD, Capitan A, Pausch H, Stothard P, Van Binsbergen R, Brøndum RF, et al. Whole-genome sequencing of 234 bulls facilitates mapping of monogenic and complex traits in cattle. Nat Genet. 2014;46:858–67.
10. Raven L, Cocks B, Hayes B. Multibreed genome wide association can improve precision of mapping causative variants underlying milk production in dairy cattle. BMC Genomics. 2014;15:62.
11. Ducrocq V. Genekit, BLUP software. Version June 2011.
12. Sargolzaei M, Chesnais JP, Schenkel FS. A new approach for efficient genotype imputation using information from relatives. BMC Genomics. 2014;15:478.
13. van Binsbergen R, Bink MC, Calus MP, van Eeuwijk FA, Hayes BJ, Hulsegge I, et al. Accuracy of imputation to whole-genome sequence data in Holstein Friesian cattle. Genet Sel Evol. 2014;46:41.
14. Bouwman AC, Veerkamp RF. Consequences of splitting whole-genome sequencing effort over multiple breeds on imputation accuracy. BMC Genet. 2014;15:105.

15. Boussaha M, Michot P, Letaief R, Hoze C, Fritz S, Grohs C, et al. Construction of a large collection of small genome variations in French dairy and beef breeds using whole-genome sequences. Genet Sel Evol. 2016;48:87.

16. Li H, Handsaker B, Wysoker A, Fennell T, Ruan J, Homer N, et al. The sequence alignment/map format and SAMtools. Bioinformatics. 2009;25:2078–9.

17. McLaren W, Pritchard B, Rios D, Chen Y, Flicek P, Cunningham F. Deriving the consequences of genomic variants with the Ensembl API and SNP effect predictor. Bioinformatics. 2010;26:2069–70.

18. Kumar P, Henikoff S, Ng PC. Predicting the effects of coding non-synonymous variants on protein function using the SIFT algorithm. Nat Protoc. 2009;4:1073–82.

19. Yang J, Lee SH, Goddard ME, Visscher PM. GCTA: a tool for genome-wide complex trait analysis. Am J Hum Genet. 2011;88:76–82.

20. Fu WX, Liu Y, Lu X, Niu XY, Ding XD, Liu JF, et al. A genome-wide association study identifies two novel promising candidate genes affecting Escherichia coli F4ab/F4ac susceptibility in swine. PLoS One. 2012;7:e32127.

21. Yang j, Ferreira T, Morris AP, Medland SE, Genetic Investigation of ANthropometric Traits (GIANT) Consortium, DIAbetes Genetics Replication and Meta-analysis (DIAGRAM) Consortium, et al. Conditional and joint multiple-SNP analysis of GWAS summary statistics identifies additional variants influencing complex traits. Nat Genet. 2012;44:369–75.

22. McLaren W, Gil L, Hunt SE, Riat HS, Ritchie GR, Thormann A, et al. The Ensembl variant effect predictor. Genome Biol. 2016;17:122.

23. Szklarczyk D, Franceschini A, Wyder S, Forslund K, Heller D, Huerta-Cepas J, et al. STRING v10: protein–protein interaction networks, integrated over the tree of life. Nucl Acids Res. 2015;43:D447–52.

24. Gautier M, Laloe D, Moazami-Goudarzi K. Insights into the genetic history of French cattle from dense SNP data on 47 worldwide breeds. PLoS One. 2010;5:e13038.

25. Hoze C, Fouilloux MN, Venot E, Guillaume F, Dassonneville R, Fritz S, et al. High-density marker imputation accuracy in sixteen French cattle breeds. Genet Sel Evol. 2013;45:33.

26. Chamberlain AJ, Vander Jagt CJ, Hayes BJ, Khansefid M, Marett LC, Millen CA, et al. Extensive variation between tissues in allele specific expression in an outbred mammal. BMC Genomics. 2015;16:993.

27. Kemper KE, Reich CM, Bowman PJ, vander Jagt CJ, Chamberlain AJ, Mason BA, et al. Improved precision of QTL mapping using a nonlinear Bayesian method in a multi-breed population leads to greater accuracy of across-breed genomic predictions. Genet Sel Evol. 2015;47:29.

28. Kemper KE, Littlejohn MD, Lopdell T, Hayes BJ, Bennett LE, Williams RP, et al. Leveraging genetically simple traits to identify small-effect variants for complex phenotypes. BMC Genomics. 2016;17:858.

29. Raven LA, Cocks BG, Kemper KE, Chamberlain AJ, Vander Jagt CJ, Goddard ME, et al. Targeted imputation of sequence variants and gene expression profiling identifies twelve candidate genes associated with lactation volume, composition and calving interval in dairy cattle. Mamm Genome. 2016;27:81–97.

30. Iso-Touru T, Sahana G, Guldbrandtsen B, Lund MS, Vilkki J. Genome-wide association analysis of milk yield traits in Nordic red cattle using imputed whole genome sequence variants. BMC Genet. 2016;17:55.

31. van den Berg I, Boichard D, Lund MS. Comparing power and precision of within-breed and multibreed genome-wide association studies of production traits using whole-genome sequence data for 5 French and Danish dairy cattle breeds. J Dairy Sci. 2016;99:8932–45.

32. Littlejohn MD, Tiplady K, Fink TA, Lehnert K, Lopdell T, Johnson T, et al. Sequence-based association analysis reveals an MGST1 eQTL with pleiotropic effects on bovine milk composition. Sci Rep. 2016;6:25376.

33. Cohen-Zinder M, Seroussi E, Larkin DM, Loor JJ, Everts-van der Wind A, Lee JH, et al. Identification of a missense mutation in the bovine ABCG2 gene with a major effect on the QTL on chromosome 6 affecting milk yield and composition in Holstein cattle. Genome Res. 2005;15:936–44.

34. Grosclaude F. Le polymorphisme génétique des principales lactoprotéines bovines. INRA Prod Anim. 1988;1:5–17.

35. Caroli AM, Chessa S, Erhardt GJ. Invited review: milk protein polymorphisms in cattle: effect on animal breeding and human nutrition. J Dairy Sci. 2009;92:5335–52.

36. Ganai NA, Bovenhuis H, van Arendonk JA, Visker MH. Novel polymorphisms in the bovine beta-lactoglobulin gene and their effects on beta-lactoglobulin protein concentration in milk. Anim Genet. 2009;40:127–33.

37. Grisart B, Coppieters W, Farnir F, Karim L, Ford C, Berzi P, et al. Positional candidate cloning of a QTL in dairy cattle: identification of a missense mutation in the bovine DGAT1 gene with major effect on milk yield and composition. Genome Res. 2002;12:222–31.

38. Littlejohn MD, Tiplady K, Lopdell T, Law TA, Scott A, Harland C, et al. Expression variants of the lipogenic AGPAT6 gene affect diverse milk composition phenotypes in Bos taurus. PLoS One. 2014;9:e85757.

39. Harold D, Abraham R, Hollingworth P, Sims R, Gerrish A, Hamshere ML, et al. Genome-wide association study identifies variants at CLU and PICALM associated with Alzheimer's disease. Nat Genet. 2009;41:1088–93.

10

Analysis of the genetic variation in mitochondrial DNA, Y-chromosome sequences, and *MC1R* sheds light on the ancestry of Nigerian indigenous pigs

Adeniyi C. Adeola[1,2], Olufunke O. Oluwole[3], Bukola M. Oladele[3], Temilola O. Olorungbounmi[3], Bamidele Boladuro[3], Sunday C. Olaogun[4], Lotanna M. Nneji[1,2,11], Oscar J. Sanke[5], Philip M. Dawuda[6], Ofelia G. Omitogun[7], Laurent Frantz[8,9], Robert W. Murphy[1,10], Hai-Bing Xie[1,2], Min-Sheng Peng[1,2,11]* and Ya-Ping Zhang[1,2,11,12]*

Abstract

Background: The history of pig populations in Africa remains controversial due to insufficient evidence from archaeological and genetic data. Previously, a Western ancestry for West African pigs was reported based on loci that are involved in the determination of coat color. We investigated the genetic diversity of Nigerian indigenous pigs (NIP) by simultaneously analyzing variation in mitochondrial DNA (mtDNA), Y-chromosome sequence and the *melanocortin receptor 1* (*MC1R*) gene.

Results: Median-joining network analysis of mtDNA D-loop sequences from 201 NIP and previously characterized loci clustered NIP with populations from the West (Europe/North Africa) and East/Southeast Asia. Analysis of partial sequences of the Y-chromosome in 57 Nigerian boars clustered NIP into lineage HY1. Finally, analysis of *MC1R* in 90 NIP resulted in seven haplotypes, among which the European wild boar haplotype was carried by one individual and the European dominant black by most of the other individuals (93%). The five remaining unique haplotypes differed by a single synonymous substitution from European wild type, European dominant black and Asian dominant black haplotypes.

Conclusions: Our results demonstrate a European and East/Southeast Asian ancestry for NIP. Analyses of *MC1R* provide further evidence. Additional genetic analyses and archaeological studies may provide further insights into the history of African pig breeds. Our findings provide a valuable resource for future studies on whole-genome analyses of African pigs.

Background

The origins of African pig breeds are highly controversial owing to a paucity of archaeological and genetic data for hypothesis testing [1, 2]. Previous genetic analyses of West African pigs revealed that they shared maternal and paternal haplotypes with European wild boars and pigs, but not with Near Eastern wild boars [3]. The limited size of West African pig samples did not allow discriminating them from pigs domesticated in North Africa and/ or from pigs introduced by the European colonizers during the 15th–19th centuries. Early Portuguese sailors circumnavigated Africa and, in doing so, they may have introduced a European gene pool into some West African pigs [1]. However, this hypothesis has not been formally tested through genetic analysis. Some Iberian pigs are classified as black hairy and this pattern is common

*Correspondence: pengminsheng@mail.kiz.ac.cn; zhangyp@mail.kiz.ac.cn
[1] State Key Laboratory of Genetic Resources and Evolution, Yunnan Laboratory of Molecular Biology of Domestic Animals, Kunming Institute of Zoology, Chinese Academy of Sciences, Kunming, China
Full list of author information is available at the end of the article

in indigenous West African pigs [1]. However, the causal genetic variants that underlie the color phenotype in the latter pigs remain largely unexplored. Melanocortin receptor 1 (*MC1R*) is a major determinant in color phenotype [4]. Functional mutations in *MC1R* result in different coat colors in domestic animals, such as cattle [5], horses [6], goats [7], sheep [8–10] and pigs [11–13]. Research on *MC1R* has provided valuable insights into the evolution of domesticated animals [13–15]. For instance, Linderholm et al. [13] showed that, among the alleles of *MC1R*, there is a novel black allele unique to Polynesian pigs. Therefore, we used this gene to investigate the genetic diversity and origin of hairy black Nigerian indigenous pigs (NIP) as well as data from mitochondrial DNA (mtDNA) and Y-chromosomes of NIP to provide insights into the origin of NIP.

Methods
Animals
Peripheral blood samples were collected from 204 NIP distributed in six Nigeria states after receiving appropriate permission from their owners (see Additional file 1: Table S1).

Analysis of mtDNA D-loop sequences
Our data involved the amplification and sequencing of 630-base pair (bp) fragments of mtDNA D-loop (the methods are detailed in Additional file 2; GenBank accession numbers: KU561971–KU562068 and KY055561–KY055663). The final dataset for analysis comprised 201 NIP (de novo) and 722 mtDNA D-loop sequences of pigs retrieved from GenBank (see Additional file 3: Table S2). All 923 sequences were aligned and trimmed to 464 bp, which corresponded to nucleotide positions between 112 and 575 of the reference sequence EF545567 [16]. A median-joining network of 923 pig sequences was constructed using NETWORK 5.0 [17].

Y-chromosome analysis
Paternal genetic data were also obtained from 57 Nigerian indigenous sires (see Additional file 1: Table S1) by sequencing 370 bp of intron 1 and part of the flanking exons 1 and 2 of the Y-linked gene *UTY* (*ubiquitously transcribed tetratricopeptide repeat*), which contains repeats (see methods in Additional file 2; GenBank accession numbers: KU561941–KU561970 and KY234314–KY234340). Single nucleotide polymorphisms (SNPs) in the *UTY* amplicon were used to diagnose Y-chromosome lineages HY1 and HY2 versus HY3 [3].

Analysis of *MC1R* sequences
Finally, we analyzed sequence variation in *MC1R* for 90 NIP (see Additional file 1: Table S1) by sequencing the entire *MC1R*-coding region i.e. 963 bp (see methods in Additional file 2; GenBank accession numbers: KX264504–KX264593).

Results and discussion
Mitochondrial DNA
NIP individuals clustered with pig individuals from both the West (Europe/North Africa) and East/Southeast Asia (Fig. 1). These results were consistent with previous analyses of West African pigs [3]. The early introduction of unimproved Iberian swine by the Portuguese into West Africa may have influenced NIP [1]. Ubiquitous standard European breeds, such as Large White and Landrace, which are white pigs, are widespread in Africa because of their excellent productivity, which often overcomes that of local populations [1]. Previously, genetic analyses of indigenous and commercially-developed crossbred pigs from southwestern Nigeria raised concerns about the possibility of genetic erosion in the locally-adapted pigs [18]. Introgression of the Asian matrilineal haplotype into European commercial pigs might have resulted in the clustering of some NIP with East/Southeast Asian pigs. It is also possible that the observed Asian haplotypes in NIP were inherited directly through female Asian introgression due to a low frequency of the European haplotype in NIP that carried the Asian haplotype (Fig. 1).

Y-chromosome
All of the 57 analyzed Nigerian sires were assigned to the HY1 haplotype only (data not shown), which occurs widely in both Europe and Asia [3]. None of the NIP were assigned to HY3, which is unique to Asia and was detected at considerable high frequency in Kenyan pigs (35%) and Zimbabwean Mukota pigs (100%) [3]. This might be due to the influence of East/Southeast Asian pigs on African pigs and, particularly the Mukota pigs from Zimbabwe, which closely resemble the Chinese lard pig, in terms of morphology. Our finding agrees with that of an earlier study that reported Western ancestry for West African pigs [3].

MC1R variation
Analyses using PHASE version 2.1.1 [19, 20] on the NIP samples led to the construction of seven haplotypes for *MC1R* (see Additional file 4: Table S3). The median joining network (Fig. 2) and Additional file 5: Table S4 show that there was one individual with the E^+ European wild type *MC1R* haplotype [14]. Although this homozygous individual carried the European wild type, it displayed a variable coat color phenotype. Within the tested sample of NIP, the E^{D2} (European-dominant black) was the most frequent *MC1R* haplotype at 93% (see Additional file 3:

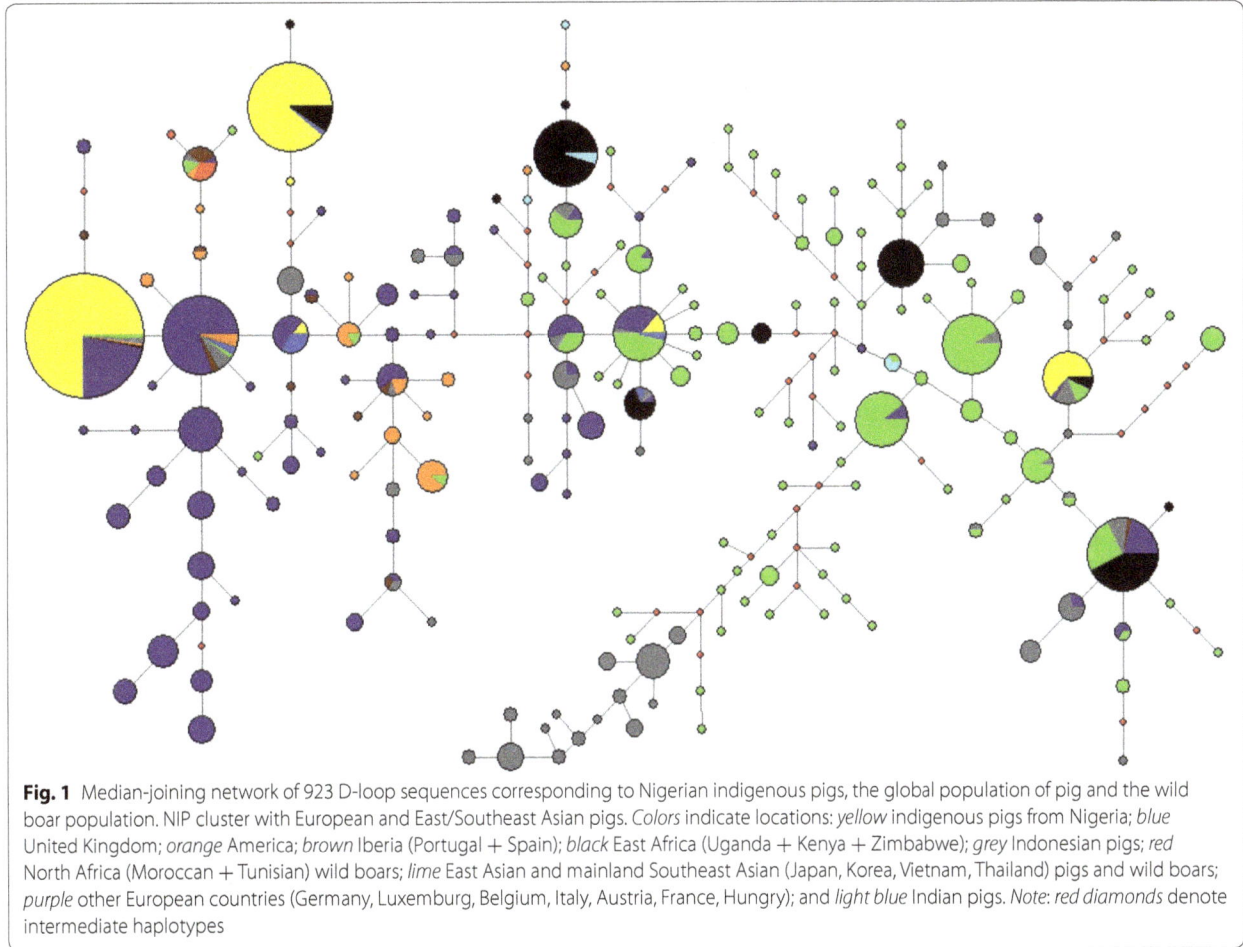

Fig. 1 Median-joining network of 923 D-loop sequences corresponding to Nigerian indigenous pigs, the global population of pig and the wild boar population. NIP cluster with European and East/Southeast Asian pigs. *Colors* indicate locations: *yellow* indigenous pigs from Nigeria; *blue* United Kingdom; *orange* America; *brown* Iberia (Portugal + Spain); *black* East Africa (Uganda + Kenya + Zimbabwe); *grey* Indonesian pigs; *red* North Africa (Moroccan + Tunisian) wild boars; *lime* East Asian and mainland Southeast Asian (Japan, Korea, Vietnam, Thailand) pigs and wild boars; *purple* other European countries (Germany, Luxemburg, Belgium, Italy, Austria, France, Hungry); and *light blue* Indian pigs. *Note: red diamonds* denote intermediate haplotypes

Table S2). The remaining five unique haplotypes differed by a single synonymous substitution from the E^+ (European wild type), E^{D2} (European dominant black) and Asian E^{D1} (dominant black) haplotypes (Fig. 2).

Direct selection for non-camouflage patterns was proposed to be an essential component of the selection of coat color loci in domestic animals [14], which may have been fostered by animal husbandry. Independent selective sweeps have been identified in Chinese and European pigs resulting in the dominant black color. For instance, in Polynesia and Europe, selection of pigs for the D124N substitution in *MC1R* resulted in a dominant black color, whereas selection for the L102P substitution in *MC1R* was responsible for the dominant black color in Chinese pigs [13–15]. These mutations have been used to differentiate Polynesian, European and Asian black pigs. Therefore, the high frequency of the European dominant black color haplotype in NIP suggests the occurrence of gene flow from local European pig breeds. The

NIP individuals that carry the East Asian *MC1R* haplotype might have originated from European black pigs, which agrees with findings from a genome-wide analysis that showed that introgression of Asian haplotypes via anthropogenic hybridization and selection has influenced the genomic architecture of European pigs [21]. Similarly, another possibility is a direct Asian introgression in NIP. Future investigations based on evidence from whole-genome sequence data should test these possibilities (Additional file 2).

Conclusions

In summary, this study reveals that NIP have mainly a European ancestry with some East/Southeast Asian ancestry, which may be due to direct introgression or through introgression from European pig breeds, themselves derived from introgression with Asian breeds. It also provides a first glimpse on *MC1R* variation across populations of indigenous pigs in one West Africa

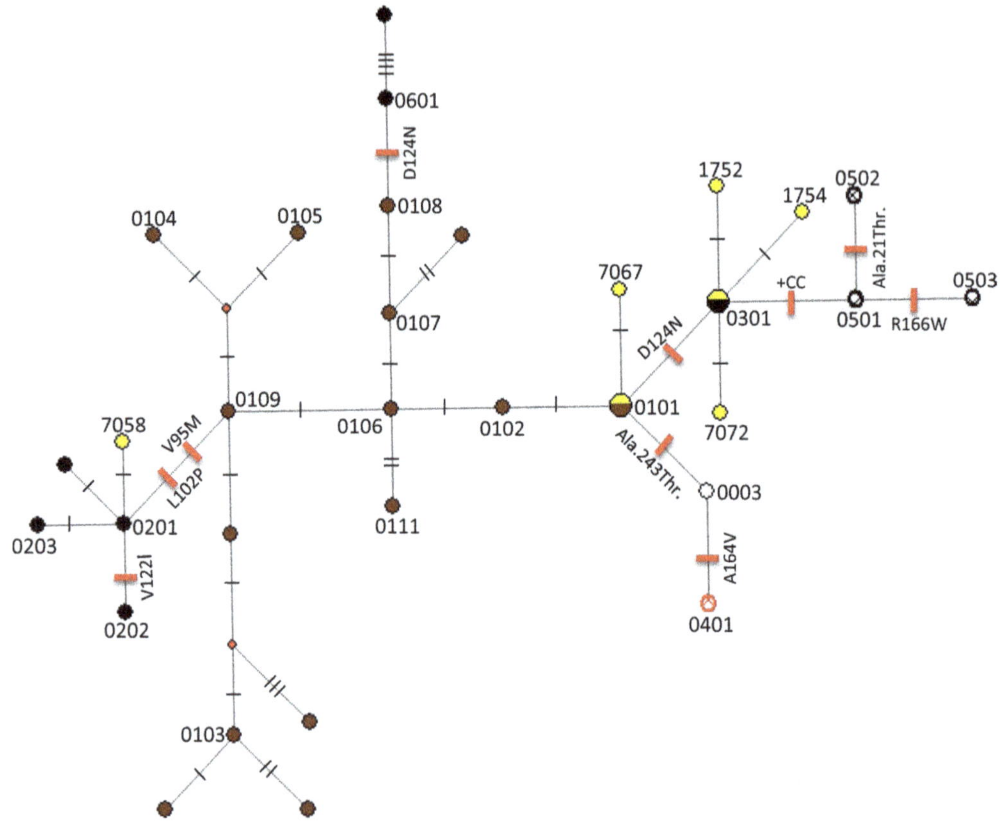

Fig. 2 Median joining network of *MC1R* haplotypes in Nigerian, Polynesian, Asian and European pigs. All known haplotypes are represented by *circles*. *Colors* inside the *circles* indicate the type and nomenclature as follows [13, 14]: *brown* (E+ European and Asian—wild type); *yellow* Nigerian indigenous pigs (NIP); *red* (e—recessive red—European); *black* and *white* (E^P—spotted black—European); and *black* (E^D2 and E^D1—dominant black—European, Polynesian and Asian). Differences in sequences are noted on each of the *branches* and *the small dash lines* represent the number of steps. *Red ticks* perpendicular to each branch represent non-synonymous mutations that change the protein sequence. Note: *red diamond symbols* represent intermediate haplotypes

country. This study was designed to provide a valuable resource for future studies on whole-genome analyses of African pigs.

Additional files

Additional file 1: Table S1. Data on the 204 hairy black NIP sampled from six states in Nigeria. Information on the mtDNA, Y-chromosome and *MC1R* of the 204 NIP samples.

Additional file 2. Details on blood sampling of NIP individuals and sequencing of mtDNA D-loop, Y-chromosome and *MC1R* sequences [15–17, 23]. This is comprehensive information on sampling and sequencing procedure for the 204 NIP.

Additional file 3: Table S2. Data on the 722 pig D-loop sequences analyzed in this study. European, African and Asian mitochondrial control region sequences retrieved from GenBank were used for the median joining network analysis (Fig. 1). Assignments to sub-haplogroups and variants conform to the pig and wild boar mtDNA tree obtained from DomeTree [22].

Additional file 4: Table S3. Mutations in the *MC1R* coding region defining seven haplotypes and their frequencies in Nigerian indigenous pigs. 0301 is the European dominant black pig haplotype; column NIP provides copy number of each haplotype among the samples; dots indicate identity with the previously reported [14] European wild boar haplotype (0101).

Additional file 5: Table S4. *MC1R* alleles in the global population of pigs. Thirty-four *MC1R* haplotypes (seven NIP *de novo* plus 27 downloaded from GenBank) were used to construct the *MC1R* network.

Authors' contributions
ACA, MSP, and YPZ conceived the work. ACA, OOO, BMO, TOO, BB, SCO, and OJS performed animal sampling, ACA and LMN performed the experiment. ACA and MSP performed data analysis. LMN provided technical assistance. ACA, LF, RWM, HBX, MSP, and YPZ were involved in the writing of the paper. All authors read and approved the final manuscript.

Author details
[1] State Key Laboratory of Genetic Resources and Evolution, Yunnan Laboratory of Molecular Biology of Domestic Animals, Kunming Institute of Zoology, Chinese Academy of Sciences, Kunming, China. [2] Sino-Africa Joint Research Center, Chinese Academy of Sciences, Kunming, China. [3] Institute of Agricultural Research and Training, Obafemi Awolowo University, Ibadan, Nigeria.

[4] Department of Veterinary Medicine, University of Ibadan, Ibadan, Nigeria.
[5] Taraba State Ministry of Agriculture and Natural Resources, Jalingo, Nigeria.
[6] Department of Veterinary Surgery and Theriogenology, College of Veterinary Medicine, University of Agriculture Makurdi, Makurdi, Nigeria. [7] Department of Animal Sciences, Obafemi Awolowo University, Ile-Ife, Nigeria. [8] The Palaeogenomics and Bio-Archaeology Research Network, Research Laboratory for Archaeology, University of Oxford, Oxford, UK. [9] School of Biological and Chemical Sciences, Queen Mary University of London, London, UK. [10] Centre for Biodiversity and Conservation Biology, Royal Ontario Museum, Toronto, Canada. [11] Kunming College of Life Science, University of Chinese Academy of Sciences, Kunming, China. [12] State Key Laboratory for Conservation and Utilization of Bio-Resources in Yunnan, Yunnan University, Kunming, China.

Acknowledgements

We are grateful to all volunteers who assisted in sampling. This work was supported by the Sino-Africa Joint Research Center, Chinese Academy of Sciences (SAJC201611 and SAJC201306) and the Animal Branch of the Germplasm Bank of Wild Species, Chinese Academy of Sciences (the Large Research Infrastructure Funding). The Youth Innovation Promotion Association, Chinese Academy of Sciences provided support to MSP. In addition, this work was also supported, in part, by the Chinese Academy of Sciences President's International Fellowship Initiative (2017VBA0003), and the manuscript preparation by a Natural Sciences and Engineering Research Council of Canada Discovery Grant A3148 to R.W.M.

Competing interests

The authors declare that they have no competing interests.

References

1. Blench RM. A history of pigs in Africa. In: Blench RM, Mac Donald KC, editors. The origins and development of African livestock: archaeology, genetics, linguistics and ethnography. London: Routledge; 2000. p. 355–67.
2. Amills M, Ramírez O, Galman-Omitogun O, Clop A. Domestic pigs in Africa. Afr Archaeol Rev. 2013;30:73–82.
3. Ramirez O, Ojeda A, Tomas A, Gallardo D, Huang LS, Folch JM, et al. Integrating Y-chromosome, mitochondrial, and autosomal data to analyze the origin of pig breeds. Mol Biol Evol. 2009;26:2061–72.
4. Lin JY, Fisher DE. Melanocyte biology and skin pigmentation. Nature. 2007;445:843–50.
5. Rouzaud F, Martin J, Gallet PF, Delourme D, Goulemot-Leger V, Amigues Y, et al. A first genotyping assay of French cattle breeds based on a new haplotype of the extension gene encoding the melanocortin-1 receptor (MC1R). Genet Sel Evol. 2000;32:511–20.
6. Marklund L, Moller MJ, Sandberg K, Andersson L. A missense mutation in the gene for melanocyte-stimulating hormone receptor (MC1R) is associated with the chestnut coat color in horses. Mamm Genome. 1996;7:895–9.
7. Fontanesi L, Beretti F, Riggio V, Dall'Olio S, Gonzalez EG, Finocchiaro R, et al. Missense and nonsense mutations in melanocortin 1 receptor (MC1R) gene of different goat breeds: association with red and black coat colour phenotypes but with unexpected evidences. BMC Genet. 2009;10:47.
8. Vage DI, Klungland H, Lu D, Cone RD. Molecular and pharmacological characterization of dominant black coat color in sheep. Mamm Genome. 1999;10:39–43.
9. Vage DI, Fleet MR, Ponz R, Olsen RT, Monteagudo LV, Tejedor MT, et al. Mapping and characterization of the dominant black colour locus in sheep. Pigment Cell Res. 2003;16:693–7.
10. Fontanesi L, Dall'Olio S, Beretti F, Portolano B, Russo V. Coat colours in the Massese sheep breed are associated with mutations in the agouti signalling protein (ASIP) and melanocortin 1 receptor (MC1R) genes. Animal. 2011;5:8–17.
11. Kijas JM, Wales R, Tornsten A, Chardon P, Moller M, Andersson L. Melanocortin receptor 1 (MC1R) mutations and coat color in pigs. Genetics. 1998;150:1177–85.
12. Kijas JM, Moller M, Plastow G, Andersson L. A frameshift mutation in MC1R and a high frequency of somatic reversions cause black spotting in pigs. Genetics. 2001;158:779–85.
13. Linderholm A, Spencer D, Battista V, Frantz L, Barnett R, Fleischer RC, et al. A novel MC1R allele for black coat colour reveals the Polynesian ancestry and hybridization patterns of Hawaiian feral pigs. R Soc Open Sci. 2016;3:160304.
14. Fang M, Larson G, Ribeiro HS, Li N, Andersson L. Contrasting mode of evolution at a coat color locus in wild and domestic pigs. PLoS Genet. 2009;5:e1000341.
15. Li J, Yang H, Li JR, Li HP, Ning T, Pan XR, et al. Artificial selection of the melanocortin receptor 1 gene in Chinese domestic pigs during domestication. Heredity (Edinb). 2010;105:274–81.
16. Wu GS, Yao YG, Qu KX, Ding ZL, Li H, Palanichamy MG, et al. Population phylogenomic analysis of mitochondrial DNA in wild boars and domestic pigs revealed multiple domestication events in East Asia. Genome Biol. 2007;8:R245.
17. Bandelt HJ, Forster P, Röhl A. Median-joining networks for inferring intraspecific phylogenies. Mol Biol Evol. 1999;16:37–48.
18. Adeola AC, Omitogun OG. Characterization of indigenous pigs in Southwestern Nigeria using blood protein polymorphism. Anim Genet Resour. 2012;51:125–30.
19. Stephens M, Smith NJ, Donnelly P. A new statistical method for haplotype reconstruction from population data. Am J Hum Genet. 2001;68:978–89.
20. Stephens M, Scheet P. Accounting for decay of linkage disequilibrium in haplotype inference and missing-data imputation. Am J Hum Genet. 2005;76:449–62.
21. Bosse M, Lopes MS, Madsen O, Megens HJ, Crooijmans RP, Frantz LA, et al. Artificial selection on introduced Asian haplotypes shaped the genetic architecture in European commercial pigs. Proc Biol Sci. 2015;282:20152019.
22. Peng MS, Fan L, Shi NN, Ning T, Yao YG, Murphy RW, et al. DomeTree: a canonical toolkit for mitochondrial DNA analyses in domesticated animals. Mol Ecol Resour. 2015;15:1238–42.
23. Tamura K, Peterson D, Peterson N, Stecher G, Nei M, Kumar S. MEGA5: molecular evolutionary genetics analysis using maximum likelihood, evolutionary distance, and maximum parsimony methods. Mol Biol Evol. 2011;28:2731–9.

11

Contributions of linkage disequilibrium and co-segregation information to the accuracy of genomic prediction

Xiaochen Sun[*], Rohan Fernando and Jack Dekkers

Abstract

Background: Traditional genomic prediction models using multiple regression on single nucleotide polymorphisms (SNPs) genotypes exploit associations between genotypes of quantitative trait loci (QTL) and SNPs, which can be created by historical linkage disequilibrium (LD), recent co-segregation (CS) and pedigree relationships. Results from field data analyses show that prediction accuracy is usually much higher for individuals that are close relatives of the training population than for distantly related individuals. A possible reason is that historical LD between QTL and SNPs is weak and, for close relatives, prediction accuracy of SNP models is mainly contributed by pedigree relationships and CS. Information from pedigree relationships decreases fast over generations and only contributes to within-family prediction. Information from CS is affected by family structures and effective population size, and can have a substantial contribution to prediction accuracy when modeled explicitly.

Results: In this study, a method to explicitly model CS was developed by following the transmission of putative QTL alleles using allele origins at SNPs. Bayesian hierarchical models that combine information from LD and CS (LD-CS model) were developed for genomic prediction in pedigree populations. Contributions of LD and CS information to prediction accuracy across families and generations without retraining were investigated in simulated half-sib datasets and deep pedigrees with different recent effective population sizes, respectively. Results from half-sib datasets showed that when historical LD between QTL and SNPs is low, accuracy of the LD model decreased when the training data size is increased by adding independent sire families, but accuracies from the CS and LD-CS models increased and plateaued rapidly. Results from deep pedigree datasets show that the LD model had high accuracy across generations only when historical LD between QTL and SNPs was high. Modeling CS explicitly resulted in higher accuracy than the LD model across generations when the mating design generated many close relatives.

Conclusions: Our results suggest that modeling CS explicitly improves accuracy of genomic prediction when historical LD between QTL and SNPs is low. Modeling both LD and CS explicitly is expected to improve accuracy when recent effective population size is small, or when the training data include many independent families.

Background

The feasibility of obtaining genotypes of dense single nucleotide polymorphisms (SNPs) with genome-wide coverage has improved accuracy of estimated breeding values by genomic prediction [1–6]. To date, most statistical models for genomic prediction are based on multiple regression of phenotypes on SNP genotype covariates (SNP models). The estimated SNP effects are used to predict genomic estimated breeding values (GEBV) for selection candidates, which are usually progeny of individuals in the training population [7]. Linkage disequilibrium (LD) between quantitative trait loci (QTL) and SNPs was initially thought to be the only source of genetic information that contributes to accuracy of genomic prediction using SNP models, until [8] and [9] showed that co-segregation (CS) of QTL with SNPs and pedigree relationships that are implicitly captured by SNP genotypes also contribute to prediction accuracy.

*Correspondence: xsun1120@iastate.edu
Department of Animal Science and Center for Integrated Animal Genomics, Iowa State University, Ames, IA 50011, USA

Co-segregation is an important source of information that contributes to accuracy of genomic prediction [9, 10]. Alleles co-segregate when they originate from the same parental chromosome. Thus, in this study, CS is defined as a non-random association between the grand-parental allele origins of two linked loci. For instance, the maternal alleles of an individual at two loci co-segregate when both alleles originate from the same grand maternal chromosome [11, 12]. With high-density genotyping, the probability that a QTL allele co-segregates with its two adjacent SNP alleles is high. For example, the average distance between adjacent SNPs on the Illumina Bovine SNP50 BeadChip is 50 kb [13, 14], and the average recombination rate between two adjacent SNPs is only around 0.0005 per meiosis, assuming a typical crossover rate of 1 % per million base pair.

In analyses of field datasets using SNP models, high accuracy of genomic prediction has been mainly observed among close relatives [1, 2, 15, 16], and prediction accuracy decreases rapidly when the validation individuals are separated from training individuals by more generations [5, 16–18]. The latter does not agree with results from simulation studies in which the LD between QTL and SNPs was high [7–9, 19]. These results suggest that LD between QTL and SNPs is low in current livestock populations, and that prediction accuracy of the SNP model mainly comes from CS and pedigree relationships that are implicitly captured by SNP genotypes [8, 10, 16, 17, 20].

Simulation studies have shown that both LD and CS information contribute to prediction accuracy of the genomic best linear unbiased prediction (GBLUP) model [9]. Information from historical LD was persistent across generations and contributed to prediction accuracy across families and across validation generations. CS information that is captured implicitly by GBLUP was not persistent across families or generations, and its contribution to prediction accuracy decreased when the number of unrelated families increased in the training population [9]. Simulation studies of an aquaculture breeding program [21] showed that the contribution of CS information to prediction accuracy was similar across a wide range of SNP densities, while the contribution of LD information dropped significantly with decreasing SNP density, indicating that the accuracy due to CS was not affected by the level of LD. In a study using data from Italian Brown Swiss bulls, the effect of LD and CS information on prediction accuracy was investigated for the GBLUP model, with the covariance structure of genomic breeding values constructed either using LD or CS information at SNPs [10]. The GBLUP model that fitted both LD and CS had a similar accuracy as that fitting only CS, which was slightly higher than that fitting only LD

[10]. Their results also suggest that when historical LD between QTL and SNPs is low, prediction accuracy for closely related individuals mainly comes from CS instead of LD.

Although LD between SNPs has been shown to be sizable in livestock populations [14, 22–24], LD between SNPs and unobservable QTL can be much lower than LD between SNPs, which is probably due to the difference in minor allele frequencies (MAF) of SNPs and QTL. QTL for economically important traits are likely to have low MAF either because the traits have been subject to directional selection for a long time [25–27], or because some QTL are the result of mutations that occurred more recently than the mutations that caused SNPs [25, 27, 28]. SNPs included on SNP chips usually have moderate to high MAF due to ascertainment bias from sequencing and prototype genotyping of reference samples [13]. Since LD between loci that have different MAF is low, LD information contributes little to prediction accuracy when most QTL have much lower MAF than SNPs (Detailed discussion is provided in Additional file 1). Modeling CS explicitly can increase accuracy when historical LD is low because CS information follows transmission of QTL alleles among related individuals, which is independent of the level of LD between QTL and SNPs.

Explicit modeling of CS information for genomic prediction was proposed and developed by Luan et al. [10]. In Luan et al. [10], CS was modeled at each SNP locus using the method in Fernando and Grossman [29], and a realized relationship matrix using CS information was constructed by averaging across all SNPs with equal weights. The method for modeling CS in Luan et al. [10] can be improved in two aspects. First, since CS signals span long genomic distances, modeling CS across multiple SNPs is expected to capture the same amount of CS information as modeling at each SNP, but modeling across multiple loci can substantially improve computational efficiency. Second, Luan et al. [10] assumed that the contribution of CS information at each SNP was the same. The variance at QTL can vary due to differences in allele frequencies and effects at the QTL, and these variances can be treated as unknowns and marginalized in a Bayesian analysis. In this study, a new method is developed to model CS explicitly. The CS model follows the transmission of putative QTL within 1-cM genomic windows. A detailed description of a Bayesian hierarchical model for genomic prediction using CS is provided, and a Gibbs sampling algorithm for prediction of breeding values is derived.

Persistence of prediction accuracy across validation generations without retraining (long-term accuracy) is an important criterion to evaluate contributions of LD and CS information to prediction accuracy. Habier et al. [9] showed that LD information was more persistent than the CS information that is implicitly captured by SNP genotypes because CS information decays across generations due to recombination within large chromosome segments. Modeling CS explicitly at small putative QTL regions is expected to improve accuracy across generations because recombination is less likely to happen within small chromosome segments. The contribution of CS information to accuracy across generations by modeling CS explicitly has not been studied.

Recent effective population size (N_e) is another important factor that affects the contribution of CS information to long-term accuracy. For a given size of the training and validation populations, individuals are more closely related when recent N_e is smaller and CS information is expected to contribute more to long-term accuracy in that case for two reasons. First, CS is generated as associations between loci over long chromosome regions. Smaller recent N_e causes higher CS due to stronger drift and selection of alleles in recent generations. Second, with smaller recent N_e, fewer founder alleles are each inherited by relatively more offspring, and, thus, the values of founder alleles can be estimated more accurately when more data is available for each allele. In livestock populations, recent N_e is affected by the mating design and can vary greatly between breeding programs. Therefore, it is important to study the effect of recent N_e on the contribution of CS information to long-term accuracy of genomic prediction.

The objectives of this study were (1) to develop a Bayesian statistical method to model CS explicitly by following the transmission of putative QTL alleles of pedigree founders, (2) to investigate contributions of LD and CS information to the accuracy of genomic prediction across unrelated families and validation generations without re-training, and (3) to investigate the effects of historical LD, recent N_e, and MAF of QTL on the advantage of modeling LD and CS explicitly to improve prediction accuracy.

Methods
Genomic prediction models using LD, CS and combined LD-CS information
Definitions of LD, CS, and pedigree relationship are presented for a population with known pedigree information following Habier et al. [12]. LD is defined as all non-random associations between allele states in pedigree founders. These associations result in greater similarity between individuals that have the same marker allele states due to QTL that are in LD with the markers. CS is defined as non-random association between allele origins in pedigree generations. These associations result in higher covariances between relatives that have the same

marker allele origins than those conditional only on pedigree relationships due to CS between QTL and markers. Similarly, relatives that have different marker allele origins will have lower covariances than those conditional only on pedigree relationships. Once LD is defined within pedigree founders, all the "new" associations created in pedigree generations can be explained by co-segregation and pedigree relationships. Since pedigree relationships quickly dissipate with generations, their contribution to long-term accuracy is small [8, 12]. Therefore, this study only focuses on the contributions of LD and CS information.

Following Meuwissen et al. [7], the statistical model that uses LD information for prediction of GEBV for a quantitative trait is written as:

$$\mathbf{y} = \mathbf{X}\boldsymbol{\beta} + \mathbf{Z}\boldsymbol{\alpha} + \mathbf{e}, \qquad (1)$$

where $\mathbf{y}$ is an $n \times 1$ vector of trait phenotypes of n training individuals, $\boldsymbol{\beta}$ is a vector of non-genetic fixed effects, $\mathbf{X}$ is the design matrix for fixed effects, $\mathbf{Z}$ is an $n \times m$ matrix with each row containing genotypes (coded as 0/1/2) at m SNPs of each training individual, $\boldsymbol{\alpha}$ is an $m \times 1$ vector of allele substitution effects of the m SNPs, and $\mathbf{e}$ is an $n \times 1$ vector of residuals. Informative prior distributions are usually given to $\boldsymbol{\alpha}$ to allow simultaneous estimation of all SNP effects.

In the LD model (1), QTL effects are not explicitly fitted but SNPs are used as surrogates for QTL due to LD. The genotypic value at the QTL that is captured by surrounding SNP genotypes can be viewed as the conditional expectation of this genotypic value given SNP genotypes. When LD between QTL and SNPs is not complete, the true genotypic value at the QTL deviates from its conditional expectation. Therefore, under low LD, the LD model can only capture part of the genetic variance at QTL.

The CS model is given by

$$\mathbf{y} = \mathbf{X}\boldsymbol{\beta} + \sum_{j=1}^{n_q} \mathbf{W}_j \mathbf{v}_j + \mathbf{e}, \qquad (2)$$

where $\mathbf{y}$ is an $n \times 1$ vector of trait phenotypic values of n training individuals, $\boldsymbol{\beta}$ and $\mathbf{X}$ are the same as in the LD model (1), $\mathbf{v}_j$ is a vector of the values of founder alleles at the jth putative QTL, with n_q the number of putative QTL, $\mathbf{W}_j$ is the covariate matrix for $\mathbf{v}_j$, and $\mathbf{e}$ is an $n \times 1$ vector of residuals. As in the LD model, informative prior distributions are given to $\mathbf{v}_j$'s to allow simultaneous estimation of the value of founder alleles. Definitions of the value of founder alleles $\mathbf{v}_j$ and their covariates $\mathbf{W}_j$ are given in the next section.

The model that fits both LD and CS (LD-CS model) includes the LD and CS terms from models (1) and (2),

$$\mathbf{y} = \mathbf{X}\boldsymbol{\beta} + \mathbf{Z}\boldsymbol{\alpha} + \sum_{j=1}^{n_q} \mathbf{W}_j\mathbf{v}_j + \mathbf{e}. \qquad (3)$$

In the LD-CS model (3), the conditional expectation of the genotypic value at a putative QTL is captured by surrounding SNP genotypes in the LD term, while the genotypic value at a putative QTL are explicitly fitted in the CS term. When LD between QTL and SNPs is not complete, deviations between QTL genotypic values and their conditional expectations on SNP genotypes are captured by the CS term. Therefore the LD-CS model (3) is expected to capture most genetic variance at QTL under incomplete LD.

Statistical modeling of CS information

Co-segregation of alleles at two loci means that these alleles share identical grand-parental allele origins, i.e. they both originate from the same chromosome of a parent. The indicator of parental allele origin at one locus is a Bernoulli variable. In this study, the allele origin indicator equals 0 if it originates from its grand-maternal allele, and 1 if it originates from its grand-paternal allele. When allele origins of parents and offspring at a SNP are known, the probability that the allele origin of a putative QTL linked to the SNP is grand-paternal (equals 1) can be calculated using recombination rates between QTL and SNPs, which is termed the probability of descent of the QTL allele (PDQ). Suppose that allele origins are known for an individual's maternal alleles at two SNPs M_1 and M_2. Then, assuming no interference, the PDQ at the putative QTL is calculated as follows when the origins of both SNP alleles are the mother's maternal allele, i.e. $O_1^m = 0$ and $O_2^m = 0$,

$$
\begin{aligned}
&\Pr(O_Q^m = 0 | O_1^m = 0, O_2^m = 0) \\
&= \frac{\Pr(O_1^m = 0, O_Q^m = 0, O_2^m = 0)}{\Pr(O_1^m = 0, O_2^m = 0)} \\
&= \frac{(1 - r_1)(1 - r_2)}{1 - r_{12}},
\end{aligned}
\qquad (4)
$$

where O_i^m is the maternal allele origin at M_i for $i = 1, 2$, O_Q^m is the maternal allele origin at the QTL, r_1 is the recombination rate between M_1 and QTL, r_2 is the recombination rate between QTL and M_2, and r_{12} is the recombination rate between M_1 and M_2. Recombination rates r_1, r_2 and r_{12} can be calculated from the map distance between M_1 and M_2 using mapping functions [30].

In most cases, positions of QTL are not known, and genome-wide CS information is modeled for putative QTL within each non-overlapping genomic window of a certain length. In this study, putative QTL were fitted at genome windows of lengths 0.5, 1 and 2 cM, respectively,

to study the impact of window length on the accuracy of the CS and LD-CS models. Within each genomic window, the putative QTL is assumed to be located at the midpoint of SNPs M_1 and M_2 that are flanking the window. It follows that $r_1 = r_2 = 0.5[1 - \exp(-d_{M_1M_2})]$ and $r_{12} = 0.5[1 - \exp(-2d_{M_1M_2})]$, where $d_{M_1M_2}$ is the distance between M_1 and M_2 in Morgan. Transmission of putative QTL alleles is followed by PDQ calculated using allele origins of M_1 and M_2. When there is no recombination between M_1 and M_2, the PDQ of the putative QTL is either 0 or 1, indicating the transmission of grand maternal or paternal QTL allele, respectively. When recombination occurs between M_1 and M_2, the PDQ of the putative QTL is 0.5, meaning that the value of the recombinant QTL allele is the average of the values of the two parental QTL alleles.

The method for modeling CS uses PDQ to follow the transmission of founder alleles at putative QTL. The true breeding value (TBV) of an individual is the summation of the value of its maternal and paternal alleles (denoted as v^m and v^p, respectively) of putative QTL across the genome. At each putative QTL, the values of founder QTL alleles were assumed independent. The values of non-founder QTL alleles are linear combinations of the values of founder QTL alleles, with covariates determined by the PDQ. The covariates of maternal and paternal QTL alleles of a non-founder individual i ($\mathbf{w}'^m_i$ and $\mathbf{w}'^p_i$, respectively) were calculated recursively as

$$\mathbf{w}'^m_i = \text{PDQ}^m_i \mathbf{w}'^m_{\text{dam}} + (1 - \text{PDQ}^m_i)\mathbf{w}'^p_{\text{dam}}, \qquad (5)$$

and

$$\mathbf{w}'^p_i = \text{PDQ}^p_i \mathbf{w}'^m_{\text{sire}} + (1 - \text{PDQ}^p_i)\mathbf{w}'^p_{\text{sire}}, \qquad (6)$$

where PDQ^m_i (PDQ^p_i) is the maternal (paternal) PDQ for non-founder i, and $\mathbf{w}'^m_{\text{dam}}$ ($\mathbf{w}'^p_{\text{sire}}$) is the dam's maternal (sire's paternal) covariates for founder QTL alleles. Vectors $\mathbf{w}'^m_{\text{dam}}$ and $\mathbf{w}'^p_{\text{sire}}$ at each QTL have dimension equal to twice the number of pedigree founders.

Vectors $\mathbf{w}'^m_i$ and $\mathbf{w}'^p_i$ comprise the rows in the incidence matrix $\mathbf{W}_H$ that relates the values of QTL alleles of all individuals with those of founders. The incidence matrix that relates TBV with the values of founder QTL alleles ($\mathbf{W}$) was derived by the summation of every two rows in $\mathbf{W}_H$ that correspond to the paternal and maternal QTL alleles of the same individual, i.e.

$$\mathbf{W} = (\mathbf{I}_n \otimes [1, 1]) \times \mathbf{W}_H,$$

where $\mathbf{I}_n$ is the identity matrix with dimension n. The vector of TBV of all individuals ($\mathbf{g}_{CS}$) can be written as

$$\mathbf{g}_{CS} = \mathbf{W}\mathbf{v},$$

where $\mathbf{v}$ is the vector of values of founder alleles at all putative QTL.

Prediction of breeding values

Method BayesA and BayesB were used to estimate SNP allele substitution effects and values of founder QTL alleles. Details of Bayesian inference using BayesA and BayesB are given in Additional file 2. The value of π_{SNP} for BayesB in the LD and LD-CS models was calculated as $1 - \frac{\text{Number of QTL}}{\text{Number of SNPs}}$. The value of π_{CSE} for BayesB in the CS and LD-CS models was 0.95, indicating the proportion of founder alleles that have an ignorable effect on TBV. The Gibbs sampler was run for 21,000 iterations, with the first 1000 discarded as burn-in. Point estimates of SNP effects and values of founder alleles at putative QTL were posterior means calculated from the MCMC samples.

Simulation of the base population

Contributions of LD and CS information to prediction accuracy were investigated using simulated datasets of paternal half-sib designs with different numbers of independent sire families, and extended pedigrees with three mating designs that differed in recent N_e. Parents of half-sibs and founders of extended pedigrees were random samples from the same base population, following closely the simulations of Habier et al. [8, 16] and Sun et al. [31].

The simulated genome comprised two chromosomes, each 1 Morgan long. Each chromosome was evenly covered by 4000 SNPs. Fifty candidate QTL were randomly positioned within each cM of the genome. The mutation rate for QTL and SNPs was 2.5×10^{-5} per meiosis per locus. The number of crossovers per chromosome was sampled from a Poisson distribution with mean 1.0, and the positions of the crossovers were sampled from a uniform distribution.

Two scenarios were simulated for historical LD between SNPs in the base population. In the scenario of high historical LD between SNPs, the base population was generated as follows. The initial generations comprised a population with $N_e = 500$ that was randomly mated for 500 generations to generate LD between closely linked loci, after which the population was shrunk to $N_e = 200$ and randomly mated for another 100 generations to create LD over longer genetic distances. In the next 10 generations, the population was linearly scaled up to an actual size of 2000 as the base population. In the scenario with no historical LD between SNPs, a population of actual size 2000 was generated as base population with SNP and QTL alleles randomly sampled with frequency 0.5. This resulted in a population that was both in linkage equilibrium and in Hardy-Weinberg equilibrium.

In the base population, 2000 SNPs with MAF higher than 0.05 on each chromosome and one segregating QTL within each cM of the genome were sampled according to their MAF, depending on the scenario (Table 1). In the Common QTL scenario, all QTL had MAF between 0.01 and 0.5. In the Rare QTL scenario, all QTL had MAF

between 0.01 and 0.06. Additive QTL effects were randomly sampled from a standard normal distribution. The TBV were obtained as the summation of all QTL allele values for a given individual. Allele substitution effects of QTL were scaled in the base population to achieve a genetic variance equal to 4.29. Normally distributed random errors with mean 0 and variance 10.0 were added to TBV to generate phenotypes for a quantitative trait with a narrow sense heritability of 0.3.

For each pedigree, 50 replicated datasets were independently simulated for each scenario in Table 1. All replicated datasets used the same initial SNP positions but had different randomly sampled QTL effects and, after selection of loci based on MAF, had different positions of QTL and SNPs.

Simulation of half-sib datasets

To study contributions of LD and CS information to prediction accuracy across unrelated families, paternal half-sib families from different numbers of sires were simulated. From the base population, s sires and $20 \times s$ dams were randomly sampled without replacement as the parents of half-sib offspring. Each of the s sires was mated with 20 dams, with each dam producing one offspring. Within each sire family, 10 random half-sib offspring were used in the training population and the other 10 for validation. Independent datasets were generated for different numbers of sire families, $s = 1, 2, 5, 10, 50, 100$ and 200, corresponding to training population sizes of 10, 20, 50, 100, 500, 1000 and 2000, respectively.

Simulation of pedigree population designs

To study the effect of recent N_e on long-term accuracy due to LD or CS information, three mating designs with

Table 1 Minor allele frequencies (MAF) of QTL and SNPs and the level of historical linkage disequilibrium (LD) in the base population of simulated scenarios

Scenario	Common QTL[a]		Rare QTL[b]	
	High LD[c]	No LD[d]	High LD	No LD
MAF of QTL	0.01–0.5		0.01–0.06	
MAF of SNPs	0.06–0.5			
LD between QTL and SNPs	>0	=0	≈0	=0
LD between SNPs	>0	=0	>0	=0

[a] The scenario with MAF of QTL between 0.01 and 0.50, and MAF of SNPs between 0.06 and 0.50

[b] The scenario with MAF of QTL between 0.01 and 0.06, and MAF of SNPs between 0.06 and 0.50

[c] The scenario with high LD in the base population created by historical generations

[d] The scenario with linkage equilibrium in the base population by independently sampling genotypes of QTL and SNPs

different recent N_e were simulated. The mating designs were represented by three pedigrees with 13 non-overlapping generations but different numbers of parents and offspring per mating. The founders (first generation) of all three pedigrees comprised five sires, each mated with 10 dams. Sires and dams of the first generation were randomly sampled from a base population of size 2000. Every mating in the first generation produced six male and six female progeny (second generation).

In pedigree 1, five sires and 50 dams were randomly selected in each generation starting from the 2nd generation. Each sire was mated with 10 dams, each producing six male and six female progeny. Pedigree 1 represents a balanced nested design where a small number of sires was selected in each generation and each sire on average contributed equally to the next generation. The N_e for pedigree 1 was calculated as $N_e = \frac{4 \times 5 \times 50}{5+50} = 18.2$ [32].

In pedigree 2, all 300 sires and 300 dams from generation 2 were used as parents. Each sire was mated with one dam, producing one male and one female progeny. Pedigree 2 represents an outbred population where all individuals survived, but each individual had a relatively limited contribution to future generations. Since each individual contributes an equal number of gametes to the next generation, and the variance of family sizes is zero, the N_e of pedigree 2 is approximated by $2N = 1200$, where N is the actual population size of each generation (600 for pedigree 2) [32].

In pedigree 3, five sires and 70 dams were randomly selected in each generation starting from the second generation. One sire was mated with 50 dams, each dam producing five male and five female progeny, representing an influential sire family. Each of the other four sires was mated with five dams, each dam producing two male and three female progeny, representing four small sire families. Pedigree 3 represents an unbalanced nested design, where the genetics of one individual dominates future generations. The N_e of pedigree 3 was much less than 18.2.

The first five pedigree generations, with size 2455 (pedigree 1 and 2) and 2475 (pedigree 3), were used for training. Each of the following eight generations, with size 600, were used for validation. Prediction accuracy was calculated as the correlation between GEBV and TBV in each validation generation.

In all simulated datasets, allele origins were assumed known without error at all SNPs. Putative QTL were fitted within each 1-cM genome and PDQ were calculated using the simulated true allele origins. In addition, for pedigree 1, allele origins at all SNPs were either assumed known without error, or imputed using the LDMIP software, with default parameter settings [33]. Putative QTL were fitted within every 0.5, 1 and 2 cM and PDQ were

calculated using either true or imputed allele origins, to investigate the impacts of length of genomic windows to model CS and unknown allele origins.

Results

Half-sib designs

In the Common QTL scenario with high historical LD, prediction accuracy of the LD model increased from less than 0.2 with one half-sib family and quickly plateaued around 0.8 when the number of half-sib families exceeded 50, which corresponds to a training size of 500 (Fig. 1). Accuracy of the LD-CS model could not be distinguished from that of the LD model at all training sizes. Accuracy of the CS model increased from 0.2 and plateaued around 0.4 when the training size exceeded 500, which was much lower than accuracies of the LD or LD-CS model (Fig. 1). These results suggest that when LD between QTL and SNPs is high, the LD model has high accuracy by capturing information from both LD and CS, and modeling CS explicitly in addition to LD does not improve prediction accuracy.

In the Rare QTL scenario with high historical LD between SNPs, the actual level of historical LD between QTL and SNPs was low due to all QTL having much lower MAF than SNPs. Accuracy of the LD model increased with training size from 0.1 to about 0.45, which was much lower than accuracy with the Common QTL scenario (Fig. 1). Accuracy of the CS model also increased with training size and plateaued around 0.4, which was similar to the Common QTL scenario. Accuracy of the LD-CS model increased and became significantly higher than accuracy of both the LD and CS models when the training size exceeded 100 (10 half-sib families) (Fig. 1). These results suggest that when LD between QTL and SNPs is low, the contribution of CS information is more important than when LD between QTL and SNPs is high, and that modeling CS explicitly in addition to LD improves prediction accuracy across unrelated families.

In the Common QTL and Rare QTL scenarios without historical LD, the LD, CS and LD-CS models had similar accuracies when the training size was less than 500 (Fig. 2). The CS and LD-CS models had higher accuracy than the LD model when the training size exceeded 500. Accuracy from the LD model decreased from 0.35 to 0.25 when the training size exceeded 500 (Fig. 2). These results suggest that when there is no historical LD between QTL and SNPs, accuracy of the LD model comes from implicitly capturing CS information, but the ability to capture CS information decreases when a large number of unrelated families are included in the training population. Without historical LD, the CS model has much higher accuracy than the LD model due to explicitly capturing CS information.

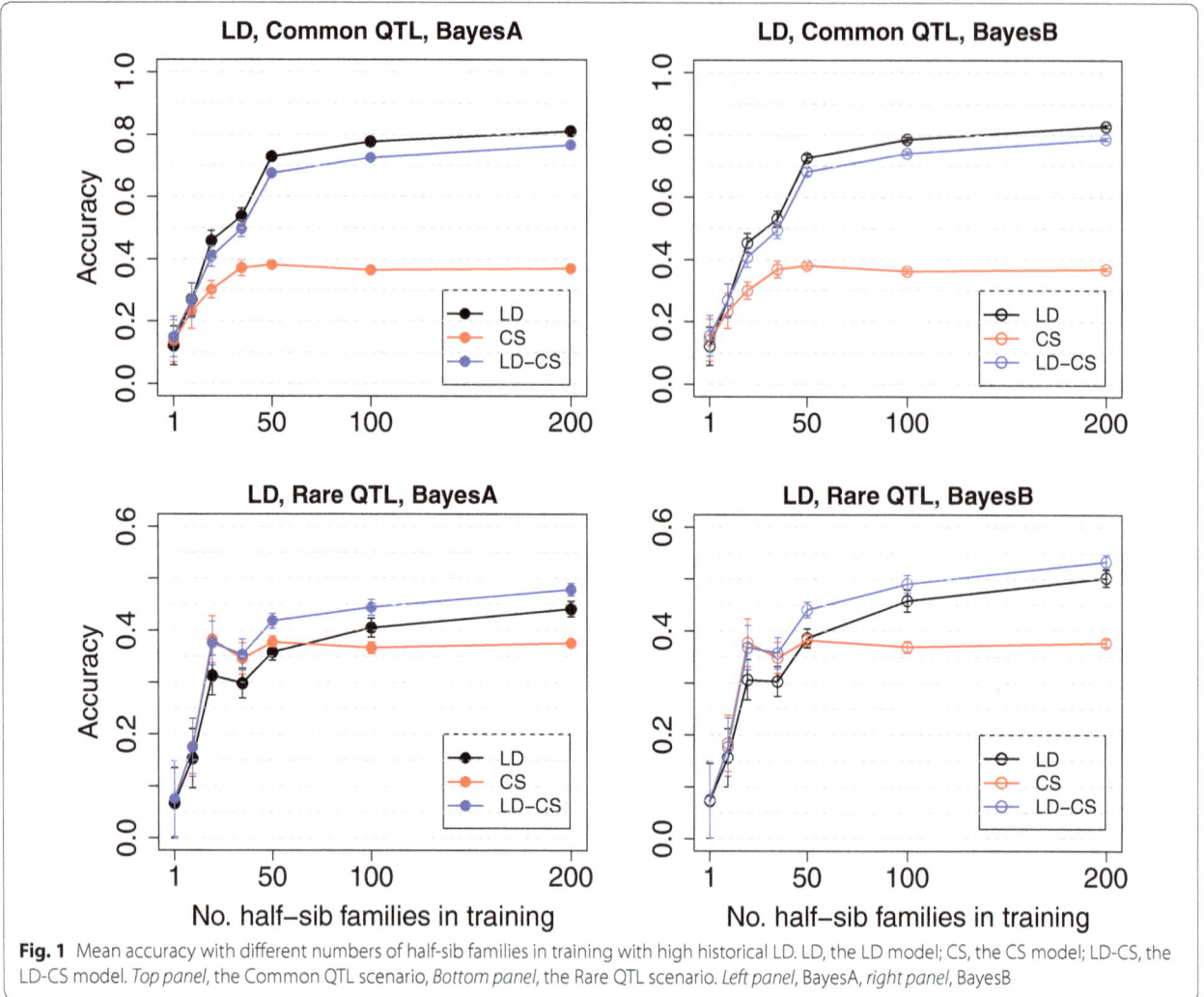

Fig. 1 Mean accuracy with different numbers of half-sib families in training with high historical LD. LD, the LD model; CS, the CS model; LD-CS, the LD-CS model. *Top panel*, the Common QTL scenario, *Bottom panel*, the Rare QTL scenario. *Left panel*, BayesA, *right panel*, BayesB

Pedigree mating designs

In the Common QTL scenario with high historical LD, the LD model had higher accuracy than either the CS or LD-CS model (Fig. 3). For all three pedigrees, the LD-CS model had slightly lower accuracy than the LD model, but the CS model had much lower accuracy than the LD model. Accuracies from the LD and LD-CS models only decreased marginally across the eight validation generations, but the accuracy of the CS model decreased rapidly (Fig. 3). These results suggest that when historical LD between QTL and SNPs is high, the LD model has persistently high accuracy across validation generations without retraining by accurately capturing QTL effects. The CS model estimates only the values of founder alleles, and the values of recombinant alleles that are generated across generations cannot be accurately estimated. Accuracies of the LD and LD-CS models were similar for all three pedigrees, because accuracy was mostly

contributed by LD that was generated historically, which was not eroded within a limited number of generations.

Reductions in accuracy for the CS model across validation generations were less severe in pedigrees 1 and 3 compared with pedigree 2 (Fig. 3). The reason is that the number of founder alleles was much smaller in pedigrees 1 and 3 than in pedigree 2 and, thus, the values of founder alleles could be estimated more accurately due to more data available per founder allele. Similar trends in prediction accuracy were observed for BayesB compared with BayesA, except that the difference in accuracy between the LD and LD-CS models was smaller for BayesB than for BayesA, especially in pedigrees 1 and 3 (Fig. 3).

In the Rare QTL scenario with high historical LD, the CS and LD-CS models had higher accuracy than the LD model (Fig. 4). The reduction in accuracy across validation generations was larger for the LD model than for the CS and LD-CS models, especially for pedigrees 1

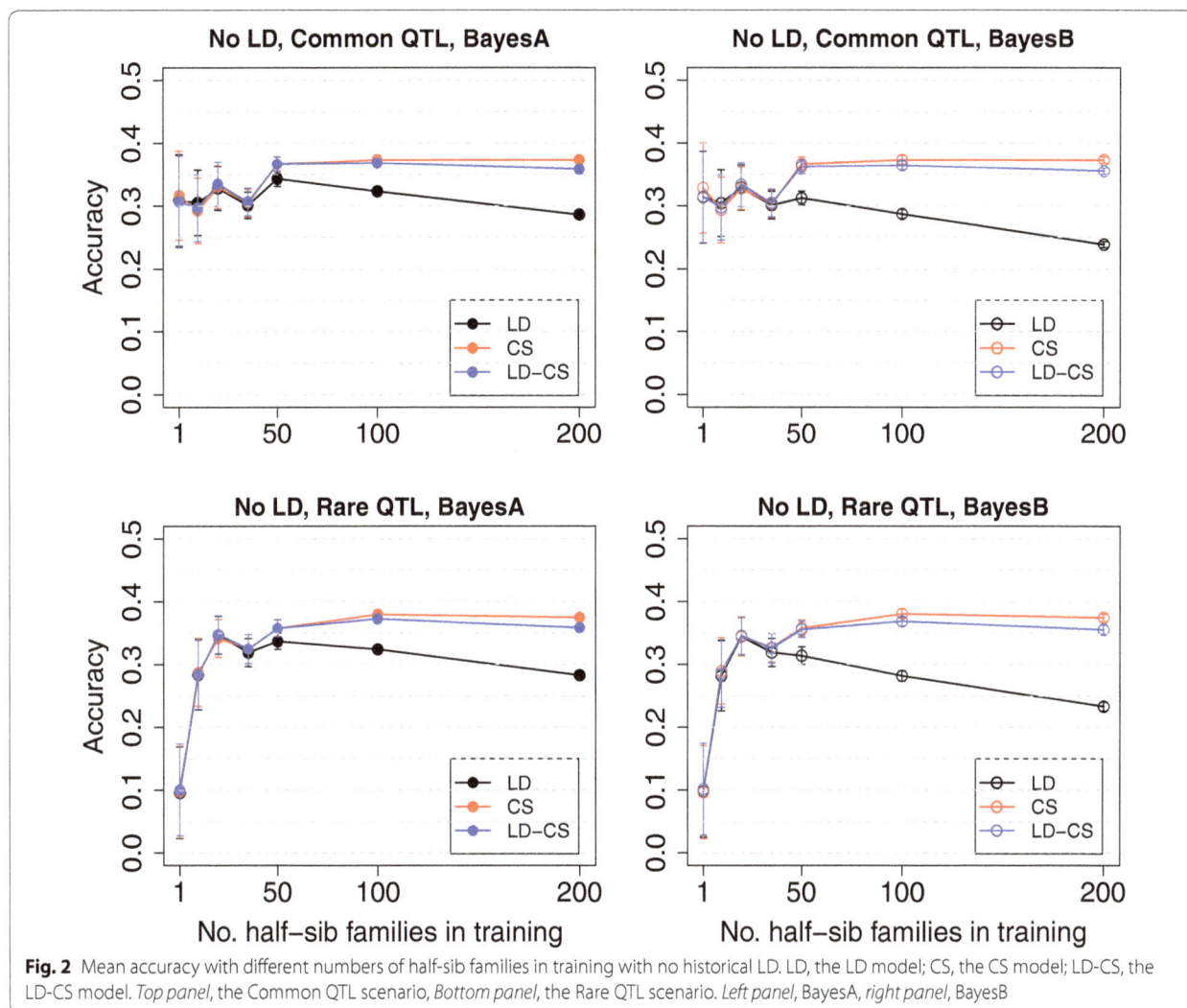

Fig. 2 Mean accuracy with different numbers of half-sib families in training with no historical LD. LD, the LD model; CS, the CS model; LD-CS, the LD-CS model. *Top panel*, the Common QTL scenario, *Bottom panel*, the Rare QTL scenario. *Left panel*, BayesA, *right panel*, BayesB

and 3. These results suggest that when LD between QTL and SNPs is low, the accuracy of the LD model mostly comes from capturing CS information, and this information decreases across validation generations due to recombination. In pedigrees 1 and 3, the LD-CS model had slightly higher accuracy than the CS model when using BayesA, but accuracies of the CS and LD-CS models were almost the same when using BayesB (Fig. 4). In pedigree 2, the LD-CS model had significantly higher accuracy than the CS model for both BayesA and BayesB. This is because when recent N_e is large, as in pedigree 2, the CS model has a disadvantage due to a large number of segregating alleles, each with relatively little data that contribute to the estimation of its value. Modeling LD in addition to CS improves prediction accuracy by implicitly capturing extra CS information. In conclusion, the contribution of CS information to prediction accuracy is more important in pedigrees with few parents than in pedigrees with many parents, because with few parents,

the value of founder QTL alleles can be estimated more accurately due to more data being available.

In the Rare QTL scenario with high historical LD, accuracy of BayesB was higher than that of BayesA for the CS model (Fig. 4). This is because when QTL alleles have low MAF, the proportion of founder alleles that carry the favorable QTL allele is low. BayesB is more effective than BayesA to accurately estimate the value of the small proportions of founder alleles that carry favorable QTL alleles.

In the Common QTL scenario without historical LD, the LD, CS and LD-CS models had almost the same accuracy in either pedigree 1 or 3; while in pedigree 2, the LD model had much lower accuracy than the CS and LD-CS models (Fig. 5). When there is no historical LD, only CS information contributes to prediction accuracy. Recent LD between linked QTL and SNPs was created quickly within several generations in pedigrees 1 and 3 due to high genetic drift, in which case the LD model

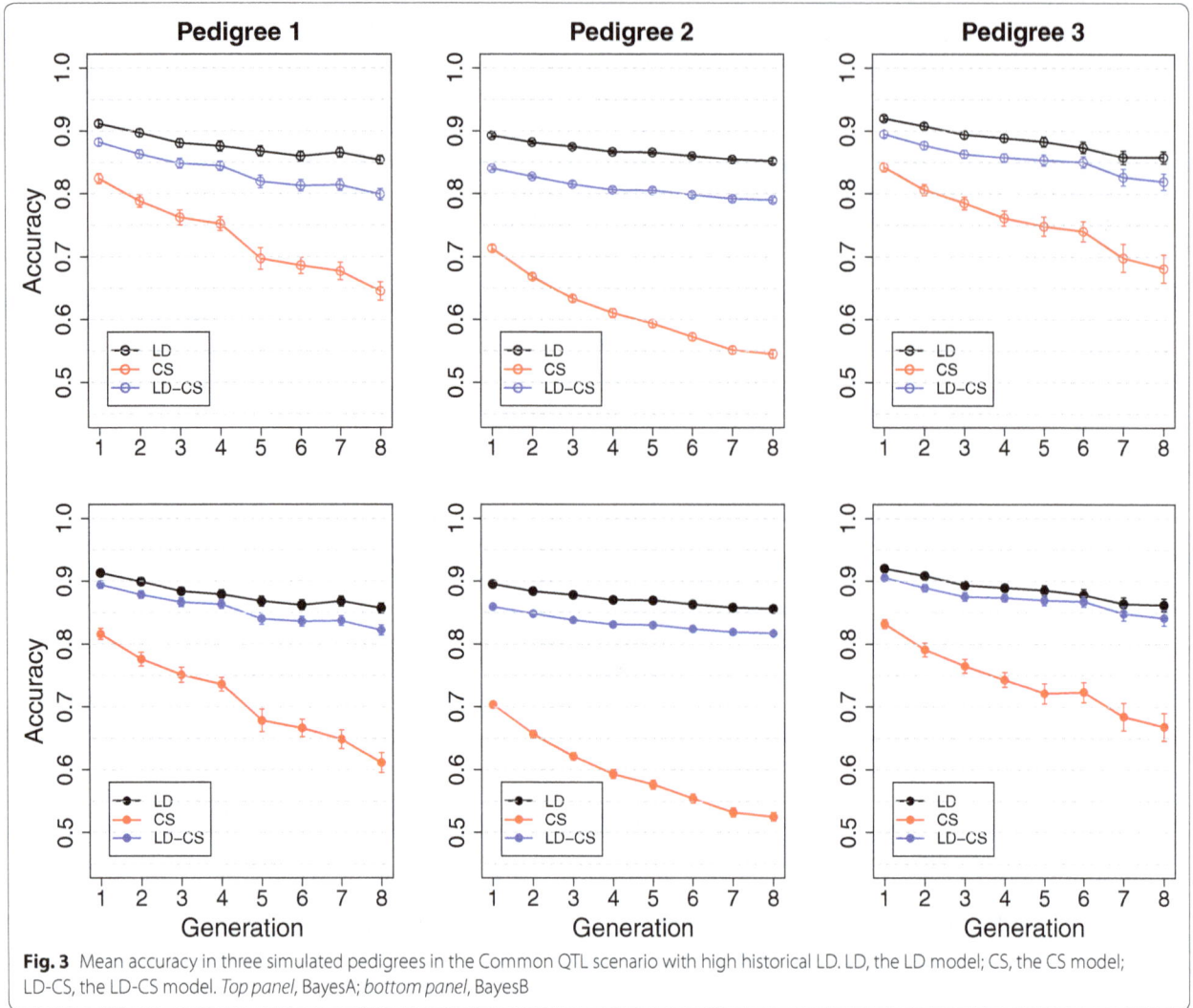

Fig. 3 Mean accuracy in three simulated pedigrees in the Common QTL scenario with high historical LD. LD, the LD model; CS, the CS model; LD-CS, the LD-CS model. *Top panel*, BayesA; *bottom panel*, BayesB

can capture as much CS information as the CS model. The creation of recent LD was slower in pedigree 2 due to much less drift compared with pedigrees 1 and 3, and hence the LD model could only capture part of the CS information.

In the Rare QTL scenario without historical LD, the CS and LD-CS models had similar accuracies, which was higher than the accuracy of the LD model (Fig. 6). This is because high LD cannot be created within several recent generations due to the difference in MAF between QTL and SNPs. Method BayesB had higher accuracy than BayesA for the CS and LD-CS models, because the value of the small proportion of founder alleles that carry favorable QTL alleles was estimated more accurately using BayesB than BayesA. In contrast, method BayesB had lower accuracy than BayesA for the LD model,

because more SNPs were fitted by BayesA than by BayesB and hence captured more CS information [9].

Impact of unknown allele origins

To study the impact of unknown allele origins on the accuracy of the CS and LD-CS models, prediction accuracy was also compared for models using simulated true allele origins and allele origins imputed by the LDMIP software [33] in the simulated datasets of Pedigree 1 with high historical LD. In the Common QTL scenario, the CS model using imputed allele origins had lower accuracy when using imputed allele origins than when using true allele origins, and the accuracy decreased faster over generations, while the LD-CS model had similar accuracy using either imputed or true allele origins (Fig. 7). The reason is that, when there was high historical LD,

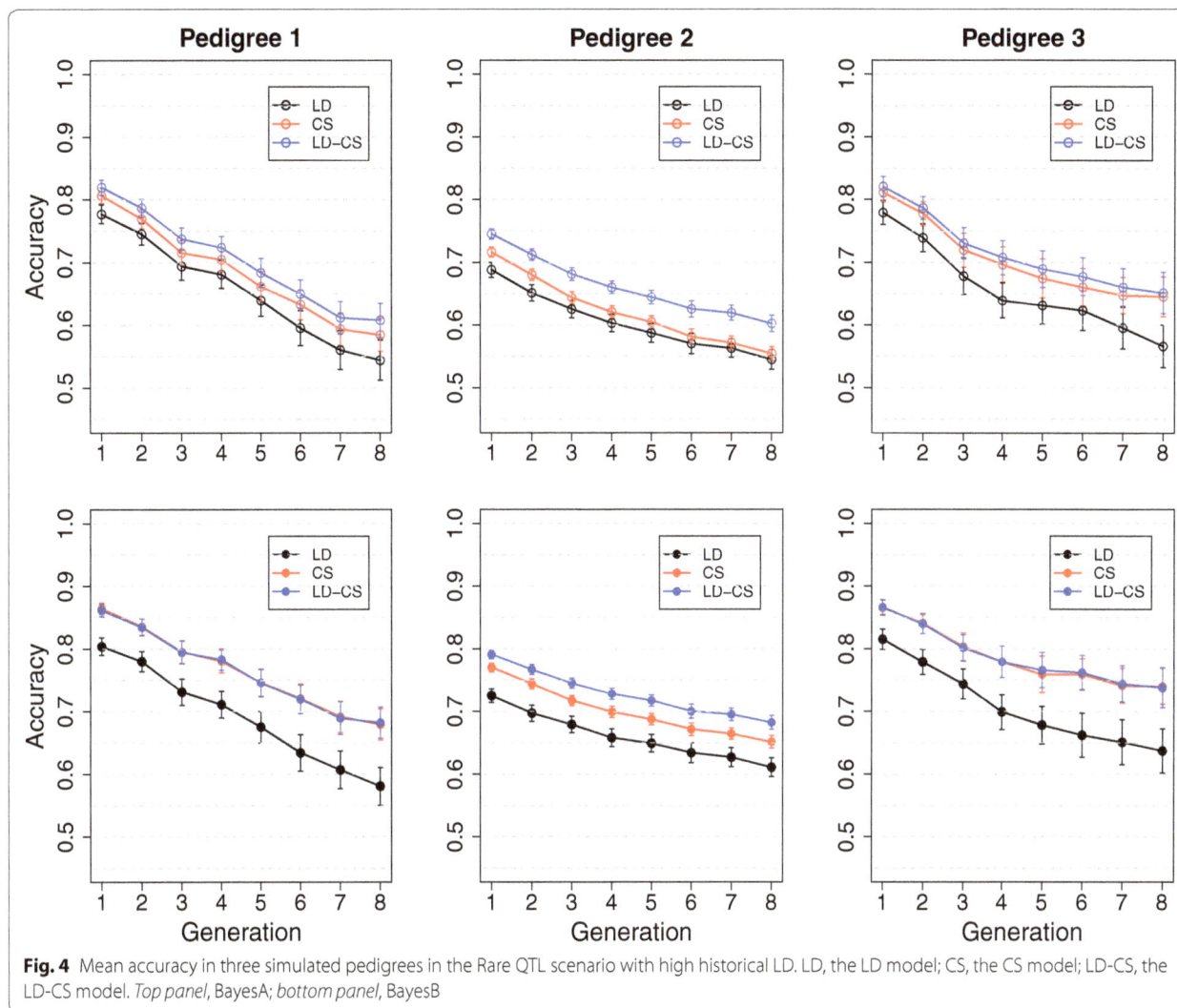

Fig. 4 Mean accuracy in three simulated pedigrees in the Rare QTL scenario with high historical LD. LD, the LD model; CS, the CS model; LD-CS, the LD-CS model. *Top panel*, BayesA; *bottom panel*, BayesB

the accuracy was mostly contributed by LD, which was hardly affected by the accuracy of allele origins. In the Rare QTL scenario, the CS and LD-CS models had lower accuracy when using imputed allele origins than when using true allele origins (Fig. 7). The reason was that, when there was no historical LD, the accuracy was mostly contributed by CS, which was significantly affected by the accuracy of allele origins.

Impact of the length of the genomic window used to model CS

To study the impact of the length of the genomic window used to model CS on the accuracy of the CS and LD-CS models, values of founder alleles at putative QTL were fitted at genome windows of lengths 0.5, 1 and 2 cM, respectively, in the simulated datasets of pedigree 1 with high historical LD. Prediction accuracies of CS models

that used different window lengths was similar, although shorter window lengths tended to have slightly higher accuracy in the Common QTL scenario but lower accuracy in the Rare QTL scenario (Fig. 8). Prediction accuracies of LD-CS models that used different window lengths was almost identical in the Common and Rare QTL scenarios (Fig. 8). Similar results were also observed for BayesA (results not shown). These results indicate that the length of the genomic window used to model CS has minimal impact on the accuracy of the CS and LD-CS models. A possible reason is that only very few windows contain recombinations and most of the non-recombinant windows are not affected by window length. Fitting putative QTL at each SNP may show a significant advantage over e.g. 1-cM bins only when a larger number of observations is also available, to compensate for the much larger number of putative QTL allele effects to estimate.

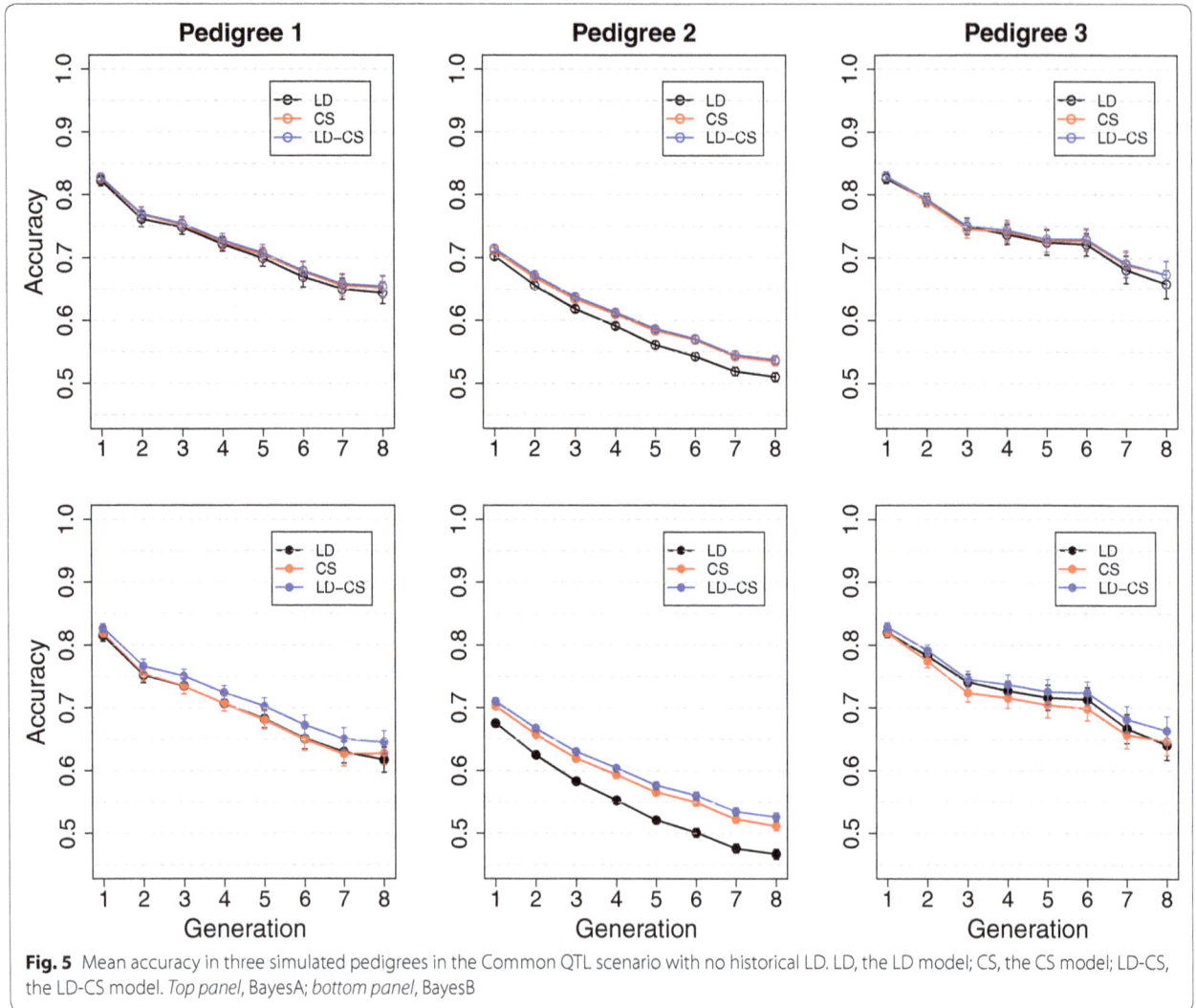

Fig. 5 Mean accuracy in three simulated pedigrees in the Common QTL scenario with no historical LD. LD, the LD model; CS, the CS model; LD-CS, the LD-CS model. *Top panel*, BayesA; *bottom panel*, BayesB

Discussion

The objectives of this study were (1) to develop a Bayesian statistical method to model CS information explicitly, (2) to study contributions of LD and CS information to accuracy of genomic prediction across unrelated families and validation generations without re-training, and (3) to study the effects of historical LD, recent N_e and MAF of QTL on the advantage of modeling LD and CS explicitly in improving prediction accuracy. The major focus of this study was a theoretical exploration of contributions of LD and CS information to prediction accuracy in pedigree populations. The new CS and LD-CS model developed in this study enables precise disentanglement of CS information from LD since it models CS explicitly using allele origins. The simulation studies that were performed assist in this disentanglement by limiting genetic information to LD, CS and pedigree relationships, ruling

out the noise that exists in real datasets (e.g. dominance, epistasis, imprinting, epigenetics). Investigation on real datasets is, however, worthy of future studies for potential applications of the CS and LD-CS models. In the following section, the mechanisms by which LD and CS information contribute to prediction accuracy across families and generations, as well as the effects of historical LD, recent N_e and MAF of QTL on prediction accuracy are discussed.

Explicit modeling of CS information

In this study, a new method that explicitly models CS information was developed for genomic prediction in pedigree populations. This method models transmission of putative QTL alleles within consecutive non-overlapping genomic windows of sufficiently small length (1 cM in this study), such that the recombination rate is

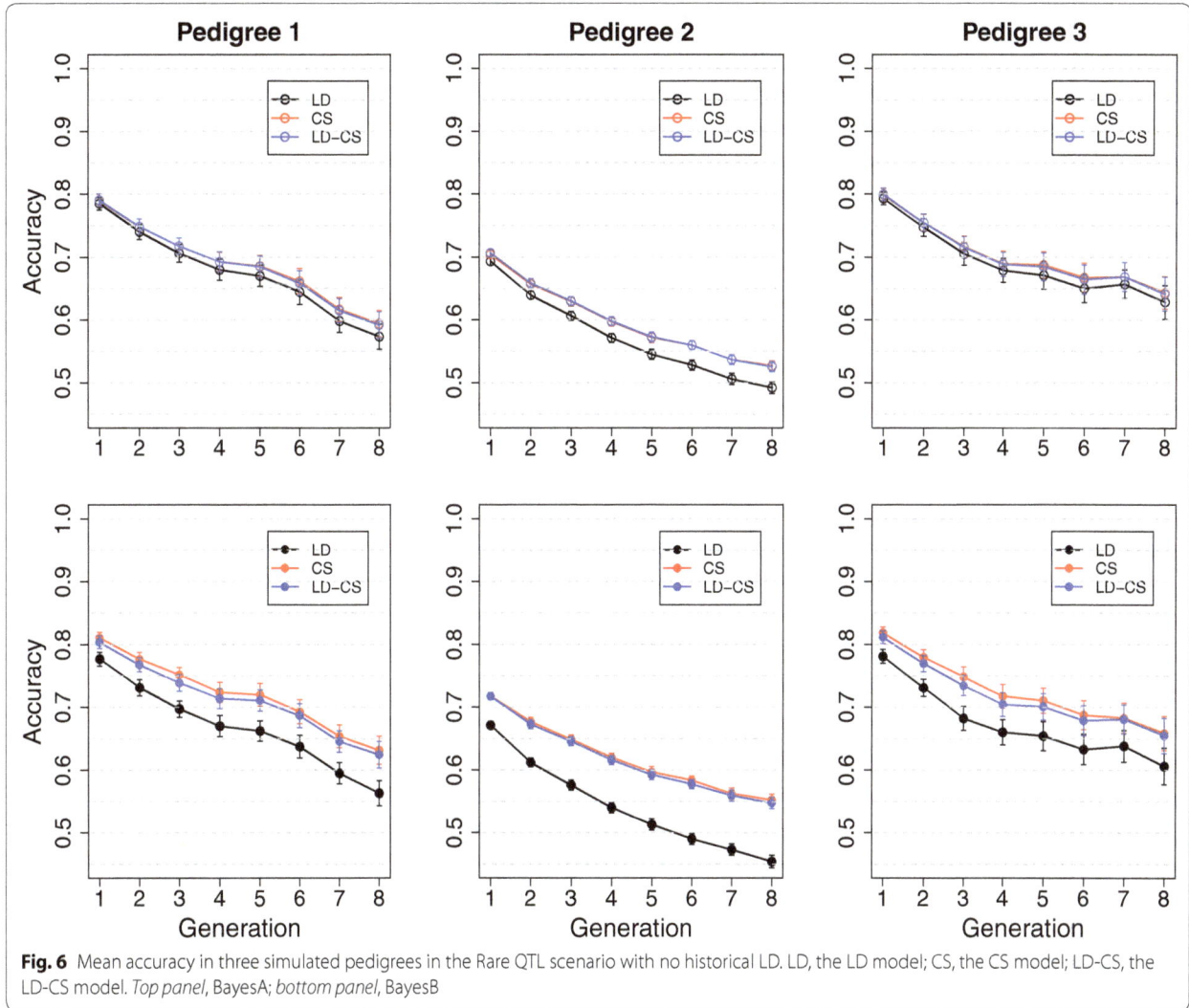

Fig. 6 Mean accuracy in three simulated pedigrees in the Rare QTL scenario with no historical LD. LD, the LD model; CS, the CS model; LD-CS, the LD-CS model. *Top panel*, BayesA; *bottom panel*, BayesB

so small that the alleles at all polymorphic loci within the window are expected to co-segregate for several generations. Co-segregation of QTL alleles was modeled using parental allele origins at SNPs that cover the genomic window, which are independent of the level of LD between QTL and SNPs. This method is applicable to any type of pedigree population provided that allele origins are available (or can be imputed) from founders to offspring. The method of modeling CS at putative QTL using allele origins at observable SNPs is similar to the method developed by Fernando and Grossman [29], but has the advantage that it allows estimation of values of founder alleles at putative QTL using single site Gibbs sampling. In Fernando and Grossman [29], the value of paternal and maternal alleles at putative QTL were fitted in the model for every pedigree individual to explain its breeding value. The value of QTL alleles is usually

correlated among individuals that are related by the pedigree, and therefore, the estimation of these values was achieved by solving mixed model equations, which requires the inverse of the covariance matrix of these values for all pedigree members. Computation of the Fernando and Grossman [29] method is manageable for marker-assisted selection, where the number of molecular markers is usually small, but would not be feasible for genomic prediction. However, for genomic prediction using dense SNP panels, the CS model developed in this study is computationally tractable because the breeding values for all pedigree members are modeled using only the QTL alleles of pedigree founders, which are assumed independent and MCMC methods can be feasibly implemented to estimate their values.

The CS model (2) developed in this study can also be written as an equivalent breeding value model

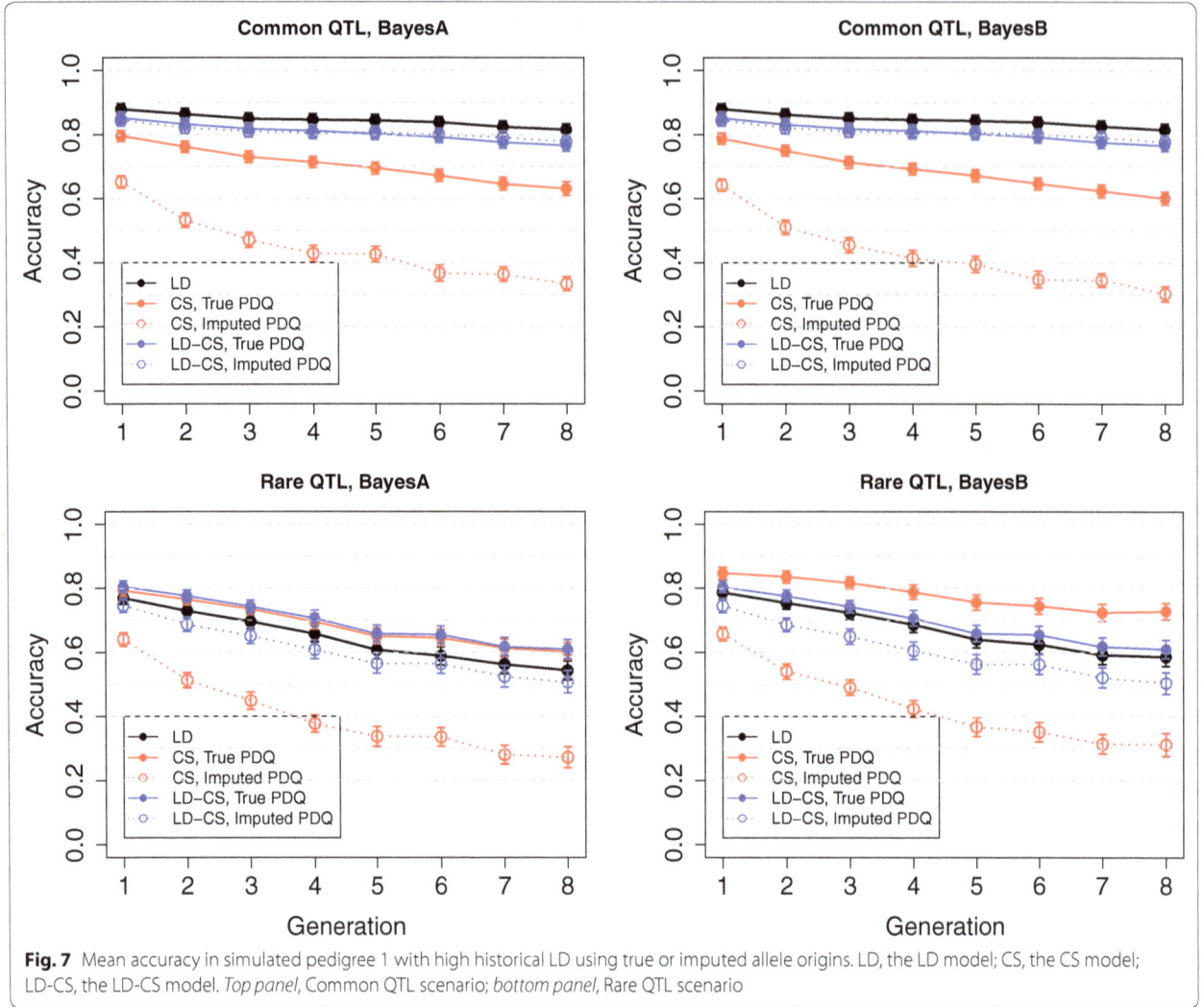

Fig. 7 Mean accuracy in simulated pedigree 1 with high historical LD using true or imputed allele origins. LD, the LD model; CS, the CS model; LD-CS, the LD-CS model. *Top panel*, Common QTL scenario; *bottom panel*, Rare QTL scenario

$$\mathbf{y} = \mathbf{X}\boldsymbol{\beta} + \mathbf{g}_{CS} + \mathbf{e}, \tag{7}$$

where

$$\mathbf{g}_{CS} = \sum_{j=1}^{n_q} \mathbf{W}_j \mathbf{v}_j,$$

and

$$\mathrm{Var}(\mathbf{g}_{CS}) = \mathbf{G}_{CS} = \sum_{j=1}^{n_q} \mathbf{W}_j \mathbf{D}_j \mathbf{W}_j', \quad \text{with}$$

$$\mathbf{D}_j = \mathrm{diag}\{\sigma_{jk}^2\}_{k=1}^{n_j}.$$

The covariance matrix of breeding values due to CS ($\mathbf{g}_{CS}$), $\mathbf{G}_{CS}$, quantifies the genetic covariance among individuals due to co-segregation at putative QTL. The genetic covariance between two individuals depends on the number of common founder alleles that the two

individuals share through identity-by-descent, averaged across n_q QTL, with the corresponding QTL effect variances as weights. This equivalent breeding value model (7) was used by Luan et al. [10] in their study on the contributions of CS and LD information to prediction accuracy in Italian Brown Swiss bulls. In Luan et al. [10], CS was modeled at each SNP locus of the Bovine SNP 50K chip, which were assumed to be surrogates of QTL. The covariance matrix $\mathbf{G}_{CS}$ was constructed independently at each SNP, and then averaged across all SNPs using equal weights. Compared to Luan et al. [10], the CS model developed herein has three advantages. First, the CS model (2) fits putative QTL within short genomic windows, which are much fewer than the number of SNPs in the 50K chip. Modeling CS at each SNP is not necessary because CS information is based on linkage and is conserved over longer genomic distances. Second, the CS model (2) allows different variances of QTL effects depending on their allele frequencies and sizes of the

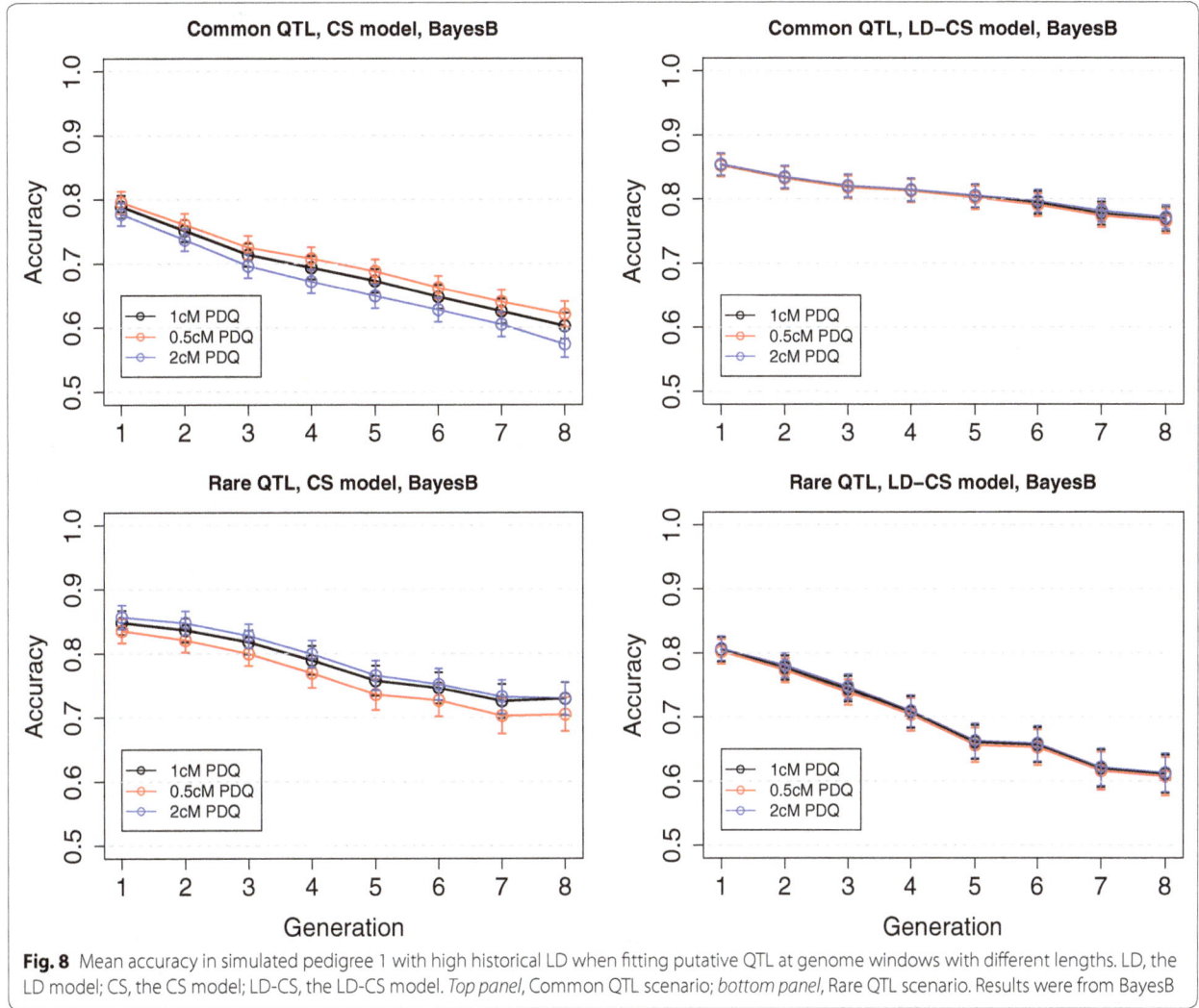

Fig. 8 Mean accuracy in simulated pedigree 1 with high historical LD when fitting putative QTL at genome windows with different lengths. LD, the LD model; CS, the CS model; LD-CS, the LD-CS model. *Top panel*, Common QTL scenario; *bottom panel*, Rare QTL scenario. Results were from BayesB

QTL effects. Larger QTL effects are estimated with less shrinkage, or equivalently larger weights in $\mathbf{G}_{CS}$. Third, the computation time for model (2) increases linearly with the number of individuals (n) times the number of founder QTL alleles ($\sum_{j=1}^{n_q} n_j$), while for the mixed model approach in Luan et al. [10], computation time increases cubically with n because it requires the inverse of a dense matrix, $\mathbf{G}_{CS}$.

It is generally accepted that LD between QTL and SNPs, co-segregation of QTL with SNP alleles, and pedigree relationships at QTL captured by SNPs are the three main sources of information that contribute to the accuracy of genomic prediction [8–10, 16, 17, 34]. Most previous studies aimed at disentangling these three sources of information were based on multiple regression models on SNP genotypes (the LD model). The LD model only allows the part of CS information that is implicitly captured by SNP genotypes to be evaluated, which is highly variable depending on the number and density of SNPs,

the level of historical LD, population structure and pedigree relationships. The CS model developed in this study enables precise disentanglement of CS from LD information because explicit modeling of CS information using parental allele origins does not depend on the level of LD between QTL and SNPs. As an intriguing consequence, results in this study are in contrast to some typical findings in several previous studies based on the LD model. For example, Habier et al. [9] showed that CS information captured by SNP genotypes contributed little to prediction accuracy across half-sib families and prediction accuracy decreased rapidly with increasing training size. In this study, prediction accuracy from the CS model persisted with increasing training size regardless of historical LD. This difference is mainly because Habier et al. [9] only considered the part of CS information that is implicitly captured by GBLUP, while the CS model in this study captures most of CS information due to modeling CS explicitly.

Contributions of LD and CS information in half-sib designs

In the simulated datasets with different numbers of unrelated half-sib families, both LD and CS information contributed to the accuracy of genomic prediction. Accuracy of the LD model relies on the level of LD between QTL and SNPs in the base population. Accuracy of the CS model relies on accurate estimation of the value of founder QTL alleles that are transmitted to half-sib offspring within the same family. Accuracy of the LD model increased rapidly with increasing training size when LD between QTL and SNPs was high, because high LD is conserved across families and increasing training size brings in more data to improve estimation of SNP effects. However, when historical LD is low or zero, accuracy of the LD model mainly comes from capturing CS information, which only exists within the same half-sib family. Consequently, with more unrelated half-sib families included in the training population, accuracy of the LD model decreased and became lower than accuracy of the CS model. This is because only half-sibs from the same family contribute to prediction accuracy. In the CS model, the value of founder QTL alleles is estimated using only information from the same half-sib family, while the LD model estimates SNP effects by pooling CS information across all families, which is erroneous because linkage phase and LD is highly variable across a large number of unrelated families.

Contributions of LD and CS information in extended pedigrees with different recent N_e

Three mating designs were simulated that differed in the number of parents per generation and the number of progeny per mating. Pedigrees 1 and 3 resemble the breeding program where a few sires are selected and intensively used for breeding in each generation. CS information made significant contributions to prediction accuracy in pedigrees 1 and 3, because a limited number of sire alleles segregated among a large number of their progeny, and the value of the sires' alleles can be estimated accurately based on the amount of data available. Pedigree 1 is a balanced nested design with identical family sizes, which is similar to the structure of nucleus herds in swine [35] or poultry [5] breeding programs. Pedigree 3 is an unbalanced design with an influential sire in each generation that has more than 80 % of the total progeny, which resembles a dairy cattle population, where artificial insemination is widely used [36]. CS information had a larger contribution in pedigree 3 than in pedigree 1, because most progeny in a cohort inherited alleles from only one sire in pedigree 3. In contrast, pedigree 2 resembles an outbred population where all individuals survive and each mating has very few progeny. The number of unique parental alleles is large but each allele

is transmitted only to very few progeny. As a result, the value of founder QTL alleles in pedigree 2 cannot be estimated accurately because each allele has only limited data available.

The difference between the three mating designs can be quantified by recent N_e. The N_e of pedigrees 1 and 3 was less than 20, while that of pedigree 2 was close to 200. CS information has a larger contribution to prediction accuracy in a population with a smaller recent N_e because individuals tend to be more closely related and share more founder alleles at QTL. The importance of CS information in the three mating designs was clearly illustrated in the scenario without historical LD among founders, where the long-term accuracy only stems from CS information. As shown in Figs. 5 and 6, the long-term accuracy by modeling CS explicitly was most persistent for pedigree 3, followed by pedigree 1, and least persistent for pedigree 2. A similar trend was also observed for the CS model when both LD and CS information contributed to prediction accuracy (Figs. 5 and 6).

The contribution of LD information to prediction accuracy should not depend on the mating design because high historical LD between QTL and SNPs is mostly between closely linked loci and hardly erodes within several recent generations. However, because the LD model also implicitly captures information from CS and pedigree relationships [9], accuracy of the LD model is affected by the mating design. For example, in the scenario with high historical LD, accuracy of the LD model was higher for pedigrees 1 and 3 than for pedigree 2 (Fig. 5). These results agree with Muir [19], who found that prediction accuracy of the GBLUP model decreased when recent N_e increased, and this reduction was larger when QTL and SNPs were in linkage equilibrium (LE) than when they were in LD.

In general, when historical LD is high between QTL and SNPs, long-term accuracy is mostly contributed by LD information and CS information has little contribution regardless of recent N_e. However, when historical LD is low, CS information contributes most to long-term accuracy, especially when the mating design creates a very small recent N_e.

The effect of MAF of QTL on contributions of LD and CS information

LD quantifies the correlation between allele states at QTL and SNPs. The LD model captures this correlation using multiple regression on SNP genotypes. Strong correlations can only exist when QTL and SNPs have similar MAF, as in the Common QTL scenario. Correlations are expected to be low when most QTL have low MAF, while SNPs have moderate MAF, as in the Rare QTL scenario. In the simulated datasets, the correlation between allele states exists in the form of historical LD and recent CS. Historical LD between closely linked loci is hardly eroded by recombination. The correlation generated by CS can exist between loci over long chromosome regions, which erodes fast with recombination. Both forms of correlation can be captured by the LD model, however, the LD model has persistent long-term accuracy only when historical LD between QTL and SNPs is high, which requires similar MAF between QTL and SNPs. When historical LD is low due to most QTL having low MAF, prediction accuracy of the LD model mainly comes from implicitly capturing CS information, which decreases rapidly across validation generations because CS information across long chromosome regions erodes fast with recombination. Similar results have been observed by Habier et al. [9] for the GBLUP model.

CS follows the transmission of QTL alleles among related individuals, which is independent of historical LD. As a result, prediction accuracy from the CS model is not affected by the level of historical LD. Accuracy due to CS information depends on (1) the length of the founder haplotypes that are used to follow transmission of putative QTL alleles, which determines the rate of erosion of CS due to recombination; and (2) the accuracy with which the value of founder alleles can be estimated, which depends on the amount of phenotype data for the progeny that inherited the same founder haplotype. The CS model that used haplotypes of length 1 cM is expected to have only a few recombinations within haplotypes, and therefore CS information contributes to long-term accuracy provided that the value of founder QTL alleles can be estimated correctly with sufficient data.

The effect of prior distributions on prediction accuracy

Prior distributions in BayesA and BayesB [7] were used to allow simultaneous estimation of SNP effects α in the LD model and the value of founder QTL alleles v_j in the CS model. GBLUP was also used in this study and had similar results as BayesA (results not shown). BayesA represents a method of shrinkage regression without variable selection. In BayesA, independent t prior distributions are given to α_l and v_{jk}. When using the posterior mode as the point estimate of a parameter β, the amount of shrinkage imposed by a scaled t prior distribution with degrees of freedom v and scale parameter S^2, $t(0, v, S^2)$, is proportional to $\log\left(1 + \frac{\beta^2}{vS^2}\right)$ [37]. The posterior mean used in this study is expected to be close to the mode due to the almost symmetric posterior distribution at convergence of the MCMC [31]. This means that the estimates of small β are heavily shrunk towards zero but large β are less shrunk. BayesB represents a variable selection method, in which the prior for each α_l and v_{jk} is a mixture of a point mass at zero and a t distribution elsewhere.

BayesB results in much heavier shrinkage towards zero than BayesA, and consequently the effective number of loci fitted in the model is larger in BayesA than in BayesB [9].

The effect of using alternate prior distributions on the LD model is twofold. When historical LD is high between QTL and SNPs, high prediction accuracy is usually achieved by the LD model, with nearly unbiased estimates of large SNP effects, while effectively shrinking small SNP effects towards zero. In the simulated datasets of this study, BayesB resulted in higher accuracy than BayesA because the number of QTL was much smaller than the number of SNPs. When historical LD is low, prediction accuracy mainly comes from implicitly capturing CS information, which depends on the effective number of SNPs fitted in the model. Then, BayesA results in higher accuracy than BayesB due to fitting relatively more SNPs, which can capture more CS information than BayesB [9].

The effect of prior distributions on the CS model does not depend on historical LD, but does depend on the MAF of QTL. When the MAF of QTL is high, BayesA tends to result in higher accuracy than BayesB because of fitting more founder alleles that co-segregate with common QTL alleles. When the MAF of QTL is low, BayesB tends to result in higher accuracy than BayesA, because only a small proportion of founder alleles carry QTL and their values can be estimated more accurately by variable selection of BayesB.

Implementation of LD-CS model in field datasets

In real livestock populations, persistently high accuracy across validation generations using the LD model has rarely been observed [5, 16–18], which suggests that historical LD between QTL and SNPs is not perfect, and prediction accuracy relies more on CS than on LD information [10, 20]. The LD-CS model is recommended to improve long-term accuracy due to capturing both LD and CS information explicitly. Simulation results in this study suggest that the LD-CS model tends to result in the highest prediction accuracy of predictions in almost all scenarios. Using a similar LD-CS model as model (7), Luan et al. [10] show that in a pedigree population of Italian Brown Swiss bulls, LD information does not contribute to accuracy beyond that due to CS information. Recent studies in pig [28] and Atlantic salmon [38] breeding populations showed that the CS model of Luan et al. [10] had similar or lower accuracy than the GBLUP model, depending on trait heritability, SNP density, and the number of generations of pedigree data used to infer SNP allele origins. Results from field data suggest that in livestock populations, both CS and LD information contribute to prediction accuracy, and modeling CS explicitly can achieve almost the same accuracy as fitting SNP genotypes. Furthermore, the simulation results in this study suggest that modeling LD and CS explicitly improves prediction accuracy compared to modeling either LD or CS, when historical LD between QTL and SNPs is low due to most QTL being rare, as represented by the Rare QTL scenario.

There are several computational issues in implementing the CS or LD-CS models (2) in field datasets. First, obtaining parental allele origins from SNP genotypes for all pedigree members can be computationally prohibitive. This is usually achieved in two steps. SNP genotypes are first phased into haplotypes, which are then used to infer parental allele origins using pedigree information [12, 33]. Our results show that the advantage of modeling CS information explicitly in improving prediction accuracy depends strongly on the accuracy of allele origins, and errors in phasing and allele origin imputation can reduce or even nullify the advantage of the CS or LD-CD models. This problem can become less demanding with the availability of higher SNP density, genome re-sequencing, and identification of multi-allelic markers such as copy number variants and insertions/deletions. Second, the computation time for the MCMC algorithm of the CS model (2) increases with the number of pedigree founders, because the number of alleles at each putative QTL (n_j) is twice the number of founders. It is suggested that putative QTL be modeled at every 1 cM of the genome to reduce the total number of QTL alleles, as justified by the fact that recombination occurs very rarely within a 1-cM genomic window over several consecutive generations. Furthermore, instead of treating the values of QTL alleles as independent, they can be clustered according to their probability of identity-by-descent with respect to some historical common ancestors beyond the pedigree founders [33, 39, 40]. However, when the number of founders is large, the equivalent breeding value model (7) is recommended since the mixed model equations have the number of genotyped individuals as dimension.

Conclusions
In this study, a new method that explicitly models co-segregation information was developed for genomic prediction of breeding values. Breeding values in this CS model were modeled as the sum of independent values of putative QTL alleles among pedigree founders, which were traced down in the pedigree using SNP haplotypes. When the training size was increased by adding unrelated half-sib families, accuracy of the CS model increased and plateaued, but accuracy of the LD model that fits SNP genotypes dropped when historical LD between QTL and SNPs was low. Modeling both LD and CS information improved prediction accuracy compared to modeling either LD or CS, especially when historical LD was low and recent CS information contributed substantially to prediction accuracy among families, which is probably the case for recent genomic evaluation in most livestock populations. The effects of recent N_e, historical LD, and MAF of QTL on persistence of accuracy across validation generations without retraining were investigated when explicitly modeling LD and CS information. The LD model had persistently high accuracy across validation generations only when historical LD between QTL and SNPs was high, which requires that QTL and SNPs have similar MAF. When historical LD between QTL and SNPs was low, accuracy of the LD model came mostly from capturing CS information, which was much lower and less persistent than that of the CS and LD-CS models. The contribution of CS information increased with smaller recent N_e, because the number of segregating QTL alleles of pedigree founders was smaller and their value could be estimated more accurately with sufficient data. Since the recent N_e of most livestock populations is small and historical LD between QTL and SNPs tends to be low, modeling CS explicitly in addition to LD has potential to improve long-term accuracy provided that the allele origins can be accurately imputed.

Additional files

Additional file 1. Boundaries of linkage disequilibrium between a QTL and a SNP with different MAF. This file provides derivation and simulation results for boundaries of linkage disequilibrium between a QTL and a SNP with different MAF.

Additional file 2. Bayesian inference for the LD-CS model. This file provides derivation of full conditional distributions of parameters in the LD-CD model, and MCMC algorithm to get point estimates of parameter values.

Authors' contributions
Perceived the idea: RLF and JCMD. Designed the study: XS, RLF and JCMD. Developed the statistical methods: XS and RLF. Performed the analyses: XS. Interpreted the results: XS, RLF and JCMD. Wrote the manuscript: XS, RLF and JCMD. All authors read and approved the final manuscript.

Acknowledgements
This work was supported by the US Department of Agriculture, Agriculture and Food Research Initiative, National Institute of Food and Agriculture Competitive Grant 2010-65205-20341 and by National Institutes of Health Grant R01GM099992. Comments from two reviewers were greatly acknowledged.

Competing interests
The authors declare that they have no competing interests.

References

1. Hayes B, Bowman P, Chamberlain A, Goddard M. Invited review: genomic selection in dairy cattle—progress and challenges. J Dairy Sci. 2009;92:433–43.

2. VanRaden P, van Tassell C, Wiggans G, Sonstegard T, Schnabel R, Taylor J, Schenkel F. Invited review: reliability of genomic predictions for North American Holstein bulls. J Dairy Sci. 2009;92:16–24.

3. Daetwyler H, Hickey J, Henshall J, Dominik S, Gredler B, van der Werf J, Hayes B. Accuracy of estimated genomic breeding values for wool and meat traits in a multi-breed sheep population. Anim Prod Sci. 2010;50:1004–10.

4. Garrick D. The nature, scope and impact of genomic prediction in beef cattle in the United States. Genet Sel Evol. 2011;43:17.

5. Wolc A, Stricker C, Arango J, Settar P, Fulton J, O'Sullivan N, Preisinger R, Habier D, Fernando R, Garrick D, Lamont S, Dekkers J. Breeding value prediction for production traits in layer chickens using pedigree or genomic relationships in a reduced animal model. Genet Sel Evol. 2011;43:5.

6. Ostersen T, Christensen O, Henryon M, Nielsen B, Su G, Madsen P. Deregressed EBV as the response variable yield more reliable genomic predictions than traditional EBV in pure-bred pigs. Genet Sel Evol. 2011;43:38.

7. Meuwissen T, Hayes B, Goddard M. Prediction of total genetic value using genome-wide dense marker maps. Genetics. 2001;157:1819–29.

8. Habier D, Fernando R, Dekkers J. The impact of genetic relationship information on genome-assisted breeding values. Genetics. 2007;177:2389–97.

9. Habier D, Fernando R, Garrick D. Genomic BLUP decoded: a look into the black box of genomic prediction. Genetics. 2013;194:597–607.

10. Luan T, Woolliams J, Odegard J, Dolezal M, Roman-Ponce S, Bagnato A, Meuwissen T. The importance of identity-by-state information for the accuracy of genomic selection. Genet Sel Evol. 2012;44:28.

11. He W, Fernando R, Dekkers J, Gilbert H. A gene frequency model for QTL mapping using Bayesian inference. Genet Sel Evol. 2010;42:21.

12. Habier D, Totir L, Fernando R. A two-stage approximation for analysis of mixture genetic models in large pedigrees. Genetics. 2010;185:655–70.

13. Matukumalli L, Lawley C, Schnabel R, Taylor J, Allan M, Heaton M, O'Connell J, Moore S, Smith T, Sonstegard T, van Tassell C. Development and characterization of a high density SNP genotyping assay for cattle. PLoS One. 2009;4:5350.

14. Qanbari S, Pimentel E, Tetens J, Thaller G, Lichtner P, Sharifi A, Simianer H. The pattern of linkage disequilibrium in German Holstein cattle. Anim Genet. 2010;41:346–56.

15. Luan T, Woolliams J, Lien S, Kent M, Svendsen M, Meuwissen T. The accuracy of genomic selection in Norwegian Red cattle assessed by cross-validation. Genetics. 2009;183:1119–26.

16. Habier D, Tetens J, Seefried F, Lichtner P, Thaller G. The impact of genetic relationship information on genomic breeding values in German Holstein cattle. Genet Sel Evol. 2010;42:5.

17. Wientjes Y, Veerkamp R, Calus M. The effect of linkage disequilibrium and family relationships on the reliability of genomic prediction. Genetics. 2013;193:621–31.

18. Weng Z, Wolc A, Shen X, Fernando R, Dekkers J, Arango J, Settar P, Fulton J, O'Sullivan N, Garrick D. Effects of number of training generations on genomic prediction for various traits in a layer chicken population. Genet Sel Evol. 2016;48:22.

19. Muir W. Comparison of genomic and traditional BLUP-estimated breeding value accuracy and selection response under alternative trait and genomic parameters. J Anim Breed Genet. 2007;124:342–55.

20. Daetwyler H, Kemper K, van der Werf J, Hayes B. Components of the accuracy of genomic prediction in a multi-breed sheep population. J Anim Sci. 2012;90:3375–84.

21. Vela-Avitua S, Meuwissen T, Luan T, Odegard J. Accuracy of genomic selection for a sib-evaluated trait using identity-by-state and identity-by-descent relationships. Genet Sel Evol. 2015;47:9.

22. Hayes B, Visscher P, McPartlan H, Goddard M. Novel multilocus measure of linkage disequilibrium to estimate past effective population size. Genome Res. 2003;13:635–43.

23. de Roos A, Hayes B, Spelman R, Goddard M. Linkage disequilibrium and persistence of phase in Holstein–Friesian. Jersey and Angus cattle. Genetics. 2008;179:1503–12.

24. Espigolan R, Baldi F, Boligon A, Souza F, Gordo D, Tonussi R, Cardoso D, Oliveira H, Tonhati H, Sargolzaei M, Schenkel F, Carvalheiro R, Ferro J, Albuquerque L. Study of whole genome linkage disequilibrium in Nellore cattle. BMC Genomics. 2013;14:305.

25. Hayes B, Pryce J, Chamberlain A, Bowman P, Goddard M. Genetic architecture of complex traits and accuracy of genomic prediction: coat colour, milk-fat percentage, and type in Holstein cattle as contrasting model traits. PLoS Genet. 2010;6:1001139.

26. Daetwyler H, Capitan A, Pausch H, Stothard P, van Binsbergen R, Brondum R, Liao X, Djari A, Rodriguez S, Grohs C, Esquerre D, Bouchez O, Rossignol M, Klopp C, Rocha D, Fritz S, Eggen A, Bowman P, Coote D, Chamberlain A, Anderson C, VanTassell C, Hulsegge I, Goddard M, Guldbrandtsen B, Lund M, Veerkamp R, Boichard D, Fries R, Hayes B. Whole-genome sequencing of 234 bulls facilitates mapping of monogenic and complex traits in cattle. Nat Genet. 2014;46:858–65.

27. Druet T, Macleod I, Hayes B. Toward genomic prediction from whole-genome sequence data: impact of sequencing design on genotype imputation and accuracy of predictions. Heredity. 2014;112:39–47.

28. Meuwissen T, Odegard J, Andersen-Ranberg I, Grindflek E. On the distance of genetic relationships and the accuracy of genomic prediction in pig breeding. Genet Sel Evol. 2014;46:49.

29. Fernando R, Grossman M. Marker assisted selection using best linear unbiased prediction. Genet Sel Evol. 1989;21:467–77.

30. Haldane J. The combination of linkage values, and the calculation of distances between the loci of linked factors. J Genet. 1919;8:299–309.

31. Sun X, Qu L, Garrick D, Dekkers J, Fernando R. A fast EM algorithm for Bayes A-like prediction of genomic breeding values. PLoS One. 2012;7:49157.

32. Falconer D, Mackay T. Introduction to quantitative genetics, 4th ed.: Pearson Education Limited; 1996.

33. Meuwissen T, Goddard M. The use of family relationships and linkage disequilibrium to impute phase and missing genotypes in up to whole-genome sequence density genotypic data. Genetics. 2010;185:1441–9.

34. de los Campos G, Vazquez A, Fernando R, Klimentidis Y, Sorensen D. Prediction of complex human traits using the genomic best linear unbiased predictor. PLoS Genet. 2013;9:1003608.

35. Cleveland M, Hickey J, Forni S. A common dataset for genomic analysis of livestock populations. G3 (Bethesda). 2012;2:35–429.

36. Schaeffer L. Strategy for applying genome-wide selection in dairy cattle. J Anim Breed Genet. 2006;123:218–23.

37. Sorensen D, Gianola D. Likelihood, Bayesian, and MCMC Methods in quantitative genetics. New York: Springer; 2002.

38. Odegard J, Moen T, Santi N, Korsvoll S, Kjoglum S, Meuwissen T. Genomic prediction in an admixed population of atlantic salmon (salmo salar). Front Genet. 2014;5:402.

39. Meuwissen T, Goddard M. Prediction of identity by descent probabilities from marker-haplotypes. Genet Sel Evol. 2001;33:605–34.

40. Meuwissen T, Goddard M. Multipoint identity-by-descent prediction using dense markers to map quantitative trait loci and estimate effective population size. Genetics. 2007;176:2551–60.

12

Multi-breed genomic prediction using Bayes R with sequence data and dropping variants with a small effect

Irene van den Berg[1][*][iD], Phil J. Bowman[2,3], Iona M. MacLeod[2], Ben J. Hayes[2,4], Tingting Wang[2], Sunduimijid Bolormaa[2] and Mike E. Goddard[1,2]

Abstract

Background: The increasing availability of whole-genome sequence data is expected to increase the accuracy of genomic prediction. However, results from simulation studies and analysis of real data do not always show an increase in accuracy from sequence data compared to high-density (HD) single nucleotide polymorphism (SNP) chip genotypes. In addition, the sheer number of variants makes analysis of all variants and accurate estimation of all effects computationally challenging. Our objective was to find a strategy to approximate the analysis of whole-sequence data with a Bayesian variable selection model. Using a simulated dataset, we applied a Bayes R hybrid model to analyse whole-sequence data, test the effect of dropping a proportion of variants during the analysis, and test how the analysis can be split into separate analyses per chromosome to reduce the elapsed computing time. We also investigated the effect of imputation errors on prediction accuracy. Subsequently, we applied the approach to a dataset that contained imputed sequences and records for production and fertility traits for 38,492 Holstein, Jersey, Australian Red and crossbred bulls and cows.

Results: With the simulated dataset, we found that prediction accuracy was highly increased for a breed that was not represented in the training population for sequence data compared to HD SNP data. Either dropping part of the variants during the analysis or splitting the analysis into separate analyses per chromosome decreased accuracy compared to analysing whole-sequence data. First, dropping variants from each chromosome and reanalysing the retained variants together resulted in an accuracy similar to that obtained when analysing whole-sequence data. Adding imputation errors decreased prediction accuracy, especially for errors in the validation population. With real data, using sequence variants resulted in accuracies that were similar to those obtained with the HD SNPs.

Conclusions: We present an efficient approach to approximate analysis of whole-sequence data with a Bayesian variable selection model. The lack of increase in prediction accuracy when applied to real data could be due to imputation errors, which demonstrates the importance of developing more accurate methods of imputation or directly genotyping sequence variants that have a major effect in the prediction equation.

Background

The increasing availability of whole-sequence data, which should contain causative mutations for complex traits, is expected to increase the accuracy of genomic prediction and to aid in the identification of these causative mutations. There are two advantages of using sequence data over single nucleotide polymorphism (SNP) chip genotypes. First, if the SNP chip does not explain all of the genetic variance explained by the sequence, prediction accuracy will be limited regardless of the prediction method used. Second, if there is no single SNP that is in complete linkage disequilibrium (LD) with a quantitative trait locus (QTL), prediction accuracy using SNP chip genotypes will decrease. In particular, the latter

---
*Correspondence: irene.vandenberg@unimelb.edu.au
[1] Faculty of Veterinary and Agricultural Science, University of Melbourne, Parkville, VIC, Australia
Full list of author information is available at the end of the article

influences Bayesian prediction methods, which work best when they identify a single SNP with a large effect. Both of these reasons concern the LD between causative mutations and SNPs. In dairy cattle, LD is extensive within a breed but the phase of LD varies between breeds [1], which is expected to decrease across-breed prediction. Use of sequence data is expected to increase the accuracy of multi-breed and across-breed prediction, which would be beneficial for breeds with small reference population sizes [2].

However, results from both simulation studies and analysis of real data do not always show an increase in accuracy from sequence data compared to SNP chip genotypes. The large number of variants makes analysis of all sequence variants and accurate estimation of all effects computationally challenging. Furthermore, the higher rate of genotype errors due to imputation errors in sequence data compared to SNP chip data [3], may limit the benefit of sequence data over SNP chips. Studies using whole-sequence data in dairy cattle [4] and chicken [5] showed no or very little increase in prediction accuracy compared to high-density SNP data, using either genomic best linear unbiased prediction (GBLUP) or a Bayesian variable selection model. Several stimulation studies [6, 7] indicate that, rather than analysing all sequence variants together, preselecting variants that are close to the causative mutations can lead to increased prediction accuracy. In dairy cattle [8, 9] and *Drosophila* [10], substantial increases in accuracy were obtained when several tens, hundreds or thousands variants were selected based on a genome-wide association study (GWAS) and used for prediction in addition to genome-wide SNPs.

On the contrary, other studies show that preselecting sequence variants can lead to an increase in bias and, thus, an increase in accuracy is not evident. Calus et al. [11] used split-and-merge Bayesian selection, where the analysis was split into several subsets that were analysed in a first step to select the most informative variants. Subsequently, selected variants were analysed together. This resulted in a prediction accuracy that is slightly lower or equal to that obtained with the 50 K SNP chip, and increased the bias. Similar results were obtained by Veerkamp [12], using a conditional and joint GWAS. Both Calus et al. [11] and Veerkamp et al. [12] used data on one breed only, Holstein, and the long distance over which LD is conserved within Holstein populations [1] reduces the potential benefit of sequence data over medium- or high-density SNP data [13]. Another approach is preselection of variants based on their functional annotations, which results in small increases in accuracy in dairy cattle [14] and chickens [15], although

Heidaritabar et al. [5] found no increases in prediction accuracy using a similar approach in chickens.

While promising results were obtained by selecting variants based on a GWAS [9], it required testing a large number of scenarios to find a set of variants that increased prediction accuracy. Furthermore, because a GWAS generally tests only one SNP at a time, it does not account for LD between SNPs, which results in the selection of many variants associated with the same QTL. Limiting the number of variants per QTL resulted in a higher accuracy than selecting all variants with a p value below a certain threshold. Therefore, a model that analyses multiple SNPs simultaneously may be more efficient in identifying sequence variants that increase prediction accuracy than a GWAS, which tests SNPs one at a time. Bayesian variable selection models are effective for the identification of causative mutations [16]. However, analysing all sequence variants simultaneously is computationally expensive. To speed up the analysis, Wang et al. [17] developed a hybrid version of the Bayes R variable selection model, which substantially decreases the computing time by first running an expectation–maximization (EM) module, followed by a reduced number of Monte Carlo Markov chain (MCMC) iterations. To further decrease computing time, a proportion of the variants can be dropped either directly after the EM module, or after a number of MCMC iterations.

While the Bayes R hybrid model decreases computing time substantially compared to Bayes R [17], estimating effects for millions of sequence variants simultaneously remains computationally challenging. An approximation to analysing all sequence variants simultaneously could be achieved by splitting up the analysis per chromosome, which makes it feasible to analyse all variants on a chromosome with a Bayesian variable selection model, such as the Bayes R hybrid model.

Our objective was to find a strategy to approximate multi-breed and across-breed prediction, by analysing whole-sequence data with a Bayesian variable selection model. First, we used a simulated dataset that consisted of a filtered set of whole-genome sequence variants to test the accuracy of the Bayes R hybrid model. We also considered the effect of dropping variants with little or no effect during the analysis and tested how the analysis can be split into chromosomes to reduce the elapsed computing time. Furthermore, we investigated the effect of imputation errors on the prediction accuracy. Subsequently, we applied the tested approach to a dataset that contains imputed sequences and records for production and fertility traits for a large number of Holstein, Jersey, Australian Red and crossbred bulls and cows.

Methods

For this study, we used two datasets: a small dataset, with a reduced number of variants and simulated phenotypes, to speed up initial comparisons of different scenarios and a second dataset to test the scenarios in practice, which contained a much larger number of sequence variants and individuals, with daughter trait deviations (DTD) for bulls and trait deviations (TD) for cows for milk, fat, protein and fertility.

Simulated data

The simulated dataset was the AUS-Sim set that is described in more detail by Macleod et al. [14]. This dataset consisted of realised imputed sequence variants for 3047 Holstein bulls, 4942 Holstein cows, 770 Jersey bulls, 1553 Jersey cows, 869 Red Holstein bulls, 741 Australian Red cows and 114 Australian Red bulls. All Holstein and Jersey individuals were used as reference population and the Red Holstein and Australian Red individuals as validation population.

All individuals were genotyped with the Illumina BovineSNP50 chip [18], or custom 50 K chips, and either genotyped with or imputed to the 800 K Illumina BovineHD beadChip. For part of the analysis, the 600,641 SNPs on the HD chip were used (HD). In addition, genotypes for approximately two million sequence variants in gene coding regions and variants that were 5000 bp up- and down-stream of genes were imputed. Annotations for the sequence variants were collated using NGS-SNP [19]. After filtering out variants with a minor allele frequency (MAF) lower than 0.0002 and variants in complete LD, this dataset (SEQ) contained 994,019 variants, including 45,026 non-synonymous coding (NSC) variants, 578,734 variants located within 5 kb upstream and downstream of genes, or in 3/5' untranslated genic regions (REG), and 370,259 variants on the HD chip.

QTL were randomly sampled from all SEQ variants. In total, 4000 causative mutations were simulated, of which 3485, 500 and 15 were categorised as having small, medium and large effects on the trait. Effects were sampled from three normal distributions, with a mean of 0 and variances of $0.0001\sigma_g^2$, $0.001\sigma_g^2$ and $0.01\sigma_g^2$ for small, medium and large QTL respectively, where σ_g^2 is the additive genetic variance. The true breeding value (TBV) of individual j was computed as $TBV_j = \sum_{i=1}^{4000} x_{ij}a_i$, where x_{ij} is the genotype of individual j for QTL i, and a_i the additive effect of QTL i. To obtain a phenotype with a heritability (h^2) of 0.6, an environmental effect was sampled from a normal distribution and added to the TBV. A Holstein breed effect was sampled from $N(10, 1)$ and added to the phenotype for all Holstein individuals.

To investigate the effect of imputation errors on prediction accuracy, errors were added to the SEQ variants for the reference population, the validation population or both populations. For each allele, the probability of an error (e) was simulated as $e = \frac{r}{\sqrt{MAF}}$, where r was equal to 0.0013, 0.0027, 0.0066, 0.0132 or 0.0264 to simulate an average e of 0.005, 0.0101, 0.025, 0.050 or 0.100, respectively. Each imputation error scenario was replicated 10 times.

Pedigree information for all individuals was obtained from the Australian Dairy Herd Improvement Scheme (ADHIS) and Interbull.

Real data

The second dataset contained daughter trait deviations (DTD) or trait deviations (TD) for milk, fat, protein and fertility for 38,540 animals. Animals were genotyped with the Illumina BovineSNP50 chip [18] and imputed to or directly genotyped with the Illumina 800 K BovineHD bead chip. Subsequently, sequences of Holstein, Jersey and Australian Red bulls and cows from Run 5 of the 1000 bulls genome project [20] were used as the reference set to impute sequence genotypes for all individuals using FImpute [21]. During the imputation process, FImpute failed to impute parts of chromosomes 12 and 23, and for these regions, only the HD genotypes were available. This was the case between 25 and 30 Mb on chromosome 12 and between 62 and 70.5 Mb and between 72.5 and 75 Mb on chromosome 23. These regions contained a large number of structural variants and had a low density of HD SNPs, which may have hindered the imputation process. After imputation, the dataset contained 21,379,438 variants, of which 90,010 NSC, 1459,566 REG, 5520,343 intronic variants, 77,299 synonymous variants and 14,232,221 intergenic variants. The HD SNP chip contained 3977 and 360,816 of the synonymous and intergenic variants, respectively. The number of variants used for the analysis was substantially smaller after removing variants with a MAF lower than 0.002 and LD pruning. LD pruning was performed using PLINK [22] to remove variants in high LD ($r^2 > 0.9$). For LD pruning, variants were divided into four groups based on their functional annotations: NSC variants, REG variants, variants on the HD chip and all other variants. Annotations for the sequence variants were collated using the NGS-SNP software [19]. LD pruning was first performed within each group, followed by removal of REG variants with an r^2 higher than 0.9 with a NSC variant, HD variants with an r^2 higher than 0.9 with a NSC variant or a REG variant and other variants with an r^2 higher than 0.9 with a NSC variant. After filtering based on MAF and LD, 4812,745 variants were retained for further analysis.

The dataset was split up into a reference population with Holstein and Jersey bulls born before 2005, and Holstein, Jersey and crossbred cows, and a validation population with Holstein and Jersey bulls born in 2005 and after, and Australian Red bulls and cows. Animals in the reference population that had sons in the validation population were removed from the dataset. Furthermore, seven animals were removed from the dataset because their sequence differed for less than 10,000 variants from another individual in the dataset. Because of the presence of crossbred individuals, a principal component analysis (PCA) was used to divide the Holstein and Jersey animals in five different clusters, as shown in Figure S1 (see Additional file 1: Figure S1). Clusters 1, 2 and 3 contained mainly Holstein individuals, while clusters 4 and 5 contained mainly Jersey individuals. The crossbred individuals were present in all clusters. Three Jersey cows were removed from the analysis because they were assigned to clusters 1 and 2, and one Holstein cow and one Holstein bull were removed from the analysis because they were assigned to cluster 5. The clusters were set as fixed effect to account for breed differences. In total, the reference population for production traits contained 35,775 individuals, including 22,868 Holstein cows, 3124 Holstein bulls, 6144 Jersey cows, 787 Jersey bulls and 2852 crossbred cows. An overview of the reference population is in Table 1. In the validation population, the number of individuals in clusters 3 and 4 was small, i.e. 28 and 20 individuals, respectively. Therefore, the individuals in these clusters were not used in the analysis. In total, the validation population contained 2717 individuals, including 799 Holstein bulls, 200 Jersey bulls, 1579 Australian Red cows and 139 Australian Red bulls. Table 2 summarizes the validation population.

Statistical analysis

We used the hybrid version of the Bayes R mixture model described by Wang et al. [23] for our analyses:

$$\mathbf{y} = \mathbf{Xb} + \mathbf{Za} + \mathbf{Wv} + \mathbf{e},$$

where $\mathbf{y}$ is a vector of phenotypes (TD or DTD), $\mathbf{X}$ a design matrix that allocates phenotypes to vector $\mathbf{b}$ with fixed effects, fitting the overall mean, breed and sex as fixed effects, $\mathbf{Z}$ is a design matrix that allocates phenotypes to vector $\mathbf{a}$ with polygenic breeding values distributed as $N(0, \mathbf{A}\sigma_a^2)$, where $\mathbf{A}$ the pedigree-based relationship matrix, σ_a^2 is the polygenic variance, $\mathbf{W}$ is a design matrix of genotypes, $\mathbf{v}$ a vector of variant effects, and $\mathbf{e}$ a vector of residual errors distributed as $N(0, \mathbf{E}\sigma_e^2)$, where $\mathbf{E}$ is a diagonal matrix with diagonals $1/w_j$, where the weighting coefficient w_j is based on the number of records available for individual j [24], and σ_e^2 is the residual variance. Variant effects ($\mathbf{v}$) were drawn from one of four normal distributions with $N(0, 0\sigma_g^2)$, $N(0, 0001\sigma_g^2)$, $N(0, 001\sigma_g^2)$, and $N(0, 01\sigma_g^2)$, respectively, where σ_g^2 is the additive genetic variance. The prior distribution for the

Table 1 Reference population

Cluster	Breed	Sex	Production	Fertility
1	Holstein	Cows	8757	7853
	Holstein	Bulls	1246	1230
	Crossbred	Cows	447	401
2	Holstein	Cows	12,140	10,926
	Holstein	Bulls	1607	1551
	Crossbred	Cows	824	735
3	Holstein	Cows	1936	1831
	Holstein	Bulls	271	229
	Jersey	Cows	10	10
	Jersey	Bulls	1	1
	Crossbred	Cows	738	684
4	Holstein	Cows	35	30
	Jersey	Cows	609	584
	Jersey	Bulls	190	145
	Crossbred	Cows	710	668
5	Jersey	Cows	5525	5281
	Jersey	Bulls	596	551
	Crossbred	Cows	133	109

The number of individuals in the reference population is split up per cluster, breed and sex for production traits and fertility

Table 2 Number of individuals in the validation population

Cluster	Breed	Sex	Production	Fertility
1	Holstein	Bulls	357	294
2	Holstein	Bulls	442	338
5	Jersey	Bulls	200	167
–	Australian Red	Cows	1579	1507
–	Australian Red	Bulls	139	133

The number of individuals in the validation population is split up per cluster, breed and sex for production traits and fertility

proportion of variants in each of these distributions was $\mathbf{P} \sim \text{Dirichlet}(\alpha), \alpha = [1, 1, 1, 1]$.

The hybrid variant of Bayes R uses first an expectation–maximization (EM) module to estimate $\mathbf{a}$, $\mathbf{P}$, $\mathbf{b}$, $\mathbf{v}$, and σ_e^2. Then, the estimates of these parameters are used as starting values for the subsequent Monte Carlo Marcov chain (MCMC) module, for 10,000 iterations, without burn-in.

The accuracy of prediction was defined as the correlation of the predicted breeding value with the TD (cows) or DTD (bulls) between validation animals.

Dropping of variants

To speed up the analysis, it is possible to drop some of the variants during the different stages of analysis (e.g. after the EM step or after a certain number of MCMC iterations). Variants were ranked based on their posterior inclusion probability (PIP) to be included in any of the distributions with a non-zero variance, and the variants with the lowest PIP were dropped in order to drop the desired proportion of variants. After dropping, the mixing proportions at the time of dropping were added to the prior for the rest of the analysis, to account for the dropped variants.

Scenarios in the simulated dataset

Using the simulated dataset, we tested several strategies to analyse sequence data, which are summarized in Table 3: all sequence variants analysed together (S_FULL_D0), all variants analysed per chromosome (S_CHR_Dd), variants selected based on their PIP from each chromosome (CHR) reanalysed with all chromosomes together (S_KEPT_Dd), and variants selected by CHR and all HD variants reanalysed with all chromosomes together (S + HD_KEPT + HD_Dd). As a comparison to the S_FULL_D0 scenario, we analysed all HD genotypes (HD_FULL_D0). In the S_FULL_Dd scenarios, the sequence variants were analysed simultaneously with $d = 0, 0.25, 0.5, 0.7$ or 0.9 as the target proportion of variants dropped during the analysis. Variants were dropped after the EM step, after 200 MCMC iterations, or after 10,000 MCMC iterations.

In scenarios S_CHR_Dd with $d = 0, 0.7$ or 0.9, the sequence variants were split up and analysed per chromosome. The effects of variants that were estimated during HD_FULL_D0 were used to correct the DTD and TD for all other chromosomes except the chromosome that was analysed. After analysing all the chromosomes, the estimated effects of variants of all the chromosomes were used to estimate a genome-wide breeding value.

Using the variant effects estimated by S_CHR_Dd directly to compute breeding values assumes that effects are estimated independently between chromosomes. Therefore, in scenarios S_KEPT_Dd with $d = 0.7$ or 0.9, variants that were retained in the model by S_CHR_Dd were reanalysed in a genome-wide analysis to re-estimate effects of variants and GEBV.

The approach used in scenarios S_KEPT + HD_Dd, with $d = 0.7$ or 0.9, was the same as S_KEPT_Dd, except that in addition to the variants that were retained in the model for the analyses per chromosome, the HD variants were added into the model.

For scenarios HD_FULL_D0, S_FULL_Dd and S_CHR_Dd, the prior for the number of variants per distribution was $\alpha = [1, 1, 1, 1]$, whereas for S_KEPT_Dd and S_KEPT + HD_Dd, this was set to the posterior estimate of the number of variants per distribution obtained by S_FULL_Dd.

Scenarios with the real data

Using the real dataset, we compared scenarios HD_FULL_D0, S_CHR_D0.9, S_KEPT_D0.9 and S_KEPT + HD_D0.9. For HD_FULL_D0, the prior for the number of variant per distribution was $\alpha = [1, 1, 1, 1]$, and the posterior estimate of the HD_FULL_D0 scenario was used as prior for S_CHR_D0.9 and S_KEPT_D0.9.

Animal ethics statement

No ethical approval was required for this study.

Table 3 Overview of scenarios

Scenario	Data	Strategy	DropIter	DropProp	Simulation	Real
HD_FULL_D0	HD	FULL	–	0	Y	Y
S_FULL_D0	SEQ	FULL	–	0	Y	N
S_FULL_D0.25	SEQ	FULL	0, 200 or 10,000	0.25	Y	N
S_FULL_D0.50	SEQ	FULL	0, 200 or 10,000	0.50	Y	N
S_FULL_D0.7	SEQ	FULL	0, 200 or 10,000	0.70	Y	N
S_FULL_D0.9	SEQ	FULL	0, 200 or 10,000	0.90	Y	N
S_CHR_D0	SEQ	CHR	0	0	Y	N
S_CHR_D0.7	SEQ	CHR	10,000	0.70	Y	N
S_CHR_D0.9	SEQ	CHR	10,000	0.90	Y	N
S_KEPT_D0.7	SEQ	KEPT	10,000	0.70	Y	N
S_KEPT_D0.9	SEQ	KEPT	10,000	0.90	Y	Y
S + HD_KEPT + HD_D0.7	SEQ + HD	KEPT + HD	10,000	0.70	Y	N
S + HD_KEPT + HD_D0.9	SEQ + HD	KEPT + HD	10,000	0.90	Y	N

HD = HD genotypes used for prediction, S = sequence variants used for prediction, FULL = all variants analysed together, CHR = all variants analysed per chromosome, KEPT = variants selected by CHR reanalysed with all chromosomes together, KEPT + HD = variants selected by CHR and all HD variants reanalysed with all chromosomes together, dropProp = proportion of variants that is dropped after dropIter MCMC iterations, simulation and real indicate whether the scenario was analysed in the simulated and real datasets

Results

Simulation

The accuracy and bias of the different strategies are in Figs. 1 and 2, respectively. Differences between scenarios were more pronounced for Australian Red than for Red Holstein. For both breeds, the accuracy was higher using sequence data than HD data. The S_FULL_D0 scenario resulted in accuracies of 0.60 and 0.66 for Australian Red and Red Holstein individuals, respectively, while HD_FULL_D0 yielded accuracies of 0.45 and 0.64. Dropping 70 or 90% of the variants after 10,000 MCMC iterations resulted in accuracies that were similar or slightly reduced compared to those with S_FULL_D0. Dropping variants directly after the EM module or after 200

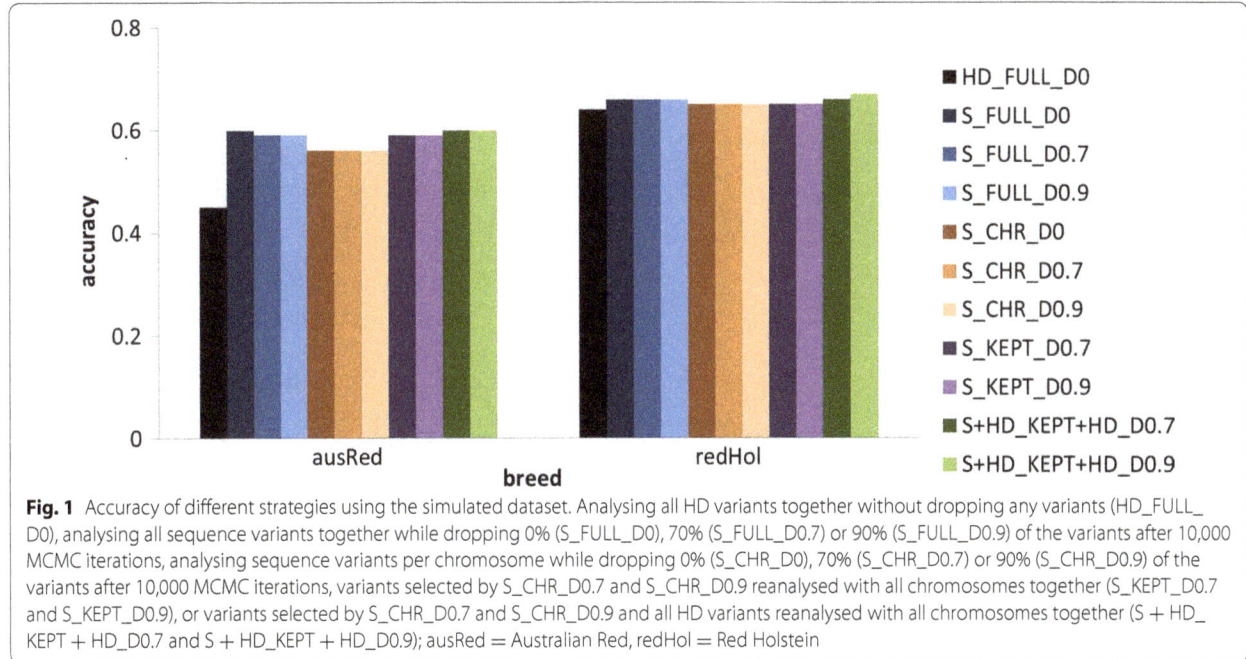

Fig. 1 Accuracy of different strategies using the simulated dataset. Analysing all HD variants together without dropping any variants (HD_FULL_D0), analysing all sequence variants together while dropping 0% (S_FULL_D0), 70% (S_FULL_D0.7) or 90% (S_FULL_D0.9) of the variants after 10,000 MCMC iterations, analysing sequence variants per chromosome while dropping 0% (S_CHR_D0), 70% (S_CHR_D0.7) or 90% (S_CHR_D0.9) of the variants after 10,000 MCMC iterations, variants selected by S_CHR_D0.7 and S_CHR_D0.9 reanalysed with all chromosomes together (S_KEPT_D0.7 and S_KEPT_D0.9), or variants selected by S_CHR_D0.7 and S_CHR_D0.9 and all HD variants reanalysed with all chromosomes together (S + HD_KEPT + HD_D0.7 and S + HD_KEPT + HD_D0.9); ausRed = Australian Red, redHol = Red Holstein

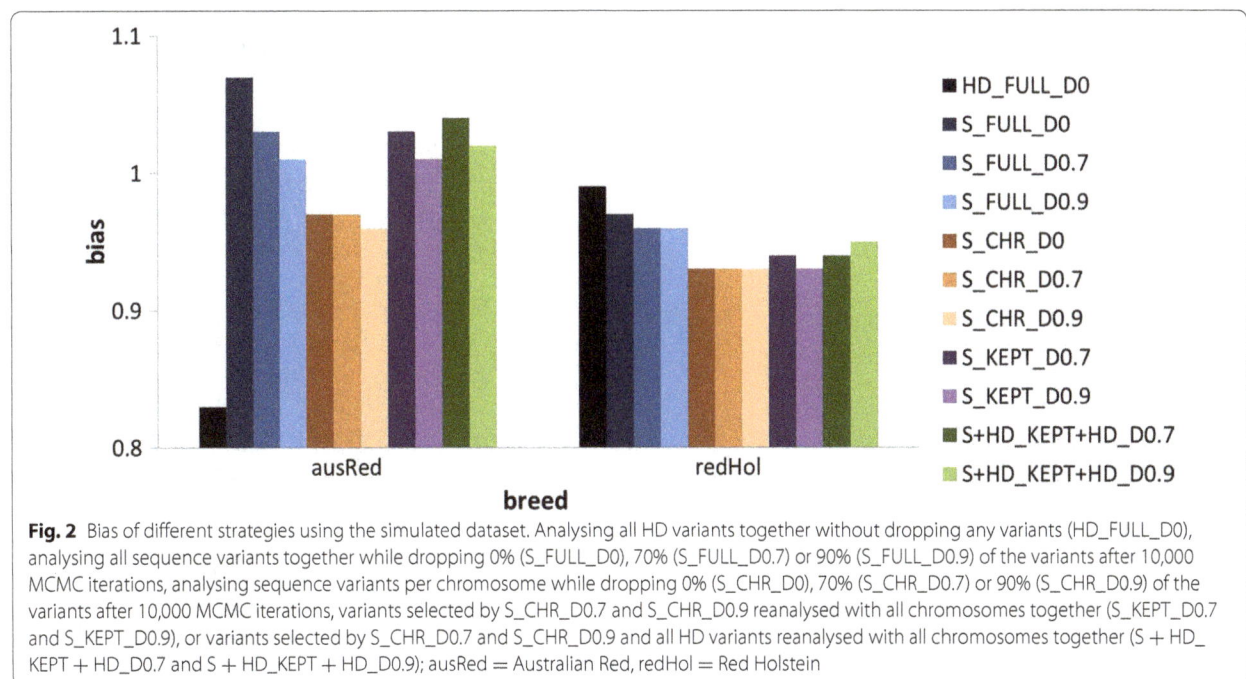

Fig. 2 Bias of different strategies using the simulated dataset. Analysing all HD variants together without dropping any variants (HD_FULL_D0), analysing all sequence variants together while dropping 0% (S_FULL_D0), 70% (S_FULL_D0.7) or 90% (S_FULL_D0.9) of the variants after 10,000 MCMC iterations, analysing sequence variants per chromosome while dropping 0% (S_CHR_D0), 70% (S_CHR_D0.7) or 90% (S_CHR_D0.9) of the variants after 10,000 MCMC iterations, variants selected by S_CHR_D0.7 and S_CHR_D0.9 reanalysed with all chromosomes together (S_KEPT_D0.7 and S_KEPT_D0.9), or variants selected by S_CHR_D0.7 and S_CHR_D0.9 and all HD variants reanalysed with all chromosomes together (S + HD_KEPT + HD_D0.7 and S + HD_KEPT + HD_D0.9); ausRed = Australian Red, redHol = Red Holstein

MCMC iterations decreased accuracy, as shown in Fig. 3. Accuracy decreased as the proportion of dropped variants increased and increased as the number of MCMC iterations increased before deciding which variants to drop. Figure 4 shows the bias as a function of the proportion of dropped variants. There was no consistent increase or decrease in bias across breeds when more variants were dropped.

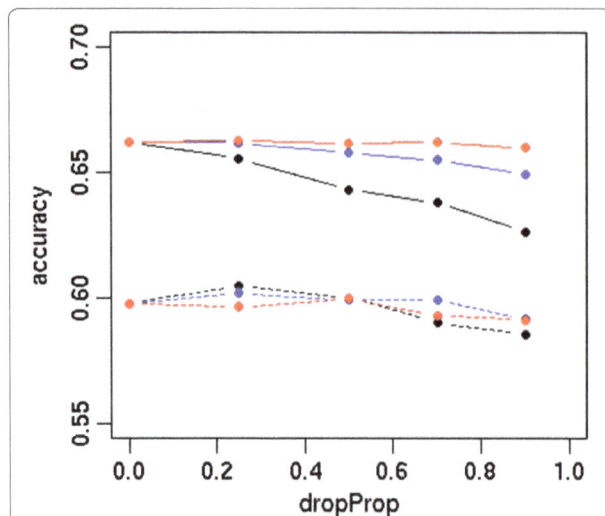

Fig. 3 Prediction accuracy as a function of the proportion of dropped variants. Variants were dropped after EM (black), 200 MCMC iterations (blue) or 10,000 MCMC iterations (red), line = Red Holstein, dashed line = Australian Red, dropProp = proportion of dropped variants

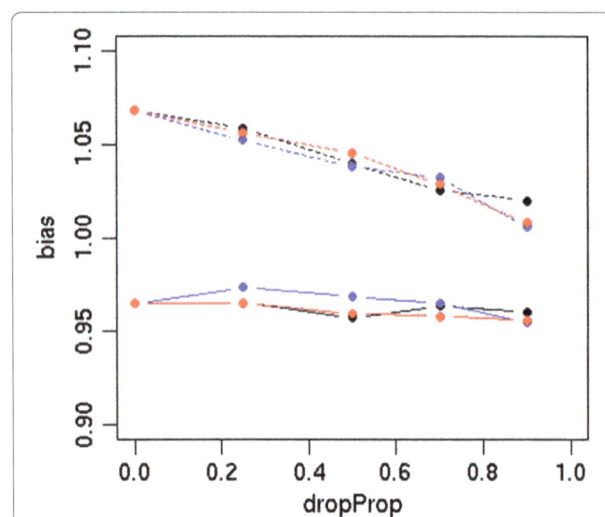

Fig. 4 Bias as a function of the proportion of dropped variants. Variants were dropped after EM (black), 200 MCMC iterations (blue) or 10,000 MCMC iterations (red), continuous line = Red Holstein, dashed line = Australian Red, dropProp = proportion of dropped variants

Splitting up the analyses per chromosome and analysing all chromosomes in parallel decreased the computing time from 55 h for S_FULL_D0 to between 1.9 and 4.5 h per chromosome. However, the accuracy was lower than that obtained by S_FULL_D0. The reduction in accuracy was up to 0.04 in Australian Red and 0.01 in Red Holstein. Combining the process of dropping 70 or 90% of the variants with splitting up the analysis per chromosome did not decrease accuracy furthermore.

Contrary to the S_CHR_Dd scenarios, reanalysing the variants that were kept in the model in a genome-wide analysis in the S_KEPT_Dd scenarios resulted in accuracies that were similar or only slightly lower than those obtained with S_FULL_Dd. Adding the HD variants in the S_KEPT + HD_Dd scenarios resulted in accuracies that were similar to those obtained with S_FULL_Dd.

Table 4 compares the number of variants assigned to each of the four distributions for the different scenarios. Generally, sequence data resulted in a larger number of variants with effects drawn from the distributions with small and large variances compared to HD data. Compared to the number of simulated QTL (3485 small, 500 medium and 15 large QTL), the number of variants included in these distributions tended to be overestimated, especially the number of variants with effects drawn from the distribution with the largest variance. While only 15 QTL had an effect size that corresponded to the distribution with the largest variance, the number of variants assigned to this distribution varied from 21 for HD_FULL_D0 to 71 for S_CHR_D0. Overestimation of the number of variants in the fourth distribution was largest in the S_CHR_D0 scenario.

Table 5 shows the proportion of variance explained by prediction markers (h_M^2) and the polygenic component (h_A^2), and the heritability computed as: $h^2 = h_M^2 + h_A^2$. In the HD_FULL_D0 scenario, h^2 was equal to 0.59 and thus, was close to the simulated heritability of 0.60. In the scenarios using sequence data, h^2 was highest when all sequence variants were used (0.64). When variants were dropped, h_A^2 increased slightly, while h^2 and h_M^2 decreased. In the S_KEPT_D0.7 scenario, h^2, h_M^2 and h_A^2 were equal to 0.60, 0.59 and 0.01, respectively. The largest h_A^2 i.e. 0.05 was obtained with the S_KEPT_D0.9 scenario. The highest h^2 were obtained with the S_KEPT + HD_Dd scenarios, i.e. 0.65 and 0.63 for S_KEPT + HD_D0.7 and S_KEPT + HD_D0.9, respectively.

Table 6 shows the number of simulated QTL dropped or retained with their respective posterior inclusion probability (PIP). For all scenarios, the majority of QTL had a PIP between 0 and 0.01. In the scenarios in which variants were dropped, the majority of QTL were dropped, and the number of dropped QTL increased as

Table 4 Average number of variants per distribution over the number of iterations in the simulated dataset

Data	Analysis	Drop	Number of variants per distribution			
			$0\,\sigma_g^2$	$0.0001\,\sigma_g^2$	$0.001\,\sigma_g^2$	$0.01\,\sigma_g^2$
HD	FULL	0.0	592,931	3286	898	21
S	FULL	0.0	914,767	5053	666	48
		0.7	369,322	2464	471	43
		0.9	172,007	1350	407	42
S	CHR	0.0	915,665	4279	519	71
		0.7	371,643	2279	396	65
		0.9	171,554	1350	358	61
S	KEPT	0.7	298,238	2118	499	44
		0.9	98,494	908	390	45
S + HD	KEPT + HD	0.7	759,835	4459	663	44
		0.9	650,252	3813	616	45

HD = HD genotypes used for prediction, S = sequence variants used for prediction, FULL = all variants analysed together, CHR = all variants analysed per chromosome, KEPT = variants selected by CHR reanalysed with all chromosomes together, KEPT + HD = variants selected by CHR and all HD variants reanalysed with all chromosomes together, dropProp = proportion of variants that is dropped after 10,000 MCMC iterations, σ_g^2 = additive genetic variance

Table 5 Proportion of variance explained by markers (h_M^2) and polygenic effect (h_A^2) in the simulated dataset

Data	Analysis	Drop	h_M^2	h_A^2	h^2
HD	FULL	0.0	0.57	0.02	0.59
S	FULL	0.0	0.63	0.01	0.64
		0.7	0.60	0.02	0.62
		0.9	0.58	0.02	0.60
S	KEPT	0.7	0.59	0.01	0.60
		0.9	0.52	0.05	0.57
S + HD	KEPT + HD	0.7	0.64	0.01	0.65
		0.9	0.62	0.01	0.63

$h^2 = h_M^2 + h_A^2$, HD = HD genotypes used for prediction, S = sequence variants used for prediction, FULL = all variants analysed together, CHR = all variants analysed per chromosome, KEPT = variants selected by CHR reanalysed with all chromosomes together, KEPT + HD = variants selected by CHR and all HD variants reanalysed with all chromosomes together, dropProp = proportion of variants that is dropped after 10,000 MCMC iterations

the proportion of dropped variants increased. The number of QTL in the classes with a PIP higher than 0.01 varied between scenarios. The number of variants with a PIP between 0.5 and 1 was largest in the S_CHR_Dd scenarios.

Figures 5 and 6 show the prediction accuracy and bias as a function of the imputation error. The prediction accuracy decreased as the number of imputation errors increased but there was no clear pattern for bias and this decrease was larger for Australian Red than for Red Holstein. It was larger when imputation errors were added only to the validation population than when they were added to the training population or to both the training and validation populations.

Real data

The accuracy and bias of the scenarios tested with real data are in Figs. 7 and 8. For all traits, S_KEPT_D0.9 and S_KEPT + HD_D0.9 tended to result in reduced accuracy and increased bias compared to HD_FULL_D0. Sequence data resulted in substantially increased accuracies only for Australian Red Bulls. Holstein bulls were grouped in two clusters, and accuracies were higher for the bulls in the HOL2 cluster that was closest to Jersey individuals in the PCA. Averaged across traits, the difference in accuracy of the S_KEPT_D0.9 scenario compared to the HD_FULL_D0 scenario was equal to −0.03, −0.01, −0.02, −0.03 and 0.11 for HOL1, HOL2, JER, RCOW and RBULL, respectively. Adding the HD variants improved the accuracy slightly with, averaged across traits, a difference compared to HD_FULL_D0 of −0.02, 0.01, −0.02, −0.02 and 0.11 for HOL1, HOL2, JER, RCOW and RBULL, respectively. Decreases in accuracy were smallest for fertility and largest for fat yield. The bias of the prediction was larger with the S_KEPT_D0.9 and S_KEPT + HD_D0.9 scenarios than with HD_FULL_D0 for HOL1, JER and RCOW. For HOL2, the bias was similar in all three scenarios, although with HD_FULL_D0, regression coefficients were above 1, while for S_KEPT_D0.9 and S_KEPT + HD_D0.9, regression coefficients were below 1. For RBULL, the bias was large for all scenarios and not consistently better in any one.

The number of variants assigned to each of the four distributions is in Table 7. S_KEPT_D0.9 resulted in fewer variants in the distribution with zero effect, more variants in the distribution with a small variance, and generally fewer or a similar number of variants in the distributions with medium and large variances. In the

Table 6 Number of simulated QTL dropped or retained with their respective posterior inclusion probability (PIP)

Analysis	Drop	Dropped	PIP					
			0–0.01	0.01–0.05	0.05–0.1	0.1–0.2	0.2–0.5	0.5–1
FULL	0.0	0	3368	435	40	15	19	20
	0.7	2159	1179	463	35	19	19	23
	0.9	2981	293	520	36	24	18	25
CHR	0.0	0	3337	465	36	12	21	26
	0.7	2177	1164	460	36	14	20	26
	0.9	3025	306	461	42	15	19	29
KEPT	0.7	2177	1088	531	38	21	22	20
	0.9	3025	251	514	41	22	19	25
KEPT+	0.7	2177	1193	446	32	17	15	17
HD	0.9	3025	347	440	34	18	16	17

FULL = all variants analysed together, CHR = all variants analysed per chromosome, KEPT = variants selected by CHR reanalysed with all chromosomes together, KEPT + HD = variants selected by CHR and all HD variants reanalysed with all chromosomes together, drop = proportion of variants that are dropped after 10,000 MCMC iterations

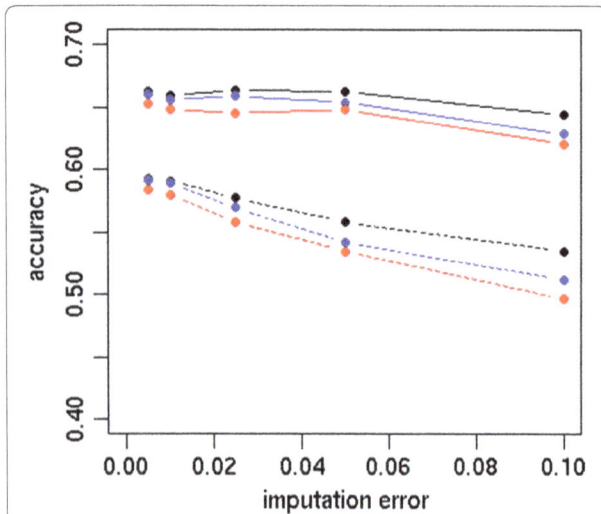

Fig. 5 Prediction accuracy as a function of imputation error. Imputation errors were added to both reference and validation population (black), only the reference population (blue) or only the validation population (red), continuous line = Red Holstein, dashed line = Australian Red

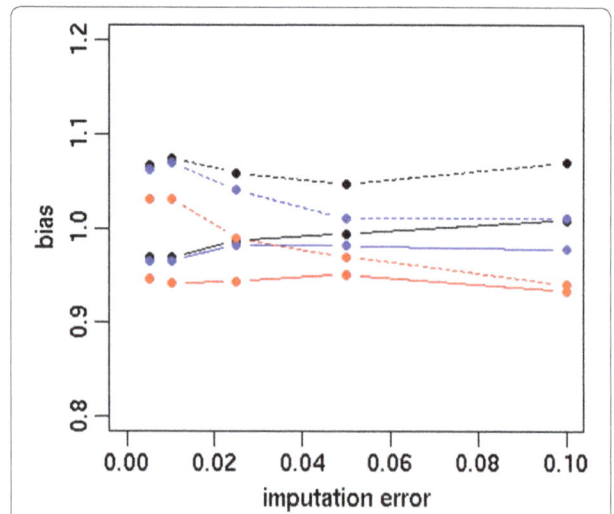

Fig. 6 Bias as a function of imputation error. Imputation errors were added to both reference and validation populations (black), only the reference population (blue) or only the validation population (red), continuous line = Red Holstein, dashed line = Australian Red

S_KEPT + HD_D0.9 scenario, there were more variants in both the distributions with zero effect and with a small variance than in the other scenarios, and generally fewer or a similar number of variants in the distribution with medium and large variances. Contrary to the milk production traits, for fertility, both S_KEPT_D0.9 and S_KEPT + HD_D0.9 resulted in more variants in the distribution with a medium variance than HD_FULL_D0.

Table 8 shows h^2_M, h^2_A and h^2 obtained with real data. h^2_M and h^2 were lowest in the HD_FULL_D0 scenario and highest in the S_KEPT + HD_D0.9 scenario, and h^2_A was highest in the HD_FULL_D0 scenario and lowest in the S_KEPT + HD_D0.9 scenario. For milk production traits, differences between S_KEPT + HD_D0.9 and HD_FULL_D0 varied between 0.22 and 0.24 for h^2_M, 0.10 and 0.12 for h^2, and were equal to 0.12 for h^2_A. Differences between S_KEPT + HD_D0.9 and S_KEPT_D0.9 were smaller, varying between 0.04 and 0.05 for h^2_M, 0.03 and 0.05 for h^2, and were equal to −0.01 for h^2_A. For fertility, h^2 was much lower, which resulted in smaller differences between scenarios, although the overall trend was the same as for production traits.

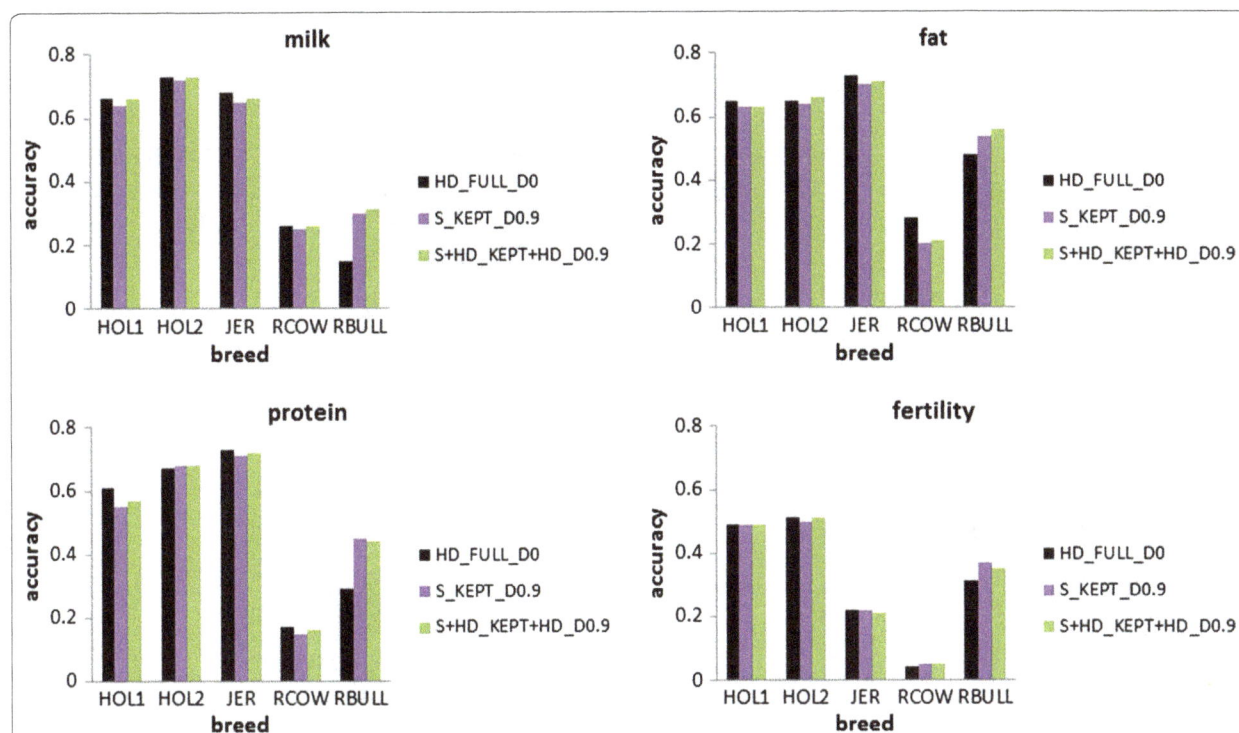

Fig. 7 Accuracy of different scenarios using real data. Analysing all HD variants together without dropping any variants (HD_FULL_D0), dropping 90% of the variants per chromosome and reanalysing the remaining variants with all chromosomes together (S_KEPT_D0.9), or dropping 90% of variants per chromosome and reanalysing the remaining variants and all HD variants with all chromosomes together (S + HD_KEPT + HD_D0.9); HOL1 = Holstein cluster 1, HOL2 = Holstein cluster 2, JER = Jersey, RCOW = Australian Red cows, RBULL = Australian Red bulls

Discussion

We focus the discussion on two points, i.e. (1) on the ability to reduce the computing time needed for analysis of whole-genome sequence data by using an EM-MCMC hybrid approach, dropping some variants from the analysis and processing chromosomes in parallel, and (2) on the reasons why genome sequence data may or may not result in higher accuracies than HD SNP genotypes.

Approximate analysis of full sequence data with Bayes R

The simulated datasets were previously analysed by Macleod et al. [14], using Bayes R. We obtained the same accuracy using full sequence data with the hybrid version of Bayes R. This is in agreement with Wang et al. [23] who show that the accuracy with the hybrid model was equal to that with Bayes R, which confirms that the hybrid model is an efficient alternative to Bayes R.

We tested a new option of the hybrid model, which drops a proportion of the variants during the analysis to decrease the required computing time even more. The dropping of variants was tested at different stages of the analysis, and the proportion of variants that were dropped varied. While dropping variants reduced computing time, it resulted in a decrease in accuracy. The decrease in accuracy became smaller as fewer variants

were dropped, and when variants were dropped after a large number of MCMC iterations. However, the goal of dropping variants is to reduce computing time, and the gain in computing time is smaller when fewer variants are dropped. Running the full MCMC chain before dropping any variants resulted in an accuracy that was similar to that in the analyses that did not drop any variants. However, if the analysis was run first for 10,000 iterations before dropping the variants, and subsequently run for another 10,000 iterations with the dropped variants computing time increased rather than decreased compared to analysing all the variants for 10,000 iterations without dropping any variants. Therefore, if the goal is to increase the speed of the analysis, it is better to use all the variants. However, if it is necessary to select variants that are associated with the trait, the results of the hybrid model can be used to select variants, but a large number of MCMC iterations is advisable. We note that dropping variants from the analysis can lead to bias. We prevented this by recording the mixing proportions for the four distributions immediately before any SNPs were dropped and adding this to the prior.

Analysing a few millions of sequence variants simultaneously is computationally challenging and would take a long time to complete. Therefore, we tested if it

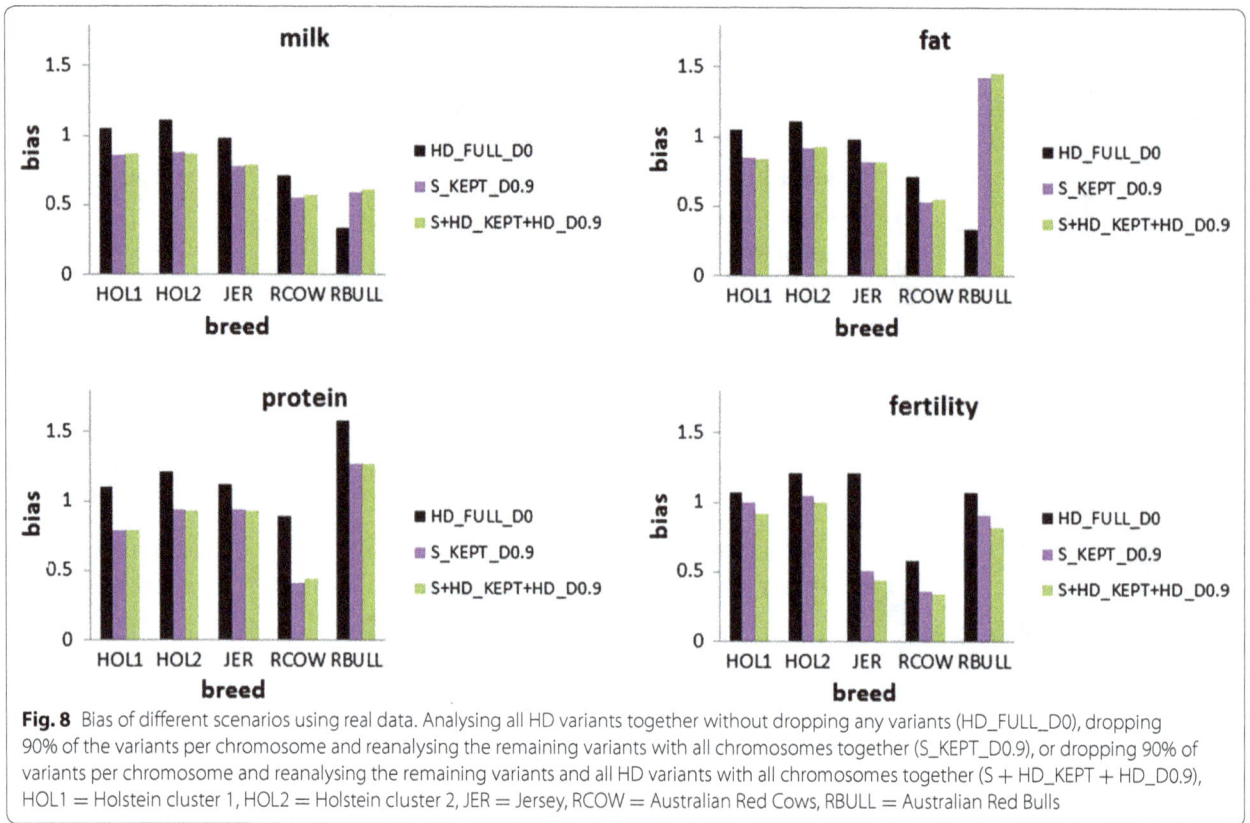

Fig. 8 Bias of different scenarios using real data. Analysing all HD variants together without dropping any variants (HD_FULL_D0), dropping 90% of the variants per chromosome and reanalysing the remaining variants with all chromosomes together (S_KEPT_D0.9), or dropping 90% of variants per chromosome and reanalysing the remaining variants and all HD variants with all chromosomes together (S + HD_KEPT + HD_D0.9), HOL1 = Holstein cluster 1, HOL2 = Holstein cluster 2, JER = Jersey, RCOW = Australian Red Cows, RBULL = Australian Red Bulls

Table 7 Average number of variants per distribution over the number of iterations in the real dataset

Trait	Data	Analysis	DropProp	Number of variants per distribution			
				$0\ \sigma_g^2$	$0.0001\ \sigma_g^2$	$0.001\ \sigma_g^2$	$0.01\ \sigma_g^2$
Milk	HD	FULL	0	627,503	4299	22	10
	S	KEPT	0.9	483,521	6603	17	6
	S + HD	KEPT + HD	0.9	1076,643	8927	22	6
Fat	HD	FULL	0	627,510	4312	9	4
	S	KEPT	0.9	483,614	6307	7	4
	S + HD	KEPT + HD	0.9	1080,078	8890	9	3
Prot	HD	FULL	0	627,347	4476	9	3
	S	KEPT	0.9	482,957	6352	8	3
	S + HD	KEPT + HD	0.9	1078,647	9025	9	2
Fert	HD	FULL	0	625,899	5668	260	8
	S	KEPT	0.9	548,382	5715	310	5
	S + HD	KEPT + HD	0.9	1135,712	10,569	436	7

Prot = protein, fert = fertility, HD = HD genotypes used for prediction, S = sequence variants used for prediction, FULL = all variants analysed together, KEPT = variants selected per chromosome reanalysed with all chromosomes together, KEPT + HD = variants selected per chromosome and all HD variants reanalysed with all chromosomes together, dropProp = proportion of variants dropped after 10,000 MCMC iterations, σ_g^2 = additive genetic variance

is possible to split up the analysis per chromosome. However, using the effects of SNPs estimated per chromosome directly to estimate breeding values resulted in a decreased accuracy compared to S_FULL_D*d*. Our

approach is somewhat similar to that tested by Calus et al. [11]. Calus et al. [11] split up the variants, but in their approach, the LD between variants in a subset was minimized. By splitting up the analysis per chromosome,

Table 8 Proportion of variance explained by prediction markers (h_M^2) and polygenic effect (h_A^2)

Trait	Data	Analysis	DropProp	h_M^2	h_A^2	h^2
Milk	HD	FULL	0.0	0.29	0.16	0.45
	S	KEPT	0.9	0.49	0.05	0.53
	S + HD	KEPT + HD	0.9	0.53	0.04	0.57
Fat	HD	FULL	0.0	0.22	0.15	0.37
	S	KEPT	0.9	0.39	0.04	0.42
	S + HD	KEPT + HD	0.9	0.44	0.03	0.47
Protein	HD	FULL	0.0	0.21	0.16	0.37
	S	KEPT	0.9	0.39	0.05	0.43
	S + HD	KEPT + HD	0.9	0.44	0.04	0.48
Fertility	HD	FULL	0.0	0.02	0.00	0.02
	S	KEPT	0.9	0.03	0.00	0.03
	S + HD	KEPT + HD	0.9	0.04	0.00	0.04

$h^2 = h_M^2 + h_A^2$, HD = high density SNP, HD = HD genotypes used for prediction, S = sequence variants used for prediction, FULL = all variants analysed together, KEPT = variants selected per chromosome reanalysed with all chromosomes together, KEPT + HD = variants selected per chromosome and all HD variants reanalysed with all chromosomes together, drop = proportion of variants dropped after 10,000 MCMC iterations

we maximised the LD. Calus et al. [11] observed that the performance of the model decreased when subsets contained variants in very high LD. This could explain why we found a reduced accuracy for the S_CHR_D*d* scenarios compared to S_FULL_D*d*, although we tried to address this issue by pruning out variants in high LD with each other. Furthermore, the dataset used by Calus et al. [11] contained only Holstein individuals, while our dataset contained individuals from multiple breeds, and LD is conserved over much longer distances within breeds than across breeds [1]. Our approach differs from that described by Calus et al. [11], in that we used the HD estimated effects to correct for all chromosomes except the chromosome being analysed. This would be the same as analysing full sequence data if the prediction based on full sequence data for other chromosomes was the same as the prediction based on HD SNPs. It appears that since the analysis based on sequence data changes the estimated effects of sequence variants, it is necessary to analyse the retained variants from all chromosomes together to maximise accuracy. Therefore, the analyses per chromosome were used to select variants rather than directly to predict breeding values. Rerunning the selected variants from all the chromosomes combined together increased the accuracy to a value that was equal or close to that obtained with S_FULL_D*d* in the simulation. However, this required to drop a large number of variants, which resulted in a decrease in accuracy even for the S_FULL_D*d* scenarios. The vast majority of variants that were dropped would probably have very small effects, and therefore were not likely to be linked to major QTL. They could, however, be used to explain part of the polygenic effects. Therefore, we added the

HD variants to the analysis, which further increased the accuracy.

Potential advantage of sequence data over HD SNP genotypes

Using the simulated data, analysis of sequence data resulted in a higher accuracy than analysis of HD SNP genotypes, i.e. there was a large advantage of S_FULL_D0 over HD_FULL_D0, and consequently, the accuracy of any scenario using sequence data was higher than HD_FULL_D0, even for the scenarios with an accuracy lower than that of S_FULL_D0. For the Red Holstein validation population, the advantage of S_FULL_D0 over HD_FULL_D0 was much smaller than for the Australian Red validation population. This is likely because the Red Holstein is much more closely related to the Holstein individuals in the reference populations. Because LD is conserved over much shorter distances across breeds than within breeds, sequence data is thought to be especially beneficial for multi-breed and across-breed prediction [2].

There are two reasons for the use of sequence data resulting in higher accuracy: it might capture more of the genetic variance and it might include QTL with large effects when there are no HD SNPs in complete LD with these QTL. However, the variance not explained by SNPs (h_A^2) was only 0.01 to 0.02 higher than when sequence data was analysed. Therefore, this does not explain the large increase in accuracy observed, and it appears that the prediction equation based on HD SNPs used SNPs in LD with the QTL and that the phase of LD differed in the validation and training populations. By comparison, the prediction based on sequence data must have emphasised variants that were closer to the QTL (or even were the QTL) and this LD was better conserved in the validation population. In reality, the missing heritability in HD SNP chip data is likely to be higher than 0.02, and consequently, the analysis of simulated data may under-estimate the advantage of sequence data in this respect. Indeed, in the analysis of the real data, h^2_A was much higher than in the simulation study. In the S_KEPT_D0.9 and S_KEPT + HD_D0.9 scenarios using real data, h^2_M and h^2 were substantially higher than in HD_FULL_D0, while h^2_A was lower, which suggests that using sequence data reduced the amount of missing heritability.

Difference between results obtained with simulated and real data

When real data was analysed, sequence data resulted in an accuracy that was similar to that of HD_FULL_D0. This is in line with several studies that reported little or no advantage of sequence data over HD or 50 K geno-types, especially within breed [4, 11, 12], but it differs from results obtained in our simulation study. These dif-ferences may have been caused by differences between the simulated and real data. In the simulated data, we simulated a moderate number of QTL, which were pre-sent in the sequence data but not in HD data. In the real data, it is possible that the number of QTL was larger but that fewer QTL had medium to large effects, which made it more difficult for Bayes R to distinguish between vari-ants in high LD with the causative mutations and variants that have no effect on the trait. In the simulated data, we assumed that all causative mutations had the same effect in all breeds, but in reality, breed x QTL interactions may result in different effects. In addition, not all sequence variants were included in the data analysed and it is likely that some causal mutations were absent. Furthermore, while an Australian Red validation population was used in both the simulated and real data, the Red Holstein bulls used as validation population were more distantly related to the Holstein individuals in the reference pop-ulation than the Holstein bulls used as validation in the real dataset. Sequence data is expected to be more advan-tageous for multi-breed and across-breed prediction than for within-breed prediction, and therefore, using two relatively distantly related validation populations likely resulted in the sequence data to be more advantageous in the simulated dataset than in the within-breed scenarios.

Another potential cause of lack of accuracy in pre-diction using sequence data is that most sequence data are obtained by imputation rather than direct sequenc-ing, and consequently, imputation errors are introduced. Because the genotypes used in the simulated dataset were obtained by imputation, it is likely that the imputa-tion errors in this dataset are similar to those in the real

dataset. However, in the simulation, the estimation of the effects of the causative mutations was based on the imputed genotypes, while in reality, the effects are based on the true genotypes. Therefore, the effect of imputation errors in the real data is expected to be larger. To inves-tigate this, additional imputation errors were simulated, either in all individuals, only in the training population or only in the validation population. As expected, increasing the number of imputation errors decreased the accuracy, and the largest decrease in accuracy was observed when the errors were present in the validation population. In the training population, the effect of imputation errors is likely less marked, because the genotype errors can dif-fer between individuals, and if the genotype is correct in the majority of animals, it may not have a large influence on the estimated effect. In contrast, errors in the geno-types of the validation population directly influence their estimated breeding value, and thereby the accuracy. For a few chromosomes, the correlation and concordance rate between imputed and true sequence genotypes were computed (see Additional file 2: Table S1). The expected reduction in prediction accuracy based on these corre-lations and concordance rates, is even greater than the observed reduction in accuracy in the scenarios using sequence data compared to HD_FULL_D0.

The correlation ranged from 0.92 for chromosome 5 to 0.94 for chromosomes 1 and 20, and the concordance from 0.94 to 0.95 (see Additional file 2: Table S1). While some imputation software programs provide a measure of imputation accuracy for each variant, this is not the case for FImpute, and we only filtered variants based on MAF. Filtering out incorrectly imputed variants may increase prediction accuracy.

In the real data, the only large increase in accuracy with sequence data was observed for Australian Red bulls. Because LD is conserved over shorter distances across breeds than within breeds [1], sequence data is expected to be especially beneficial for across-breed prediction [7]. However, for the Australian Red cows, the accuracy of the scenarios using sequence data was at most similar to that using HD data. While the vast majority of the Aus-tralian Red bulls were genotyped at HD, most cows were genotyped at lower densities. Consequently, imputation accuracy may be lower for the cows than for the bulls, which could be a possible explanation for the reduced accuracy observed in the Australian Red cows.

Conclusions

We present an efficient approach to approximate analy-sis of full sequence data with a Bayesian variable selection model. While the simulation study provided promising results, when we applied the method to a real dataset, the accuracy obtained was at most similar to that obtained with HD genotypes, and bias increased. The lack of increase in prediction accuracy could be due to errors introduced in the genotypes by imputation. Therefore, it is necessary to develop more accurate methods of imputation or to directly genotype sequence variants that have an important effect in the prediction equation.

Additional files

Additional file 1: Figure S1. Principal component analysis of Holstein and Jersey individuals. Description: *PC1* principal component 1, *PC2* principal component 2. Left graph shows the different clusters based on PC1 (dark blue = HOL1, medium blue = HOL2, light blue = HOL3, dark green = JER1, light green = JER2), right graph breeds based on the pedigree (dark blue = purebred Holstein, medium blue crossbreds with more Holstein than Jersey ancestors, light blue = Holstein × Jersey crossbreds, dark green = crossbreds with more Jersey than Holstein ancestors, light green = purebred Jersey).

Additional file 2: Table S1. Correlation and concordance between true and imputed sequence genotypes for variants on chromosomes 1, 5, 20 and 25.

Authors' contributions
IB carried out the main statistical analysis, wrote the manuscript and participated in the design of the study. IMM simulated the data. PB wrote the latest version of the EM Bayes R hybrid software. BJH carried out the imputation. SB computed the imputation accuracy. TW developed the EM Bayes R hybrid algorithm. MEG designed the study and helped draft the manuscript. All authors read and approved the final manuscript.

Author details
[1] Faculty of Veterinary and Agricultural Science, University of Melbourne, Parkville, VIC, Australia. [2] Agriculture Victoria, AgriBio, Centre for AgriBioscience, Bundoora, VIC 3083, Australia. [3] School of Applied Systems Biology, La Trobe University, Bundoora, VIC 3083, Australia. [4] Queensland Alliance for Agriculture and Food Innovation, Centre for Animal Science, University of Queensland, St Lucia, QLD, Australia.

Acknowledgements
This research was supported by the Center for Genomic Selection in Animals and Plants (GenSAP) funded by The Danish Council for Strategic Research. We acknowledge DataGene and CRV Netherlands for providing access to data used in this study. We acknowledge our partners in the 1000 Bull Genomes Project for access to the reference genomes. We acknowledge Dr Paul Stothard and team at the University of Alberta for collating annotation information of sequence variants used in this study.

Competing interests
The authors declare that they have no competing interests.

References
1. de Roos APW, Hayes BJ, Spelman RJ, Goddard ME. Linkage disequilibrium and persistence of phase in Holstein-Friesian, Jersey and Angus cattle. Genetics. 2008;179:1503–12.
2. Lund MS, van den Berg I, Ma P, Brøndum RF, Su G. Review: how to improve genomic predictions in small dairy cattle populations. Animal. 2016;10:1042–9.
3. Brøndum RF, Guldbrandtsen B, Sahana G, Lund MS, Su G. Strategies for imputation to whole genome sequence using a single or multi-breed reference population in cattle. BMC Genomics. 2014;15:728.
4. van Binsbergen R, Calus MPL, Bink MCAM, van Eeuwijk FA, Schrooten C, Veerkamp RF. Genomic prediction using imputed whole-genome sequence data in Holstein Friesian cattle. Genet Sel Evol. 2015;47:71.
5. Heidaritabar M, Calus MPL, Megens HJ, Vereijken A, Groenen MAM, Bastiaansen JWM. Accuracy of genomic prediction using imputed whole-genome sequence data in white layers. J Anim Breed Genet. 2016;133:167–79.
6. Pérez-Enciso M, Rincón JC, Legarra A. Sequence-vs. chip-assisted genomic selection: accurate biological information is advised. Genet Sel Evol. 2015;47:43.
7. van den Berg I, Boichard D, Guldbrandtsen B, Lund MS. Using sequence variants in linkage disequilibrium with causative mutations to improve across-breed prediction in dairy cattle: a simulation study. G3 (Bethesda). 2016;6:2553–61.
8. Brøndum RF, Su G, Janss L, Sahana G, Guldbrandtsen B, Boichard D, et al. Quantitative trait loci markers derived from whole genome sequence data increases the reliability of genomic prediction. J Dairy Sci. 2015;98:4107–16.
9. van den Berg I, Boichard D, Lund MS. Sequence variants selected from a multi-breed GWAS can improve the reliability of genomic predictions in dairy cattle. Genet Sel Evol. 2016;48:83.
10. Ober U, Huang W, Magwire M, Schlather M, Simianer H, Mackay TF. Accounting for genetic architecture improves sequence based genomic prediction for a Drosophila fitness trait. PLoS One. 2015;10:e0126880.
11. Calus MPL, Bouwman AC, Schrooten C, Veerkamp RF. Efficient genomic prediction based on whole-genome sequence data using split-and-merge Bayesian variable selection. Genet Sel Evol. 2016;48:49.
12. Veerkamp RF, Bouwman AC, Schrooten C, Calus MPL. Genomic prediction using preselected DNA variants from a GWAS with whole-genome sequence data in Holstein-Friesian cattle. Genet Sel Evol. 2016;48:95.
13. MacLeod IM, Hayes BJ, Goddard ME. The effects of demography and long-term selection on the accuracy of genomic prediction with sequence data. Genetics. 2014;198:1671–84.
14. MacLeod IM, Bowman PJ, Vander Jagt CJ, Haile-Mariam M, Kemper KE, Chamberlain AJ, et al. Exploiting biological priors and sequence variants enhances QTL discovery and genomic prediction of complex traits. BMC Genomics. 2016;17:144.
15. Ni G, Cavero D, Fangmann A, Erbe M, Simianer H. Whole-genome sequence-based genomic prediction in laying chickens with different genomic relationship matrices to account for genetic architecture. Genet Sel Evol. 2017;49:8.
16. Kemper KE, Reich CM, Bowman P, vander Jagt CJ, Chamberlain AJ, Mason BA, et al. Improved precision of QTL mapping using a nonlinear Bayesian method in a multi-breed population leads to greater accuracy of across-breed genomic predictions. Genet Sel Evol. 2015;47:29.
17. Wang T, Chen YP, Goddard ME, Meuwissen THE, Kemper KE, Hayes BJ. A computationally efficient algorithm for genomic prediction using a Bayesian model. Genet Sel Evol. 2015;47:34.
18. Matukumalli LK, Lawley CT, Schnabel RD, Taylor JF, Allan MF, Heaton MP, et al. Development and characterization of a high density SNP genotyping assay for cattle. PLoS One. 2009;4:e5350.
19. Grant JR, Arantes AS, Liao X, Stothard P. In-depth annotation of SNPs arising from resequencing projects using NGS-SNP. Bioinformatics. 2011;27:2300–1.
20. Daetwyler HD, Capitan A, Pausch H, Stothard P, van Binsbergen R, Brøndum RF, et al. Whole-genome sequencing of 234 bulls facilitates mapping of monogenic and complex traits in cattle. Nat Genet. 2014;46:858–65.
21. Sargolzaei M, Chesnais JP, Schenkel FS. A new approach for efficient genotype imputation using information from relatives. BMC Genomics. 2014;15:478.
22. Purcell S, Neale B, Todd-Brown K, Thomas L, Ferreira MAR, Bender D, et al. PLINK: a tool set for whole-genome association and population-based linkage analyses. Am J Hum Genet. 2007;81:559–75.

23. Wang T, Chen Y-PP, Bowman PJ, Goddard ME, Hayes BJ. A hybrid expecta-
    tion maximisation and MCMC sampling algorithm to implement Bayes-
    ian mixture model based genomic prediction and QTL mapping. BMC
    Genomics. 2016;17:744.

24. Garrick DJ, Taylor JF, Fernando RL. Deregressing estimated breeding val-
    ues and weighting information for genomic regression analyses. Genet
    Sel Evol. 2009;41:55.

13

Genomic selection for crossbred performance accounting for breed-specific effects

Marcos S. Lopes[1,2,3*], Henk Bovenhuis[2], André M. Hidalgo[2], Johan A. M. van Arendonk[2], Egbert F. Knol[1] and John W. M. Bastiaansen[2]

Abstract

Background: Breed-specific effects are observed when the same allele of a given genetic marker has a different effect depending on its breed origin, which results in different allele substitution effects across breeds. In such a case, single-breed breeding values may not be the most accurate predictors of crossbred performance. Our aim was to estimate the contribution of alleles from each parental breed to the genetic variance of traits that are measured in crossbred offspring, and to compare the prediction accuracies of estimated direct genomic values (DGV) from a traditional genomic selection model (GS) that are trained on purebred or crossbred data, with accuracies of DGV from a model that accounts for breed-specific effects (BS), trained on purebred or crossbred data. The final dataset was composed of 924 Large White, 924 Landrace and 924 two-way cross (F1) genotyped and phenotyped animals. The traits evaluated were litter size (LS) and gestation length (GL) in pigs.

Results: The genetic correlation between purebred and crossbred performance was higher than 0.88 for both LS and GL. For both traits, the additive genetic variance was larger for alleles inherited from the Large White breed compared to alleles inherited from the Landrace breed (0.74 and 0.56 for LS, and 0.42 and 0.40 for GL, respectively). The highest prediction accuracies of crossbred performance were obtained when training was done on crossbred data. For LS, prediction accuracies were the same for GS and BS DGV (0.23), while for GL, prediction accuracy for BS DGV was similar to the accuracy of GS DGV (0.53 and 0.52, respectively).

Conclusions: In this study, training on crossbred data resulted in higher prediction accuracy than training on purebred data and evidence of breed-specific effects for LS and GL was demonstrated. However, when training was done on crossbred data, both GS and BS models resulted in similar prediction accuracies. In future studies, traits with a lower genetic correlation between purebred and crossbred performance should be included to further assess the value of the BS model in genomic predictions.

Background

In pig breeding, selection takes place in purebred lines and genetic evaluations are performed mainly with information that is collected on purebreds, in high-health environments, although the final product of the pig industry is a crossbred animal. This strategy may not be optimal when the objective is to improve crossbred performance. Genetic progress realized at the purebred level may not fully translate to improved crossbred performance (under field conditions) when the genetic correlation between purebred and crossbred performance is less than 1 [1, 2]. Low genetic correlations between purebred and crossbred performance in pigs have been reported for many production traits [3, 4] and can be caused by genotype-by-environment interaction, non-additive biological (or functional) effects (such as dominance and epistasis) or breed-specific effects of (genetic marker) alleles. Therefore, if the goal is to improve crossbred performance by selection in purebreds, effects that

*Correspondence: Marcos.Lopes@topigsnorsvin.com
[1] Topigs Norsvin Research Center, P.O. Box 43, 6640 AA Beuningen, The Netherlands
Full list of author information is available at the end of the article

influence the genetic correlation between purebred and crossbred performance must be evaluated. In addition, the use of crossbred data in genetic evaluations must be considered [1, 5–8].

Using either real or simulated data, several studies have investigated the relevance of genotype-by-environment interactions and dominance effects for pig breeding [9–14]. However, to date, breed-specific effects of genetic marker alleles have not been extensively studied. Breed-specific effects are observed when the same allele, say allele *A*, of a given marker has a different effect on the crossbred phenotype depending on its breed origin. Breed-specific effects at genetic markers may occur when the linkage disequilibrium (LD) between markers and quantitative trait loci (QTL) differ between breeds or when the allele frequencies of the QTL vary across breeds [6]. When breed-specific effects are present, allele substitution effects will differ between breeds, and therefore, the breeding values that are estimated by using only data from one of the purebred parental line may not accurately predict crossbred performance.

With the recent availability of high-density marker genotypes on both purebred and crossbred animals, we can now include crossbred data in genomic evaluations. Breed origin of alleles in crossbreds can also be determined and used to build breed-specific relationship matrices, as proposed by Christensen et al. [15]. Replacing the genomic relationship of a traditional genomic selection model (GS model) by the breed-specific relationship matrices (BS model), allows us to quantify the contributions of each parental breed to the additive genetic variance of the trait in crossbreds. In addition, we can also estimate breed-specific breeding values that can be backsolved for breed-specific marker effects [16]. Breed-specific marker effects could then be used to predict direct genomic values of purebred animals for crossbred performance, which would make it possible to benefit from training on crossbred field data.

In simulation studies, Ibanez-Escriche et al. [6] concluded that the BS model may not be required to effectively select purebreds for crossbred performance, while Esfandyari et al. [7] concluded that accounting for the breed origin of alleles can substantially improve accuracy of genomic prediction if the size of the training population is sufficiently large and the parental breeds are not very closely related. Applying the method proposed by Christensen et al. [15] to real pig data, Xiang et al. [17] concluded that a BS model is a good method for selecting purebreds for crossbred performance, resulting in higher prediction accuracy than the GS model. However, further studies using real data are still necessary to determine the relevance of breed-specific effects for genomic prediction. In this study, we investigated the value of breed-specific effects for predicting crossbred performance using real data. First, the contribution of each parental breed to the genetic variance was quantified for traits that were measured in a two-way crossbred population. Second, prediction accuracies were estimated with the GS and BS model, using either purebred or crossbred training data.

Methods
Ethics statement
The data used for this study were obtained as part of routine data recording in a commercial breeding program. Samples collected for DNA extraction were only used for routine diagnostic purposes of the breeding program. Data recording and sample collection were conducted strictly in line with the rules given by Dutch Animal Research Authorities.

Data
Phenotypic and genotypic data were available for pigs from two purebred populations: Large White (LW) and Landrace (LR), and from a two-way crossbred population (F1) that consisted of animals produced by reciprocal crosses of the purebred populations (LW♂ × LR♀ and LR♂ × LW♀). Phenotypic data were available for litter size (LS, sum of piglets born alive and stillborn in the same litter) and gestation length (GL, number of days between insemination and farrowing). Both traits were recorded from parities 2 to 7. Records from the first parity were excluded because LS and GL measured in the first versus later parities have been described as different traits based on low genetic correlation [18, 19].

Phenotypic data on both traits were available for 22,597 LW, 27,035 LR, and 29,847 F1 animals (Table 1). The F1 population consisted of 14,964 animals from the LW♂ × LR♀ cross, and 14,883 animals from the LR♂ × LW♀ cross. On average, data from 3.8, 3.6, and 2.5 parities per animal were available in the LW, LR, and F1 populations, respectively. Data from the purebred populations were recorded on genotyped animals (3723 LW and 3291 LR) and their non-genotyped contemporaries (i.e. animals from the same breed and farm as the genotyped animals; 18,874 LW and 23,744 LR). The purebred animals were located on 18 (LW) and 20 (LR) farms and were born between 2004 and 2014. Data from the F1 population were recorded on 1126 genotyped animals and their 1120 non-genotyped contemporaries. These genotyped F1 animals and their contemporaries were located on six farms. Finally, data were also recorded on 27,601 non-genotyped F1 offspring of the genotyped purebred animals. This additional group of F1 animals was located across 111 farms and was only used to increase the size of the crossbred population to estimate the genetic correlation between purebred and crossbred performance. All F1 animals were born between 2010 and 2014.

Phenotypes of genotyped animals were pre-adjusted for fixed effects using the larger dataset, i.e. including contemporaries, such that fixed effects were accounted for more accurately. Fixed effects were estimated by fitting a single-trait, pedigree-based mixed linear model for each population, using ASReml v3.0 [20]. The model used for LS included the fixed effects of parity, interval between weaning and pregnancy (days), whether more than one insemination was performed or not (yes or no), litter type (whether the boar used for the inseminations was from the same breed as the sow, i.e. purebred litter, or from a different breed i.e. crossbred litter), and herd-year-season, and the random effects of service sire, permanent environmental effects, and additive genetic effects. The model used for GL included the fixed effects of parity, whether more than one insemination procedure was performed or not (yes or no), litter type (purebred or crossbred), herd-year-season, and the covariate LS. The random effects of the model for GL were the same as for LS. When evaluating the performance of the F1 animals, the fixed effect of litter type was not included in the model for either trait.

Genetic correlations between purebred and crossbred performance

Genetic correlations between purebred and crossbred performance for both traits were estimated using a three-trait model, as described by Lutaaya et al. [21]. Performance in each population (two purebreds and one crossbred) was considered as a different trait. The three-trait model was implemented in ASReml 3.0 [20]. Effects accounted for in the three-trait model were the same as those accounted for in the single-trait model that was used for the pre-adjustment of the phenotypes, except that a fixed effect for breed reciprocity was added for crossbred performance (LR♂ × LW♀ or LW♂ × LR♀).

Genotyping

Genotyping was performed mainly using the Illumina Porcine SNP60 Beadchip, but some animals from all populations were genotyped using the Illumina Porcine SNP60 v2 Beadchip. Genotypic data were available on 3723 LW, 3291 LR, and 1126 F1 animals (Table 1). In the purebred populations, both males and females were genotyped. In the F1 population, only females were genotyped. Genotypes of all animals were imputed to the SNP60 Beadchip for all SNPs that passed the quality control. The quality control excluded SNPs with a GenCall lower than 0.15, a call rate lower than 0.95, a minor allele frequency lower than 0.01, and SNPs that deviated significantly from Hardy–Weinberg equilibrium ($\chi^2 > 600$). SNPs located on the sex chromosomes and unmapped SNPs were also excluded. Positions of the SNPs were based on the Sscrofa10.2 assembly of the reference genome [22]. All genotyped animals had a frequency of missing genotypes below the threshold of 0.05 in order to exclude poorly genotyped animals. After quality control and imputation, 39,788 SNPs for LW, 41,299 SNPs for LR, and 45,515 SNPs for F1 were available for further analyses.

Imputation and phasing of the genotype data

Imputation and phasing of the genotype data were performed using AlphaImpute [23], combining genomic and pedigree information to determine the parental origin of alleles. Imputation of missing genotypes of the purebred populations was performed within populations using all SNPs that passed quality control. For the F1 population, imputation of missing genotypes and phasing of the data were performed by combining the F1 data with the imputed purebred data but using only the 36,733 SNPs that segregated (minor allele frequency >0.01) in each population.

To ensure the use of accurately phased haplotypes for determining breed origin of alleles, a threshold was applied to the F1 phased data. For each SNP genotype of

Table 1 Summary statistics

Population[a]	Phenotypes[b]		Genotypes[c]	Genotypes and phenotypes[d]		Mean ± standard deviation[e]	
	Animals	Records	Animals	Animals	Records	LS	GL
LW	22,597	84,837	3723	924	3358	15.91 ± 3.71	115.38 ± 1.63
LR	27,035	96,431	3291	924	3319	15.40 ± 3.56	116.10 ± 1.61
F1	29,847	75,143	1126	924	3771	15.93 ± 3.59	115.12 ± 1.49

[a] Large White (LW), Landrace (LR), and two-way crossbred (F1)

[b] Number of animals with phenotypic information and total number of phenotypic records for these animals

[c] Number of genotyped animals used for imputation and phasing procedures

[d] Number of genotyped animals and number of phenotypic records for these animal used for estimating the variance components and SNP effects

[e] Mean ± standard deviation of litter size (LS) and gestation length (GL) of the populations used for estimating the variance components and SNP effects

each individual, AlphaImpute [23] generates two probabilities: P_1 is the probability that a specific allele was inherited from the father, e.g. allele G of a G/C genotype, and P_2 is the probability that the same allele was inherited from the mother. For a heterozygous animal (CG), for which allele C was inherited with certainty from the father (and therefore allele G from the mother), the probabilities would be $P_1 = 0$ and $P_2 = 1$. When the phasing cannot be performed with certainty, these probabilities will have values between 0 and 1. Values of P_1 or P_2 between 0.1 and 0.9 were considered to have poor phasing. SNPs that were considered poorly phased in more than 95% of animals were excluded from the dataset. Then, animals that had more than 5% poorly phased genotypes were excluded. After this quality control, 924 F1 animals with genotypes for 31,930 SNPs were available to estimate variance components and SNP effects. The same set of SNPs was also used to estimate variance components and SNP effects for the purebred populations.

After phasing of the genotype data, the breed origin of alleles was easily determined because the breeds of the parents of the F1 individuals were known. The final F1 population included 414 animals from the LW♂ × LR♀ cross and 510 animals from the LR♂ × LW♀ cross.

Estimation of variance components and SNP effects

The number of animals with both phenotypes and genotypes was larger in the purebred than in the crossbred population. Because the size of the training population influences the estimation of SNP effects and consequently prediction accuracy [24], we randomly selected 924 animals born between 2010 and 2014 from each purebred population for use in estimation of variance components and SNP effects. Our aim was to conduct a fair comparison, since the size of the training population and range of birth years were the same for each population (Table 1). In order to have independent datasets for validation analyses (discussed below), the purebred animals used as training population had no offspring or sibs in the F1 population.

Variance components and SNP effects were estimated within each population using a traditional genomic selection model (GS model) and a model that accounts for breed-specific effects (BS model). The GS model was applied to both purebred and crossbred data, while the BS model was applied only to the crossbred data. These models were implemented in ASReml [20], as follows:

$$\mathbf{y} = \mathbf{1}\mu + \mathbf{Ss} + \mathbf{Pp} + \mathbf{Z}\mathbf{u}_{GS} + \mathbf{e} \quad \text{(GS model)}$$

$$\mathbf{y} = \mathbf{1}\mu + \mathbf{Ss} + \mathbf{Pp} + \mathbf{Z}_{LW}\mathbf{u}_{BS|LW}$$
$$+ \mathbf{Z}_{LR}\mathbf{u}_{BS|LR} + \mathbf{e} \quad \text{(BS model)}$$

where $\mathbf{y}$ is a vector of phenotypes pre-adjusted for fixed effects; μ is the mean of the populations and $\mathbf{1}$ a vector of 1s; $\mathbf{S}$ is the design matrix for service sire effects; $\mathbf{s}$ is an unknown vector of service sire effects; $\mathbf{P}$ is the design matrix for permanent environmental effects; $\mathbf{p}$ is an unknown vector of permanent environmental effects; $\mathbf{Z}$, $\mathbf{Z}_{LW}$ and $\mathbf{Z}_{LR}$ are design matrices for the additive genetic effects; $\mathbf{u}_{GS}$ is an unknown vector of additive genetic effects (i.e. breeding values); $\mathbf{u}_{BS|LW}$ and $\mathbf{u}_{BS|LR}$ are unknown vectors of breed-specific additive genetic effects (i.e. breed-specific breeding values). Assumed distributions were $\mathbf{s} \sim N\left(\mathbf{0}, \mathbf{I}\sigma_s^2\right)$, $\mathbf{p} \sim N\left(\mathbf{0}, \mathbf{I}\sigma_r^2\right)$, $\mathbf{u}_{GS} \sim N\left(\mathbf{0}, \mathbf{G}\sigma_{a_{GS}}^2\right)$, $\mathbf{u}_{BS|F1_{LW}} \sim N\left(\mathbf{0}, \mathbf{B}_{LW}\sigma_{a_{BS|F1_{LW}}}^2\right)$, and $\mathbf{u}_{BS|F1_{LR}} \sim N\left(\mathbf{0}, \mathbf{B}_{LR}\sigma_{a_{BS|F1_{LR}}}^2\right)$, where $\mathbf{I}$ is an identity matrix, σ_s^2 is the service sire variance, σ_r^2 is the permanent environmental variance, $\mathbf{G}$ is the traditional genomic additive relationship matrix, $\sigma_{a_{GS}}^2$ is the additive genetic variance, $\mathbf{B}_{LW}$ and $\mathbf{B}_{LR}$ are breed-specific genomic relationship matrices, and $\sigma_{a_{BS|F1_{LW}}}^2$ and $\sigma_{a_{BS|F1_{LR}}}^2$ are breed-specific additive genetic variances. Heritability was defined as $\sigma_{a_{GS}}^2/\sigma_P^2$ for the GS model and as $(\sigma_{a_{BS|F1_{LW}}}^2 + \sigma_{a_{BS|F1_{LR}}}^2)/\sigma_P^2$ for the BS model, where σ_P^2 is the total phenotypic variance (sum of all variances from each model). The $\mathbf{G}$ matrix was built according to VanRaden [25]:

$$\mathbf{G} = \frac{\mathbf{MM}'}{2\sum_{i=1}^{n} p_i q_i},$$

where p_i and q_i are the allele frequencies of the ith genetic marker, and $\mathbf{M}$ is a matrix of centered genotype codes ($0 - 2p_i$, $1 - 2p_i$, $2 - 2p_i$). The $\mathbf{B}_{LR}$ and $\mathbf{B}_{LW}$ matrices were built according to the genomic gametic relationship matrices described by Christensen et al. [15] and Nishio and Satoh [26]:

$$\mathbf{B} = \frac{\mathbf{LL}'}{\sum_{i=1}^{n} p_i^* q_i^*},$$

where p_i^* and q_i^* are the frequencies of the allele codes from either LW ($\mathbf{B}_{LW}$) or LR ($\mathbf{B}_{LR}$) in the F1 population, and $\mathbf{L}$ is a matrix of centered allele codes ($0 - p_i^*$, $1 - p_i^*$) from either breed LW ($\mathbf{B}_{LW}$) or LR ($\mathbf{B}_{LR}$). After obtaining the estimated breeding values (EBV) from the GS model ($\hat{\mathbf{u}}_{GS}$) and the BS model ($\hat{\mathbf{u}}_{BS|F1_{LW}}$ and $\hat{\mathbf{u}}_{BS|F1_{LR}}$), we backsolved these EBV to obtain estimates of SNP effects, which were used to estimate the direct genomic values (DGV) of the validation animals. Backsolving of EBV from the GS models to obtain SNP effect estimates ($\hat{\boldsymbol{a}}_{GS}$) was performed as described by Wang et al. [16]:

$$\hat{\boldsymbol{a}}_{GS} = \frac{\mathbf{M}'\mathbf{G}^{-1}\hat{\boldsymbol{u}}_{GS}}{2\sum_{i=1}^{n} p_i q_i},$$

and backsolving of EBV from the BS model to obtain breed-specific SNP effect estimates ($\hat{\boldsymbol{a}}_{BS|F1_{LW}}$ and $\hat{\boldsymbol{a}}_{BS|F1_{LR}}$, for LW and LR breed, respectively), was performed by extending the method described by Wang et al. [16]:

$$\hat{\boldsymbol{a}}_{BS} = \frac{\mathbf{L}'\mathbf{B}^{-1}\hat{\boldsymbol{u}}_{BS}}{\sum_{i=1}^{n} p_i^* q_i^*}.$$

Predicting crossbred performance

A schematic representation of the steps involved in the prediction analyses is in Fig. 1. Individual performance of genotyped crossbred sows was predicted with the SNP effects estimated by the GS model and the BS model. A validation using a 40-fold random training-validation populations was performed to evaluate prediction accuracies. For each replicate, 10% of the genotyped F1 animals (N = 92) were randomly assigned to the validation population and the other 90% (N = 832) were assigned to the F1 training population. For the purebred training populations, 90% (N = 832) of the animals that were used to estimate variance components were randomly

assigned to the training population in each replicate. Within each replicate, traditional (GS) DGV of validation animals were estimated as: $\hat{\boldsymbol{u}}_{GS|j} = \mathbf{M}_{F1}\hat{\boldsymbol{a}}_{GS|j}$, where $\mathbf{M}_{F1}$ is a matrix of centered genotypes of the F1 validation animals and $\hat{\boldsymbol{a}}_{GS|j}$ is a vector of SNP effects estimated using the GS model on the training animals, where subscript j indicates the breed of the animals included in the training population (LW, LR or F1).

In addition, within each replicate, two types of breed-specific (BS) DGV of the validation animals were estimated. The first type of BS DGV used SNP effects estimated within the parental purebred populations as: $\hat{\boldsymbol{u}}_{BS|j} = \mathbf{L}_{F1j}\hat{\boldsymbol{a}}_{GS|j}$, where $\mathbf{L}_{F1j}$ is a matrix of centered allele codes that the F1 validation animals inherited from the jth parental purebred populations (LW or LR) and $\hat{\boldsymbol{a}}_{GS|j}$ is defined as above. Thus, separate BS DGV were estimated, one for each parental purebred population. In addition, total the BS DGV ($\hat{\boldsymbol{u}}_{BS|LW} + \hat{\boldsymbol{u}}_{BS|LR}$) was calculated for the validation animals. The second type of BS DGV used SNP effects that were estimated within the crossbred population as: $\hat{\boldsymbol{u}}_{BS|F1j} = \mathbf{L}_{F1j}\hat{\boldsymbol{a}}_{BS|F1j}$, where $\hat{\boldsymbol{a}}_{BS|F1j}$ is a vector of SNP effects that were estimated using the BS

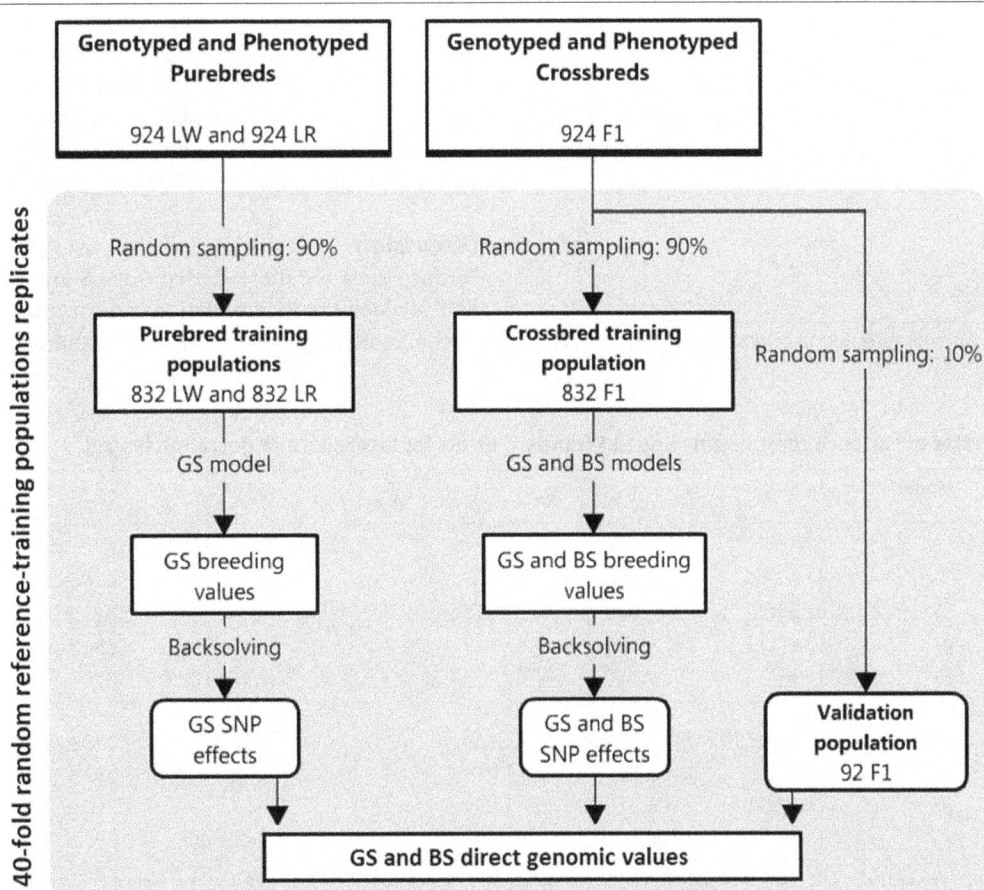

Fig. 1 Schematic representation of the steps involved in the prediction analyses. *LW* Large-White, *LR* landrace population, *F1* two-way crossbred, *GS model* traditional genomic selection model, *BS model* model that accounts for breed-specific effects

model on the alleles that the F1 training animals inherited from the jth parental purebred populations, and $\mathbf{L}_{F1_j}$ is defined as above. Total BS DGV ($\hat{\mathbf{u}}_{BS|F1_{LW}} + \hat{\mathbf{u}}_{BS|F1_{LR}}$) were also calculated for the validation animals.

Prediction accuracy was defined as the correlation of the GS DGV, the BS DGV, or the total BS DGV with pre-adjusted phenotypes in the validation population. Prediction accuracies presented are averages over 40-fold random reference-training population replicates.

Results

The estimate of the pedigree-based genetic correlation between purebred LW and crossbred performance was 0.91 ± 0.04 for LS and 0.92 ± 0.02 for GL (Table 2). The estimate of the pedigree-based genetic correlation between purebred LR and crossbred performance was slightly lower, i.e. 0.89 ± 0.04 for LS and 0.88 ± 0.03 for GL. Estimates of the pedigree-based heritability for all traits and populations are provided in Table 2.

Table 2 Estimates of pedigree-based heritability (h^2) and genetic correlation between purebred and crossbred (r_{pc}) populations from a three-trait model

Population[a]	h^2	r_{pc}
Litter size		
LW	0.18 ± 0.01	0.91 ± 0.04
LR	0.14 ± 0.01	0.89 ± 0.04
F1	0.14 ± 0.01	
Gestation length		
LW	0.39 ± 0.01	0.92 ± 0.02
LR	0.39 ± 0.01	0.88 ± 0.03
F1	0.37 ± 0.01	

[a] Populations used in the analyses were Large White (LW), Landrace (LR), and two-way crossbred (F1)

Using the GS model, estimates of the heritability of 0.15 ± 0.03 and 0.12 ± 0.03 were obtained for LS in the LW and LR populations, respectively (Table 3). Using the GS and BS models, estimates of the heritability for LS in the F1 population were similar (0.12 ± 0.03). Using the GS model, estimates of the heritability of 0.34 ± 0.04 and 0.33 ± 0.04 were obtained for GL in the LW and LR populations, respectively (Table 3). For GL in the F1 population, estimates of the heritability of 0.39 ± 0.04 and 0.40 ± 0.04 were obtained with the GS and BS models, respectively. For both traits, the estimate of the breed-specific additive genetic variance was slightly larger for alleles that were inherited from the LW population compared to alleles that were inherited from the LR population, although the standard errors were high (0.74 ± 0.23 and 0.56 ± 0.24 for LS, and 0.42 ± 0.08 and 0.40 ± 0.08 for GL, respectively).

The highest accuracy for predicting the performance of crossbred sows was observed when training was done on crossbred data (Table 4), for which the GS and BS models resulted in similar prediction accuracies. For LS, when training was done on crossbred data, prediction accuracy was the same for the GS DGV and the total BS DGV (0.23 ± 0.08). For GL, when training was done on crossbred data, similar prediction accuracies were obtained for the total BS DGV and the GS DGV (0.53 ± 0.08 and 0.52 ± 0.08, respectively). For both traits, the BS DGV based on the LW alleles resulted in higher prediction accuracies than the BS DGV based on the LR alleles (0.21 ± 0.08 vs. 0.12 ± 0.09 for LS; 0.43 ± 0.08 vs. 0.34 ± 0.09 for GL).

Discussion

In this study, we showed, that, for LS and GL, the same SNP allele in the F1 population can contribute differently to the additive genetic variance depending on its breed

Table 3 Estimates of variance components and ($\pm$) standard errors for litter size and gestation length

Population[a]	Model[b]	σ_s^2	σ_r^2	σ_a^2	$\sigma_{a_{LW}}^2$	$\sigma_{a_{LR}}^2$	σ_e^2	h^2
Litter size								
LW	GS	0.43 ± 0.13	1.70 ± 0.30	1.81 ± 0.36	–	–	8.26 ± 0.25	0.15 ± 0.03
LR	GS	0.02 ± 0.09	1.22 ± 0.28	1.30 ± 0.31	–	–	8.80 ± 0.27	0.12 ± 0.03
F1	GS	0.10 ± 0.10	1.38 ± 0.28	1.42 ± 0.34	–	–	8.54 ± 0.24	0.12 ± 0.03
F1	BS	0.11 ± 0.10	1.37 ± 0.29	–	0.74 ± 0.23	0.56 ± 0.245	8.52 ± 0.24	0.12 ± 0.03
Gestation length								
LW	GS	0.21 ± 0.03	0.33 ± 0.05	0.68 ± 0.09	–	–	0.78 ± 0.02	0.34 ± 0.04
LR	GS	0.23 ± 0.03	0.26 ± 0.05	0.64 ± 0.09	–	–	0.82 ± 0.03	0.33 ± 0.04
F1	GS	0.16 ± 0.02	0.23 ± 0.06	0.81 ± 0.11	–	–	0.90 ± 0.03	0.39 ± 0.04
F1	BS	0.16 ± 0.02	0.17 ± 0.06	–	0.42 ± 0.08	0.40 ± 0.08	0.90 ± 0.03	0.40 ± 0.04

Variance components: service sire (σ_s^2), permanent environment (σ_r^2), additive (σ_a^2), additive for the alleles of the F1 population inherited from the LW ($\sigma_{a_{LW}}^2$) and LR ($\sigma_{a_{LR}}^2$) populations, and error (σ_e^2). h^2: heritability

[a] Large White (LW), Landrace (LR), two-way crossbred (F1)

[b] Traditional genomic selection model (GS) and a model that accounts for breed-specific effects (BS)

Table 4 Prediction accuracy of performance of crossbred sows for gestation length and litter size

Model[a]	Training[b]	Accuracy[c]	SD[d]
Litter size			
GS	LW	0.06	0.10
	LR	0.07	0.11
	F1	**0.23**	0.08
BS	LW	0.06	0.11
	LR	0.06	0.13
	LW and LR*	0.09	0.12
	F1$_{LW}$	0.21	0.08
	F1$_{LR}$	0.12	0.09
	F1$_{LW}$ and F1$_{LR}$*	**0.23**	0.08
Gestation length			
GS	LW	0.42	0.08
	LR	0.30	0.09
	F1	**0.52**	0.08
BS	LW	0.39	0.08
	LR	0.23	0.10
	LW and LR*	0.45	0.08
	F1$_{LW}$	0.43	0.08
	F1$_{LR}$	0.34	0.09
	F1$_{LW}$ and F1$_{LR}$*	**0.53**	0.08

* Predicted direct genomic value was the "total direct genomic value" (sum of the breed-specific direct genomic values)

[a] GS, traditional genomic selection model; BS, model that accounts for breed-specific effects

[b] LW, Large White; LR, Landrace; F1, two-way crossbred; F1$_{LW}$, alleles of the F1 population inherited from the LW population; F1$_{LR}$, alleles of the F1 population inherited from the LR population. Training populations in each replicate were defined as a random set of 90% (N = 832) of the animals used for the estimation of variance components

[c] Average of the 40 replicates; accuracy was defined as the correlation between the direct genomic values of the validation population [random set of 10% (N = 92) of the crossbred animals used for the estimation of variance components] and their average pre-adjusted phenotypes in each replicate

[d] Standard deviation over replicates; the highest accuracies for each model and trait are indicated in bold

origin (Table 3) and that accounting for this difference only has a small impact on prediction accuracy (Table 4). The standard errors of the breed-specific variance estimates were rather high, as expected with a small dataset (N = 924), especially for LS, for which it reached 43% of the variance estimate (Table 3) and, thus, these results must be considered carefully. While standard errors can increase due to inaccurate determination of the breed origin of alleles, this increase is expected to be limited because a stringent quality control was applied to the phased genotypes.

Predicting performance of genotyped crossbred sows was more accurate when training was on crossbred data instead of purebred data (Table 4). The superiority of training on crossbred data compared to purebred data for predicting crossbred performance is in line with previous studies that were carried out using both simulated and real data. In simulation studies, training on crossbred data has been reported to yield either similar [27] or higher accuracies [7, 17] compared to training on purebred data, while using real data, training on crossbred data has been found to yield the highest prediction accuracies [2, 28].

The GS and BS models resulted in similar prediction accuracies when training was on crossbred data. Greater benefits of using the BS model over the GS model are expected when crossbred populations are larger and more distant parental breeds are crossed [7]. In the current study, we evaluated a small F1 population (N = 924) that was obtained by crossing two dam lines. If a sire line was used instead as one of the parental line, the parental breeds may, depending on the lines chosen, be more distant [29] and the BS model could have a larger impact on prediction accuracy. In pig breeding, the cross between sire and dam lines is typically done by mating F1 sows to boars from a sire line in the next generation. Therefore, we expect that applying the BS model to such a three-way crossbred population may result in larger benefits over the GS model. However, to evaluate three-way crossbred populations, even larger crossbred populations may be required because each crossbred will only carry two (grand) parental alleles.

Predictions of the performance of genotyped crossbred sows using BS DGV based on alleles of the LW breed resulted in higher accuracies than using BS DGV from alleles of the LR breed. This advantage of the LW breed compared to the LR breed when training was on crossbred data is consistent with the larger amount of variance that is explained by alleles of the LW breed (Table 3) and also with the higher estimate of the pedigree-based genetic correlation between the LW and F1 populations compared to that between the LR and F1 populations (Table 2). Furthermore, when training was on purebred data, the performance of genotyped crossbred sows was more accurately estimated when total BS DGV were used than when GS DGV based on SNP effects estimated in each purebred were used (Table 4). This suggests that determining the breed origin of the alleles in crossbred sows is beneficial, even if the training is on purebred data.

In this study, the BS model was applied to a training population composed of crossbred animals only. As a further step, the benefits of accounting for breed-specific effects could be evaluated under combined crossbred and purebred selection (CCPS), which has been described as an efficient way of increasing genetic progress in both purebred and crossbred populations [5, 30]. Such an approach was proposed by Christensen et al. [15] and further evaluated by Christensen et al. [31] and consists of performing genomic evaluation using a combination of the genomic relationship of the purebred populations and the breed-specific relationship matrices of the crossbred population. One of the limitations of applying CCPS using pedigree-based models is that it also resulted in increased rates of inbreeding [32]. However, with genomic-based models this increased inbreeding from CCPS is expected to be limited, or even absent, because the genomic information allows estimation of Mendelian sampling and therefore reduces the emphasis on family information in selection [1].

A CCPS approach to estimate BS DGV could be applied for both two-way and three-way crossbred populations. In the current study, we evaluated a two-way crossbred population and determination of the breed origin of alleles depended on pedigree information. In three-way crossbred populations, pedigree information is not commonly recorded, and thus, different strategies would be required to determine the breed origin of alleles. Recently, Bastiaansen et al. [33] proposed a method to determine breed origin of alleles in crossbreds using long-range phasing that can be applied to crossbred populations where pedigree information is lacking and this has been further evaluated by Sevillano et al. [34] and Vandenplas et al. [35]. With this method, close relationships between the crossbred and purebred genotyped animals would not be required because long-range phasing will work even with distant purebred relatives of the crossbreds. Therefore, future studies on practical applications of BS models in CCPS should evaluate a combination of the methods proposed by Christensen et al. [15] and Bastiaansen et al. [33]. When crossbred genotypes are not available, an alternative strategy would be to apply models that account for purebred genotypes and crossbred phenotypes only, as proposed by Tusell et al. [36]. These authors showed that such a strategy improves the theoretical accuracy of selection for crossbred performance without crossbred genotypes. However, we must keep in mind that with the fast developments of genotyping platforms and techniques, the major cost limitation for the use of crossbred data in genomic evaluation may come from obtaining phenotypes rather than genotypes.

In this study, we investigated the relevance of breed-specific effects when genomic selection is applied to real data for two reproductive traits in pigs. In future studies, traits that have a lower genetic correlation between purebred and crossbred performance should be included because benefits of BS models are expected to be larger in those cases. In addition to studying less correlated traits, investigating breed-specific effects in crosses of more distant purebred populations may result in larger benefits of the BS model. Furthermore, evaluation of larger datasets than those used in the current study is also required

for more conclusive results and to quantify the benefits of accounting for breed-specific effects in prediction models.

Conclusions

In this study, we provide evidence of breed-specific SNP effects for litter size and gestation length in a two-way crossbred population. Predicting performance of crossbred sows was shown to be more accurate when training was performed on crossbred instead of purebred data. However, when training was done on crossbred data, the GS and BS models resulted in similar prediction accuracies. In future studies, traits with lower genetic correlations between purebred and crossbred performance should be evaluated to confirm the potential benefit of BS models in genomic predictions.

Authors' contributions
MSL, HB, JAMvA, AMH, EFK, and JWMB conceived and designed the experiments. MSL and EFK performed the experiments. MSL performed the data analysis. MSL wrote the paper, with input from HB, JAMvA, AMH, EFK and JWMB. All authors read and approved the final manuscript.

Author details
[1] Topigs Norsvin Research Center, P.O. Box 43, 6640 AA Beuningen, The Netherlands. [2] Animal Breeding and Genomics Centre, Wageningen University, 6708 PB Wageningen, The Netherlands. [3] Topigs Norsvin, Curitiba, PR 80.420-210, Brazil.

Acknowledgements
Financial support from the Breed4Food Consortium is gratefully acknowledged.

Competing interests
The authors declare that they have no competing interests.

References
1. Dekkers JCM. Marker-assisted selection for commercial crossbred performance. J Anim Sci. 2007;85:2104–14.
2. Hidalgo AM, Bastiaansen JW, Lopes MS, Harlizius B, Groenen MA, de Koning DJ. Accuracy of predicted genomic breeding values in purebred and crossbred pigs. G3 (Bethesda). 2015;5:1575–83.
3. Zumbach B, Misztal I, Tsuruta S, Holl J, Herring W, Long T. Genetic correlations between two strains of Durocs and crossbreds from differing production environments for slaughter traits. J Anim Sci. 2007;85:901–8.
4. Cecchinato A, de los Campos G, Gianola D, Gallo L, Carnier P. The relevance of purebred information for predicting genetic merit of survival at birth of crossbred piglets. J Anim Sci. 2010;88:481–90.
5. Wei M, van der Steen HAM. Comparison of reciprocal recurrent selection with pure-line selection systems in animal breeding (a review). Anim Breed Abstr. 1991;59:281–98.
6. Ibanez-Escriche N, Fernando RL, Toosi A, Dekkers JC. Genomic selection of purebreds for crossbred performance. Genet Sel Evol. 2009;41:12.
7. Esfandyari H, Sørensen AC, Bijma P. A crossbred reference population can improve the response to genomic selection for crossbred performance. Genet Sel Evol. 2015;47:76.
8. Van Grevenhof IE, van der Werf JH. Design of reference populations for genomic selection in crossbreeding programs. Genet Sel Evol. 2015;47:14.
9. Lopes MS, Bastiaansen JW, Janss L, Knol EF, Bovenhuis H. Estimation of additive, dominance, and imprinting genetic variance using genomic data. G3 (Bethesda). 2015;5:2629–37.

10. Esfandyari H, Sørensen AC, Bijma P. Maximizing crossbred performance through purebred genomic selection. Genet Sel Evol. 2015;47:16.
11. Silva FF, Mulder HA, Knol EF, Lopes MS, Guimarães SE, Lopes PS, et al. Sire evaluation for total number born in pigs using a genomic reaction norms approach. J Anim Sci. 2014;92:3825–34.
12. Knap PW, Su G. Genotype by environment interaction for litter size in pigs as quantified by reaction norms analysis. Animal. 2008;2:1742–7.
13. Lopes MS, Bastiaansen JW, Janss L, Knol EF, Bovenhuis H. Genomic prediction of growth in pigs based on a model including additive and dominance effects. J Anim Breed Genet. 2016;133:180–6.
14. Vitezica ZG, Varona L, Elsen JM, Misztal I, Herring W, Legarra A. Genomic BLUP including additive and dominant variation in purebreds and F1 crossbreds, with an application in pigs. Genet Sel Evol. 2016;48:6.
15. Christensen OF, Madsen P, Nielsen B, Su G. Genomic evaluation of both purebred and crossbred performances. Genet Sel Evol. 2014;46:23.
16. Wang H, Misztal I, Aguilar I, Legarra A, Muir W. Genome-wide association mapping including phenotypes from relatives without genotypes. Genet Res (Camb). 2012;94:73–83.
17. Xiang T, Nielsen B, Su G, Legarra A, Christensen OF. Application of single-step genomic evaluation for crossbred performance in pig. J Anim Sci. 2016;94:936–48.
18. Irgang R, Fávero JA, Kennedy BW. Genetic parameters for litter size of different parities in Duroc, Landrace, and large white sows. J Anim Sci. 1994;72:2237–46.
19. Hanenberg E, Knol E, Merks J. Estimates of genetic parameters for reproduction traits at different parities in Dutch Landrace pigs. Livest Prod Sci. 2001;69:179–86.
20. Gilmour AR, Gogel B, Cullis B, Thompson R, Butler D. ASReml user guide release 3.0. Hemel Hempstead: VSN International Ltd.; 2009.
21. Lutaaya E, Misztal I, Mabry J, Short T, Timm H, Holzbauer R. Genetic parameter estimates from joint evaluation of purebreds and crossbreds in swine using the crossbred model. J Anim Sci. 2001;79:3002–7.
22. Groenen MA, Archibald AL, Uenishi H, Tuggle CK, Takeuchi Y, Rothschild MF, et al. Analyses of pig genomes provide insight into porcine demography and evolution. Nature. 2012;491:393–8.
23. Hickey JM, Kinghorn BP, Tier B, Wilson JF, Dunstan N, van der Werf JH. A combined long-range phasing and long haplotype imputation method to impute phase for SNP genotypes. Genet Sel Evol. 2011;43:12.
24. Daetwyler HD, Pong-Wong R, Villanueva B, Woolliams JA. The impact of genetic architecture on genome-wide evaluation methods. Genetics. 2010;185:1021–31.
25. VanRaden PM. Efficient methods to compute genomic predictions. J Dairy Sci. 2008;91:4414–23.
26. Nishio M, Satoh M. Genomic best linear unbiased prediction method including imprinting effects for genomic evaluation. Genet Sel Evol. 2015;47:32.
27. Toosi A, Fernando RL, Dekkers JCM. Genomic selection in admixed and crossbred populations. J Anim Sci. 2010;88:32–46.
28. Lourenco DA, Tsuruta S, Fragomeni BO, Chen CY, Herring WO, Misztal I. Crossbreed evaluations in single-step genomic best linear unbiased predictor using adjusted realized relationship matrices. J Anim Sci. 2016;94:909–19.
29. Veroneze R, Bastiaansen JW, Knol EF, Guimarães SE, Silva FF, Harlizius B, et al. Linkage disequilibrium patterns and persistence of phase in purebred and crossbred pig (Sus scrofa) populations. BMC Genet. 2014;15:126.
30. Bijma P, van Arendonk JAM. Maximizing genetic gain for the sire line of a crossbreeding scheme utilizing both purebred and crossbred information. Anim Sci. 1998;66:529–42.
31. Christensen OF, Legarra A, Lund MS, Su G. Genetic evaluation for three-way crossbreeding. Genet Sel Evol. 2015;47:98.
32. Bijma P, Woolliams JA, van Arendonk JAM. Genetic gain of pure line selection and combined crossbred purebred selection with constrained inbreeding. Anim Sci. 2001;72:225–32.
33. Bastiaansen JWM, Bovenhuis H, Lopes MS, Silva F, Megens HJWC, Calus MPL. SNP effects depend on genetic and environmental context. In: Proceedings of the 10th world congress on genetics applied to livestock production, 17–22 August 2014. Vancouver; 2014.
34. Sevillano CA, Vandenplas J, Bastiaansen JW, Calus MPL. Empirical determination of breed-of-origin of alleles in three-breed cross pigs. Genet Sel Evol. 2016;48:55.
35. Vandenplas J, Calus MPL, Sevillano CA, Windig JJ, Bastiaansen JW. Assigning breed origin to alleles in crossbred animals. Genet Sel Evol. 2016;48:61.
36. Tusell L, Gilbert H, Riquet J, Mercat MJ, Legarra A, Larzul C. Pedigree and genomic evaluation of pigs using a terminal-cross model. Genet Sel Evol. 2016;48:32.

14

From cow to cheese: genetic parameters of the flavour fingerprint of cheese investigated by direct-injection mass spectrometry (PTR-ToF-MS)

Matteo Bergamaschi[1], Alessio Cecchinato[1*], Franco Biasioli[2], Flavia Gasperi[2], Bruno Martin[3,4] and Giovanni Bittante[1]

Abstract

Background: Volatile organic compounds determine important quality traits in cheese. The aim of this work was to infer genetic parameters of the profile of volatile compounds in cheese as revealed by direct-injection mass spectrometry of the headspace gas from model cheeses that were produced from milk samples from individual cows.

Methods: A total of 1075 model cheeses were produced using raw whole-milk samples that were collected from individual Brown Swiss cows. Single spectrometry peaks and a combination of these peaks obtained by principal component analysis (PCA) were analysed. Using a Bayesian approach, we estimated genetic parameters for 240 individual spectrometry peaks and for the first ten principal components (PC) extracted from them.

Results: Our results show that there is some genetic variability in the volatile compound fingerprint of these model cheeses. Most peaks were characterized by a substantial heritability and for about one quarter of the peaks, heritability (up to 21.6%) was higher than that of the best PC. Intra-herd heritability of the PC ranged from 3.6 to 10.2% and was similar to heritabilities estimated for milk fat, specific fatty acids, somatic cell count and some coagulation parameters in the same population. We also calculated phenotypic correlations between PC (around zero as expected), the corresponding genetic correlations (from −0.79 to 0.86) and correlations between herds and sampling-processing dates (from −0.88 to 0.66), which confirmed that there is a relationship between cheese flavour and the dairy system in which cows are reared.

Conclusions: This work reveals the existence of a link between the cow's genetic background and the profile of volatile compounds in cheese. Analysis of the relationships between the volatile organic compound (VOC) content and the sensory characteristics of cheese as perceived by the consumer, and of the genetic basis of these relationships could generate new knowledge that would open up the possibility of controlling and improving the sensory properties of cheese through genetic selection of cows. More detailed investigations are necessary to connect VOC with the sensory properties of cheese and gain a better understanding of the significance of these new phenotypes.

Background

Volatile organic compounds (VOC) are important molecules that determine the distinct flavours of cheeses and,

consequently, their perceived quality [1, 2]. The development of flavour in cheese depends on the origin and gross composition of milk [3]. Milk provides the main components for the cheese-making process as well as microorganisms that release proteases and lipases, which catalyse the breakdown of lipids and proteins and lead to flavour development in cheese [4]. It is well known that cheese types are characterized by different aroma

*Correspondence: alessio.cecchinato@unipd.it
[1] Department of Agronomy, Food, Natural Resources, Animals and Environment (DAFNAE), University of Padua, Viale dell'Università 16, 35020 Legnaro, PD, Italy
Full list of author information is available at the end of the article

profiles [5, 6] and several studies have focused on the relationships between the sensory properties of cheese and the dairy system used, the cows' feeding regime and milk quality [7–9]. Moreover, sensory appraisal can have a huge impact on the economic value of cheese [10, 11]. Given the subjectivity, high cost and limited repeatability of sensory evaluation, and the need to better understand its chemical and biological basis, in recent years several techniques have been used to determine the qualitative characteristics of cheese flavour compounds [12–14]. Gas-chromatography combined with headspace extraction has been commonly used to investigate the link between VOC and the flavour of cheese [15–17]. Solid-phase micro-extraction and gas-chromatography mass spectrometry have been used to extract VOC from individual full-fat ripened cheeses in order to study the effects of dairy system, herd, and the cows' parity, stage of lactation and milk yield on these quality traits [18]. Recently, a model cheese procedure was used to produce a large number (more than 1000) of individual model cheeses [19] that were used to estimate the genetic parameters of cheese yields and nutrient recovery [20]. In addition, the direct-injection spectrometry method (proton transfer reaction-time of flight-mass spectrometry, PTR-ToF-MS) was used for the first time to obtain the fingerprints of volatile compounds in the same model cheeses [21]. Two hundred and forty peaks were detected from which the principal components (PC) were extracted which showed that dairy systems and individual cow characteristics had an effect on these new phenotypes.

In spite of the centrality and importance of VOC, which are potentially related to sensory properties, to date, no research has been carried out to estimate the heritability and genetic correlations of their concentrations in cheese. Given the economic importance of the perceived flavour in the cheese industry, a detailed knowledge of the genetic parameters of the VOC profile is fundamental to be able to evaluate the possibility of modifying cheese flavour in the future through breeding programmes using direct or indirect prediction of these traits (e.g., using infrared technology). Our objective was to estimate the genetic parameters of spectrometry peaks obtained by PTR-ToF-MS and of their PC to characterize the volatile compound fingerprint of model cheeses obtained from the milk of individual Brown Swiss cows.

Methods

Field data

This work is a part of the "Cowability-Cowplus projects", which involve collection of milk samples from a large number of Brown Swiss cows (n = 1075) from different herds (n = 72) located in northern Italy (Trento

province). The production environment was previously described in [22]. On each day, only one herd was visited and 15 cows from the herd were individually sampled once during evening milking. The herds were sampled over a full year to cover all seasons and rearing conditions. In the experimental area, cows on the permanent farms are not grazed and their feeding regime is almost constant all year around. Part of the herds are moved to Alpine pastures during summer, but samples were not taken from them during transhumance. Detailed descriptions of the herds, the cows' characteristics, and the sampling procedure are available in previous papers on cheese VOC [18, 21]. Briefly, milk samples (without preservative) were immediately refrigerated (4 °C) and transferred to the Cheese Making Laboratory of the Department of Agronomy, Food, Natural Resources, Animals and Environment (DAFNAE) of the University of Padua (Legnaro, Padua, Italy). All milk samples were collected as routine collection and thus no ethical approval was necessary. Data on individual cows and herds were provided by the Superbrown Consortium of Bolzano and Trento (Italy), and pedigree information was supplied by the Italian Brown Swiss Cattle Breeders Association (ANARB, Verona, Italy). The analysis included cows with phenotypic records on the investigated traits and all their known ancestors. Each sampled cow had at least four generations of known ancestors, and the pedigree file included 8845 animals. There were 1326 sires in the whole pedigree, among which 264 had progeny with records in the dataset (each sire had between 2 and 80 daughters).

Individual cheese-making procedure

Gross milk composition was measured using a MilkoScan FT6000 (Foss Electric A/S, Hillerød, Denmark). Somatic cell count was obtained from the Fossomatic FC counter (Foss) then converted to somatic cell score (SCS) by logarithm transformation [23]. All raw whole-milk samples were transformed into cheeses within 20 h of collection. The cheese-making procedure was designed to produce a laboratory "model-cheese" under the normal laboratory conditions for testing the coagulation properties of milk [19]. Briefly, 1500 mL of milk were heated at 35 °C in a stainless steel micro-vat, to which was added a thermophilic starter culture to reduce the effects of the microflora of the milk samples, and then rennet. On average, milk rennet coagulation time (RCT) was 20.3 min. Commercial rennet [Hansen standard 160 with 80 ± 5% chymosin and 20 ± 5% pepsin; 160 international milk clotting units (IMCU) × mL^{-1}; Pacovis Amrein AG, Bern, Switzerland] was diluted 20:1 with distilled water, and 9.6 mL of rennet solution was added to each vat to obtain a final concentration of 51.2 IMCU $\times$ L^{-1} of milk. The resulting curd from each vat was cut, drained, shaped into wheels, pressed, salted and weighed. All model cheeses were ripened for 60 days at 15 °C before sampling for the VOC analyses.

Descriptive statistics on daily milk yield and fat and protein content of milk from the Brown Swiss cows selected for the study, and fat and protein content of the model cheeses are in Table 1.

PTR-ToF-MS analysis

A cylindrical sample (1.1 × 3.5 cm) of each cheese was kept at −80 °C until VOC analysis. The headspace gas of each model cheese (n = 1075) was measured using a commercial PTR-ToF-MS 8000 instrument supplied by Ionicon Analytik GmbH, Innsbruck (Austria) following a modified version of the procedure described in [24]. Details of the analytical procedures and peak selection are in [18]. Briefly, cheese samples chosen randomly from the set of 1075 samples were thawed and kept at room temperature (about 20 °C) for 6 h. Sub-samples (3 g) from each cheese were placed in glass vials (20 mL; Supelco, Bellefonte, USA) equipped

with PTFE/Silicone septa (Supelco) and were measured every day. Internal calibration and peak extraction were performed as described in [25], which made it possible to assign, in some cases, a chemical formula to relevant spectrometry peaks. Absolute headspace VOC concentrations, expressed as parts per billion by volume (ppb$_v$), were calculated from peak areas using the formula described in the literature [26] with a constant reaction rate coefficient of the proton transfer reaction of 2×10^{-9} cm^3/s.

PTR-ToF-MS data

As discussed in detail in [21], 619 peaks describing VOC were obtained from the headspace gas of 1075 individual model cheeses using PTR-ToF-MS. Data compression was performed by selecting the peaks that displayed a spectrometry area greater than 1 part per billion by volume, which yielded 240 peaks after elimination of interfering ions. In addition, tentative interpretation of the spectrometry peaks was made based on the fragmentation patterns of the 61 most important volatile compounds in terms of spectrometry area that were retrieved from the available solid-phase micro-extraction gas chromatography mass spectrometry data on the same model cheeses, or from the literature, representing about 80% of the total spectral intensity. The strongest peaks detected by PTR-ToF-MS were at m/z 43.018 and 43.054, tentatively attributed to alkyl fragments, and at m/z 61.028 and 45.033, tentatively attributed to acetic acid and ethanol, respectively [18, 21].

Multivariate analysis of VOC

Multivariate data treatment (PCA) was carried out on the standardized spectrometry peaks using Statistica 7.1 (StatSoft, Paris) in order to summarize the information and provide a new set of ten PC. The statistical methodology is described in detail in [21]. The descriptive statistics of these ten PC, which represented 73.6% of the total variance of all VOC, are in Table 1.

Genetic parameters of VOC and their PC

Non-genetic effects analysed in a previous phenotypic study on the same dataset [21] were considered for the estimation of the genetic parameters of VOC and of their PC, but the effects of the micro-vats that were used on each sampling-processing date were not included in the statistical model because the adopted model cheese-making procedure showed good repeatability and reproducibility [19, 21].

All genetic models accounted for the effects of herd/sampling-processing date (72 levels) and the cows' days in milk (DIM; class 1: <50 days, class 2: 51–100 days, class 3: 101–150 days; class 4: 151–200 days; class 5:

Table 1 Descriptive statistics for milk production, cheese composition and the first principal components characterizing the volatile compound fingerprint of 1075 individual model cheeses analysed by PTR-ToF-MS

Traits	Mean	CV (%)
Milk yield (kg × day^{-1})	24.6	32.1
Milk composition		
Fat (%)	4.4	20.5
Protein (%)	3.8	10.5
Fat/protein	1.18	21.2
Casein/protein	0.769	2.34
SCS (U)	3.03	1.86
Cheese composition		
Fat (%)	38.2	11.5
Protein (%)	27.1	15.1
Cheese volatile fingerprint	Total phenotypic variance (%)	Cumulative phenotypic variance (%)
PC1	28.30	28.30
PC2	10.90	39.20
PC3	8.59	47.79
PC4	7.61	55.40
PC5	6.06	61.46
PC6	3.74	65.19
PC7	2.68	67.87
PC8	2.26	70.14
PC9	1.85	71.98
PC10	1.58	73.56

SCS = log$_2$(SCC/100,000) + 3, where SCC is somatic cells per mL

201–250 days; class 6: 251–300 days; class 7: >300 days) and parity (1–4 or more) for all traits.

Univariate models were fitted to estimate variance components and heritabilities for the traits analyzed.

The model assumed for VOC and PC was:

$$\mathbf{y} = \mathbf{Xb} + \mathbf{Z}_1\mathbf{h} + \mathbf{Z}_2\mathbf{a} + \mathbf{e}, \tag{1}$$

where $\mathbf{y}$ is the vector of phenotypic records with dimension n; $\mathbf{X}$, $\mathbf{Z}_1$, and $\mathbf{Z}_2$ are appropriate incidence matrices for systematic effects $\mathbf{b}$, herd/sampling-processing date effects $\mathbf{h}$, and polygenic additive genetic effects $\mathbf{a}$, respectively; and $\mathbf{e}$ is the vector of residual effects. More specifically, $\mathbf{b}$ included the non-genetic effects of DIM and parity.

All models were analysed using a standard Bayesian approach. Joint distribution of the parameters in a given model was proportional to:

$$p\left(\mathbf{b}, \mathbf{h}, \mathbf{a}, \sigma_e^2, \sigma_h^2, \sigma_a^2 | \mathbf{y}\right) \propto p\left(\mathbf{y} | \mathbf{b}, \mathbf{h}, \mathbf{a}, \sigma_e^2\right) p\left(\sigma_e^2\right) p(\mathbf{b})$$
$$\times\, p\left(\mathbf{h} | \sigma_h^2\right) p\left(\sigma_h^2\right) p\left(\mathbf{a} | \mathbf{A}, \sigma_a^2\right) p\left(\sigma_a^2\right),$$

where $\mathbf{A}$ is the numerator relationship matrix between individuals, and σ_e^2, σ_h^2 and σ_a^2 are the residual, herd/sampling-processing date and additive genetic variances, respectively. The a priori distribution of $\mathbf{h}$ and $\mathbf{a}$ were assumed to be multivariate normal, as follows:

$$p\left(\mathbf{h} | \sigma_h^2\right) \sim N\left(\mathbf{0}, \mathbf{I}\sigma_h^2\right),$$

$$p\left(\mathbf{a} | \sigma_a^2\right) \sim N\left(\mathbf{0}, \mathbf{A}\sigma_a^2\right),$$

where $\mathbf{I}$ is an identity matrix with dimensions equal to the number of elements in $\mathbf{h}$. Flat priors were assumed for $\mathbf{b}$ and the variance components.

To estimate the genetic correlations between VOC, PC and milk composition, we conducted a set of bivariate analyses that implemented model (1) in its multivariate version. In this case, the traits involved were assumed to jointly follow a multivariate normal distribution along with the additive genetic, herd and residual effects. The corresponding prior distributions of these effects were:

$$\mathbf{a} | \mathbf{G}_0, \mathbf{A} \sim MVN(0, \mathbf{G}_0, \otimes \mathbf{A}),$$

$$\mathbf{h} | \mathbf{H}_0, \sim N(0, \mathbf{H}_0, \otimes \mathbf{I}_n),$$

$$\text{and}\quad \mathbf{e} | \mathbf{R}_0, \sim N(0, \mathbf{R}_0, \otimes \mathbf{I}_m),$$

where $\mathbf{G}_0$, $\mathbf{H}_0$ and $\mathbf{R}_0$ are the corresponding variance–covariance matrices between the involved traits, and $\mathbf{a}$, $\mathbf{h}$ and $\mathbf{e}$ are vectors with dimensions equal to the number

of animals in the pedigree (n and m) times the number of traits considered.

Bayesian inference

Marginal posterior distributions of all unknowns were estimated using the Gibbs sampling algorithm [27]. The TM program (http://snp.toulouse.inra.fr/~alegarra) was used for all Gibbs sampling procedures. Chain lengths and burn-in period were assessed by visual inspection of the trace plots and by the diagnostic tests described in [28, 29]. After preliminary analysis, chains of 850,000 samples were used, with a burn-in period of 50,000. One in every 200 successive samples was retained. The lower and upper bounds of the highest 95% probability density regions (HPD 95%) for the parameters of concern were obtained from the estimated marginal densities. The posterior mean was used as the point estimate for all parameters.

Across-herd heritability was computed as:

$$h_{AH}^2 = \frac{\sigma_a^2}{\sigma_a^2 + \sigma_h^2 + \sigma_e^2},$$

where σ_a^2, σ_h^2, and σ_e^2 are additive genetic, herd/sampling-processing date and residual variances, respectively.

Intra-herd heritability was computed as:

$$h_{IH}^2 = \frac{\sigma_a^2}{\sigma_a^2 + \sigma_e^2},$$

where σ_a^2 and σ_e^2 are additive genetic and residual variances, respectively.

Additive genetic correlations (r_a) were computed as:

$$r_a = \frac{\sigma_{a1,a2}}{\sigma_{a1} \cdot \sigma_{a2}},$$

where $\sigma_{a1,a2}$ is the additive genetic covariance between traits 1 and 2, and σ_{a1} and σ_{a2} are the additive genetic standard deviations for traits 1 and 2, respectively.

The herd/sampling-processing date correlations (r_h) were computed as:

$$r_h = \frac{\sigma_{h1,h2}}{\sigma_{h1} \cdot \sigma_{h2}},$$

where $\sigma_{h1,h2}$ is the herd/sampling-processing date covariance between traits 1 and 2, and σ_{h1} and σ_{h2} are the herd/sampling-processing date standard deviations for traits 1 and 2, respectively.

The residual correlations (r_e) were computed as:

$$r_e = \frac{\sigma_{e1,e2}}{\sigma_{e1} \cdot \sigma_{e2}}.$$

where $\sigma_{e1,e2}$ is the residual covariance between traits 1 and 2, and σ_{e1} and σ_{e2} are the residual standard deviations for traits 1 and 2, respectively.

Results and discussion

Variance components and heritability of individual spectrometry peaks of the volatile compound fingerprint of cheese

A univariate Bayesian animal model was applied to each of the 240 individual spectrometry peaks. The variance components and heritability estimates are in Table S1 (see Additional file 1: Table S1). Table 2 shows the distribution of the intra-herd heritability estimates for the individual peaks related to the VOC of the cheese samples. Only a few peaks are characterized by a very low heritability (six peaks with a heritability lower than 3.5%).

Table 2 shows that there is a tendency towards a decrease in concentration with increasing heritability (note that the concentration is expressed on a logarithmic scale). This can be interpreted as a decrease in primary substrates, which are involved in a large number of potential metabolic pathways involved in the production of VOC. Compounds with lower concentrations are sometimes characterized by a proportional increase in instrumental error and, then, a decrease in their heritability is expected. This is not true for the spectrometry peaks that were examined in this study, although a large number of peaks with very low concentrations (<1 ppb$_v$) were not included here. This is an indirect indication of the enormous potential of the PTR-ToF-MS method for evaluating the volatile compound fingerprint of cheese. The ten VOC that had the highest estimated heritability among the VOC that were tentatively identified or unidentified are in Tables 3 and 4, respectively. The results confirm that several individual spectrometry peaks are characterized by heritability estimates of the same magnitude as those for milk yield, some milk quality traits [30, 31] and also some technological parameters [32].

It is interesting to note that some peaks are related to PC and correspond with specific odours and aromas detected in many cheese varieties [2, 13]. For instance, among the masses that were most positively correlated with PC, Bergamaschi et al. [21] detected m/z 117.091 and m/z 145.123, and their isotopes m/z 118.095 and 146.126, which in our study had heritabilities of 12 and 13%, respectively. The same authors tentatively attributed these peaks to ethyl butanoate, ethyl-2-methylpropanoate and ethyl hexanoate [21]. Esters originate from the interaction between free fatty acids and alcohols that are produced by microorganisms and are responsible for fruity-floral notes in cheese aroma [13]. In addition, the peak with a theoretical mass m/z 95.017 that is associated with methyldisulfanylmethane had a heritability of 12.5% and characterized PC4 (Table 3). This sulphur compound is either derived from the diet or formed from

Table 2 Average concentrations and estimates of phenotypic (σ_P), residual (σ_E), herd (σ_H), and additive genetic (σ_A) standard deviations, and of intra-herd heritability (h^2) categories for 240 spectrometric peaks from PTR-ToF-MS analysis of 1075 individual model cheeses made from Brown Swiss cows' milk

h^2 (%)	Peaks numbers	Concentration[a]		Average of the SD				Average h^2
		Average In ppb$_v$[b]	CV (%)[c]	σ_P	σ_E	σ_H	σ_A	
<2	0							
2–4	11	6.26	29.1	1.008	0.874	0.468	0.166	0.035
4–6	60	6.25	28.3	1.006	0.823	0.480	0.188	0.050
6–8	69	6.05	31.4	1.004	0.864	0.402	0.234	0.068
8–10	38	5.30	20.5	1.004	0.817	0.459	0.253	0.087
10–12	22	5.19	22.1	1.006	0.795	0.450	0.276	0.108
12–14	20	6.73	27.8	0.995	0.725	0.579	0.280	0.130
14–16	13	5.28	12.3	0.994	0.795	0.488	0.328	0.146
16–18	4	4.91	5.2	1.001	0.788	0.498	0.358	0.172
18–20	1	5.17		0.986	0.721	0.572	0.352	0.192
>20	2	5.33	1.2	1.020	0.666	0.691	0.344	0.211
All	240	5.90	28.5	1.003	0.822	0.462	0.239	0.082

SD standard deviation

[a] Mean value of each peak of the various classes

[b] Data expressed in natural log-transformed (ln) parts per billion by volume

[c] Coefficient of variation of the mean value of each peak calculated by dividing the standard deviation by the mean of the ppb$_v$ concentration of the PTR spectrometry peaks within each intra-herd heritability class

Table 3 Spectrometry peaks with the highest heritability (h^2) with tentative identification of volatile compounds from PTR-ToF-MS analysis of 1075 model cheeses; their phenotypic (σ_P), residual (σ_E), herd (σ_H) and additive genetic (σ_A) SD

Measured mass (m/z)	Theoretical mass (m/z)	Tentative identification	Sum formula	In ppb$_v^a$	CV (%)	SD σ_P	σ_E	σ_H	σ_A	h^2
49.011	49.0106	Methanethiol	CH_5O^+	6.82	9.8	0.994	0.800	0.507	0.302	0.125
57.033	57.0335	3-Methyl-1-butanol	$C_3H_5O^+$	6.70	10.0	1.007	0.805	0.516	0.317	0.134
75.080	75.0810	Butan-1-ol, pentan-1-ol, heptan-1-ol	$C_4H_{11}O^+$	7.70	14.7	0.990	0.763	0.554	0.302	0.136
81.070	81.0699	Alkyl fragment (terpenes)	$C_6H_9^+$	5.37	8.6	1.021	0.669	0.692	0.340	0.206
83.086	83.0855	Hexanal, nonanal	$C_6H_{11}^+$	6.13	11.3	0.998	0.825	0.453	0.333	0.140
95.017	95.0161	Methyldisulfanylmethane	$C_2H_7O_2S^+$	5.04	14.1	1.013	0.847	0.415	0.370	0.161
117.091	117.0910	Ethyl butanoate, ethyl-2-methylpropanoate	$C_6H_{13}O_2^+$	9.14	6.6	0.986	0.780	0.526	0.295	0.125
118.095	118.0940	Ethyl butanoate, ethyl-2-methylpropanoate	$C5^{[13]}CH_{13}O_2^+$	6.53	8.7	0.986	0.775	0.531	0.298	0.129
145.123	145.1220	Ethyl hexanoate, octanoic acid	$C_8H_{17}O_2^+$	7.43	10.5	0.971	0.777	0.495	0.304	0.133
146.126	146.1260	Ethyl hexanoate, octanoic acid	$C_7^{[13]}CH_{17}O_2^+$	5.30	11.5	0.972	0.774	0.498	0.310	0.138

SD standard deviation

[a] Data expressed in natural log-transformed (ln) parts per billion by volume

the amino acid methionine that is released during cheese ripening [3]. We found that m/z 81.070, which is associated with terpene fragments, had a relatively high heritability (20.6%). Terpenes are products of the degradation of carotenoids [33] and are listed as cheese odorants that have a fresh, green odour [34]. It is well known that volatile terpene in cheese is a biomarker of the area of production, and the type and phenological stage of forage [35, 36]. We found that the amount of these molecules is also affected by the genetic background of the animals (Table 3). As discussed by Bugaud et al. [37], the abundance of plants such as *Gramineae* or dicotyledons can influence the concentration of plasmin and terpenes in milk.

These particular peaks should be the first to be studied in terms of their effect on the flavour and acceptability of cheese, because they display exploitable genetic variation.

Variance components and heritability of PC extracted from volatile profiles of cheese

The proportions of variance explained by the first ten PC are in Table 1. A list of the tentatively identified individual VOC that were found as the most highly correlated with each PC is in Additional file 2: Table S2. Estimates of the marginal posterior densities for the additive genetic, herd/sampling-processing date and residual variances are in Table 5. The herd/sampling-processing date variance was always larger than the genetic variance of each PC, but was smaller than the residual variance in all but two cases i.e. PC3 had a greater herd/sampling-processing date variance than the residual variance, while for PC5

they were of the same magnitude. These data show that the proportions of the main sources of variation in individual PC (genetic, herd and individual/residual components) differ.

All ten PC that were extracted from the volatile compound fingerprint of the model cheeses had a heritability higher than 0 (Table 5). Across-herd heritability of the PC ranged from 2.4 to 8.6%, while intra-herd heritability ranged from 3.6 to 10.2%. The marginal posterior distributions of the intra-herd heritability of these ten PC are in Fig. 1a, b.

Notably, the two most important PC, which explained about 40% of the overall variability, were characterized by similar heritabilities (h_{IH}^2: 8.4% for PC1 and 8.5% for PC2).

Herds were previously classified according to dairy system, i.e. traditional or modern, which vary in terms of milk yield, destination of milk, facilities, feed management and use of maize silages [21, 22]. Modern dairy farms had modern facilities, loose animals, milking parlours and used total mixed rations with or without silage, whereas traditional dairy farms had small buildings, animals were tied and milked at the stall, and the feed was mainly composed of hay and compound feed. It is worth noting that PC1 was not affected by dairy system, while PC2 was higher in the cheese samples from milk that was produced by cows reared in modern facilities and fed on total mixed rations, whether with or without silage [21] than by cows reared in the traditional system. Both PC varied during lactation, but in opposite directions: PC1 decreased curvilinearly, while PC2 increased linearly.

Table 4 Unidentified spectrometry peaks with the highest heritability (h^2) from PTR-ToF-MS analysis of 1075 model cheeses; their phenotypic (σ_P), residual (σ_E), herd (σ_H) and additive genetic (σ_A) SD

m/z	ln ppb$_v^a$	CV (%)	SD				h^2
			σ_P	σ_E	σ_H	σ_A	
50.056	4.58	9.6	0.992	0.754	0.562	0.313	0.147
93.431	4.54	10.0	0.998	0.759	0.542	0.354	0.179
95.095	5.44	18.1	0.991	0.766	0.534	0.331	0.157
111.104	5.08	18.5	0.991	0.774	0.528	0.322	0.147
120.092	4.95	14.5	0.998	0.797	0.478	0.363	0.172
121.122	5.17	15.8	0.986	0.721	0.572	0.352	0.192
136.140	4.26	8.9	0.990	0.759	0.549	0.319	0.150
137.132	5.28	7.3	1.019	0.663	0.691	0.348	0.216
171.173	4.87	15.7	1.004	0.905	0.202	0.384	0.152
173.153	5.20	9.3	0.997	0.825	0.440	0.345	0.149

SD standard deviation

[a] Data expressed in natural log-transformed (ln) parts per billion by volume

Table 5 Features of marginal posterior densities of additive genetic (σ_A^2), herd/sampling-processing date (σ_H^2), and residual (σ_E^2) variances, and across-herd (h_{AH}^2) and intra-herd (h_{IH}^2) heritabilities for principal components derived from the volatile fingerprint of 1075 individual model cheeses analysed by PTR-ToF-MS

Traits	σ_A^2		σ_H^2		σ_E^2		h_{AH}^2		h_{IH}^2	
	Mean	HPD 95%	Mean	HPD 95%	Mean	HPD 95%	Mean	HPD 95%	Mean	HPD 95%
PC1	4.49	0.46; 11.70	15.14	9.79; 22.59	48.81	41.85; 55.09	0.065	0.01; 0.17	0.084	0.01; 0.21
PC2	1.48	0.09; 4.04	9.12	6.21; 13.24	15.95	13.53; 18.04	0.056	0.01; 0.15	0.085	0.01; 0.22
PC3	0.71	0.05; 1.90	11.77	8.22; 16.60	9.31	8.09; 10.44	0.033	0.01; 0.08	0.071	0.01; 0.18
PC4	1.62	0.15; 3.96	2.80	1.69; 4.37	14.27	12.05; 16.28	0.086	0.01; 0.21	0.102	0.01; 0.24
PC5	0.43	0.02; 1.27	7.30	5.08; 10.43	7.21	6.32; 8.04	0.029	0.01; 0.09	0.057	0.01; 0.16
PC6	0.22	0.01; 0.73	2.93	1.97; 4.23	5.97	5.36; 6.59	0.024	0.01; 0.08	0.036	0.01; 0.11
PC7	0.47	0.04; 1.27	1.61	1.06; 2.39	4.50	3.77; 5.11	0.071	0.01; 0.19	0.095	0.01; 0.25
PC8	0.48	0.06; 1.14	0.64	0.36; 1.04	4.43	3.79; 5.04	0.086	0.01; 0.20	0.098	0.01; 0.23
PC 9	0.21	0.01; 0.65	0.47	0.25; 0.77	3.86	3.40; 4.30	0.047	0.01; 0.16	0.053	0.01; 0.14
PC10	0.30	0.01; 0.86	0.47	0.26; 0.76	3.08	2.56; 3.50	0.078	0.01; 0.22	0.089	0.01; 0.25

Mean = mean of the marginal posterior density of the parameter; HPD 95% = lower and upper bound of the 95% highest posterior density region

The third PC was characterized by a slightly lower intra-herd heritability than PC1 and PC2 (7.1%), and much lower across-herd heritability (3.3%) due to the large effect of herd/sampling-processing date on this component of cheese flavour (Table 5). It is worth noting that this sizeable environmental variability is not due to the dairy system but to the large variability among herds within each dairy system [21]. This PC, which explains less than 9% of the total volatile fingerprint variation, was found to increase with daily milk yield of the cow, but was not affected by parity and DIM. The fourth PC, which explained almost 8% of the total cheese volatile fingerprint, was the most heritable ($h_{AH}^2 = 8.6\%$ and $h_{IH}^2 = 10.2\%$). This PC was not much affected by dairy system or individual herd, but was found to increase during lactation [21]. Among the other PC, PC5, PC6 and PC9 were characterized by a low heritability (<6%) and PC7, PC8 and PC10 had heritabilities that ranged from 7.1 to 9.8% and were intermediate between those of PC1 and PC2 and those of PC4 (Table 5). To our knowledge, this is the first report on heritability estimates for phenotypes that describe the profile of volatile compounds in cheese. As discussed above, slightly more than one third of the spectrometry peaks had estimated heritabilities that were similar to those of the three PC of the volatile fingerprint with low heritabilities, about one third of the spectrometry peaks had estimated heritabilities similar to those of the other seven PC, and about a quarter had estimated heritabilities that were higher than those of the most heritable PC (PC4).

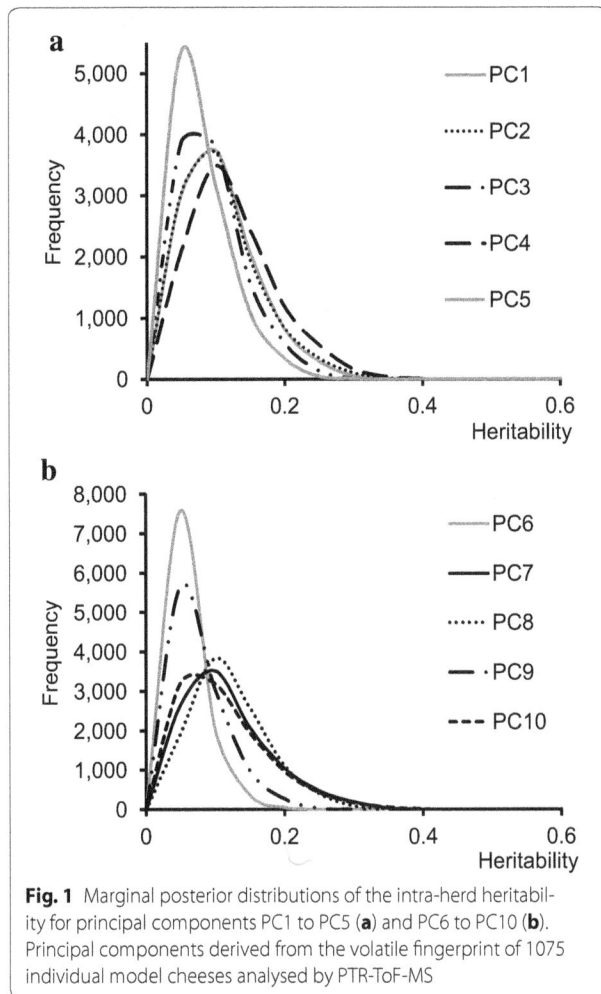

Fig. 1 Marginal posterior distributions of the intra-herd heritability for principal components PC1 to PC5 (**a**) and PC6 to PC10 (**b**). Principal components derived from the volatile fingerprint of 1075 individual model cheeses analysed by PTR-ToF-MS

Given that no other data are available in the literature, it was interesting to compare the heritability of the PC of the volatile fingerprint of the model cheeses with the heritability of other traits that were studied in the same project with the same cows, or at the population level with the same breed and in the same area. Estimated heritability of daily milk yield (18.2%) [20, 38] in individual cows was about double that of most of the cheese VOC PC, while the estimated heritability reported by Cecchinato et al. [39] was similar to that of the PC with the highest heritability. Regarding milk quality, heritabilities of fat content (12.2%) and SCS (9.6%) [38] were similar to those of the PC of the cheese volatile fingerprint with the highest heritability, while milk protein (28%), casein (28%), casein number (i.e. the ratio between casein and total protein) (15.1%), lactose (17.0%) and urea (35.6%) were much more heritable at both the experimental and population levels. The estimated heritabilities of the detailed fatty acid profile of the milk samples were in the same range as those of the PC of the volatile fingerprint of cheeses

obtained from the same milk, with the exception of the saturated odd-numbered fatty acids and a few others [40]. A possible explanation for such similar ranges of heritabilities could be related to the origin of the VOC, since many molecules may be produced from milk fat via different biosynthetic pathways [3]. For example, beta-oxydation and decarboxylation of milk fat produce methyl ketones and secondary alcohols, and esterification of hydroxy fatty acids produces lactones. Fatty acids can also react with alcohol groups to form esters such as ethyl butanoate and ethyl hexanoate, which are correlated with PC1.

As for the cheese-making process, the traditional milk coagulation properties were also much more heritable than the PC of the cheese volatile fingerprint, with the exception of curd firmness recorded 45 min after rennet addition [41]. Modelling of curd firming [42] in the same milk samples yielded estimated heritabilities that were higher than those of the PC of the cheese volatile fingerprint for rennet coagulation time and for the curd firming instant rate constant, and that were of the same magnitude as those for potential curd firmness and the syneresis instant rate constant [43].

The technological traits (three cheese yields and four milk nutrient recoveries in the curd) measured in the fresh model cheeses were also much more heritable than the PC of the volatile fingerprint of the same model cheeses after 2 months of ripening [20]. The same traits predicted by Fourier transform infrared spectrometry using the calibration proposed by Ferragina et al. [44] on the same milk samples [45] and at the population level [39] were characterized by heritability estimates of the same size.

In summary, the PC of the volatile compound fingerprints of cheeses obtained by using the PTR-ToF-MS procedure are not only heritable, their heritability estimates are similar to those of several milk quality traits (fat content, content in many fatty acids, and SCS) and of some coagulation properties (potential curd firmness and syneresis instant rate constant) that are already selected for or for which genetic selection has been proposed. How can animal genetic characteristics affect cheese VOC has never been studied. However, it should be emphasized that the majority of VOC in ripened cheese originate from the breakdown of fresh cheese components by (1) milk native enzymes produced by the cow, or (2) enzymatic activity of cheese micro-organisms.

For example, proteolysis releases amino acids via the Strecker reaction, which are the precursors of a wide variety of volatile compounds including aldehydes, such as 2-methylbutanal, 3-methylbutanal, hexanal and nonanal, that may be responsible for green and herbaceous aromas in cheese [13] and that are correlated with PC (see Additional file 2: Table S2). The presence and the activity of milk native enzymes in relation to cheese flavour and VOC need to be further studied, but it is known that some of these activities are under the genetic control of the lactating cow. Lipoprotein lipase (LPL), which has been well described in humans, has the potential to hydrolyse the greater part of milk fats, but this is prevented by the membrane of the fat globule [46]. Also, the casein cleavage by plasmin is a well-known reaction that leads to particular sensory characteristics of cheese depending on the milk plasmin activity [47]. Hydrolysis of lactoproteins by enzymes varies with the presence of polymorphisms that induce amino acid substitutions or deletions that change the site of cleavage of the enzymes, thereby generating further modifications of the cheese characteristics. For example, the cleavage of β-casein by plasmin differs greatly between the β-casein variants A1 and C [48, 49]. However, the growth and activity of the micro-organisms present in the milk may also be directly related to milk components with anti-microbial activity, such as lactoferrin, which has been shown to be heritable [50]. Another hypothesis is that very indirect relationships may occur between the technological properties of milk, such as curd firming and syneresis, i.e., water expulsion from the curd [51] and the growth and activity of micro-organisms in cheeses, and the main compounds of milk, such as caseins. Indeed, the content in milk caseins and their genetic variants drive coagulation and draining kinetics and may modify the water content of fresh cheese [52], which, in turn, may modify the growth or activity of micro-organisms during cheese ripening. Other similar indirect effects may also occur for other milk compounds (with variable heritabilities), such as soluble proteins, urea, SCS, carotenoids, etc.

Phenotypic, genetic, herd/sampling-processing date and residual correlations among VOC

Figure 2 shows that, unlike the PC, the correlations between the ten VOC with the highest heritabilities varied, but were generally positive. The residual correlations were often moderate to high and positive. Only the peaks relative to methanethiol (theoretical mass m/z 49.011) and methyldisulfanylmethane (theoretical mass m/z 95.017) had very low residual correlations with the other eight VOC. These two VOC also had negative genetic and herd/sampling-processing date correlations with the others, which were generally positively correlated with each other. As discussed above, microorganisms are considered to be the key agents in the production of these volatile compounds in ripened cheese. Analysis of individual VOC concentrations is quantitative, thus it is logical that an increase in the global quantity of VOC in the cheese would result in an increase in most of the individual VOC, and especially those with the highest concentrations and therefore in positive phenotypic correlations with each other.

It is likely that methanethiol and methyldisulfanylmethane are determined by genetic and/or herd pathways that differ from those characterizing the global quantity of cheese odorants, which could explain the residual independence from other VOC. The reasons for the negative genetic and herd correlations need to be investigated in future research.

Phenotypic, genetic, herd/sampling-processing date and residual correlations among PC

As expected with PC, the phenotypic correlations among PC were always close to 0 (−0.02 to 0.02) and the 0 value was always included in their HPD 95% (data not shown). These are the results from the multivariate data treatment that generates a few variables, either uncorrelated or with a low level of correlation, which may be desirable indicators of the volatile compound fingerprint of cheese. However, this phenotypic independence is the result of additive genetic, herd/sampling-processing date and residual correlations, which are sometimes very different from zero but opposite in sign. Figure 2 reports the genetic, herd/sampling-processing date and residual correlations among the first ten PC of the volatile compound fingerprint of cheese.

For example, the first three PC, which together explain about half of the variability of all the 240 spectrometry peaks obtained with PTR-ToF-MS, are negatively correlated with each other from a genetic point of view (Fig. 2). However, PC2 and PC3 were positively correlated for herd/sampling-processing date but had a low negative residual correlation. Other high additive genetic correlations were found between PC3 and PC6 (positive), and between PC10 and both PC5 and PC7 (negative). Some high correlations between herd-sampling and processing date were found among the ten PC, while the residual correlations were generally much lower (Fig. 2).

The correlations between PC in 60-day old cheeses could be the result of many different (and independent) metabolic pathways (i.e., lipolysis vs. proteolysis) that involve many ripening agents. Thus, correlations between PC seem to present a more qualitative picture (proportions among groups) of the volatile compound fingerprint of cheese, while the correlations between individual spectrometry peaks describe the quantitative relationships among them. Since these results cannot be compared with the literature because of lack of data, further research on these correlations is necessary to assess their importance and reveal the significance of these new phenotypes, especially in relation to the sensorial properties of cheeses. Moreover, research on the various herd and animal factors that affect the PC [21] and their

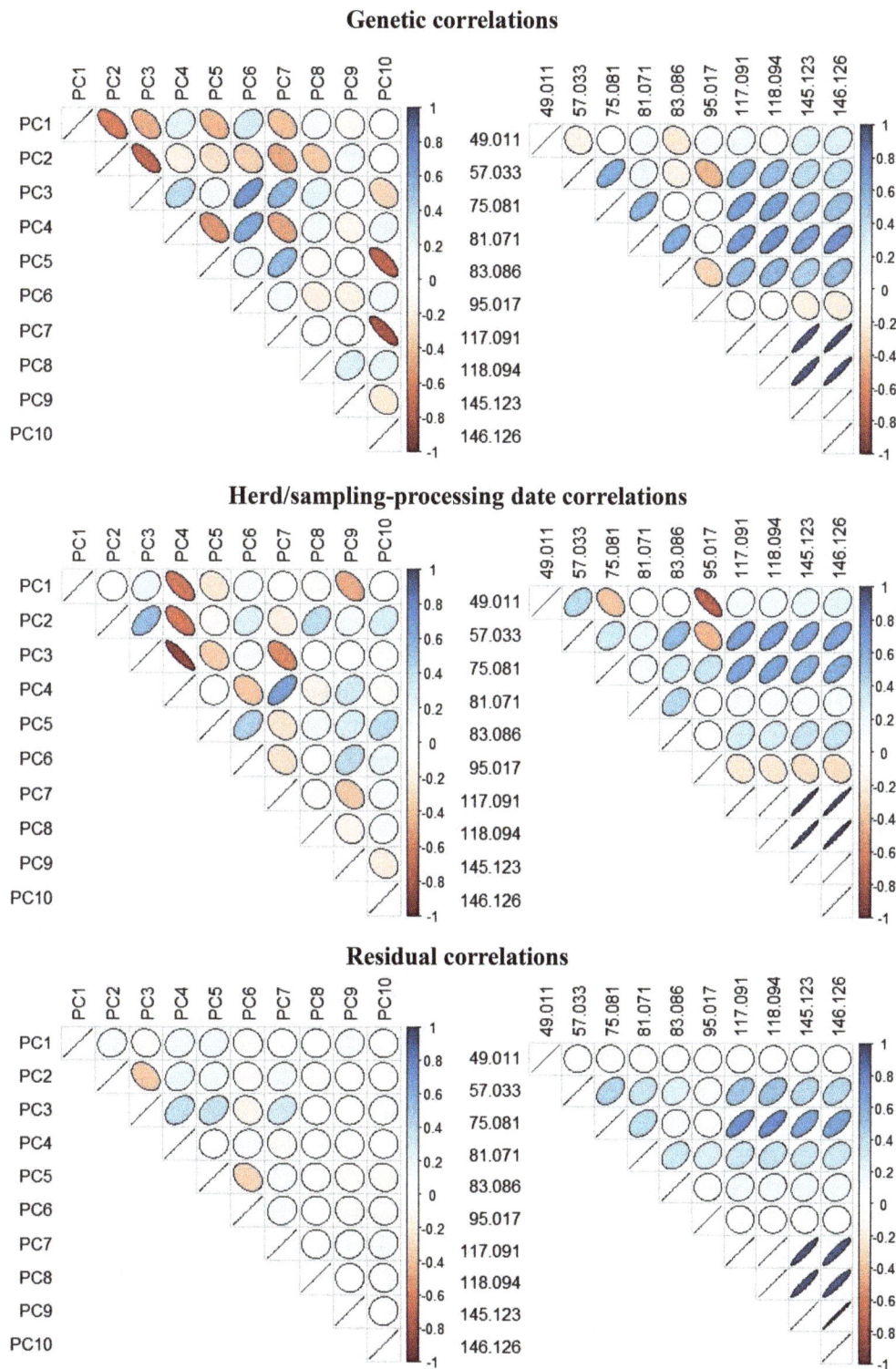

Fig. 2 Genetic, herd/sampling-processing date and residual correlations among the first ten principal components and among the ten identified individual VOC with the highest heritability. Correlations among the first ten principal components (*left triangles*) and among the ten individual identified VOC (*right triangles*); all estimates (expressed as the mean of the marginal posterior distribution of the parameter) ranged from no correlation (*uncoloured circles*) to high correlations (*thin, dark-coloured ovals*); negative correlations: *reddish ovals* from *top left* to *bottom right*; positive correlations: *bluish ovals* from *top right* to *bottom left*

Genetic correlations

	PC1	PC2	PC3	PC4	PC5	PC6	PC7	PC8	PC9	PC10	49.011	57.033	75.081	81.071	83.086	95.017	117.091	118.095	145.123	146.126
Fat	-0.55	0.4	-0.21	0.07	-0.03	-0.2	-0.09	0	-0.33	0.01	0.22	0.66	0.56	0.19	-0.03	0.13	0.26	0.27	0.15	0.17
Protein	-0.53	0.49	-0.14	-0.21	0.14	-0.29	0.14	-0.19	-0.12	-0.43	0.43	0.11	0.51	0.55	-0.16	0.22	0.34	0.33	0.34	0.33
Fat / Protein	-0.26	0.11	-0.13	0.19	-0.17	-0.22	-0.32	0.19	-0.34	0.3	-0.03	0.68	0.28	-0.22	0.04	-0.09	0.06	0.08	-0.06	-0.05
Casein	-0.43	0.59	-0.22	-0.26	0.36	-0.34	0.23	-0.14	-0.22	-0.51	0.48	-0.01	0.46	0.48	-0.19	0.24	0.19	0.18	0.18	0.17
Casein / Protein	0.65	-0.2	-0.24	-0.09	0.25	-0.31	-0.4	0.17	-0.32	0.22	-0.46	-0.3	-0.63	-0.71	-0.06	-0.31	-0.67	-0.66	-0.68	-0.67
Lactose	0.08	0.22	0.31	0.42	0.18	0.13	0.07	0.15	0.43	-0.02	-0.14	-0.06	0.05	0.05	0.38	-0.46	0.13	0.15	0.13	0.14
Total solids	-0.58	0.5	-0.14	0.12	0.09	-0.24	0.06	-0.1	-0.28	-0.21	0.28	0.56	0.59	0.3	-0.07	0.03	0.29	0.3	0.2	0.22
pH	0.31	-0.02	-0.14	-0.24	0.05	-0.47	0.12	0	0.18	0.07	0.22	0.06	-0.14	-0.11	0.06	0.38	-0.34	-0.36	-0.36	-0.42
SCS	0.4	-0.48	0.07	0.02	-0.45	0.15	0.14	0.37	0.45	-0.1	0.01	-0.25	-0.56	0.04	-0.05	-0.17	-0.21	-0.2	-0.06	-0.04

Herd/sampling-processing date correlations

	PC1	PC2	PC3	PC4	PC5	PC6	PC7	PC8	PC9	PC10	49.011	57.033	75.081	81.071	83.086	95.017	117.091	118.095	145.123	146.126
Fat	-0.01	0.08	0.04	-0.07	-0.03	0.06	0.05	0.07	-0.09	0.03	0.13	0.13	0	-0.04	0.12	0	-0.09	-0.09	-0.03	-0.04
Protein	-0.11	0.19	-0.17	0.15	0.03	-0.02	0.19	0.17	0.31	-0.46	0.01	0.01	0.26	0.09	0.23	0.05	0	-0.01	0.13	0.13
Fat / Protein	0.04	0.01	0.14	-0.18	-0.06	0.07	-0.04	0.02	-0.22	0.23	0.14	0.15	-0.11	-0.09	0.04	-0.02	-0.1	-0.1	-0.09	-0.11
Casein	-0.11	0.2	-0.12	0.15	-0.06	-0.06	0.14	0.23	0.3	-0.53	0.07	0.13	0.25	0.08	0.22	0	0.04	0.04	0.18	0.19
Casein / Protein	-0.06	0.1	0.16	0.03	-0.42	-0.14	-0.15	0.32	0.05	-0.48	0.33	0.52	0.01	0.02	0.02	-0.21	0.2	0.2	0.27	0.28
Lactose	-0.08	-0.12	0.38	-0.06	-0.11	-0.26	-0.06	-0.37	-0.19	-0.06	0.13	0.06	-0.1	-0.05	0.03	-0.54	0.32	0.32	0.27	0.27
Total solids	-0.06	0.07	0.04	-0.02	-0.03	0.01	0.1	0.03	-0.04	-0.13	0.15	0.11	0.05	-0.01	0.17	-0.09	-0.02	-0.02	0.06	0.05
pH	-0.12	-0.19	0	0.07	-0.05	-0.28	0.04	-0.3	-0.06	-0.05	-0.08	0.18	-0.02	0.11	0.28	-0.28	0.07	0.07	0.07	0.08
SCS	-0.17	-0.13	-0.13	0.16	0.01	-0.12	0.07	0.11	0.15	0.19	0.14	0.07	0.04	0.29	0.31	0.21	0.15	0.14	0.19	0.15

Residual correlations

	PC1	PC2	PC3	PC4	PC5	PC6	PC7	PC8	PC9	PC10	49.011	57.033	75.081	81.071	83.086	95.017	117.091	118.095	145.123	146.126
Fat	-0.05	-0.08	0.13	0	0.09	0	-0.03	0.07	0.01	-0.05	0.01	-0.05	0.04	0.04	0.06	-0.07	0.06	0.06	0.05	0.05
Protein	-0.14	0.22	-0.04	0.04	-0.06	-0.08	0.01	-0.02	0.03	0.1	0	0.15	0.25	-0.04	0.04	-0.05	0.14	0.14	0.18	0.2
Fat / Protein	0	-0.15	0.15	0	0.1	0.04	-0.02	0.06	0.01	-0.08	0.01	-0.09	-0.04	0.06	0.04	-0.05	0.01	0.01	-0.02	-0.02
Casein	-0.18	0.21	-0.02	0.04	-0.06	-0.08	-0.01	-0.03	0.06	0.1	0.01	0.19	0.3	0.05	0.05	-0.03	0.18	0.18	0.23	0.24
Casein / Protein	-0.13	-0.02	0.07	0	0.02	-0.02	0	-0.01	0.08	-0.01	0.04	0.08	0.15	0.23	0.06	0.06	0.07	0.07	0.08	0.09
Lactose	-0.12	0.08	-0.1	-0.17	0.01	-0.12	0.09	-0.01	-0.16	0.01	0	0.08	0.14	0.11	-0.04	0.15	-0.07	-0.08	-0.12	-0.12
Total solids	-0.11	0.01	0.1	-0.02	0.07	-0.04	-0.02	0.08	0.01	0	0	0.01	0.14	0.05	0.07	-0.05	0.1	0.1	0.09	0.09
pH	0.01	-0.06	-0.04	0	-0.11	0.11	-0.03	-0.01	0.04	0	-0.03	-0.09	-0.08	-0.06	-0.04	-0.03	0.06	0.06	0.12	0.13
SCS	0.05	-0.1	0.03	0.02	0	0.07	-0.12	-0.02	0.04	0.01	0.01	-0.04	-0.1	-0.1	-0.04	0.06	0.14	0.14	0.23	0.22

Fig. 3 Genetic, herd/sampling-processing date and residual correlations between the first ten principal components and the ten identified individual VOC with milk quality traits. Correlations between the first ten principal components (*left squares*) and milk quality traits, and between the ten identified individual VOC (*right squares*) with the highest heritability estimates (expressed as means of the marginal posterior distribution of the parameter) and milk quality traits, ranging from *bright red* for highly negative correlations to *no colour* for uncorrelated traits, and *bright blue* for highly positive correlations

correlations with milk quality traits would help to better characterize cheese flavour.

Phenotypic, genetic, herd/sampling-processing date and residual correlations of VOC and PC with milk quality traits

Finally, we analyzed the dependency of cheese VOC and their PC on the quality traits of the milk used for cheese making. Figure 3 shows that there are positive and negative correlation coefficients that are small at the residual level, low to moderate at the herd level and low to high at the genetic level.

We did not find any PC or VOC with patterns that clearly differed from the others. However, it is clear that some milk quality traits have a more marked effect on the ten VOC with the highest heritability. In particular, contents of fat, protein, casein and total solids in milk are generally positively correlated with all VOC, which indicates that genetic factors associated with more concentrated milk are also responsible for an increase in the content of VOC in the resulting cheese. These VOC with the highest heritability seem to reflect the initial composition (fat and casein contents, and therefore the content of VOC precursors) of the raw whole milk used in cheese making and could be considered quantitative elements.

Regarding the casein number, all the genetic correlations are negative except for two highly negative correlations. Since both the numerator and denominator of the ratio are generally positively correlated with VOC concentration, this could mean that the genetic factors responsible for an increase in the ratio between caseins and whey protein could differ from those related to the increase in total protein and casein and, thus, could be responsible for a decrease in VOC concentration.

Based on Fig. 3, milk pH and SCS may also exert some negative effects on cheese VOC, which should also be further studied.

Conclusions

In this work, we report the first estimated genetic parameters of the cheese volatile compound profile in dairy cattle. Although the sample size was rather limited, our results are very original and show, as expected, that the large majority of the 240 spectrometry peaks obtained with PTR-ToF-MS were characterized by heritability estimates that were similar to those obtained for the PC extracted from them, while about one sixth of these 240 peaks were characterized by higher heritabilities, which were similar to those for daily milk yield and several other milk and cheese traits. Only a small proportion of the peaks had a very low heritability. Heritability estimates for the ten PC were low, although for seven of them they were similar to those for other milk traits, such as fat content, many fatty acid percentages, SCS and some curd firming model parameters. Variability due to herd-sampling-processing date varied greatly among the PC. Although the phenotypic correlations among the ten PC were, as expected, close to 0, there were some sizeable genetic correlations, in particular those among herd/sampling-processing date, while the residual correlations were generally lower. The most heritable VOC and PC of cheeses had variable and generally positive genetic, herd and residual correlations with each other. The correlations between selected VOC and PC of cheeses with quality traits of milk before cheese making were also variable and were generally higher for genetic correlations, intermediate for herd correlations and lower for residual correlations. Our results demonstrate the existence of exploitable genetic variation in the factors related to cheese volatile profiles that are potentially useful for improving the flavour of cheese. This study opens new avenues of research for the characterization of the

relationships between individual peaks or PC and the sensory profile of cheese, and for the study of their use for indirect prediction using infrared technology.

Additional files

Additional file 1: Table S1. Average concentrations of spectrometry peaks and coefficients of variation (CV, %) from PTR-ToF-MS analysis of 1075 cheese samples, together with their phenotypic (σ_P), residual (σ_E), herd (σ_H) and additive genetic (σ_A) SD and intra-herd heritability (h^2). [1]Data expressed in natural log-transformed (ln) parts per billion by volume; [2]Mean = mean of the marginal posterior density of the parameters; [3]PSD = posterior standard deviation.

Additional file 2: Table S2. Measured and theoretical mass, sum formula and tentative identification of the five PTR-ToF-MS peaks that have the highest phenotypic correlation (r) with each PC. Highest correlation coefficients between the first ten PC and the volatile compounds tentatively attributed to specific spectrometric fragments of PTR-ToF-MS spectra

Authors' contributions
MB performed the phenotypic analysis and drafted the manuscript; AC performed the genetic analysis; GB supervised the experiment and directed the study; FB and FG supervised the PTR-ToF-MS analysis. All authors revised the manuscript critically for intellectual content. All authors read and approved the final manuscript.

Author details
[1] Department of Agronomy, Food, Natural Resources, Animals and Environment (DAFNAE), University of Padua, Viale dell'Università 16, 35020 Legnaro, PD, Italy. [2] Department of Food Quality and Nutrition, Research and Innovation Centre, Fondazione Edmund Mach (FEM), Via E. Mach 1, 38010 San Michele all'Adige, TN, Italy. [3] INRA, UMR Herbivores, 63122 Saint-Genès Champanelle, France. [4] Clermont Université, VetAgro Sup, BP 10448, 63000 Clermont-Ferrand, France.

Acknowledgements
The authors would like to thank the Autonomous Province of Trento (Italy) for funding the project, the Superbrown Consortium of Bolzano and Trento (Trento, Italy) for carrying out the recordings, and the Italian Brown Swiss Cattle Breeders Association (ANARB, Verona, Italy) for providing pedigree information. The authors also acknowledge Elisa Forato (DAFNAE, University of Padua, Legnaro, Italy), Luca Cappellin (Fondazione Edmund Mach, San Michele all'Adige, Trento, Italy) and Andrea Romano (Free University of Bolzano, Faculty of Science and Technology, Bolzano, Italy) for the PTR-ToF-MS analysis; Claudio Cipolat-Gotet for the cheese making; Roberto Chimetto, Lorenzo Giorato and Valter Giraldo (DAFNAE, University of Padua, Legnaro, Italy) for herd sampling.

Competing interests
The authors declare that they have no competing interests.

References
1. Molimard P, Spinnler HE. Review: compounds involved in the flavor of surface mold-ripened cheeses: origins and properties. J Dairy Sci. 1996;79:169–84.
2. Bellesia F, Pinetti A, Pagnoni UM, Rinaldi R, Zucchi C, Caglioti L, et al. Volatile components of Grana Parmigiano-Reggiano type hard cheese. Food Chem. 2003;83:55–61.
3. McSweeney PL, Sousa MJ. Biochemical pathways for the production of flavour compounds in cheeses during ripening: a review. Lait. 2000;80:293–324.

4. Fox PF, Wallace JM. Formation of flavor compounds in cheese. Adv Appl Microbiol. 1997;45:17–86.

5. Drake SL, Gerard PD, Drake MA. Consumer preferences for mild Cheddar cheese flavors. J Food Sci. 2008;73:S449–55.

6. Liggett RE, Drake MA, Delwiche JF. Impact of flavor attributes on consumer liking of Swiss cheese. J Dairy Sci. 2008;91:466–76.

7. Martin B, Verdier-Metz I, Buchin S, Hurtaud C, Coulon JB. How do the nature of forages and pasture diversity influence the sensory quality of dairy livestock products? Anim Sci. 2005;81:205–12.

8. Coppa M, Verdier-Metz I, Ferlay A, Pradel P, Didienne R, Farruggia A, et al. Effect of different grazing systems on upland pastures compared with hay diet on cheese sensory properties evaluated at different ripening times. Int Dairy J. 2011;21:815–22.

9. Romanzin A, Corazzin M, Piasentier E, Bovolenta S. Effect of rearing system (mountain pasture vs. indoor) of Simmental cows on milk composition and Montasio cheese characteristics. J Dairy Res. 2013;80:390–9.

10. Bittante G, Cecchinato A, Cologna N, Penasa M, Tiezzi F, De Marchi M. Factors affecting the incidence of first-quality wheels of Trentingrana cheese. J Dairy Sci. 2011;94:3700–7.

11. Bittante G, Cologna N, Cecchinato A, De Marchi M, Penasa M, Tiezzi F, et al. Monitoring of sensory attributes used in the quality payment system of Trentingrana cheese. J Dairy Sci. 2011;94:5699–709.

12. LeQuéré J. Cheese flavour: instrumental techniques. In: Fox PF, McSweeney PLH, Cogan TM, Guinee TP, editors. Cheese—chemistry, physics and microbiology. 3rd ed. Oxford: Elsevier Academic Press; 2004. p. 489–510.

13. Cornu A, Rabiau N, Kondjoyan N, Verdier-Metz I, Pradel P, Tournayre P, et al. Odour-active compound profiles in Cantal-type cheese: effect of cow diet, milk pasteurization and cheese ripening. Int Dairy J. 2009;19:588–94.

14. Carunchia Whetstine ME, Drake MA, Nelson BK, Barbano DM. Flavor profiles of full-fat and reduced-fat cheese and cheese fat made from aged cheddar with the fat removed using a novel process. J Dairy Sci. 2006;89:505–17.

15. Delgado FJ, González-Crespo J, Cava R, Ramírez R. Formation of the aroma of a raw goat milk cheese during maturation analysed by SPME–GC–MS. Food Chem. 2011;129:1156–63.

16. Thomsen M, Gourrat K, Thomas-Danguin T, Guichard E. Multivariate approach to reveal relationships between sensory perception of cheeses and aroma profile obtained with different extraction methods. Food Res Int. 2014;62:561–71.

17. Valdivielso I, Albisu M, de Renobales M, Barron LJR. Changes in the volatile composition and sensory properties of cheeses made with milk from commercial sheep flocks managed indoors, part-time grazing in valley, and extensive mountain grazing. Int Dairy J. 2016;53:29–36.

18. Bergamaschi M, Aprea E, Betta E, Biasioli F, Cipolat-Gotet C, Cecchinato A, et al. Effects of dairy system, herd within dairy system, and individual cow characteristics on the volatile organic compound profile of ripened model cheeses. J Dairy Sci. 2015;98:2183–96.

19. Cipolat-Gotet C, Cecchinato A, De Marchi M, Bittante G. Factors affecting variation of different measures of cheese yield and milk nutrient recovery from an individual model cheese-manufacturing process. J Dairy Sci. 2013;96:7952–65.

20. Bittante G, Cipolat-Gotet C, Cecchinato A. Genetic parameters of different measures of cheese yield and milk nutrient recovery from an individual model cheese-manufacturing process. J Dairy Sci. 2013;96:7966–79.

21. Bergamaschi M, Biasioli F, Cappellin L, Cecchinato A, Cipolat-Gotet C, Cornu A, et al. Proton transfer reaction time-of-flight mass spectrometry: a high-throughput and innovative method to study the influence of dairy system and cow characteristics on the volatile compound fingerprint of cheeses. J Dairy Sci. 2015;98:8414–27.

22. Sturaro E, Marchiori E, Cocca G, Penasa M, Ramanzin M, Bittante G. Dairy systems in mountainous areas: farm animal biodiversity, milk production and destination, and land use. Livest Sci. 2013;158:157–68.

23. Ali AKA, Shook GE. An optimum transformation for somatic cell concentration in milk. J Dairy Sci. 1980;63:487–90.

24. Fabris A, Biasioli F, Granitto PM, Aprea E, Cappellin L, Schuhfried E, et al. PTR-TOF-MS and data-mining methods for rapid characterisation of agro-industrial samples: influence of milk storage conditions on the volatile compounds profile of Trentingrana cheese. J Mass Spectrom. 2010;45:1065–74.

25. Cappellin L, Biasioli F, Schuhfried E, Soukoulis C, Märk TD, Gasperi F. Extending the dynamic range of proton transfer reaction time-of-flight mass spectrometers by a novel dead time correction. Rapid Commun Mass Spectrom. 2011;25:179–83.

26. Lindinger W, Hansel A, Jordan A. On-line monitoring of volatile organic compounds at pptv levels by means of proton-transfer-reaction mass spectrometry (PTR-MS) medical applications, food control and environmental research. Int J Mass Spectrom Ion Process. 1998;173:191–241.

27. Gelfand AE, Smith AF. Sampling-based approaches to calculating marginal densities. J Am Stat Assoc. 1990;85:398–409.

28. Geweke J. Evaluating the accuracy of sampling-based approaches to the calculation of posterior moments. In: Berger JO, Bernardo JM, Dawid AP, Smith AFM, editors. Bayesian statistics. Oxford: Oxford University Press; 1992. p. 164–93.

29. Gelman A, Rubin DB. Inference from iterative simulation using multiple sequences. Stat Sci. 1992;7:457–72.

30. Othmane MH, Carriedo JA, San Primitivo F, Fuente LDL. Genetic parameters for lactation traits of milking ewes: protein content and composition, fat, somatic cells and individual laboratory cheese yield. Genet Sel Evol. 2002;34:581–96.

31. Rosati A, Van Vleck LD. Estimation of genetic parameters for milk, fat, protein and mozzarella cheese production for the Italian river buffalo *Bubalus bubalis* population. Livest Prod Sci. 2002;74:185–90.

32. Ikonen T, Morri S, Tyrisevä A, Ruottinen O, Ojala M. Genetic and phenotypic correlations between milk coagulation properties, milk production traits, somatic cell count, casein content, and pH of milk. J Dairy Sci. 2004;87:458–67.

33. Carpino S, Mallia S, La Terra S, Melilli C, Licitra G, Acree T, et al. Composition and aroma compounds of Ragusano cheese: native pasture and total mixed rations. J Dairy Sci. 2004;87:816–30.

34. Horne J, Carpino S, Tuminello L, Rapisarda T, Corallo L, Licitra G. Differences in volatiles, and chemical, microbial and sensory characteristics between artisanal and industrial Piacentinu Ennese cheeses. Int Dairy J. 2005;15:605–17.

35. Viallon C, Verdier-Metz I, Denoyer C, Pradel P, Coulon JB, Berdagué JL. Desorbed terpenes and sesquiterpenes from forages and cheeses. J Dairy Res. 1999;66:319–26.

36. Cornu A, Kondjoyan N, Martin B, Verdier-Metz I, Pradel P, Berdagué J, et al. Terpene profiles in Cantal and Saint-Nectaire-type cheese made from raw or pasteurised milk. J Sci Food Agric. 2005;85:2040–6.

37. Bugaud C, Buchin S, Coulon JB, Hauwuy A, Dupont D. Influence of the nature of alpine pastures on plasmin activity, fatty acid and volatile compound composition of milk. Lait. 2001;81:401–14.

38. Cecchinato A, Ribeca C, Chessa S, Cipolat-Gotet C, Maretto F, Casellas J, et al. Candidate gene association analysis for milk yield, composition, urea nitrogen and somatic cell scores in Brown Swiss cows. Animal. 2014;8:1062–70.

39. Cecchinato A, Chessa S, Ribeca C, Cipolat-Gotet C, Bobbo T, Casellas J, et al. Genetic variation and effects of candidate-gene polymorphisms on coagulation properties, curd firmness modeling and acidity in milk from Brown Swiss cows. Animal. 2015;9:1104–12.

40. Pegolo S, Cecchinato A, Casellas J, Conte G, Mele M, Schiavon S, et al. Genetic and environmental relationships of detailed milk fatty acids profile determined by gas chromatography in Brown Swiss cows. J Dairy Sci. 2015;99:1315–30.

41. Cecchinato A, Cipolat-Gotet C, Casellas J, Penasa M, Rossoni A, Bittante G. Genetic analysis of rennet coagulation time, curd-firming rate, and curd firmness assessed over an extended testing period using mechanical and near-infrared instruments. J Dairy Sci. 2013;96:50–62.

42. Bittante G, Contiero B, Cecchinato A. Prolonged observation and modelling of milk coagulation, curd firming, and syneresis. Int Dairy J. 2013;29:115–23.

43. Cecchinato A, Albera A, Cipolat-Gotet C, Ferragina A, Bittante G. Genetic parameters of cheese yield and curd nutrient recovery or whey loss traits predicted using Fourier-transform infrared spectroscopy of samples collected during milk recording on Holstein, Brown Swiss, and Simmental dairy cows. J Dairy Sci. 2015;98:4914–27.

44. Ferragina A, Cipolat-Gotet C, Cecchinato A, Bittante G. The use of Fourier-transform infrared spectroscopy to predict cheese yield and nutrient recovery or whey loss traits from unprocessed bovine milk samples. J Dairy Sci. 2013;96:7980–90.

45. Bittante G, Ferragina A, Cipolat-Gotet C, Cecchinato A. Comparison between genetic parameters of cheese yield and nutrient recovery

or whey loss traits measured from individual model cheese-making methods or predicted from unprocessed bovine milk samples using Fourier-transform infrared spectroscopy. J Dairy Sci. 2014;97:6560–72.

46. Deeth HC. Lipoprotein lipase and lipolysis in milk. Int Dairy J. 2006;16:555–62.

47. Coulon JB, Delacroix-Buchet A, Martin B, Pirisi A. Relationships between ruminant management and sensory characteristics of cheeses: a review. Lait. 2004;84:221–41.

48. Marie C, Delacroix-Buchet A. Comparaison des variants A et C de la caséine β des laits de vaches tarentaises en modèle fromager de type Beaufort. II. Protéolyse et qualité des fromages. Lait. 1994;74:443–59.

49. Papoff C, Delacroix-Buchet A, Le Bars D, Campus R, Vodret A. Hydrolysis of bovine β-casein C by plasmin. Ital J Food Sci. 1995;7:157–68.

50. Soyeurt H, Colinet F, Arnould V, Dardenne P, Bertozzi C, Renaville R, et al. Genetic variability of lactoferrin content estimated by mid-infrared spectrometry in bovine milk. J Dairy Sci. 2007;90:4443–50.

51. Law B. Cheese-ripening and cheese flavour technology. In: Law BA, Tamime AY, editors. Technology of Cheesemaking. 2nd ed. Oxford: Wiley-Blackwell; 2010. p. 231–59.

52. Caroli AM, Chessa S, Erhardt GJ. Invited review: milk protein polymorphisms in cattle: effect on animal breeding and human nutrition. J Dairy Sci. 2009;92:5335–52.

15

Simultaneous fitting of genomic-BLUP and Bayes-C components in a genomic prediction model

Oscar O. M. Iheshiulor[1*], John A. Woolliams[1,2], Morten Svendsen[3], Trygve Solberg[3] and Theo H. E. Meuwissen[1]

Abstract

Background: The rapid adoption of genomic selection is due to two key factors: availability of both high-through-put dense genotyping and statistical methods to estimate and predict breeding values. The development of such methods is still ongoing and, so far, there is no consensus on the best approach. Currently, the linear and non-linear methods for genomic prediction (GP) are treated as distinct approaches. The aim of this study was to evaluate the implementation of an iterative method (called GBC) that incorporates aspects of both linear [genomic-best linear unbiased prediction (G-BLUP)] and non-linear (Bayes-C) methods for GP. The iterative nature of GBC makes it less computationally demanding similar to other non-Markov chain Monte Carlo (MCMC) approaches. However, as a Bayesian method, GBC differs from both MCMC- and non-MCMC-based methods by combining some aspects of G-BLUP and Bayes-C methods for GP. Its relative performance was compared to those of G-BLUP and Bayes-C.

Methods: We used an imputed 50 K single-nucleotide polymorphism (SNP) dataset based on the Illumina Bovine50K BeadChip, which included 48,249 SNPs and 3244 records. Daughter yield deviations for somatic cell count, fat yield, milk yield, and protein yield were used as response variables.

Results: GBC was frequently (marginally) superior to G-BLUP and Bayes-C in terms of prediction accuracy and was significantly better than G-BLUP only for fat yield. On average across the four traits, GBC yielded a 0.009 and 0.006 increase in prediction accuracy over G-BLUP and Bayes-C, respectively. Computationally, GBC was very much faster than Bayes-C and similar to G-BLUP.

Conclusions: Our results show that incorporating some aspects of G-BLUP and Bayes-C in a single model can improve accuracy of GP over the commonly used method: G-BLUP. Generally, GBC did not statistically perform better than G-BLUP and Bayes-C, probably due to the close relationships between reference and validation individuals. Nevertheless, it is a flexible tool, in the sense, that it simultaneously incorporates some aspects of linear and non-linear models for GP, thereby exploiting family relationships while also accounting for linkage disequilibrium between SNPs and genes with large effects. The application of GBC in GP merits further exploration.

Background

The rapid adoption of genomic selection (GS) is due to two key factors: (1) availability of high-throughput dense genotyping, and (2) availability of statistical methods to estimate and predict breeding values [1, 2]. The development of such methods is still ongoing and so far, there is

no consensus on the best approach. The methods available for genomic prediction (GP), can be broadly classified into two groups: linear and non-linear methods [3]. Genomic-best linear unbiased prediction (G-BLUP) is a typical example of a linear method, while the Bayesian methods such as Bayes-(A/B/C/etc.), are non-linear methods and often implemented by Markov chain Monte Carlo (MCMC) algorithms. A major difference between the linear and non-linear methods lies in their prior assumptions about the effects of the single-nucleotide

*Correspondence: oscar.iheshiulor@nmbu.no
[1] Department of Animal and Aquacultural Sciences, Norwegian University of Life Sciences, PO Box 5003, 1432 Ås, Norway
Full list of author information is available at the end of the article

polymorphisms (SNPs), which have been reviewed in detail by Neves et al. [4] and De Los Campos et al. [5]. Currently, linear and non-linear methods are treated as distinct approaches, and results from most empirical studies show that they yield similar prediction accuracies. However, in contrast, simulation studies reported significant differences between linear and non-linear methods [6, 7], an issue which was resolved by Daetwyler et al. [3] who demonstrated that the number of QTL (quantitative trait loci) in relation to the structure of the genome was a major factor in this discrepancy.

G-BLUP is commonly used for routine genetic evaluations because of its simple and less computationally demanding nature. Since Bayesian methods are often implemented by using MCMC algorithms, they are time consuming and computationally demanding when they deal with large numbers of SNPs. Hence, they are rarely used in routine genetic evaluations although they can potentially pick up and use SNPs with large effects or the actual causative variants. The need to reduce computational demands, while maintaining the features of Bayesian methods, has led to the development of iterative methods (non-MCMC-based Bayesian methods) such as the VanRaden's non-linear A/B [8], fastBayesB [9], MixP [10], or emBayesR [11] methods. These methods are iterative in nature hence computationally fast and yield prediction accuracies that are similar to those of MCMC-based Bayesian methods. However, they remain focused on exploiting linkage disequilibrium (LD) just as their MCMC-based counterparts.

GP uses two sources of information: genetic relationships among individuals and LD between SNPs and QTL [12, 13]. The emphasis put on these sources of information varies with the GP method used. G-BLUP through the genomic relationship matrix (**G**) exploits the relationship in a given population more comprehensively than the pedigree-based relationship matrix (**A**), both by quantifying the variation in relationships between sibs and the historical relationships between individuals in the base generation of **A** [12, 14, 15]. However, compared to G-BLUP, non-linear methods can better exploit the LD information gained through mapping of QTL [12, 13]. Thus, methods that could exploit both genetic relationships and LD might help to increase prediction accuracy and the persistency of the accuracy across time and genetic distance.

Our aim was to develop an iterative method (referred to as GBC) that combines relationship information using the G-BLUP approach with information on the LD between QTL and neighboring SNPs using the Bayes-C [16] approach of GP. In a sense, GBC shares the Bayes-A property of including all SNPs in the prediction [7] but implies different prior assumptions on the effects. Given

the importance of reducing computational demands when dealing with large numbers of SNPs, GBC follows the iterative approach of other non-MCMC-based methods but differs from both MCMC- and non-MCMC-based Bayesian methods by combining aspects of G-BLUP and Bayes-C methods for GP. We evaluated GBC using an imputed 50 K SNP chip dataset. Furthermore, predictions from GBC were compared to those from G-BLUP and Bayes-C, using real data from a population of genotyped bulls.

Methods

Phenotypes

Daughter yield deviations (DYD; [17]) on 3244 proven Norwegian Red bulls and their associated effective number of daughters (d_e; i.e. weighted number of daughters for each bull) were obtained from GENO SA (http://www.geno.no). These were extracted from the routine genetic evaluations of 2013 for three production traits, fat yield (Fkg), milk yield (Mkg) and protein yield (Pkg), and a health indicator, somatic cell count (SCC). The DYD is an estimate of the average performance of each bull's daughters, corrected for all fixed and non-genetic random effects of the daughters and genetic effects of the bulls' mates [17]. The minimum d_e was 108 and the average d_e was 177 with a standard deviation of ~31. The reliabilities of the DYD were calculated following Fikse and Banos [18] as $r^2_{DYD} = d_e/(d_e + K)$, where $K = (4 - h^2)$ and h^2 is the heritability of the trait used in the evaluations. The parameters used for each trait and average reliabilities for each trait are in Table 1. The average reliability between bulls ranged from 0.858 for SCC to 0.927 for Mkg.

Genotypes

Genotyping data were also provided by Geno SA for these bulls. Bulls were previously genotyped with different SNP chips: 2450 bulls with the 25 K Affymetrix chip (Affymetrix Inc., Santa Clara, CA), 1650 were genotyped with the Illumina Bovine50K BeadChip (Illumina Inc., San Diego, CA), and 856 were genotyped with both.

Quality control was carried out by CIGENE (http://www.cigene.no) and is described in detail by Solberg

Table 1 Heritability (h^2) and average reliability (r^2_{DYD}) of daughter yield deviations for the 3244 bulls

Trait	h^2	r^2_{DYD}
Somatic cell count (SCC)	0.136	0.858
Fat yield (Fkg)	0.213	0.906
Milk yield (Mkg)	0.277	0.927
Protein yield (Pkg)	0.235	0.915

$r^2_{DYD} = d_e/(d_e + K)$, where d_e is the effective number of daughters and $K = (4 - h^2)/h^2$

et al. [19]. Briefly, quality control was carried out post-genotyping within each set of SNP chip data so that animals with an individual call rate lower than 97% and SNPs with a call rate lower than 25% were removed. Pedigree relationships between parent and offspring were set to missing if they exceeded a Mendelian error threshold of 1%; following this, SNPs with an overall Mendelian error rate higher than 2.5% were deleted; and, for parent–offspring pairs with Mendelian errors less than 1%, SNP genotypes that were flagged as errors were set to missing. Finally, SNPs with a minor allele frequency lower than 0.05 were discarded.

We used genotype imputation to obtain ~50 K SNP genotypes. The genotypes of bulls obtained with the 25 K Affymetrix chip were imputed to the SNP density of the Illumina Bovine50K BeadChip. Genotype imputation was performed by CIGENE (http://www.cigene.no) using Beagle v3.3.1 [20] and other in-house developed software as described by Solberg et al. [19]. Following these procedures, the data contained 48,249 SNPs on 3244 bulls. SNPs that were not mapped to the bovine reference genome assembly UMD 3.1 [21] and those on the X chromosome were not included in the analyses.

Reference and validation sets

Bulls were divided into reference and validation sets following a standard animal breeding selection scheme, so that the validation dataset consisted of the 124 youngest sires born between January 1st 2007 and December 31st 2008 with a minimum of 100 actual daughters. The reference set included bulls born between 1964 and 2005 with all performance records contributing to the DYD collected before January 1st 2007, for a total of 3091 bulls. To check relationships between reference and validation sets following Clark et al. [22] and Daetwyler et al. [23], four measures of genomic relatedness were calculated from the genomic relationship according to VanRaden's method 1 [8]. For each bull these measures were: (1) the mean relationship with the reference population (mean-Rel); (2) the maximum relationship (Relmax); (3) the mean of the five largest absolute relationships (Rel5); and (4) the mean of the ten largest absolute relationships (Rel10).

Genomic prediction methods and data analysis

Three methods were implemented for GP: G-BLUP, Bayes-C, and GBC. Genetic and error variances used in the analyses were estimated from the dataset using ASReml v3.0 [24].

G-BLUP

The G-BLUP model [7, 8] used to predict genomic estimated breeding values (GEBV) was as follows:

$$\mathbf{y} = \mathbf{1}\mu + \mathbf{Zg} + \mathbf{e}, \tag{1}$$

where $\mathbf{y}$ is a vector of DYD for the reference set; $\mathbf{1}$ is a vector of ones; μ is the overall mean; $\mathbf{Z}$ is a design matrix that relates the records to genomic values; $\mathbf{g}$ is a vector of genomic values assumed to follow a multivariate normal distribution $MVN \sim (0, \sigma_g^2 \mathbf{G})$, where $\mathbf{G}$ is the genomic relationship matrix and σ_g^2 is the genetic variance; and $\mathbf{e}$ is the vector of residuals assumed to follow a multivariate normal distribution $MVN \sim (0, \sigma_e^2 \mathbf{I})$. $\mathbf{G}$ was calculated, following VanRaden's method 1 [8] using all bulls, as $\mathbf{G} = \mathbf{MM'}/2\sum p_j(1 - p_j)$, and $M_{ij} = x_{ij} - 2p_j$, where x_{ij} is the genotype of bull i for SNP j, with $x_{ij} = 0$, 1 or 2 for the reference homozygote, heterozygote and alternative homozygote, respectively, and p_j is the allele frequency of the alternative allele of SNP j for all bulls.

Bayes-C

Bayes-C, a sub-model of GBC (i.e. where the variance explained by the GBLUP term in GBC is set to zero), was also independently evaluated so that the relative performance of both approaches can be compared. Bayes-C assumes that a fraction $(1 - \pi)$ of the SNPs has zero effects and that the distribution of the effects for the other fraction (π) is normal [16]. Thus, the model of analysis for Bayes-C is:

$$\mathbf{y} = \mathbf{1}\mu + \mathbf{ZMQq} + \mathbf{e}, \tag{2}$$

where $\mathbf{M}$ is the design matrix of scaled SNP genotypes as in the calculation of $\mathbf{G}$ above; $\mathbf{Q}$ is a diagonal matrix with indicators on the diagonal that are 1 if the SNP has an effect (with prior probability π) and 0 if it has no such effect (with prior probability $(1 - \pi)$; $\mathbf{q}$ is a vector of SNP effects (q_j) assumed to be normally distributed, i.e. $q_j \sim N\left(0, \sigma_q^2\right)$ with probability π and 0 otherwise. All other model elements are defined as previously. The π values used were estimated from the dataset via a search between 1% and then 5 to 30% in increments of 5% to obtain the optimal π values. The GEBV for the validation animals was calculated as $\mathbf{M}_v\hat{\mathbf{q}}$ where $\mathbf{M}_v$ describes the scaled genotypes for each bull in the validation set, and $\hat{\mathbf{q}}$ is the posterior mean of the SNP effects. Bayes-C analyses were performed using the GS3 software [25]. The number of iterations was 20,000 with a burn-in of 2000 and a thinning interval of 100. Using 50,000 or 100,000 iterations with a burn-in of 10,000 or 20,000 had no impact on the accuracy of prediction but increased computing time.

GBC

This method fits a Bayes-C model [16] simultaneously with an effect due to background genes following a GBLUP model. This was achieved by using the iterative conditional expectation (ICE) algorithm [9], to which was added a correction for the uncertainty of the other effects of SNPs when deciding whether SNP j has an effect or not as described by Wang et al. [11]. The ICE algorithm uses the expectation/mean instead of the posterior mode, mainly because the posterior distribution is often bimodal, and when both modes are about equally high, the mode of the distribution is rather an arbitrary choice. The model of analysis used by GBC is:

$$\mathbf{y} = \mathbf{1}\mu + \mathbf{ZMQq} + \mathbf{Zg} + \mathbf{e}, \tag{3}$$

where $\mathbf{g}$ is a vector of residual breeding values with distributional assumptions as described above for G-BLUP. All other elements of the model are defined as previously. The π values were estimated from the dataset via a search between 1% and then 5 to 30% in increments of 5% to obtain the optimal π values.

The G-BLUP term was implemented as described in the section on G-BLUP, but here, it is called the residual breeding value because it represents the breeding value after the SNPs with the largest effects have been fitted through the Bayes-C term. In the Bayes-C term, the SNPs with a large effect were assumed to have a variance of $0.001\,\sigma_g^2$ as implemented here. Optimal π values and the fraction of genetic variance explained by the SNPs with a large effect in GBC are assessed by cross-validation.

Posterior probabilities of SNPs with a large effect in GBC

The posterior probability that a SNP j has a large effect is calculated from:

$$PostProb(Q_{jj} = 1) = \frac{PPR_j * LR_j}{PPR_j * LR_j + 1},$$

where PPR_j is the prior-probability-ratio ($=\pi(1-\pi)$); and LR_j is the likelihood ratio that SNP j has a large effect. The $\log(LR_j)$ equals the log-likelihood of a model with versus without the effect of SNP j (see Appendix for a derivation):

$$\log(LR_j) = \frac{1}{2}\log(\lambda) - \frac{1}{2}\log\left(m_j'm_j + \lambda\right)$$
$$+ \frac{1}{2}(y^{*\prime}m_jm_j'y^* + m_j'PEVm_j)\sigma_e^{-2}/\left(m_j'm_j + \lambda\right),$$

where $\lambda = \sigma_e^2/\sigma_q^2$; m_j are the scaled genotypes of SNP j for animals with records; y^* are the records corrected for all other effects in the model except that of SNP j; PEV is the prediction error variance matrix of the G-BLUP model; and the $m_j'PEVm_j/\left(m_j'm_j + \lambda\right)$ term corrects for the uncertainty about the other genetic effects in the model [11].

The effect of SNP j now becomes:

$$\hat{q}_j = PostProb(Q_{jj} = 1) * m_j'y^*/\left(m_j'm_j + \lambda\right),$$

where the $m_j'y^*/\left(m_j'm_j + \lambda\right)$ term equals the BLUP solution of the SNP effect when it has a large effect.

Predictive ability

The primary criterion for evaluating predictive ability was the accuracy of the predictions (r), calculated as the correlation between GEBV and DYD, divided by the square root of the average reliability of the DYD for the trait $\left(\sqrt{r_{DYD}^2}\right)$. The bias of predictions was calculated as the unweighted regression of DYD on the predicted values, where a regression coefficient of 1 denotes no bias, less than 1 implies that the spread of the GEBV is too large, and more than 1 implies their spread is too small.

Standard errors of the prediction accuracies and the regression coefficients on the DYD were computed using a custom bootstrapping R-script in R software [26]. The bootstrap procedure involved sampling with replacement of the GEBV 10,000 times. For each bootstrap sample, pairs of GEBV-DYD of an animal in the validation population are sampled with replacement, i.e. the connection between a specific GEBV and DYD is maintained in this sampling process. The resulting GEBV were correlated to the DYD, and standard errors were computed from the 10,000 bootstrap estimates of accuracy and bias. A Hotelling–Williams test [27] for dependent correlations was used to determine whether differences between the validation correlations using alternative methods were statistically significant.

Results

Genomic relatedness between validation and reference individuals

Table 2 shows the average genomic relatedness between reference individuals and between validation and reference individuals. Overall meanRel was equal to 0.03, while estimated Relmax between the validation and reference population was ~0.5, which suggests that nearly all the bulls in the validation population were closely related to the reference population (i.e. their sire is in the reference

Table 2 Average of four measures of genomic relatedness

Relatedness	meanRel	Relmax	Rel5	Rel10
Within reference	0.03 (0.01)	0.49 (0.04)	0.34 (0.05)	0.30 (0.05)
Between validation and reference	0.03 (0.00)	0.48 (0.09)	0.29 (0.05)	0.24 (0.05)

Standard deviations are in parentheses

Here, meanRel is the average relationship $(1/N_P)\sum_{j=1}^{N_p} rel(i,j)$, where N_P is the number of individuals in the reference population, $rel(i,j)$ is the relationship between validation i and reference individual j; Relmax is the maximum $(rel(i,j))$ for individual i over all reference individuals j; Rel5 is $(1/5)\sum_{j=1}^{N_p} x_{ij}rel(i,j)$, where $x_{ij} = 1$ if j is among the top 5 (i,j) for individual i and Rel10 is the extension to the top 10 relationships for i

population). For Rel5 and Rel10, genomic relatedness estimates of 0.29 and 0.24, respectively, were obtained.

Prediction methods

Table 3 shows the accuracies of predictions using alternative prediction methods. Accuracies across the four traits ranged from 0.602 to 0.716 for G-BLUP, from 0.604 to 0.733 for Bayes-C, and from 0.607 to 0.731 for GBC. The highest accuracy was found for Fkg across the three methods. Apart from the trait Fkg for which GBC resulted in a statistically significant higher accuracy than G-BLUP using the Hotelling–Williams test (P < 0.05), we observed that, although not significant, in most cases the accuracies obtained with GBC were higher than with G-BLUP and Bayes-C. Generally, on average across the four traits, G-BLUP yielded the lowest prediction accuracy while GBC yielded the highest prediction accuracy. GBC yielded a 0.009 and 0.006 increase in prediction accuracy over G-BLUP and Bayes-C, respectively. The regression coefficients (Table 4) ranged from 0.881 to 0.956 for SCC, from 1.259 to 1.326 for Fkg, from 1.435 to 1.530 for Mkg, and from 1.410 to 1.506 for Pkg. Regression coefficients differed slightly across methods.

Effects of SNPs: Bayes-C and GBC

The effects of SNPs estimated by Bayes-C and GBC are in Figs. 1, 2, 3 and 4. For Fkg, GBC picked up two SNPs with a large effect on chromosomes 5 and 12. The effects of the other SNPs were substantially shrunk towards 0. With Bayes-C, the same SNPs were observed to have large effects but several other SNPs with small to moderate effects were also found. For Mkg, we observed a similar trend, i.e. GBC identified SNPs on chromosomes 6, 12, and 28 with a large effect while Bayes-C also identified SNPs on chromosomes 6 and 12 as well as other SNPs.

Table 3 Accuracy (SE) of the predicted values for the youngest sires based on the different prediction methods

Trait$_{(π)}$	G-BLUP	Bayes-C	GBC
SCC $_{(20\%, 20\%)}$	0.602 (0.066)	0.604 (0.064)	0.607 (0.065)
Fkg $_{(10\%, 10\%)}$	0.716 (0.049)	0.733 (0.042)	0.731 (0.047)
Mkg $_{(10\%, 10\%)}$	0.705 (0.051)	0.701 (0.050)	0.719 (0.048)
Pkg $_{(10\%, 1\%)}$	0.695 (0.053)	0.689 (0.050)	0.696 (0.051)
Average	0.679	0.682	0.688

Accuracy $= \frac{corr(DYD, GEBV)}{\sqrt{r^2_{DYD}}}$

SE: standard errors computed from 10,000 bootstrap samples

G-BLUP: genomic BLUP using genomic-based relationship matrix; Bayes-C: a non-linear method that fits zero effects and normal distributions of effects for SNPs; GBC: an iterative method that fits a G-BLUP next to SNP effects with a Bayes-C prior

SCC, somatic cell count; Fkg, fat yield; Mkg, milk yield; Pkg, protein yield

π refers to the optimal π values (i.e. proportion of SNP having large effects) when using Bayes-C and GBC

Table 4 Bias (SE) of the predicted values for the youngest sires based on the different prediction methods

Trait	G-BLUP	Bayes-C	GBC
SCC	0.881 (0.111)	0.956 (0.120)	0.881 (0.109)
Fkg	1.275 (0.120)	1.326 (0.131)	1.259 (0.113)
Mkg	1.530 (0.146)	1.435 (0.136)	1.459 (0.136)
Pkg	1.506 (0.157)	1.410 (0.149)	1.461 (0.100)

Bias: measured as the regression of daughter yield deviation on the predicted values

SE: standard errors computed from 10,000 bootstrap samples

G-BLUP: genomic BLUP using genomic-based relationship matrix; Bayes-C: a non-linear method that fits zero effects and normal distributions of effects for SNPs; GBC: an iterative method that fits a G-BLUP next to SNP effects with a Bayes-C prior

SCC, somatic cell count; Fkg, fat yield; Mkg, milk yield; Pkg, protein yield

Chromosome 6 was also identified by both methods as a region that carries SNPs with a large effect on Pkg. In the case of SCC, there were many SNPs with (very) small effects across the genome as indicated by both methods especially with GBC.

Computing time and memory usage

Table 5 shows the computing time and memory usage for each method. With an Intel(R) Xeon(R) CPU E5-2670 0 @ 2.60 GHz, G-BLUP took on average 2.51 min with average memory usage of about 2197 MB to complete the analysis, Bayes-C took on average 1.10 h with average memory usage of about 1296 MB, while GBC took on average 4.2 min with average memory usage of 2474 MB. Generally, across the four traits studied, G-BLUP was fastest, followed closely by GBC in terms of computing time while in terms of memory usage Bayes-C used less memory compared to G-BLUP and GBC.

Discussion

GP uses mainly two sources of information: genetic relationships between individuals and LD between SNPs and QTL [12, 13]. The contribution of both information sources to prediction in a given population can vary across generations with relationships decaying across generations while LD may remains fairly persistent [12, 13]. Currently, these sources are included separately in the linear (i.e. G-BLUP) and non-linear (i.e. Bayesians) GP methods. While G-BLUP tries to exploit relationships maximally, the Bayes-(A/B/C/etc.) methods try to use LD between individual SNPs and genes maximally. To take advantage of both methods as well as to maintain short computing times, we developed and evaluated an iterative GP method, i.e. GBC that combines relationship information using the G-BLUP approach with information on LD between QTL and neighboring SNPs using

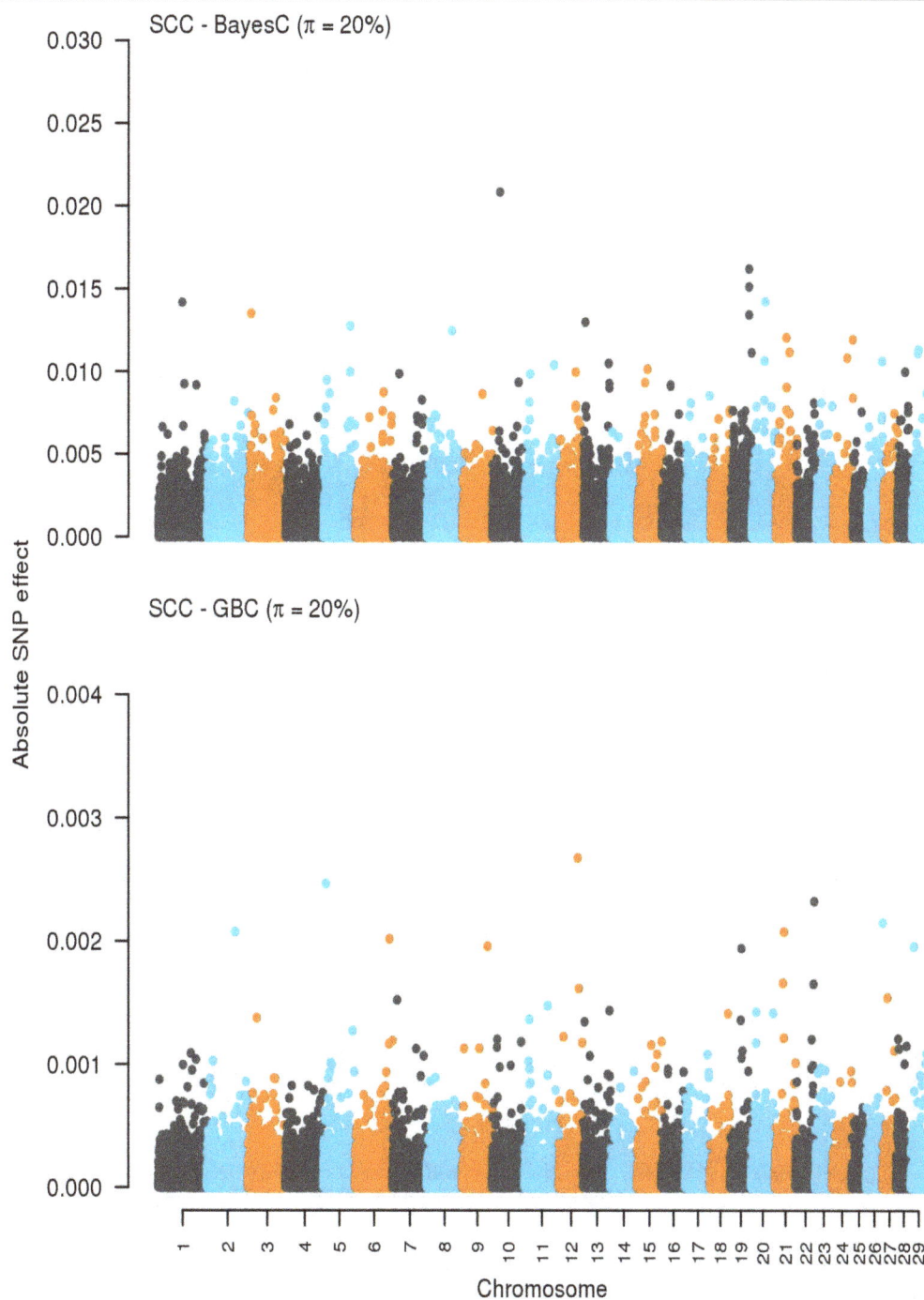

Fig. 1 Effects of SNPs estimated by using Bayes-C and GBC for somatic cell count (SCC). The absolute values of the estimates of the effects of SNPs are on the *y axis*. The *X axis* is ordered by chromosomes from 1 to 29. π refers to the optimal π value when using Bayes-C and GBC. Absolute values were standardized by $\sqrt{\sigma_g^2}$. Standardization was only for plotting purpose

the Bayes-C approach. Comparisons were made with the commonly used G-BLUP, which does not select SNPs, and Bayes-C, a non-linear method that assumes zero effects for a fraction of the SNPs and a normal distribution of the effects for the other fraction. Our results show that simultaneously fitting a GBLUP and a Bayes-C term can improve accuracy over G-BLUP and Bayes-C, alone. In terms of computational speed, GBC was much faster than a MCMC-based version of Bayes-C but used more memory compared to GBLUP and Bayes-C.

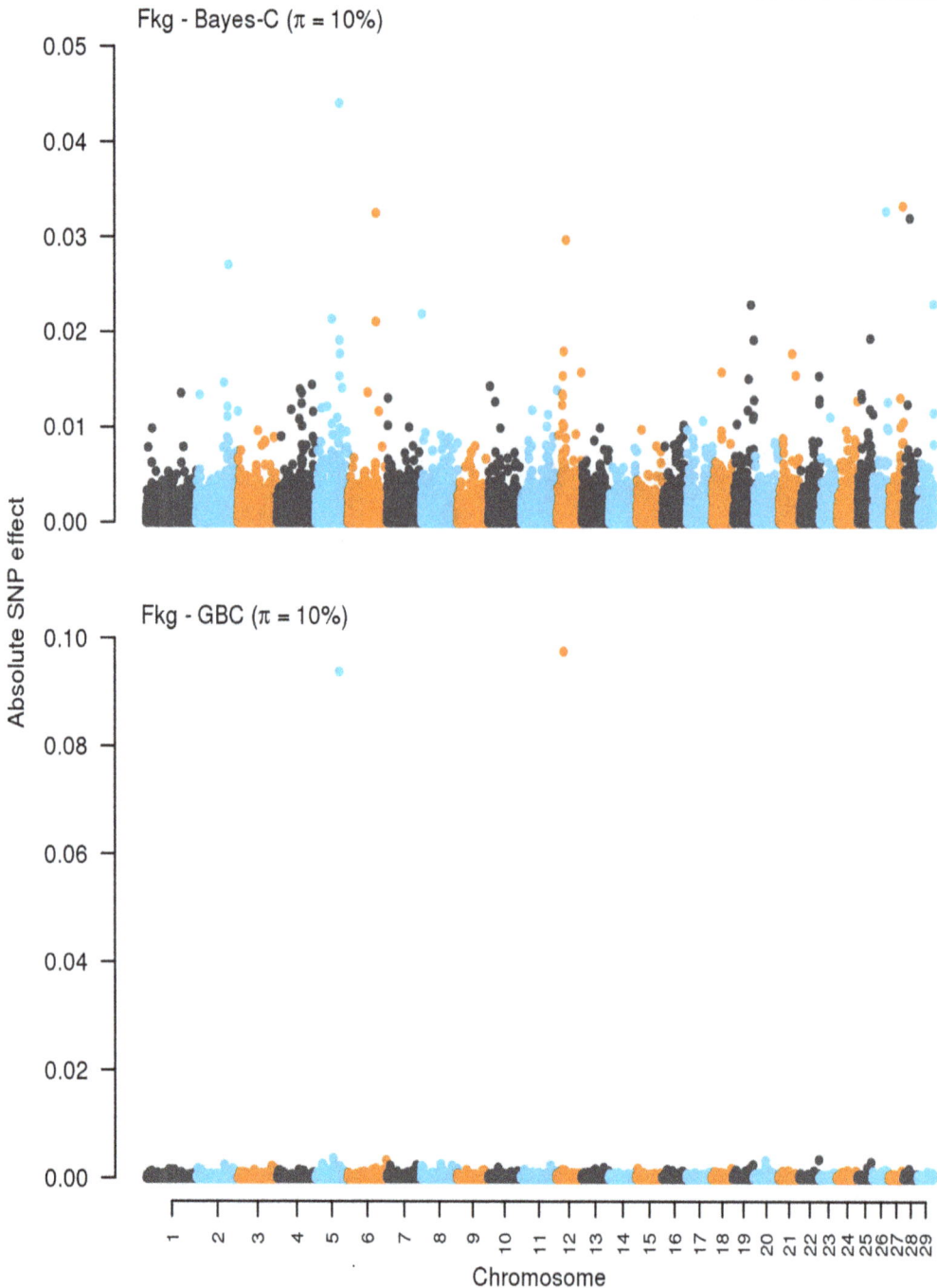

Fig. 2 Effects of SNPs estimated by using Bayes-C and GBC for fat yield (Fkg). The absolute values of the estimates of the effects of SNPs are on the *y axis*. The *X axis* is ordered by chromosomes from 1 to 29. π refers to the optimal π value when using Bayes-C and GBC. Absolute values were standardized by $\sqrt{\sigma_g^2}$. Standardization was only for plotting purpose

Prediction methods

In this study, we compared our new method GBC to two existing GP methods: G-BLUP and Bayes-C. Generally, on average across the four traits, GBC yielded a 0.009 and 0.006 increase in prediction accuracy over G-BLUP and

Bayes-C, respectively. With GBC, we anticipated that, by fitting a residual SNP term in addition to Bayes-C SNP effects, both models would complement each other: the G-BLUP term mainly picking up effects that could be explained by linkage analysis [12] and the Bayes-C term

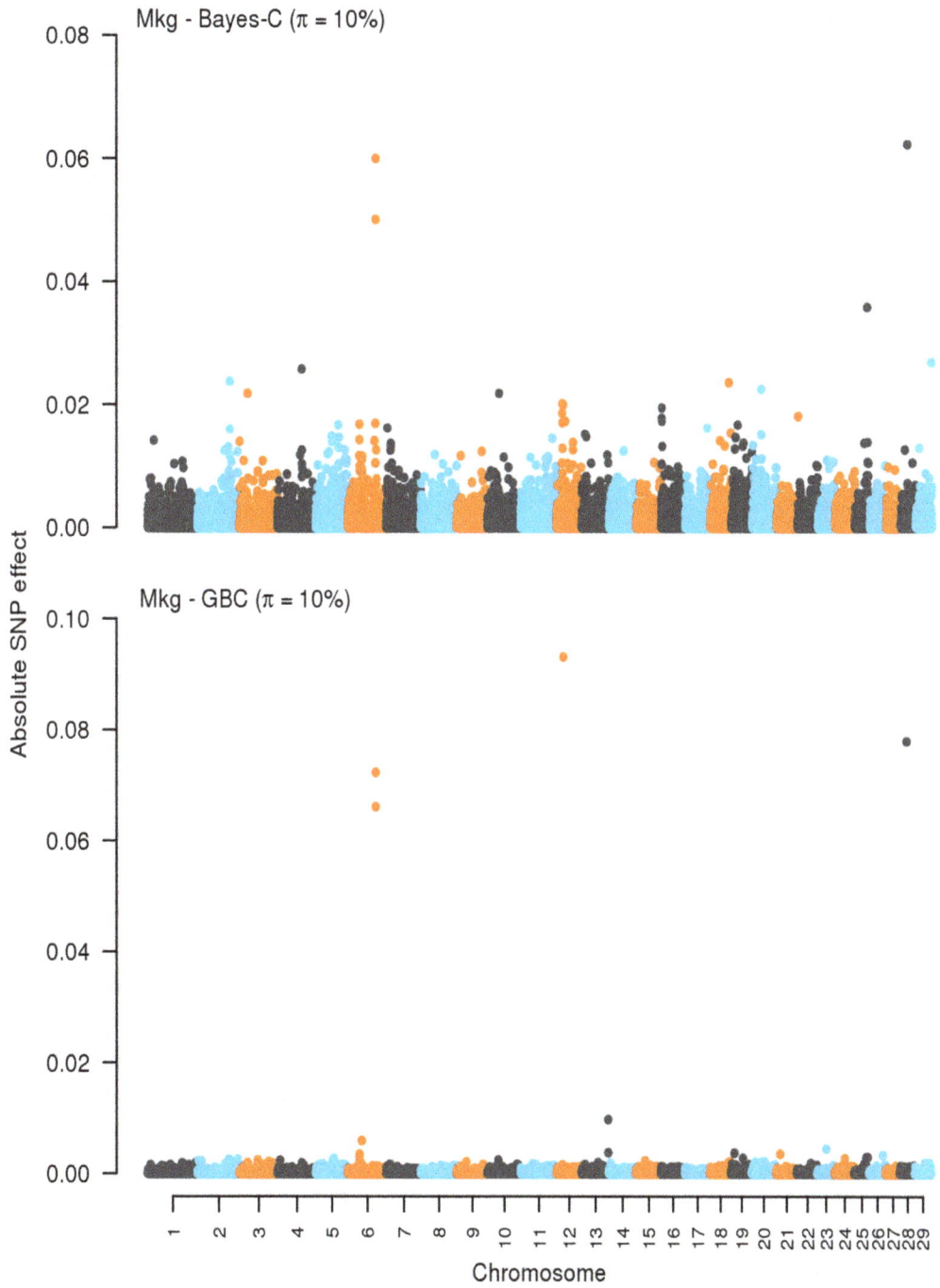

Fig. 3 Effects of SNPs estimated by using Bayes-C and GBC for milk yield (Mkg). The absolute values of the estimates of the effects of SNPs are on the y axis. The X axis is ordered by chromosomes from 1 to 29. π refers to the optimal π value when using Bayes-C and GBC. Absolute values were standardized by $\sqrt{\sigma_g^2}$. Standardization was only for plotting purpose

picking up tight LD between SNPs and genes. Consequently, we expected GBC to result in a higher accuracy of GP. Although the results agreed with this expectation, differences were small and were generally not statistically significant.

The GBC method has some similarity with Bayes-A [7], i.e. both methods fit all SNPs in the model while differentiating between SNPs with a large variance and SNPs with a small variance. Habier et al. [16] observed that Bayes-A performed marginally better than G-BLUP

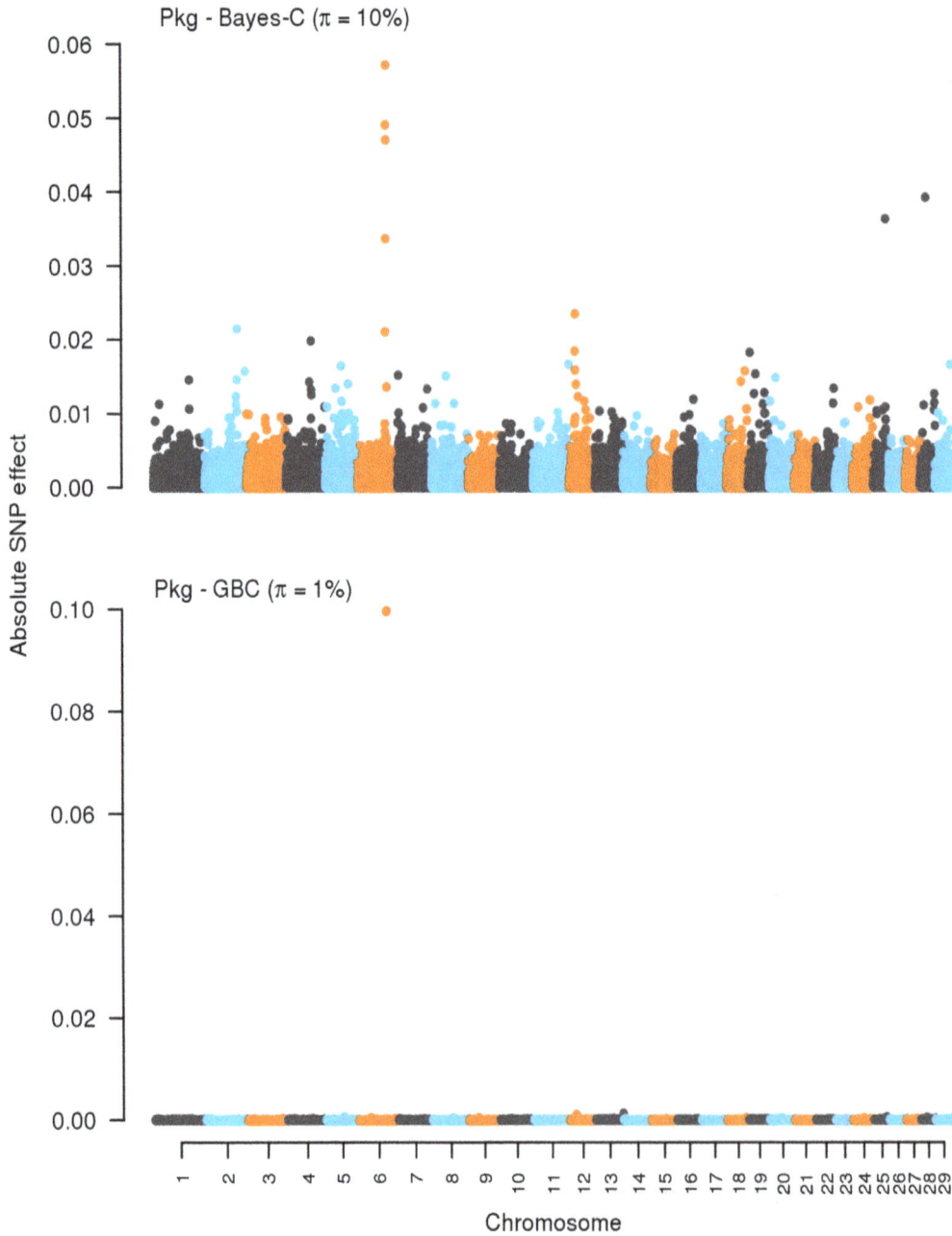

Fig. 4 Effects of SNPs estimated by using Bayes-C and GBC for protein yield (Pkg). The absolute values of the estimates of the effects of SNPs are on the *y axis*. The *X axis is* ordered by chromosomes from 1 to 29. *π* refers to the optimal *π* value when using Bayes-C and GBC. Absolute values were standardized by $\sqrt{\sigma_g^2}$. Standardization was only for plotting purpose

and Bayes-C with real data. GBC performed always marginally better than G-BLUP and Bayes-C, except for Fkg where it performed significantly better than G-BLUP. A possible explanation could be that the modeled LD blocks that surround the major genes were large and were also reasonably well captured by G-BLUP [3]. In addition, as shown in Table 2, the relationships between the animals in the validation set and those in

the reference set were generally high in our dataset. In such a case, the performance of GBC is only marginally better than that of G-BLUP and Bayes-C modeled independently. This suggests that GBC works well if the level of relationships is high. If there are no relationships or if relationships decay across generations while LD remains fairly persistent, GBC also has the potential to yield more persistent accuracies across generations since it models

Table 5 Computing time of the different prediction methods for each trait

Method	SCC	Fkg	Mkg	Pkg
G-BLUP	00:03:24 (2233.492 MB)	00:02:18 (2233.492 MB)	00:02:31 (2159.716 MB)	00:02:31 (2159.716 MB)
Bayes-C	01:04:06 (1296.312 MB)	01:10:32 (1296.312 MB)	01:14:41 (1296.312 MB)	01:10:36 (1296.312 MB)
GBC	00:03:04 (2474.432 MB)	00:04:51 (2474.436 MB)	00:05:11 (2474.436 MB)	00:04:14 (2474.436 MB)

Memory usage is in parentheses

G-BLUP: genomic BLUP using genomic-based relationship matrix; Bayes-C: a non-linear method that fits zero effects and normal distributions of effects for SNPs; GBC: an iterative method that fits a G-BLUP next to SNP effects with a Bayes-C prior

SCC somatic cell count; *Fkg*, fat yield, *Mkg*, milk yield; *Pkg*, protein yield

both information sources simultaneously. Practically, the number of QTL or major genes that underlie a trait remains largely unknown and so is the variance explained by the genes with large effects. Examples of genes with a large effect are reported in the literature, e.g. the *diacylglycerol O-acyltransferase 1* (*DGAT1*) gene involved in fat percentage in dairy cattle [28]. The GBC model assumes that some genes with a large effect can be detected based on LD while the effects of the background genes are predicted by genomic relationships, which seems to be mainly true for Fkg.

GBC, as mentioned earlier, simultaneously fits Bayes-C and G-BLUP components. This means that all SNPs are included twice in the model, first in the Bayes-C term, and second in the G-BLUP term. The G-BLUP term includes all the SNPs to explain genomic relationships between the animals, whereas the Bayes-C term answers the question whether a SNP might explain more variance than expected based on its contribution to genomic relationships (where all SNPs contribute equally). Thus, GBC opens the opportunity for SNPs with extra-large effects to be included twice in the model and thereby increasing their weight in the GP.

The regression coefficients in Table 4 are a measure of bias of the GEBV predictions. Except for the trait SCC for which the regression coefficients were lower than 1, they were above 1 for the other three traits across all methods. This implies that, for the production traits, the variance of GEBV was deflated while for SCC, it was inflated. Since, in Norwegian Red cattle, selection pressure against directly recorded mastitis is strong and mastitis is quite highly correlated to SCC, biased GP is expected for SCC (a bivariate analysis that would fit both mastitis and SCC might avoid such bias). For the production traits (Fkg, Mkg and Pkg), all investigated methods also yielded biased GEBV, which is probably due to these 124 validation bulls being under strong selection for these traits. Considering that the regression coefficients from the pedigree-BLUP (result not shown) showed similar biases, one may attribute the biases to intrinsic aspects of the data such as selection.

Effects of SNPs: Bayes-C and GBC

A key difference between Bayes-C and GBC lies in how they estimate and deal with the effects of SNPs. Bayes-C assumes a priori zero effects for a fraction $(1 - \pi)$ of the SNPs and a normal distribution of effects for the other fraction (π) [16]. GBC fits a Bayes-C like prior for the SNPs with large effects assuming that an estimated fraction π of the SNPs have a large effect with a variance of $0.001 \, \sigma_g^2$ (this proportion can differ across traits) and then it fits a G-BLUP component for all SNPs. With GBC, all SNPs have an estimated effect, thus, in a sense, GBC methods share the Bayes-A property of including all SNPs in the prediction [7] but their prior assumptions about SNP effects differ. As shown in Figs. 1, 2, 3 and 4, the methods behaved differently in terms of number of SNPs with effects and their magnitude. However, interestingly for the production traits, both methods found common SNPs with large effects on chromosomes 5, 6, and 12. We did not try to identify candidate genes in these regions, as this was outside the scope of our study. Nevertheless, several genome-wide association studies (GWAS) have reported that these chromosomes harbor QTL that affect production traits in dairy cattle [29–32]. In the case of SCC, both methods showed no clear pattern with many SNPs having very small effects.

There was a general tendency that GBC allocated large effects to (very) few SNPs while Bayes-C identified many more SNPs with moderate to large effects. This implies that, with GBC, only SNPs in high LD with the QTL tend to pick up the genes with major effects while the others are treated as residual SNP effects. Bayes-C also needs to capture SNP genetic relationships by fitting SNPs with large effects, and thus needs to fit more SNPs. Therefore, the observed differences in prediction accuracies between both methods are a reflection of how genomic regions with large and small effects are treated. The ability of GBC to not neglect any SNP effect may explain why it tended towards higher accuracies than Bayes-C. It seems that, the GBC method is also very precise in pointing towards QTL locations. This could be due to GBC showing some similarity to GWAS methods in which a single SNP and a G-BLUP term are fitted, with the G-BLUP term correcting the QTL signal for family relationships. Although the prior distribution of the effects of SNPs and the actual proportion of variance they explain remain unknown, the results of this study indicate that the assumed prior distribution for the effects of SNPs alongside the proportion of variance they explain in GBC tends to yield somewhat higher accuracy than the assumptions underlying Bayes-C.

Impact of the assumed variance for SNPs with a large effect in GBC

In this study, we assumed that, across the four traits studied, the SNPs with a large effect explained 0.1% of the genetic variance in GBC. This corresponds to the genetic variance explained by the third distribution in Bayes-R [33]. Bayes-R assumes that the effects of SNPs are derived from a mixture of four different normal distributions, each explaining 0, 0.01, 0.1, or 1% of the genetic variance, respectively. In our study, we set the genetic variance explained by SNPs to 0.1% because we considered that it was an intermediate value between that of genes with a small or a large effect. In addition, we did not want a situation where (very) few SNPs with a large effect explained a larger proportion of the genetic variance since most traits in livestock are polygenic. However, the proportion of genetic variance explained by the SNPs with a large effect might differ across traits. To investigate the impact of alternative assumptions on the variance explained for genes with a large effect, we also investigated a situation in which genes with a large effect explained 1% of the genetic variance. Assuming that SNPs with a large effect in GBC explained 1% of the genetic variance led to a 0.002 and 0.007 increase in prediction accuracy for Fkg and Mkg, respectively (result not shown), whereas for SCC and Pkg, it led to a 0.003 and 0.001 decrease in prediction accuracy, respectively. These results suggest that, in GBC, the optimal proportion of the genetic variance explained by SNPs with a large effect in GBC varies with traits. However, as also shown by the results, deviation from a 0.1% genetic variance explained by SNPs with a large effect seems to have little impact on prediction accuracy in GBC.

GBC and other non-MCMC-based Bayesian methods

On the one hand, GBC shares some similarity with other non-MCMC-based Bayesian methods in the sense that it uses an iterative approach. A key advantage of the iterative methods (non-MCMC-based methods) over MCMC-based methods is their shorter computing time. Non-MCMC-based Bayesian methods such as fastBayesB [9], MixP [10], emBayesR [11] or VanRaden's non-linear method [8] among others are computationally several orders faster than their MCMC counterparts. This is because generally, non-MCMC-based methods require much fewer iterations compared to MCMC-based methods. In agreement with the aforementioned studies, our results demonstrated the faster computing time of GBC compared to MCMC implementations of e.g. Bayes-C (Table 5). On the other hand, GBC differs from the aforementioned non-MCMC-based methods in that it simultaneously incorporates aspects of G-BLUP and Bayes-C methods for GP and thereby making it flexible for exploiting information of genomic data. In addition, unlike most other non-MCMC-based methods, GBC adds a correction for the uncertainty of other SNP effects when deciding whether a particular SNP has an effect or not as recommended by Wang et al. [11]. Not accounting for these uncertainties could result in a decline of about 8 to 9% in accuracy of prediction as demonstrated by Wang et al. [11].

Conclusions

We introduced and evaluated the GBC method for GP, which simultaneously fits G-BLUP and Bayes-C terms. The method was evaluated by using imputed 50 K SNP datasets and its relative performance was compared to G-BLUP and Bayes-C. GBC showed marginal advantages over G-BLUP and Bayes-C for most of the traits in terms of prediction accuracy. For Fkg, GBC performed significantly better than G-BLUP, which agrees with the fact that Fkg is controlled by a few genes with a large effect. Overall in our study, statistically, GBC did not significantly outperform G-BLUP and Bayes-C probably due to a high level of relationship between reference and validation individuals. However, it is a flexible tool in the sense that it simultaneously incorporates some aspects of both linear and non-linear models for GP, thereby exploiting family relationships while also accounting for LD between SNPs and genes with a large effect. Computationally, GBC was much faster than Bayes-C with a computational speed that is comparable to that of G-BLUP. The application of GBC in GP merits further exploration.

Authors' contributions
OOMI performed the study and drafted the manuscript. JAW contributed to writing the draft and revised the manuscript critically. MS extracted the data needed for the analysis and the links between information sources. TS helped in the preparation of the data. THEM coordinated the whole study, wrote the GBC program and contributed in the writing of the manuscript. All authors read and approved the final manuscript.

Author details
[1] Department of Animal and Aquacultural Sciences, Norwegian University of Life Sciences, PO Box 5003, 1432 Ås, Norway. [2] The Roslin Institute (Edinburgh), Royal (Dick) School of Veterinary Studies, University of Edinburgh, Midlothian EH25 9RG, Scotland, UK. [3] GENO SA, Holsegata 22, 2317 Hamar, Norway.

Acknowledgements
The research leading to these results has received funding from the European Union's Seventh Framework Programme for research, technological development and demonstration under Grant Agreement No. 289592 - Gene2Farm. Neither the European Commission nor the partners of the Gene2Farm project can be held responsible for views expressed in this manuscript. The authors thank Geno SA (Ås, Norway) for providing the datasets and CIGENE (Ås, Norway) for the quality control and genotype imputation. The vital comments of two anonymous reviewers and the associate editor are gratefully acknowledged.

Competing interests
The authors declare that they have no competing interests.

Appendix: Log likelihood ratio of a SNP having a normally distributed effect versus no effect

The general form of the multivariate normal probability density is:

$$P(\mathbf{y}|\mu, \mathbf{V}) \propto |\mathbf{V}|^{-\frac{1}{2}} \exp\left[-\frac{1}{2}(\mathbf{y} - \mu)'\mathbf{V}^{-1}(\mathbf{y} - \mu)\right],$$

where $\mathbf{y}$ is a vector of multivariate normally distributed variables with mean μ and (co)variance matrix $\mathbf{V}$.

First, we assume a model without a SNP effect and assume that $\mathbf{V} = \mathbf{I}\sigma_e^2$, and $\mu = 0$ for simplicity (assuming the actual data are corrected for all other effects in the model except for the putative SNP effect). The log-likelihood of this null-model becomes:

$$\mathrm{Log}L(0) = -\frac{1}{2}n \log\left(\sigma_e^2\right) - \frac{1}{2}\frac{\mathbf{y}'\mathbf{y}}{\sigma_e^2}.$$

For the alternative model with a SNP effect, we have $\mathbf{V} = \left(\mathbf{I}\sigma_e^2 + \mathbf{mm}'\sigma_q^2\right)$, where $\mathbf{m}$ is a vector of SNP genotypes of the animals with records, and σ_q^2 is the variance of the SNP effect. The $\log|\mathbf{V}|^{-1/2}$ term becomes:

$$-\frac{1}{2}\log|\mathbf{V}| = -\frac{1}{2}\log\left[(\sigma_e^2)^n\left(\sigma_q^2\right)\left(\frac{\mathbf{m}'\mathbf{m}}{\sigma_e^2} + \frac{1}{\sigma_q^2}\right)\right]$$

$$= -\frac{1}{2}n\log\left(\sigma_e^2\right) + \frac{1}{2}\log(\lambda) - \frac{1}{2}\log(\mathbf{m}'\mathbf{m} + \lambda),$$

where $\lambda = \sigma_e^2/\sigma_q^2$.

Following Woodbury [34], the inverse of $\mathbf{V}$ can be written as:

$$\mathbf{V}^{-1} = (\mathbf{I}\sigma_e^2 + \mathbf{mm}'\sigma_q^2)^{-1} = \mathbf{I}\sigma_e^{-2} - \frac{\sigma_e^{-2}\mathbf{mm}'}{(\mathbf{m}'\mathbf{m} + \lambda)}.$$

Such that $\mathbf{y}'\mathbf{V}^{-1}\mathbf{y}$ becomes:

$$\mathbf{y}'\mathbf{V}^{-1}\mathbf{y} = \frac{\mathbf{y}'\mathbf{y}}{\sigma_e^2} - \frac{\mathbf{y}'\mathbf{mm}'\mathbf{y}}{(\mathbf{m}'\mathbf{m} + \lambda)\sigma_e^2}.$$

The log-likelihood of the alternative model with a SNP effect thus becomes:

$$\mathrm{Log}L(a) = -\frac{1}{2}n\log\left(\sigma_e^2\right) + \frac{1}{2}\log(\lambda)$$
$$- \frac{1}{2}\log(\mathbf{m}'\mathbf{m} + \lambda) - \frac{1}{2}\frac{\mathbf{y}'\mathbf{y}}{\sigma_e^2} + \frac{1}{2}\frac{\mathbf{y}'\mathbf{mm}'\mathbf{y}}{(\mathbf{m}'\mathbf{m} + \lambda)\sigma_e^2}.$$

Taking the log-likelihood ratio of the alternative to the null-model yields:

$$\mathrm{Log}L(a) - \mathrm{Log}L(0)$$
$$= \frac{1}{2}\log(\lambda) - \frac{1}{2}\log(\mathbf{m}'\mathbf{m} + \lambda) + \frac{1}{2}\frac{\mathbf{y}'\mathbf{mm}'\mathbf{y}}{(\mathbf{m}'\mathbf{m} + \lambda)\sigma_e^2}.$$

Following Wang et al. [11], we account for the fact that the correction of the data $\mathbf{y}$ was not performed using the true value of all other effects in the model, but using estimates of these effects, which results in an estimate of $\mathbf{y}$ denoted by $\mathbf{y}^*$. The variance of $\mathbf{y}^*$ given the real data $\mathbf{y}$ is denoted by the prediction error covariance matrix PEV, which is assumed approximately equal to the PEV matrix from the G-BLUP model. Accounting for this uncertainty due to prediction error variances of $\mathbf{y}^*$, the expectation of the $\mathbf{y}'\mathbf{mm}'\mathbf{y}$ term is:

$$E\left(\mathbf{y}'\mathbf{mm}'\mathbf{y}\right) = \mathbf{y}^{*'}\mathbf{mm}'\mathbf{y}^* + trace\left(PEV\,\mathbf{mm}'\right)$$
$$= \mathbf{y}^{*'}\mathbf{mm}'\mathbf{y}^* + \mathbf{m}'PEV\,\mathbf{m}.$$

The expectation of the log-likelihood ratio thus becomes:

$$\mathrm{Log}(LR) = \frac{1}{2}\log(\lambda) - \frac{1}{2}\log(\mathbf{m}'\mathbf{m} + \lambda)$$
$$+ \frac{1}{2}\frac{\mathbf{y}^{*'}\mathbf{mm}'\mathbf{y}^* + \mathbf{m}'PEV\,\mathbf{m}}{(\mathbf{m}'\mathbf{m} + \lambda)\sigma_e^2}.$$

References
1. Goddard ME, Hayes BJ, Meuwissen TH. Using the genomic relationship matrix to predict the accuracy of genomic selection. J Anim Breed Genet. 2011;128:409–21.
2. Meuwissen THE, Hayes BJ, Goddard ME. Accelerating improvement of livestock with genomic selection. Annu Rev Anim Biosci. 2013;1:221–37.
3. Daetwyler HD, Pong-Wong R, Villanueva B, Woolliams JA. The impact of genetic architecture on genome-wide evaluation methods. Genetics. 2010;185:1021–31.
4. Neves HH, Carvaheiro R, Queiroz SA. A comparison of statistical methods for genomic selection in a mice population. BMC Genet. 2012;13:100.
5. De Los Campos G, Hickey JM, Pong-Wong R, Daetwyler HD, Calus MP. Whole-genome regression and prediction methods applied to plant and animal breeding. Genetics. 2013;193:327–45.
6. Meuwissen THE. Accuracy of breeding values of 'unrelated' individuals predicted by dense SNP genotyping. Genet Sel Evol. 2009;41:35.
7. Meuwissen THE, Hayes BJ, Goddard ME. Prediction of total genetic value using genome-wide dense marker maps. Genetics. 2001;157:1819–29.

8. VanRaden PM. Efficient methods to compute genomic predictions. J Dairy Sci. 2008;91:4414–23.

9. Meuwissen THE, Solberg TR, Shepherd R, Woolliams JA. A fast algorithm for BayesB type of prediction of genome-wide estimates of genetic value. Genet Sel Evol. 2009;41:2.

10. Yu X, Meuwissen THE. Using the Pareto principle in genome-wide breeding value estimation. Genet Sel Evol. 2011;43:35.

11. Wang T, Chen YP, Goddard ME, Meuwissen THE, Kemper KE, Hayes BJ. A computationally efficient algorithm for genomic prediction using a Bayesian model. Genet Sel Evol. 2015;47:34.

12. Habier D, Fernando RL, Dekkers JCM. The impact of genetic relationship information on genome-assisted breeding values. Genetics. 2007;177:2389–97.

13. Habier D, Tetens J, Seefried FR, Lichtner P, Thaller G. The impact of genetic relationship information on genomic breeding values in German Holstein cattle. Genet Sel Evol. 2010;42:5.

14. Odegard J, Meuwissen THE. Identity-by-descent genomic selection using selective and sparse genotyping. Genet Sel Evol. 2014;46:3.

15. Odegard J, Moen T, Santi N, Korsvoll SA, Kjoglum S, Meuwissen THE. Genomic prediction in an admixed population of Atlantic salmon (*Salmo salar*). Front Genet. 2014;5:402.

16. Habier D, Fernando RL, Kizilkaya K, Garrick DJ. Extension of the Bayesian alphabet for genomic selection. BMC Bioinformatics. 2011;12:186.

17. VanRaden PM, Wiggans GR. Derivation, calculation, and use of national animal model information. J Dairy Sci. 1991;74:2737–46.

18. Fikse WF, Banos G. Weighting factors of sire daughter information in international genetic evaluations. J Dairy Sci. 2001;84:1759–67.

19. Solberg TR, Heringstad B, Svendsen M, Grove H, Meuwissen THE. Genomic predictions for production and functional traits in Norwegian Red from BLUP analyses of imputed 54 K and 777 K SNP data. Interbull Bull. 2011;44:240–3.

20. Browning BL, Browning SR. A unified approach to genotype imputation and haplotype-phase inference for large data sets of trios and unrelated individuals. Am J Hum Genet. 2009;84:210–23.

21. Zimin AV, Delcher AL, Florea L, Kelley DR, Schatz MC, Puiu D, Hanrahan F, Pertea G, Van Tassell CP, Sonstegard TS, et al. A whole-genome assembly of the domestic cow, *Bos taurus*. Genome Biol. 2009;10:R42.

22. Clark SA, Hickey JM, Daetwyler HD, van der Werf JH. The importance of information on relatives for the prediction of genomic breeding values and the implications for the makeup of reference data sets in livestock breeding schemes. Genet Sel Evol. 2012;44:4.

23. Daetwyler HD, Calus MP, Pong-Wong R, de Los Campos G, Hickey JM. Genomic prediction in animals and plants: simulation of data, validation, reporting, and benchmarking. Genetics. 2013;193:347–65.

24. Gilmour AR, Gogel BJ, Cullis BR, Thompson R. ASReml User Guide Release 3.0. In. VSN International Ltd, Hemel Hempstead, HP1 1ES, UK; 2009.

25. Legarra A, Ricard A, Filangi O. GS3: genomic selection, Gibbs Sampling, Gauss Seidel (and BayesCπ). http://genoweb.toulouse.inrafr/~alegarra/gs3_folder/. (2011).

26. Core Team R. R: a language and environment for statistical computing. Vienna: Austria R Foundation for Statistical Computing; 2015.

27. Steiger JH. Tests for comparing elements of a correlation matrix. Psychol Bull. 1980;87:245–51.

28. Grisart B, Coppieters W, Farnir F, Karim L, Ford C, Berzi P, Cambisano N, Mni M, Reid S, Simon P, et al. Positional candidate cloning of a QTL in dairy cattle: identification of a missense mutation in the bovine DGAT1 gene with major effect on milk yield and composition. Genome Res. 2002;12:222–31.

29. Cole JB, Wiggans GR, Ma L, Sonstegard TS, Lawlor TJ Jr, Crooker BA, Van Tassell CP, Yang J, Wang S, Matukumalli LK, et al. Genome-wide association analysis of thirty one production, health, reproduction and body conformation traits in contemporary U.S. Holstein cows. BMC Genomics. 2011;12:408.

30. Meredith BK, Kearney FJ, Finlay EK, Bradley DG, Fahey AG, Berry DP, Lynn DJ. Genome-wide associations for milk production and somatic cell score in Holstein-Friesian cattle in Ireland. BMC Genet. 2012;13:21.

31. Nayeri S, Sargolzaei M, Abo-Ismail MK, May N, Miller SP, Schenkel F, Moore SS, Stothard P. Genome-wide association for milk production and female fertility traits in Canadian dairy Holstein cattle. BMC Genet. 2016;17:75.

32. Raven L-A, Cocks BG, Hayes BJ. Multibreed genome wide association can improve precision of mapping causative variants underlying milk production in dairy cattle. BMC Genomics. 2014;15:62.

33. Erbe M, Hayes BJ, Matukumalli LK, Goswami S, Bowman PJ, Reich CM, Mason BA, Goddard ME. Improving accuracy of genomic predictions within and between dairy cattle breeds with imputed high-density single nucleotide polymorphism panels. J Dairy Sci. 2012;95:4114–29.

34. Woodbury MA. Inverting modified matrices. In: J. Kuntzmann, editor. Memorandum report, vol. 42: Statistical Research Group. Princeton University, Princeton, NJ; 1950. P. 4.

16

Application of a Bayesian dominance model improves power in quantitative trait genome-wide association analysis

Jörn Bennewitz[1]*⬛, Christian Edel[2], Ruedi Fries[3], Theo H. E. Meuwissen[4] and Robin Wellmann[1]

Abstract

Background: Multi-marker methods, which fit all markers simultaneously, were originally tailored for genomic selection purposes, but have proven to be useful also in association analyses, especially the so-called BayesC Bayesian methods. In a recent study, BayesD extended BayesC towards accounting for dominance effects and improved prediction accuracy and persistence in genomic selection. The current study investigated the power and precision of BayesC and BayesD in genome-wide association studies by means of stochastic simulations and applied these methods to a dairy cattle dataset.

Methods: The simulation protocol was designed to mimic the genetic architecture of quantitative traits as realistically as possible. Special emphasis was put on the joint distribution of the additive and dominance effects of causative mutations. Additive marker effects were estimated by BayesC and additive and dominance effects by BayesD. The dependencies between additive and dominance effects were modelled in BayesD by choosing appropriate priors. A sliding-window approach was used. For each window, the R. Fernando window posterior probability of association was calculated and this was used for inference purpose. The power to map segregating causal effects and the mapping precision were assessed for various marker densities up to full sequence information and various window sizes.

Results: Power to map a QTL increased with higher marker densities and larger window sizes. This held true for both methods. Method BayesD had improved power compared to BayesC. The increase in power was between −2 and 8% for causative genes that explained more than 2.5% of the genetic variance. In addition, inspection of the estimates of genomic window dominance variance allowed for inference about the magnitude of dominance at significant associations, which remains hidden in BayesC analysis. Mapping precision was not substantially improved by BayesD.

Conclusions: BayesD improved power, but precision only slightly. Application of BayesD needs large datasets with genotypes and own performance records as phenotypes. Given the current efforts to establish cow reference populations in dairy cattle genomic selection schemes, such datasets are expected to be soon available, which will enable the application of BayesD for association mapping and genomic prediction purposes.

Background

With the advent of dense single nucleotide polymorphisms (SNP) panels, it has become possible to exploit linkage disequilibrium (LD) between SNPs and genes that are involved in complex or quantitative trait variation, with the aim to map genes that underlie trait variation

and to predict genomic values [1]. Genome-wide association studies (GWAS) scan the genome systematically to identify SNPs that are significantly associated with the trait of interest. Various methods to conduct GWAS are available. Single-SNP analyses are widely used, where one SNP is tested at a time for significance. The SNP genotype is usually treated as a fixed effect in a mixed linear model. Correction for the effects of population structure is done by fitting simultaneously a random polygenic term in the model [2]. For each SNP, a test statistic and

*Correspondence: j.bennewitz@uni-hohenheim.de
[1] Institute of Animal Science, University of Hohenheim, 70593 Stuttgart, Germany
Full list of author information is available at the end of the article

an error probability for the trait association is obtained in a 'frequentist' manner, which can conveniently be used for post-GWAS analyses, such as false discovery rate calculations [3], or for meta-analyses, e.g. [4]. However, with dense SNP panels, the level of multiple-testing can be enormous, which needs a very stringent significance threshold in order to prevent an inflation of type one errors. In addition, the effect of a gene can be captured only in part by a single marker due to imperfect LD, but might be better explained by using jointly the SNPs that surround the gene. To overcome these problems, multi-marker methods have been proposed, which fit all SNPs simultaneously as random effects in the model [5]. These models were originally tailored for genomic prediction or selection purposes [6], but have proven to be useful also in association analyses [7]. The simulation study of Sahana et al. [8] revealed that Bayesian multi-marker association analysis has a higher power than single-SNP analysis. These authors used also a window-based approach, where consecutive SNPs within 1 cM were used to build a window. Inference was drawn for each window by considering these window SNPs jointly. Legarra et al. [9] compared linkage and linkage disequilibrium analysis (known as LDLA), single-marker mixed model association analysis, and Bayesian whole-genome association analysis using a real data structure. They did not report a clear superiority of one method, but recommended to apply more than one method to real data.

In Bayesian analysis, inference on unknowns is drawn from their posterior distributions. Recently, Fernando and Garrick [5] and Fernando et al. [10] developed a method to control false positive results in multi-marker Bayes GWAS, which can be straightforwardly implemented in MCMC-based algorithms. This method controls the proportion of false positives by calculating the posterior probability of association of a trait with each SNP or each window of consecutive SNPs.

To the best of our knowledge, the Bayesian models mentioned above consider only additive gene effects. However, dominance is a non-negligible source of complex trait variation [11–13]. Bolormaa et al. [14] used a large-scale experiment with about 10,000 bovine individuals, which were phenotyped for 16 traits and genotyped with dense SNP panels. They conducted a GWAS using single-marker regression analysis and found many trait-associated SNPs that had a dominance effect. Moreover, the estimated dominance variance across the traits was between 0 and 42% of the phenotypic variance, with a median of 5%. It is well known that additive and dominance effects are dependent in a complicated manner [15], as described in the next section.

Verbyla et al. [16, 17] proposed a Bayesian stochastic search variable selection method, which was named BayesC by these authors. In a recent study, we extended this BayesC method by accounting for dominance, resulting in the BayesD method, with the three sub-models BayesD1, 2 and 3 [18]. These sub-models differ in the way dependencies between additive and dominance effects are modelled. Simulation studies showed that these models increased the accuracy of predicted genetic values by about 15%. Moreover, application on a real dairy cattle dataset revealed that the use of these BayesD models increased the prediction accuracy of cow's yield deviations for milk fat yield compared to a G-BLUP analysis without dominance [19]. Hence, it seems worthwhile to investigate the use of BayesD models in association analyses also. Therefore, the aim of our study was to compare the power and precision of BayesD and BayesC in a GWAS. The analysis was conducted on simulated datasets, using a protocol that simulated many segregating genes and accounted for the dependencies between additive and dominant gene effects. The models were also applied to a real dairy cattle dataset.

Methods
Simulation protocol
A forward simulation approach was used to generate a Fisher-Wright diploid population with a genome that consisted of one chromosome that was 1 Morgan (M) long. The mutation rate was 10^{-8}/bp/meiosis. In total, 1051 generations were generated. For the first 650 generations, the effective population size was Ne = 600, and then decreased to 100 in the following 350 generations, with a fast decline in the last generations. This decline was chosen in order to create an LD pattern as observed in bovine populations [20]. From generation 1000 to 1051, the Ne remained constant at 100. In the last generation, the sample size was N = 1500. According to the scaling by Ne and genome length argument given in [21], a simulated population with an N of 1500 and an Ne of 100 corresponds to a population of 45,000 individuals with a genome of 30 M, with an Ne of 100, or of 450,000 individuals with an Ne of 1000. In total, ten populations were simulated. The average number of segregating SNPs (minor allele frequency (MAF) higher than 0.01) in the last generation was about 7k. From these 7k SNPs, 2k, 1k and 0.5k SNPs were chosen based on a MAF higher than 0.03 and equal distances between SNPs. This corresponds to a marker density of 20Ne, 10Ne and 5Ne per Morgan. Scaling this to a population of individuals with a genome of 30 M and an Ne of 1000, it reflects marker densities of 600k, 300k, and 150k, respectively. If the Ne is equal to 100, this results in marker densities of 60k, 30k, and 15k, respectively. Note that these scaling arguments were derived for genomic predictions [21]. In our study, it was assumed that these scaling arguments also

approximately hold for GWAS using genomic prediction models.

For each population, five traits were simulated as follows. In the last generation, 15 of the 7k SNPs with a MAF higher than 0.05 were randomly selected as causal mutations. The minimum distance between two quantitative trait loci (QTL) was 2 cM. The additive effect (a) represents half the difference between the alternative homozygous genotypes and the dominance effect (d) represents the deviation of the heterozygous genotype from the mean of the two alternative homozygous genotypes. The distribution of the dominance coefficient $h = d/|a|$ was assumed to be $h \sim N(0.2, 0.3^2)$ [22]. The distribution of a was $a|h \sim N(0, \exp(3h))$ [15]. A scatterplot of these distributions is in Fig. 1, which shows that alleles with small additive effects had high variable dominance effects. As the size of the additive effect increased, the dominance coefficient was likely to become more positive and, on average, increased in size. Hence, on average, the genetic value of heterozygous genotypes was above the mean of the two alternative homozygous genotypes. Moreover, overdominance was a rare event. This pattern reflects the results of Caballero and Keightley [23], who found that genes with large additive effects likely have larger dominance coefficients. It is also in agreement with the well-known Kacser–Burns model [24], and with theoretical derivations about the contribution of dominance to the variation of quantitative traits [15].

From these simulated genotype values, the breeding values, dominance deviations, and genetic values of the individuals were calculated, following the derivations in Falconer and Mackay [25]. For an individual with genotype x (x representing the number of copies of the mutant allele at the causative mutation, $x = 0$, 1, or 2), the breeding value (BV) is:

$$BV(x) = \sum_{j=1}^{Q} (x_j - 2p_j)\alpha_j, \tag{1}$$

where $\alpha_j = a_j + (q_j - p_j)d_j$ is the substitution effect, p_j the frequency of the mutant allele, $q_j = 1 - p_j$, and Q is the number of simulated causative mutations (i.e. 15). The dominance deviation (DV) is:

$$DV(x) = \sum_{j=1}^{Q} -d_j x_j (x_j - 1 - 2p_j) - 2p_j^2 d_j. \tag{2}$$

The genetic value (GV) is:

$$GV(x) = \sum_{j=1}^{Q} (a_j + (2 - x_j)d_j)x_j. \tag{3}$$

The additive genetic variance was calculated as the variance of the BV and the dominance variance as the variance of the DV. The residuals were sampled from a normal distribution with mean zero and a residual variance chosen such that the narrow sense heritability was equal to 0.3 for each trait. The mean (median) of the dominance variance as the proportion of the phenotypic variance was equal to 0.1 (0.08), but varied between 0.01 and 0.29 across the simulated traits. The expected inbreeding depression was calculated as $\sum_{j=1}^{Q} 2p_j q_j d_j$ and was on average 0.023 times the phenotypic standard deviation across all simulated traits.

We simulated 50 replicates. For each replicate, we chose four marker densities, i.e. 7k, 2k, 1k, and 0.5k. The causal mutations were removed, except for the 7k dataset. For this dataset, no selection on SNPs was conducted (except for MAF), and hence, it mimics a situation where the full sequence is known and the causal mutations are among the full set of SNPs.

Bayesian models

The three BayesD sub-models of Wellmann and Bennewitz [18] differ in the way the complicated relationships between additive and dominance effects are modelled. Sub-model BayesD3 performed best in their simulation study, followed by BayesD2. However, BayesD2 showed better mixing properties in the MCMC analysis. Since the number of iterations is an issue in the application of these models, we used BayesD2 in this study, which will be thereafter named simply BayesD. BayesD1 was used in some preliminary analyses, in which it was slightly

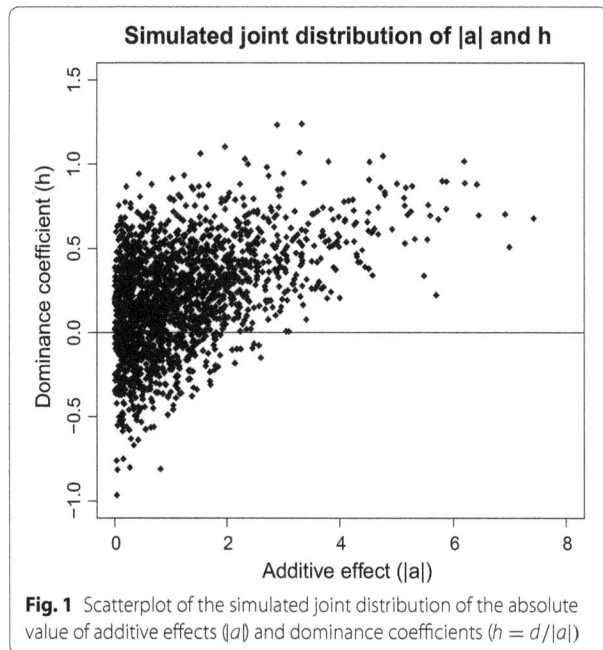

Fig. 1 Scatterplot of the simulated joint distribution of the absolute value of additive effects ($|a|$) and dominance coefficients ($h = d/|a|$)

less powerful than BayesD2 (Wellmann and Bennewitz, unpublished results). Therefore, it was not included in this study.

A full description of BayesD can be found in [18]. Only essential issues will be described here. The following general linear regression model was applied for BayesD:

$$\mathbf{y} = \mathbf{1_n}\mu + \mathbf{X\tilde{a}} + \mathbf{W\tilde{d}} + \mathbf{E},$$

where $\mathbf{y}$ is the vector of the observations of the n individuals, $\mathbf{1_n}$ is a vector of n ones and μ is the general mean. Vector $\tilde{\mathbf{a}}$ contains the random additive and vector $\tilde{\mathbf{d}}$ the random dominance effects of the M SNPs. The SNP genotypes are coded as '0 0', '0 1', and '1 1'. $\mathbf{X}$ is a known $N \times M$ matrix and contains the number of copies of 1-alleles at the SNPs for each individual, i.e. the gene content. $\mathbf{W}$ is a known $N \times M$ indicator matrix, which is 1 if the individual is heterozygous at the SNP and 0 otherwise. Errors $\mathbf{E}$ are assumed normally distributed. Let $\tilde{\theta}_j$ be the effect of SNP j, with $\tilde{\theta}_j = \left(\tilde{a}_j, \tilde{d}_j \right)$. It was assumed that the distribution of $\tilde{\theta}_j$ is a mixture of two distributions (F) that differ only by a scaling factor ε. The form of the distribution F is described in detail in [18]. Conditional on a Bernoulli-distributed indicator variable γ_j, we have $\tilde{\theta}_j | \gamma_j \sim \gamma_j F + \left(1 - \gamma_j \right) \varepsilon F$. Thus, if $\gamma_j = 1$, the marker effect comes from the distribution that has the larger variance. The prior probability of a marker being important (i.e. $\gamma_j = 1$) was pLD. The distribution of the absolute value of additive effects was assumed to follow a folded t-distribution. The distribution of the dominance effect was assumed normal conditional on the absolute additive effect. The prior distribution of the dominance effects was conditional on the additive effects and was specified such that the absolute additive effects and dominance coefficients $(h = d/|a|)$ are independent, which implies that the probability that a dominance effect is much larger in magnitude than the additive effect is small. As a result, presence of overdominance $(h > 1)$ is a rare but non negligible event. The prior probability of the sign of the additive effects depends on the allele frequency. This probability was chosen such that it is unlikely that the genetic variance of the gene is large. This is the assumption made in BayesD, because selection has shifted allele frequencies away from values for which the contribution of a gene to the genetic variance is large [18]. This prior was not used in this study, because no selection was simulated.

For the joint posterior distribution and the Markov chain, which was generated by Gibbs sampling, see [18]. BayesD is an extension of the BayesC method of Verbyla et al. [16, 17] towards accounting for dominance. Hence, the model shown above was also applied for BayesC, but without the dominance term and assuming $\tilde{\theta}_j = \tilde{a}_j$.

We chose $\varepsilon = 0.01$ and 2.5 degrees of freedom for the t-distribution of the additive effects for both BayesC and BayesD. A single MCMC chain was run for 20,000 cycles, discarding the first 10,000 as burn-in, using the R-package BayesDsamples (can be obtained from R. Wellmann). Every 100th sample, the additive and dominance (only for BayesD) effects were stored for inference purposes. The simulated variance components and expected inbreeding depression were used in the models as input parameters. Parameter pLD was chosen such that the expected number of SNPs coming from the distribution with the larger variance approached the number of QTL when the size of the SNP panel approached the total number of SNPs in the genome. We used the following calculations for pLD for analysis of the simulated data. For $t = 1, 2, 3, 4.8$, we have a marker density of $M = 250 \times 2^t$, i.e. 0.5k, 1k, 2k, and 7k. We chose $pLD = 5.5 \times 0.7^t \times 15/M$. This results in a pLD of 0.116, 0.040 and 0.014 for $t = 1, 2$, and 3 respectively, and 0.0021 for $t = 4.8$. This results in an average number of SNPs per QTL of $5.5 \times 0.7^t = 3.85$, 2.70, 1.89 and 1 for $t = 1, 2, 3$ and 4.8 respectively. The same pLD was used for both Bayes models.

Inference of association

For inference of association, we used the posterior probability of the window association (WPPA) criterion proposed by Fernando and Garrick [5] and Fernando et al. [10], using a sliding window approach. The window size was varied between 0.25, 0.5, and 1 cM. For each of the stored estimates of the SNP effects from the MCMC chain in a window w, the genomic variance was computed as follows:

$$\widehat{\sigma_{g_w}^2} = \sum_{j \in w} \left[H_j \widehat{\alpha_j^2} + H_j^2 \widehat{d_j^2} \right],$$

where $H_j = 2p_j q_j$ is the heterozygosity of SNP j, $\widehat{\alpha_j} = \widehat{a_j} + (q_j - p_j)\widehat{d_j}$ is the estimated substitution effect of SNP j, and $\widehat{a_j}$ and $\widehat{d_j}$ are the corresponding estimated additive and dominance SNP effects. For BayesC, these computations were done without including estimated dominance effects, because they were not estimated in BayesC. Note that the estimated effects can be inserted into Eqs. (1) to (3) if estimated genomic breeding values, estimated genomic dominance deviations or estimated genetic values are of interest. The summation in Eqs. (1) to (3) is then over M SNPs instead of over Q causative mutations.

For each window, the ratio q_w is calculated as:

$$q_w = \widehat{\sigma_{g_w}^2} / E\left(\sigma_{g_w}^2 \right),$$

where $E(\sigma_{g_w}^2) = \sum_{j \in w} [H_j E(\alpha^2) + H_j^2 E(d^2)]$.

Expectations $E(\alpha^2)$ and $E(d^2)$ were derived under the assumption of an equal distribution of the total genetic variance across the genome, as shown in the "Appendix". Ratios q_w were computed for both BayesD and BayesC by using the same definition and calculation of $E\left(\sigma^2_{g_w}\right)$ (i.e. the same value was used for both BayesC and BayesD). If $q_w > 1$, this indicates the presence of a causative mutation within window w with an effect greater than expected under the assumption of an equal distribution of the genetic variance across the genome. Hence, causative mutations with effects below this expectation are in general not detectable. The posterior distribution of q_w was approximated by the distribution of the q_w obtained from the stored MCMC samples of the additive and dominance effects. The WPPA was calculated by counting the number of samples for which $q_w > 1$ and dividing this by the total number of samples saved. The following three levels of WPPA were considered: 0.85, 0.95, and 0.99. According to [5, 10], this controls the proportion of false positives (PFP) at ≤ 0.15, ≤ 0.05, and ≤ 0.01, respectively.

A causal gene is mapped if at least one window within a region of 1 cM surrounding the gene shows a WPPA above the threshold. For each simulated trait, power to map a causal gene was calculated by dividing the number of mapped causative genes by the number of simulated causative genes that showed an effect greater than expected under the assumption that the genetic variance is distributed equally across the genome. Hence, very small causative genes were not counted. This guarantees that the upper bound of the power is 1, as it should be. From the definition of q_w, it becomes clear that this would not be the case if all simulated causative genes were counted. In addition, power was calculated only for causative genes that explained more than 2.5% of the simulated genetic variance, which is denoted as large-power (L-power). Mapping precision was calculated for each simulated trait as the size of the genome around a mapped causative gene with significant sliding window(s).

Application to a real dataset

The models were also applied to a Fleckvieh dairy cattle dataset, which is described in detail by Ertl et al. [12]. In brief, the dataset included 1996 cows with yield deviations (YD) for milk production and conformation traits. The YD observations were based on test-day observations adjusted for non-genetic effects, but not for the permanent environmental effect. The cows were genotyped with the Illumina BovineHD genotyping BeadChip. After quality control, 629,028 SNPs remained in the dataset. This dataset was a subset of the data used for prediction purposes [19]. In that study, milk fat yield was chosen to compare G-BLUP and BayesD with regard to their

ability to predict a cow's YD accurately. For this trait, it is known that the *DGAT1* gene has a major effect in this population and dominance is important [12]. Therefore, this trait was also chosen in our study. The narrow sense heritability was equal to 0.47 and the dominance variance was equal to 0.18 as a proportion of phenotypic variance [12]. Parameter *pLD* was set to 0.05 and 2.5 degrees of freedom were chosen for the *t*-distribution of the additive effects [19].

Results

Simulated datasets

Results of the power evaluations from the simulated datasets are in Table 1. Standard deviations are also included in this table, from which the standard errors can be calculated, if desired. Power decreased as the WPPA level increased for all simulated configurations and for both methods, as expected. L-power was substantially higher (about 0.2 across all results) than power. This indicates that it is very unlikely to find the numerous small simulated causative effects, even for low WPPA levels and with sequence data, as mimicked by the 7k dataset.

Power increased with SNP density. For example, for BayesC, a WPPA of 0.95 and a window size of 0.5, L-power was equal to 0.35, 0.39, 0.47, and 0.52 for SNP densities of 0.5k, 1k, 2k, and 7k, respectively. The same pattern was observed for BayesD.

Power also increased as window size increased. For example, L-power was equal to 0.42, 0.47, and 0.52 for window sizes of 0.25, 0.5 and 1 cM respectively (2k SNP density, 0.95 WPPA and BayesC). This pattern held true for all genetic configurations and for both methods.

Application of BayesD improved the power compared to BayesC for almost all genetic configurations. For example, for a WPPA of 0.95 and a window size of 1 cM, L-power of BayesC was equal to 0.35, 0.44, 0.52, and 0.55 and that of BayesD was equal to 0.41, 0.5, 0.53, and 0.57 for SNP densities of 0.5k, 1k, 2k and 7k, respectively. In general, it appeared that the superiority of BayesD over BayesC declined as SNP density increased. For a 7k SNP density, and a window size of 0.25, BayesC resulted in slightly greater power than BayesD.

Results of the evaluation of mapping precision are in Table 2. Because mapping precision was not affected by the WPPA level, except for some random fluctuations (not shown), they are reported as means across all WPPA levels. By definition, the lower bound of precision was the window size and, thus, window size affected the precision. For example, for BayesC (2k SNP density and 0.95 WPPA), a precision of 0.51, 0.93, and 1.75 cM was obtained for window sizes of 0.25, 0.5, and 1 cM, respectively. Precision was improved, although only slightly, with an increase in SNP density. For example, for BayesC

Table 1 Power and L-Power[a] as a function of the SNP density per M, window size in cM, window posterior probability of association (WPPA) and method (BayesC and BayesD)

SNP density (k/M)	WS	WPPA	BayesC		BayesD	
			Power	L-power[a]	Power	L-power[a]
0.5	0.25	0.85	0.261 (0.129)	0.441 (0.199)	0.282 (0.132)	0.475 (0.205)
		0.95	0.201 (0.107)	0.347 (0.169)	0.217 (0.110)	0.362 (0.154)
		0.99	0.144 (0.102)	0.249 (0.166)	0.175 (0.101)	0.300 (0.170)
	0.5	0.85	0.267 (0.138)	0.445 (0.193)	0.280 (0.120)	0.480 (0.205)
		0.95	0.205 (0.107)	0.349 (0.174)	0.230 (0.103)	0.388 (0.158)
		0.99	0.151 (0.094)	0.260 (0.151)	0.192 (0.099)	0.336 (0.182)
	1	0.85	0.279 (0.142)	0.467 (0.204)	0.298 (0.135)	0.500 (0.193)
		0.95	0.214 (0.114)	0.352 (0.156)	0.241 (0.102)	0.410 (0.161)
		0.99	0.167 (0.092)	0.285 (0.146)	0.199 (0.095)	0.343 (0.165)
1	0.25	0.85	0.265 (0.119)	0.452 (0.184)	0.290 (0.112)	0.500 (0.186)
		0.95	0.210 (0.114)	0.365 (0.194)	0.241 (0.101)	0.415 (0.173)
		0.99	0.193 (0.112)	0.337 (0.196)	0.180 (0.091)	0.315 (0.157)
	0.5	0.85	0.291 (0.121)	0.495 (0.190)	0.321 (0.123)	0.551 (0.204)
		0.95	0.228 (0.116)	0.394 (0.199)	0.265 (0.108)	0.454 (0.180)
		0.99	0.192 (0.106)	0.334 (0.174)	0.198 (0.088)	0.344 (0.151)
	1	0.85	0.316 (0.128)	0.538 (0.208)	0.341 (0.139)	0.580 (0.213)
		0.95	0.259 (0.116)	0.438 (0.171)	0.291 (0.125)	0.497 (0.204)
		0.99	0.216 (0.119)	0.369 (0.183)	0.227 (0.096)	0.388 (0.145)
2	0.25	0.85	0.285 (0.126)	0.481 (0.176)	0.299 (0.126)	0.501 (0.169)
		0.95	0.247 (0.115)	0.423 (0.172)	0.245 (0.107)	0.419 (0.155)
		0.99	0.216 (0.109)	0.367 (0.159)	0.214 (0.106)	0.325 (0.169)
	0.5	0.85	0.332 (0.118)	0.561 (0.169)	0.343 (0.129)	0.571 (0.168)
		0.95	0.276 (0.118)	0.469 (0.182)	0.284 (0.115)	0.476 (0.152)
		0.99	0.234 (0.119)	0.395 (0.172)	0.221 (0.114)	0.376 (0.176)
	1	0.85	0.348 (0.130)	0.586 (0.181)	0.371 (0.141)	0.618 (0.182)
		0.95	0.312 (0.132)	0.522 (0.179)	0.318 (0.117)	0.532 (0.147)
		0.99	0.245 (0.123)	0.406 (0.178)	0.248 (0.121)	0.420 (0.172)
7	0.25	0.85	0.307 (0.120)	0.518 (0.185)	0.301 (0.093)	0.505 (0.164)
		0.95	0.273 (0.097)	0.467 (0.156)	0.263 (0.092)	0.449 (0.166)
		0.99	0.259 (0.090)	0.435 (0.146)	0.256 (0.075)	0.393 (0.146)
	0.5	0.85	0.356 (0.107)	0.602 (0.172)	0.369 (0.099)	0.615 (0.168)
		0.95	0.309 (0.102)	0.518 (0.161)	0.312 (0.093)	0.531 (0.142)
		0.99	0.270 (0.089)	0.459 (0.165)	0.273 (0.089)	0.460 (0.135)
	1	0.85	0.378 (0.118)	0.636 (0.152)	0.394 (0.106)	0.657 (0.137)
		0.95	0.329 (0.116)	0.551 (0.156)	0.339 (0.103)	0.567 (0.152)
		0.99	0.281 (0.109)	0.465 (0.151)	0.279 (0.095)	0.472 (0.155)

Standard deviations are in parenthesis

[a] L-Power denotes the power to detect a causal gene that explains more than 2.5% of the simulated genetic variance

WS window size

and a window size of 0.5 cM, precisions of 1.1, 0.94, 0.93, and 0.91 cM were obtained for SNP densities of 0.5k, 1k, 2k, and 7k, respectively. BayesD improved mapping precision only slightly and not in all cases. In general, precision was slightly better for BayesD for smaller window sizes. However, with a window size of 1 cM, precision was slightly better for BayesC. A remarkable outcome is that even with sequence data, the precision was substantially higher than its lower bound. This held true for both methods.

Table 2 Precision as a function of SNP density per M, window size in cM, and method (BayesC and BayesD)

SNP density (k/M)	WS	BayesC	BayesD
0.5	0.25	0.653 (0.223)	0.639 (0.219)
	0.5	1.050 (0.275)	1.047 (0.245)
	1	1.779 (0.283)	1.796 (0.259)
1	0.25	0.525 (0.130)	0.517 (0.095)
	0.5	0.943 (0.138)	0.943 (0.152)
	1	1.720 (0.215)	1.744 (0.182)
2	0.25	0.506 (0.138)	0.467 (0.085)
	0.5	0.930 (0.125)	0.909 (0.121)
	1	1.746 (0.184)	1.728 (0.157)
7	0.25	0.468 (0.042)	0.449 (0.046)
	0.5	0.912 (0.074)	0.906 (0.075)
	1	1.798 (0.120))	1.802 (0.114)

Standard deviations are in parenthesis

WS window size

Real dataset

Plots of WPPA along the chromosomes are in Fig. 2 for a window size of 0.5 cM. For the two other window sizes (0.25 and 1 cM), similar plots were obtained, except that the peaks were somewhat more (less) pronounced with a window size of 1 cM (0.25 cM) (not shown). Both methods produced a clear signal on BTA14 at the chromosomal position where *DGAT1* is located and a clear signal on chromosome 10, although this was below the threshold levels used in this study. No other WPPA was above the threshold levels used in this study, which indicates that either no other genes with a larger effect are segregating in this population, or the sample size of the data set is too small, or both.

Closer inspection of the WPPA plots revealed some differences between the two methods. BayesD produced several extra signals, although far below the threshold levels used in this study. Examples are on chromosomes 6, 11, and 21. In addition, the average WPPA across the genome was slightly higher for BayesD than for BayesC. In Fig. 3, estimates of the window additive genetic variance of the window dominance variance are provided. From this figure it seems that not only the additive effects are spread across the genome, but also the dominance effects.

Discussion
Simulation protocol
Multi-marker Bayesian methods for association analyses were compared using simulated and real data. The simulation protocol regarding genetic architecture followed our current understanding of quantitative traits with regard to number of segregating causal mutations and

their additive and dominance effects. In a recent study [15], we conducted an in-depth theoretical analysis of the contribution of dominance to the variation of quantitative traits. One aim of that analysis was to develop a simulation protocol that models dominance gene effects that result in realistic genetic variance components and to validate this protocol with a sensitivity analysis. The current simulation protocol follows the recommendations given in that study.

Only one chromosome was simulated because the MCMC-based analyses were computationally demanding and replicated simulations were performed. It is possible that the power that would result from the simulation of multiple chromosomes would be somewhat reduced, but this would not alter the general findings of our study. However, the results obtained by using real data revealed that it is possible to also apply these methods in the case of genomes that consist of numerous chromosomes, and for which 630k SNPs are available. Due to the stochastic nature of the effects during the simulation of the traits, the dominance variance calculated as the proportion of the phenotypic variance ranged from 0.01 to 0.29, with a median of 0.08. These values are consistent with those reported by Bolormaa et al. [14], i.e. proportions from 0 to 0.42, with a median of 0.05, and indeed also with the estimates obtained from the dairy cow dataset that was used here. For some simulated traits, dominance was not important at all, which is also the case for real quantitative traits. We favoured this stochastic approach of simulating dominance variance proportions instead of choosing fixed proportions, because this generalised the results obtained for typical quantitative traits instead of producing results that are valid only for a defined dominance variance proportion.

BayesC versus BayesD
The results showed that method BayesD increased power to map QTL compared to BayesC in almost all analyses because it uses dominance variance as an additional source of genetic variance. This increased power can also be deduced from the results of the analyses of the real dairy cow dataset (Fig. 2), where BayesD produced several additional signals that were not found in the BayesC results.

Figure 4 shows the genetic effects and genetic variances of simulated causal mutations for a randomly chosen simulated trait for which dominance was important. Plots of the estimated window genomic variance across the chromosome and of the WPPA are also in Fig. 4. The two large causal mutations at positions 0.14 and 0.23 were detected with both methods (WPPA > 0.95), although both showed a relatively large dominance variance and the estimate of genomic variance was substantially larger

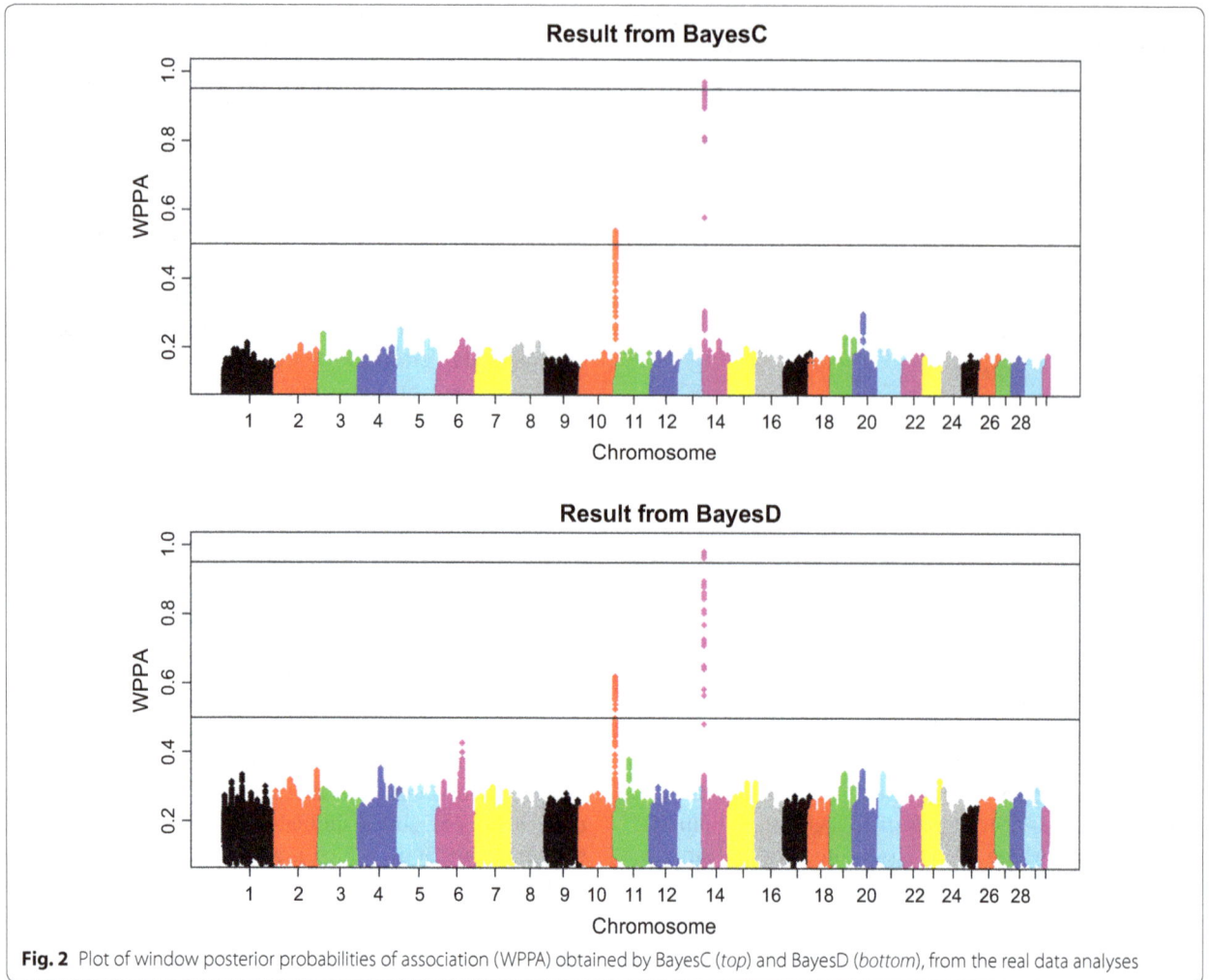

Fig. 2 Plot of window posterior probabilities of association (WPPA) obtained by BayesC (*top*) and BayesD (*bottom*), from the real data analyses

for BayesD than BayesC. For the causal mutations with a smaller effect size at positions 0.71 and 0.76, dominance was important and these mutations were only significant with BayesD (WPPA > 0.95). For the mutation with a moderate effect size at position 0.85, dominance was not important and this mutation was only detected with BayesC (WPPA > 0.95). These examples illustrate the following general findings: mutations with a large effect are likely to be detected by both methods, mutations with a moderate effect and with dominance effects are more likely to be detected by BayesD, and those without dominance by BayesC. Mutations with a small effect are not detected by either method. Hence, although the power of BayesD is generally higher than that of BayesC, the application of both models will likely improve overall power.

Since dominance is an interaction effect of the two alleles at a locus, their effects are captured in the association analysis by matched haplotype pairs, i.e. diplotypes. Diplotypes show a faster breakup around a focal point in the genome compared to haplotypes. Hence, BayesD was

expected to improve the mapping precision as well, but the observed improvement was only small and was more pronounced for smaller window sizes.

For a high SNP density and a small window size, BayesC outperformed BayesD with regard to power. This intuitively unexpected result can be explained by the fact that, in BayesC one effect is estimated per SNP, whereas in BayesD two effects are estimated per SNP. Thus, with BayesD, the effect of a causative mutation may be spread over more very closely-linked SNPs than for BayesC, and some SNPs may even be outside the window, if the window size is small.

For BayesC, only the additive genetic variance and the residual variance need to be known, which can usually be estimated from the data with high precision. In BayesD, dominance variance and inbreeding depression also need to be known. With the use of genomic data it might be possible for most traits in most populations to estimate them precisely, at least for the dominance variance [12, 13]. Common to both methods is the specification of

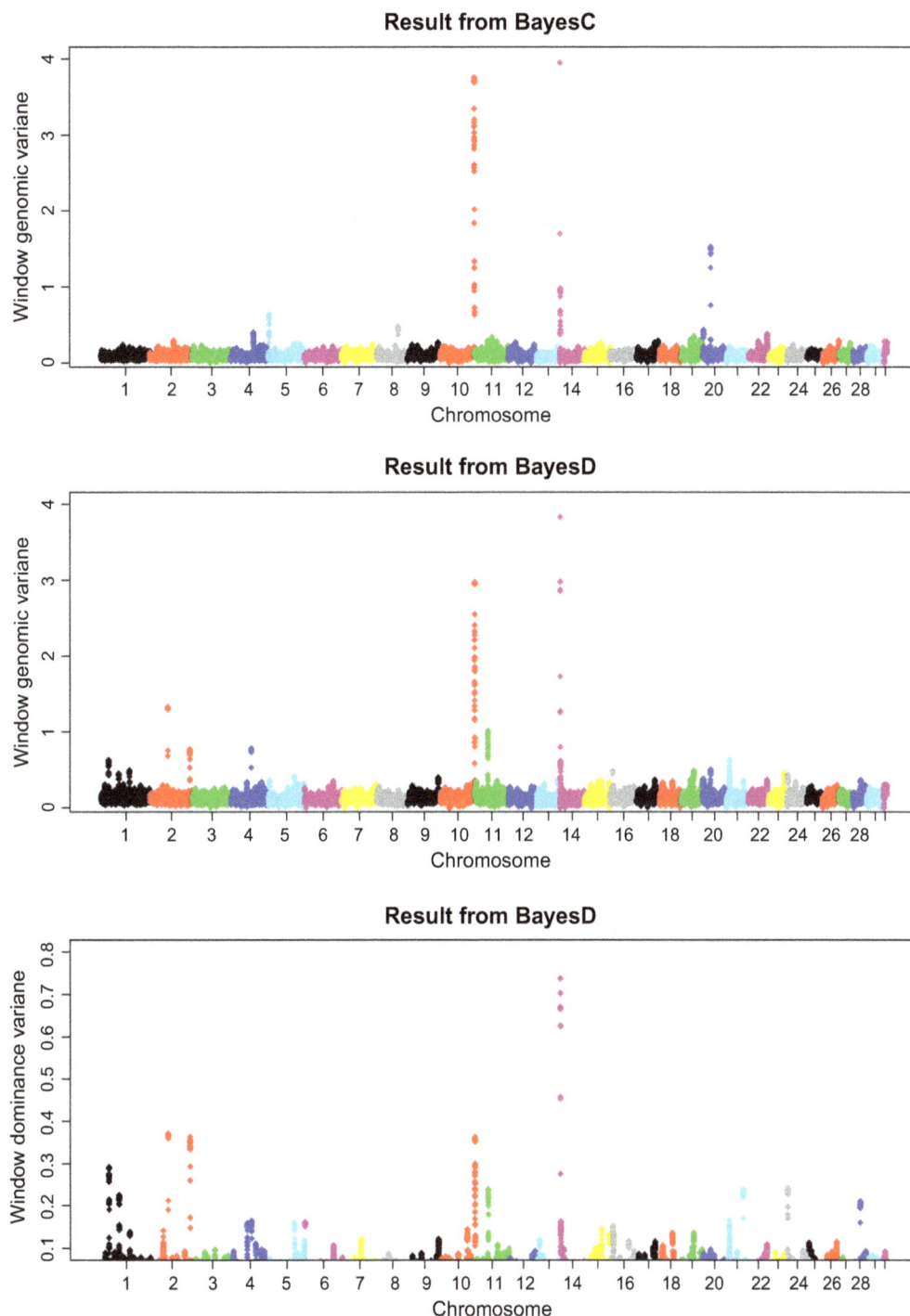

Fig. 3 Estimates of window genomic variances based on the real data analyses. The *top* and *middle panels* show the within-window estimates of genomic variances obtained by BayesC and BayesD, respectively. The *bottom panel* shows the within-window dominance variance obtained by BayesD. The window variances were multiplied by 1000. The window size was 0.5 cM

the number of degrees of the *t*-distribution of the SNP effects. We chose relatively small degrees in order to obtain a heavy-tailed distribution and thus clearer association signals. In addition, parameter *pLD*, i.e. the prior

probability that an SNP effect comes from the distribution with the large variance, needs to be specified. In this study, for the simulated datasets, this was specified under the assumption that the expected number of required

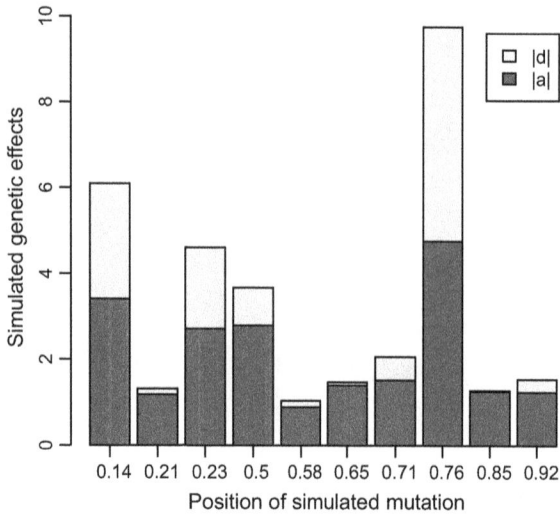

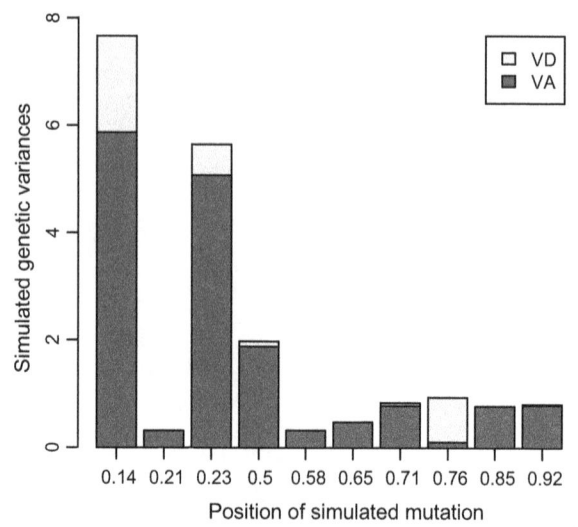

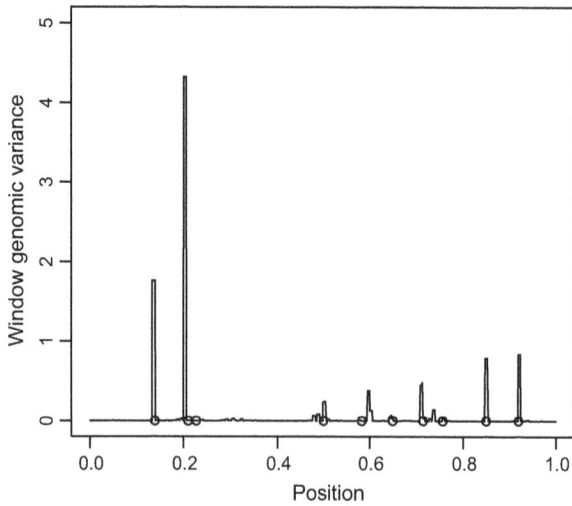

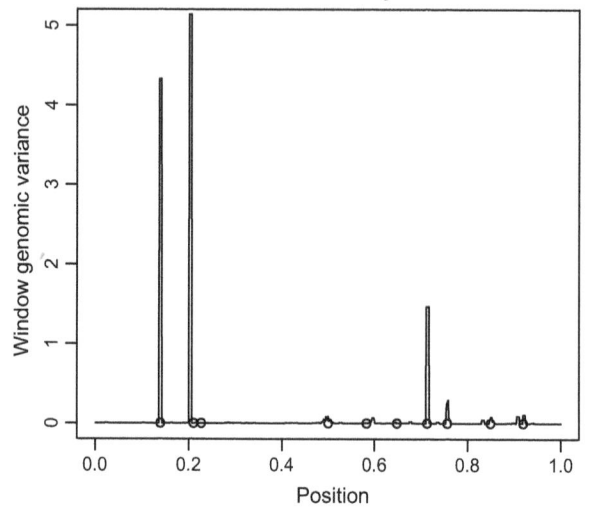

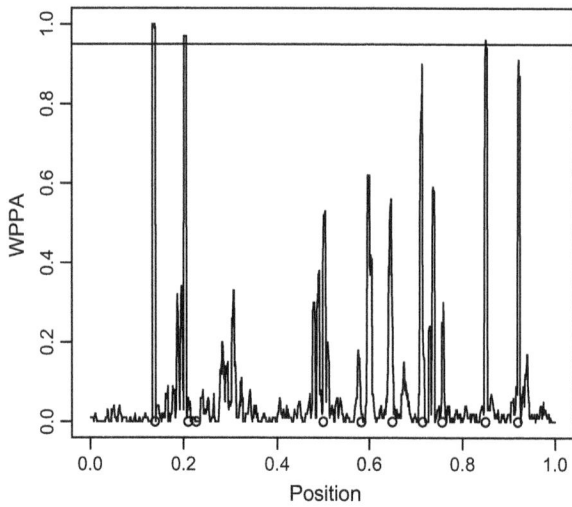

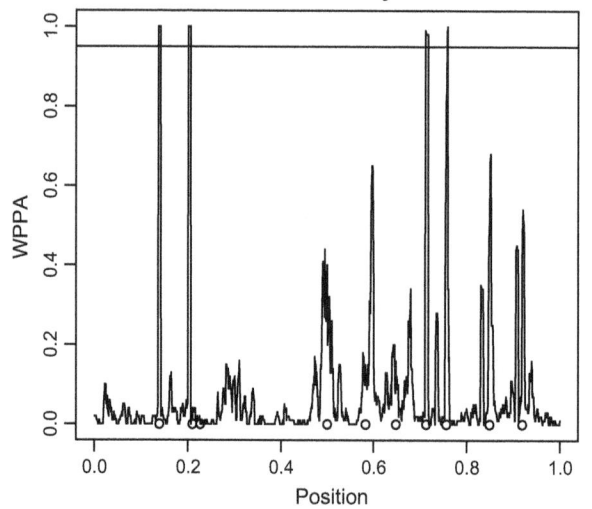

(See figure on previous page.)

Fig. 4 Simulated gene effects and BayesC and BayesD results for a single simulated trait. The *top left panel* shows the simulated additive and dominance effects of the 10 causative mutations with a non-negligible effect for a randomly chosen trait for which dominance was important. The *top right panel* shows the genetic variances of these simulated causative mutations. The *two panels in the middle* show the within-window genomic variances obtained by BayesC (*left*) and BayesD (*right*). The window posterior probability of association (WPPA) obtained from both methods are shown at the *bottom*. The positions of the 10 causative mutations are indicated by a *circle*

SNPs approaches the number of causative mutations when the size of the SNP panel approaches the total number of SNPs. Alternatively, the parameter could have been estimated from the data [26] but the current version of the MCMC algorithm does not include this option [18], or by using grid searches. In the real dataset, the input parameters were taken from an earlier study [19], in which these parameters were varied across a specified range of values. The parameters that need to be specified for the Bayesian methods are not needed in single-marker GWAS implemented in linear mixed models. This makes the application of these models more convenient and is probably one reason why these methods are more widely used.

False positive results

Our simulation study was not designed to determine whether the WPPA controls the proportion of false positives at the desired level. This hampered the comparison of WPPA between BayesC and BayesD in a formal way. Instead, we searched for elevated WPPA (at the lowest threshold level of 0.85 and even lower) in chromosomal regions of about 10 cM that had no simulated causal mutation. In very exceptional cases, an elevated WPPA was observed in these regions without any differences between BayesC and BayesD. Hence, it appears that WPPA controls the proportion of false positives in the simulation study at a low level for both methods. This might be due to the small number of degrees of freedom that was used for the t-distribution of the SNP effects, which allows large true effects to be detected, but small and spurious effects are regressed back to zero and, hence, are not detected. A larger number of degrees of freedom would probably have resulted in greater power but also in a larger number of false positives. From this, it became also obvious that the frequentist properties (power, false positive rate) of these methods remains somewhat unclear, because the WPPA criterion as implemented in this study appeared to be a poor guide for the false positive rate.

Window approach

Instead of drawing inferences from SNPs in windows, the posterior probability of the effect of a specific SNP being drawn from the distribution with the larger variance

could have been used for inference purposes. However, in the study of Sahana et al. [8], this resulted in reduced power, because the gene effects may be distributed over several consecutive SNPs and individual SNPs thus have a reduced power. Thus, the window approach is a logical consequence of applying multi-marker Bayesian methods for GWAS with dense SNP panels. Window size affected the power and precision in opposite directions, i.e. power increased with window but precision was lower for larger window sizes. Hence, there is a trade-off between these two criteria of success. An obvious solution would be to start with larger windows, e.g. of 1 cM, to find significant chromosomal regions that are associated with a trait and subsequently reduce window size to fine-map the causal mutations. With full sequence data, it can be assumed that the causal mutation is among the genotypes. Obviously, pinpointing the causal mutation within a fine-mapped region is not possible with a window approach. In this case, it might be beneficial to use posterior probabilities for individual SNPs [7], as well as other methods [27, 28]. Although complete sequence information for each individual was available in the 7k dataset, we did not attempt to detect the causal mutation within a fine-mapped region.

A sliding-window-based approach was used by moving the window boundaries one SNP forward along the chromosome. This resulted in as many windows as SNPs included in the analysis, and also in the same number of WPPA, which are highly correlated for consecutive windows. Alternatively, the chromosome could have been divided into non-overlapping windows. This would reduce the number of windows but introduces the problem of arbitrarily setting the window boundaries. It is possible that such boundaries will break a chromosomal region that harbours a causative mutation into two or more distinct windows, which would reduce the power to map the mutation. The choice of windows deserves additional investigations that take the LD structure along the genome into account [29]. Another option would be to fit haplotypes [30]. This would also extend the inferences beyond a single SNP but the SNPs would be in strong LD.

Inferences were drawn by using the WPPA criterion and using estimates of genomic variance within windows. If inference on the importance of dominance is of interest, the same criterion can be used by using estimates of dominance variance within windows. A straightforward strategy would be to map causative genes using estimates of within-window genomic variance, as done in this study, and then test the importance of dominance by using the within-window estimate of dominance variance.

Large datasets are needed

Given that the scaling argument of [21] also holds at least approximately, for GWAS, large-scale populations were simulated when scaling towards a bovine genome of 30 M and an Ne of 100 or 1000. Results show that, even with these large datasets, power to find causative genes that explain more than 2.5% of the genetic variance (L-power in Table 1) is above 0.5 only in a few configurations. Hence, our findings show that large datasets are needed if also moderate and eventually small effect associations are to be detected. Results of the real data analyses support this as well. BayesD needs genotypes and trait measurements from the same individual. In cattle, this means that cows need to be genotyped and this is becoming part of the routine genotyping strategy in many dairy cattle breeding organisations. Hence, it can be expected that such large-scale datasets that include genotyped cows will soon be available.

Conclusions

The application of BayesD for GWAS on simulated quantitative traits with realistic dominance effects resulted in increased power compared to BayesC. The increase in power was between -2 and 8% for causative genes that explain more than 2.5% of the genetic variance. This trend in increased power was also observed in the results from the real data analyses. In addition, by examining the within-window estimates of genomic dominance variance, BayesD allows for inference about the magnitude of dominance effects at significant associations, which remains hidden in BayesC analyses. Mapping precision was, however, not substantially improved by BayesD. If the aim is to map mutations with medium and small effects, large datasets that include several thousands of individuals are needed.

Authors' contributions
RW wrote the software for the population simulation. JB did the statistical analysis and wrote the paper. CE and RF contributed the Fleckvieh dairy cattle dataset. TM and RW participated in the discussion of the methods and results and revised the first draft of the paper. All authors read and approved the final manuscript.

Author details
[1] Institute of Animal Science, University of Hohenheim, 70593 Stuttgart, Germany. [2] Institute of Animal Breeding, Bavarian State Research Center for Agriculture, 85580 Grub, Germany. [3] Chair of Animal Breeding, Technical University of Munich, Liesel-Beckmann-Strasse 1, 85354 Freising, Germany. [4] Institute of Animal and Aquacultural Science, Norwegian University of Life Science, 1450 Aas, Norway.

Acknowledgements
This study was partially supported by a grant from the German Research Foundation (Deutsche Forschungsgemeinschaft, DFG). Fleckvieh cow genotyping data were generated within the AgroClustEr "Synbreed—Synergistic plant and animal breeding" (FKZ: 0315528A) supported by the German Federal Ministry of Education and Research (BMBF).

Competing interests
The authors declare that they have no competing interests.

Appendix

In this appendix, we show how to derive expectations $E(\alpha^2)$ and $E(d^2)$ under the assumption of an equal distribution of the additive genetic variance (V_A) and of dominance variance (V_D) across all SNPs M. Under this assumption, the expected additive genetic variance of a marker m can be denoted by $E(V_{A_m}) = \frac{V_A}{M}$ and the expected dominance variance by $E(V_{D_m}) = \frac{V_D}{M}$. Furthermore, $E(V_{A_m}) = E(2pq) \times E(\alpha^2)$ and $E(V_{D_m}) = E((2pq)^2) \times E(d^2)$. The expectation of $2pq$ can be approximated by the mean of the heterozygosity across all SNPs and is denoted by $\bar{H}$. Similarly, the expectation of $(2pq)^2$ can be approximated by calculating the squared heterozygosity of each SNP and subsequently by taking the mean of this value across all SNPs. This is denoted as $\overline{H^2}$. Thus, $E(\alpha^2) = \frac{V_A}{M\bar{H}}$ and $E(d^2) = \frac{V_D}{M\overline{H^2}}$, which was used for the calculation of q_w in the main text.

References

1. Goddard ME, Hayes BJ. Mapping genes for complex traits in domestic animals and their use in breeding programmes. Nat Rev Genet. 2009;10:381–91.
2. Yang J, Zaitlen NA, Goddard ME, Visscher PM, Price AL. Mixed model association methods: advantages and pitfalls. Nat Genet. 2014;46:100–6.
3. Storey JD, Tibshirani R. Statistical significance for genomewide studies. Proc Natl Acad Sci USA. 2003;100:9440–5.
4. Yang J, Ferreira T, Morris AP, Medland SE, Genetic Investigation of Anthropometric Traits (GIANT) Consortium, DIAbetes Genetics Replication And Meta-analysis (DIAGRAM) Consortium, et al. Conditional and joint multiple-SNP analysis of GWAS summary statistics identifies additional variants influencing complex traits. Nat Genet. 2012;44:369–75.
5. Fernando RL, Garrick D. Bayesian methods applied to GWAS. In: Gondro C, van der Werf J, Hayes B, editors. Genome-wide association studies and genomic prediction. Methods in molecular biology, Springer protocols. New York: Springer Press; 2013.
6. Meuwissen THE, Hayes BJ, Goddard ME. Prediction of total genetic value using genome-wide dense marker maps. Genetics. 2001;157:1819–29.
7. Goddard ME, MacLeod IM, Kemper KE, Vander Jagt CJ, Savin K, Schrooten C, et al. Research plan for the identification of QTL. In Proceedings of the 10th world congress on genetics applied to livestock production: 17–22 August 2014. Vancouver; 2014. https://asas.org/docs/default-source/wcgalp-proceedings-oral/199_paper_10348_manuscript_1635_0.pdf?sfvrsn=2.
8. Sahana G, Guldbrandtsen B, Janss L, Lund MS. Comparison of association mapping methods in a complex pedigree population. Genet Epidemiol. 2010;34:455–62.
9. Legarra A, Croiseau P, Sanchez MP, Teyssèdre S, Sallé G, Allais S, et al. A comparison of methods for whole-genome QTL mapping using dense markers in four livestock species. Genet Sel Evol. 2015;47:6.

10. Fernando RL, Toosi A, Garrick DJ, Dekkers JCM. Application of whole-genome prediction methods for genome-wide association studies: a Bayesian approach. In: Proceedings of the 10th world congress on genetics applied to livestock production: 17–22 August 2014. Vancouver; 2014. https://asas.org/docs/default-source/wcgalp-proceedings-oral/201_paper_10341_manuscript_1325_0.pdf?sfvrsn=2.

11. van Tassell CP, Misztal I, Varona L. Method R estimates of additive genetic, dominance genetic, and permanent environmental fraction of variance for yield and health traits of Holsteins. J Dairy Sci. 2000;83:1873–7.

12. Ertl J, Legarra A, Vitezica ZG, Varona L, Edel C, Emmerling R, et al. Genomic analysis of dominance effects on milk production and conformation traits in Fleckvieh cattle. Genet Sel Evol. 2014;46:40.

13. Su G, Christensen OF, Ostersen T, Henryon M, Lund MS. Estimating additive and non-additive genetic variances and predicting genetic merits using genome-wide dense single nucleotide polymorphism markers. PLoS One. 2012;7:e45293.

14. Bolormaa S, Pryce JE, Zhang Y, Reverter A, Barendse W, Hayes BJ, et al. Non-additive genetic variation in growth, carcass and fertility traits of beef cattle. Genet Sel Evol. 2015;47:26.

15. Wellmann R, Bennewitz J. The contribution of dominance to the understanding of quantitative genetic variation. Genet Res (Camb). 2011;93:139–54.

16. Verbyla KL, Hayes BJ, Bowman PJ, Goddard ME. Accuracy of genomic selection using stochastic search variable selection in Australian Holstein Friesian dairy cattle. Genet Res (Camb). 2009;91:307–11.

17. Verbyla KL, Hayes BJ, Bowman PJ, Goddard ME. Sensitivity of genomic selection to using different prior distributions. BMC Proc. 2010;4:S5.

18. Wellmann R, Bennewitz J. Bayesian models with dominance effects for genomic evaluation of quantitative traits. Genet Res (Camb). 2012;94:21–37.

19. Wellmann R, Ertl J, Emmerling R, Edel C, Götz KU, Bennewitz J. Joint genomic evaluation of cows and bulls with BayesD for prediction of genotypic values and dominance deviations. In: Proceedings of the 10th world congress on genetics applied to livestock production: 17–22 August 2014. Vancouver; 2014.

20. Villa-Angulo R, Matukumalli LK, Gill CA, Choi C, Van Tassell CP, Grefenstette JJ. High resolution haplotype block structure in the cattle genome. BMC Genet. 2009;10:19.

21. Meuwissen THE. Accuracy of breeding values of 'unrelated' individuals predicted by dense SNP genotyping. Genet Sel Evol. 2009;41:35.

22. Bennewitz J, Meuwissen THE. The distribution of QTL additive and dominance effects in porcine F2 crosses. J Anim Breed Genet. 2010;127:171–9.

23. Caballero A, Keightley PD. A pleiotropic nonadditive model of variation in quantitative traits. Genetics. 1994;138:883–900.

24. Kacser H, Burns JA. The molecular basis of dominance. Genetics. 1981;97:639–66.

25. Falconer DS, Mackay TFC. Introduction to quantitative genetics. London: Longman; 1996.

26. Habier D, Fernando RL, Kizilkaya K, Garrick D. Extension of the Bayesian alphabet for genomic selection. BMC Bioinformatics. 2011;12:186.

27. Uleberg E, Meuwissen THE. The complete linkage disequilibrium test: a test that points to causative mutations underlying quantitative traits. Genet Sel Evol. 2011;43:20.

28. Weller JI, Ron M. Invited review: quantitative trait nucleotide determination in the era of genomic selection. J Dairy Sci. 2011;94:1082–90.

29. Beissinger TM, Rosa GJ, Kaeppler SM, Gianola D, de Leon N. Defining window-boundaries for genomic analyses using smoothing spline techniques. Genet Sel Evol. 2015;47:30.

30. Calus MPL, Meuwissen THE, de Roos APW, Veerkamp RF. Accuracy of genomic selection using different methods to define haplotypes. Genetics. 2008;178:553–61.

Permissions

All chapters in this book were first published in GSE, by BioMed Central; hereby published with permission under the Creative Commons Attribution License or equivalent. Every chapter published in this book has been scrutinized by our experts. Their significance has been extensively debated. The topics covered herein carry significant findings which will fuel the growth of the discipline. They may even be implemented as practical applications or may be referred to as a beginning point for another development.

The contributors of this book come from diverse backgrounds, making this book a truly international effort. This book will bring forth new frontiers with its revolutionizing research information and detailed analysis of the nascent developments around the world.

We would like to thank all the contributing authors for lending their expertise to make the book truly unique. They have played a crucial role in the development of this book. Without their invaluable contributions this book wouldn't have been possible. They have made vital efforts to compile up to date information on the varied aspects of this subject to make this book a valuable addition to the collection of many professionals and students.

This book was conceptualized with the vision of imparting up-to-date information and advanced data in this field. To ensure the same, a matchless editorial board was set up. Every individual on the board went through rigorous rounds of assessment to prove their worth. After which they invested a large part of their time researching and compiling the most relevant data for our readers.

The editorial board has been involved in producing this book since its inception. They have spent rigorous hours researching and exploring the diverse topics which have resulted in the successful publishing of this book. They have passed on their knowledge of decades through this book. To expedite this challenging task, the publisher supported the team at every step. A small team of assistant editors was also appointed to further simplify the editing procedure and attain best results for the readers.

Apart from the editorial board, the designing team has also invested a significant amount of their time in understanding the subject and creating the most relevant covers. They scrutinized every image to scout for the most suitable representation of the subject and create an appropriate cover for the book.

The publishing team has been an ardent support to the editorial, designing and production team. Their endless efforts to recruit the best for this project, has resulted in the accomplishment of this book. They are a veteran in the field of academics and their pool of knowledge is as vast as their experience in printing. Their expertise and guidance has proved useful at every step. Their uncompromising quality standards have made this book an exceptional effort. Their encouragement from time to time has been an inspiration for everyone.

The publisher and the editorial board hope that this book will prove to be a valuable piece of knowledge for researchers, students, practitioners and scholars across the globe.

List of Contributors

Mohammad H. Ferdosi
The Centre for Genetic Analysis and Applications, School of Environmental and Rural Science, University of New England, Armidale, Australia
Animal Genetics and Breeding Unit, University of New England, Armidale, Australia

Bruce Tier
Animal Genetics and Breeding Unit, University of New England, Armidale, Australia

John Henshall
Cobb-Vantress, Siloam Springs, AR, USA
CSIRO Agriculture Flagship, FD McMaster Laboratory Chiswick, Armidale, Australia

Natalia S. Forneris
Centre for Research in Agricultural Genomics (CRAG), CSIC-IRTA-UAB-UB Consortium, 08193 Bellaterra, Barcelona, Spain
Departamento de Producción Animal, Facultad de Agronomía, Universidad de Buenos Aires, C1417DSE Buenos Aires, Argentina

Miguel Pérez-Enciso
Centre for Research in Agricultural Genomics (CRAG), CSIC-IRTA-UAB-UB Consortium, 08193 Bellaterra, Barcelona, Spain
Departament de Ciència Animal i dels Aliments, Universitat Autònoma de Barcelona, 08193 Bellaterra, Barcelona, Spain
ICREA, Passeig de Lluís Companys 23, 08010 Barcelona, Spain

Zulma G. Vitezica and Andres Legarra
GenPhySE, INRA, INPT, ENVT, Université de Toulouse, 31326 Castanet-Tolosan, France

Bruna P. Sollero and Cláudia C. G. Gomes
Embrapa Pecuária Sul, Caixa Postal 242 - BR 153 - Km 633, Bagé, Rio Grande do Sul 96.401-970, Brazil

Fernando F. Cardoso
Embrapa Pecuária Sul, Caixa Postal 242 - BR 153 - Km 633, Bagé, Rio Grande do Sul 96.401-970, Brazil
Universidade Federal de Pelotas, Capão do Leão, Rio Grande do Sul 96.000-010, Brazil

Vinícius S. Junqueira
Departamento de Zootecnia, Universidade Federal de Viçosa, Avenida Peter Henry Rolfs, s/n - Campus Universitário, Viçosa, Minas Gerais 36.570-000, Brazil

Alexandre R. Caetano
Embrapa Recursos Genéticos e Biotecnologia, Parque Estacao Biologica Final Av. W/5 Norte, Brasilia-DF, C.P. 02372, Brasília, Distrito Federal 70770-917, Brazil

Miriam Piles, Josep Ramon, Oriol Rafel, Mariam Pascual and Juan P. Sánchez
Institute for Food and Agriculture Research and Technology, Torre Marimon s/n, 08140 Caldes de Montbui, Barcelona, Spain

Mohamed Ragab
Institute for Food and Agriculture Research and Technology, Torre Marimon s/n, 08140 Caldes de Montbui, Barcelona, Spain
Poultry Production Department, Kafr El-Sheikh University, Kafr El-Sheikh 33516, Egypt

Ingrid David and Laurianne Canario
GenPhySE, INRA, Université de Toulouse, INPT, ENVT, 31326 Castanet Tolosan, France

Michael E. Goddard
Agriculture Victoria Research, AgriBio Centre, Bundoora, VIC 3083, Australia
School of Land and Environment, University of Melbourne, Parkville, VIC 3010, Australia

Hans D. Daetwyler
Agriculture Victoria Research, AgriBio Centre, Bundoora, VIC 3083, Australia
School of Applied Systems Biology, La Trobe University, Bundoora, VIC 3086, Australia
Cooperative Research Centre for Sheep Industry Innovation, Armidale, NSW 2351, Australia

Sunduimijid Bolormaa
Agriculture Victoria Research, AgriBio Centre, Bundoora, VIC 3083, Australia.
Cooperative Research Centre for Sheep Industry Innovation, Armidale, NSW 2351, Australia

Andrew A. Swan and Daniel J. Brown
Animal Genetics and Breeding Unit, University of New England, Armidale, NSW 2351, Australia
Cooperative Research Centre for Sheep Industry Innovation, Armidale, NSW 2351, Australia

Sue Hatcher
NSW Department of Primary Industries, Orange Agricultural Institute, Orange, NSW 2800, Australia
Cooperative Research Centre for Sheep Industry Innovation, Armidale, NSW 2351, Australia

Nasir Moghaddar and Julius H. van der Werf
School of Environmental and Rural Science, University of New England, Armidale, NSW 2351, Australia
Cooperative Research Centre for Sheep Industry Innovation, Armidale, NSW 2351, Australia

Rohan L. Fernando and Hao Cheng
Department of Animal Science, Iowa State University, Ames, IA 50011, USA

Dorian J. Garrick
Department of Animal Science, Iowa State University, Ames, IA 50011, USA
Institute of Veterinary, Animal and Biomedical Sciences, Massey University, Palmerston North, New Zealand

Eva M. Strucken, Hawlader A. Al-Mamun and John P. Gibson
School of Environmental and Rural Science, University of New England, Armidale 2350, Australia

Cecilia Esquivelzeta-Rabell
Pic Improvement Company (PIC), Genetic Services, Hendersonville, TN 37075, USA

Cedric Gondro
Michigan State University, Animal Science, East Lansing, Michigan 48824, USA

Okeyo A. Mwai
International Livestock Research Institute, Nairobi, Kenya

Breno O. Fragomeni, Daniela A. L. Lourenco, Yutaka Masuda and Ignacy Misztal
Edgar L. Rhodes Center for Animal and Dairy Science, University of Georgia, Athens, GA, USA

Andres Legarra
GenPhySE, INRA, INPT, INP-ENVT, Université de Toulouse, 31326 Castanet-Tolosan, France

Marie-Pierre Sanchez, Pascal Croiseau, Guy Miranda, Patrice Martin, Anne Barbat-Leterrier, Rabia Letaïef, Dominique Rocha, Mekki Boussaha and Didier Boichard
GABI, INRA, AgroParisTech, Université Paris Saclay, 78350 Jouy-en-Josas, France

Armelle Govignon-Gion
GABI, INRA, AgroParisTech, Université Paris Saclay, 78350 Jouy-en-Josas, France
Institut de l'Elevage, 75012 Paris, France

Sébastien Fritz and Chris Hozé
GABI, INRA, AgroParisTech, Université Paris Saclay, 78350 Jouy-en-Josas, France
Allice, 75012 Paris, France

Mickaël Brochard
Institut de l'Elevage, 75012 Paris, France

Adeniyi C. Adeola and Hai-Bing Xie
State Key Laboratory of Genetic Resources and Evolution, Yunnan Laboratory of Molecular Biology of Domestic Animals, Kunming Institute of Zoology, Chinese Academy of Sciences, Kunming, China
Sino-Africa Joint Research Center, Chinese Academy of Sciences, Kunming, China

Robert W. Murphy
State Key Laboratory of Genetic Resources and Evolution, Yunnan Laboratory of Molecular Biology of Domestic Animals, Kunming Institute of Zoology, Chinese Academy of Sciences, Kunming, China
Centre for Biodiversity and Conservation Biology, Royal Ontario Museum, Toronto, Canada

Lotanna M. Nneji and Min-Sheng Peng
State Key Laboratory of Genetic Resources and Evolution, Yunnan Laboratory of Molecular Biology of Domestic Animals, Kunming Institute of Zoology, Chinese Academy of Sciences, Kunming, China
Sino-Africa Joint Research Center, Chinese Academy of Sciences, Kunming, China
Kunming College of Life Science, University of Chinese Academy of Sciences, Kunming, China

Ya-Ping Zhang
State Key Laboratory of Genetic Resources and Evolution, Yunnan Laboratory of Molecular Biology of Domestic Animals, Kunming Institute of Zoology, Chinese Academy of Sciences, Kunming, China
Sino-Africa Joint Research Center, Chinese Academy of Sciences, Kunming, China
Kunming College of Life Science, University of Chinese Academy of Sciences, Kunming, China
State Key Laboratory for Conservation and Utilization of Bio-Resources in Yunnan, Yunnan University, Kunming, China

Olufunke O. Oluwole, Bukola M. Oladele, Temilola O. Olorungbounmi and Bamidele Boladuro
Institute of Agricultural Research and Training, Obafemi Awolowo University, Ibadan, Nigeria

Sunday C. Olaogun
Department of Veterinary Medicine, University of Ibadan, Ibadan, Nigeria

Oscar J. Sanke
Taraba State Ministry of Agriculture and Natural Resources, Jalingo, Nigeria

Philip M. Dawuda
Department of Veterinary Surgery and Theriogenology, College of Veterinary Medicine, University of Agriculture Makurdi, Makurdi, Nigeria

Ofelia G. Omitogun
Department of Animal Sciences, Obafemi Awolowo University, Ile-Ife, Nigeria

Laurent Frantz
The Palaeogenomics and Bio-Archaeology Research Network, Research Laboratory for Archaeology, University of Oxford, Oxford, UK
School of Biological and Chemical Sciences, Queen Mary University of London, London, UK

Xiaochen Sun, Rohan Fernando and Jack Dekkers
Department of Animal Science and Center for Integrated Animal Genomics, Iowa State University, Ames, IA 50011, USA

Irene van den Berg
Faculty of Veterinary and Agricultural Science, University of Melbourne, Parkville, VIC, Australia

Mike E. Goddard
Faculty of Veterinary and Agricultural Science, University of Melbourne, Parkville, VIC, Australia
Agriculture Victoria, AgriBio, Centre for AgriBioscience, Bundoora, VIC 3083, Australia

Iona M. MacLeod and Tingting Wang and Sunduimijid Bolormaa
Agriculture Victoria, AgriBio, Centre for AgriBioscience, Bundoora, VIC 3083, Australia

Phil J. Bowman
Agriculture Victoria, AgriBio, Centre for AgriBioscience, Bundoora, VIC 3083, Australia
School of Applied Systems Biology, La Trobe University, Bundoora, VIC 3083, Australia

Ben J. Hayes
School of Applied Systems Biology, La Trobe University, Bundoora, VIC 3083, Australia
Queensland Alliance for Agriculture and Food Innovation, Centre for Animal Science, University of Queensland, St Lucia, QLD, Australia

Egbert F. Knol
Topigs Norsvin Research Center, P.O. Box 43, 6640 AA Beuningen, The Netherlands

Marcos S. Lopes
Topigs Norsvin Research Center, P.O. Box 43, 6640 AA Beuningen, The Netherlands.
Animal Breeding and Genomics Centre, Wageningen University, 6708 PB Wageningen, The Netherlands
Topigs Norsvin, Curitiba, PR 80.420-210, Brazil

Henk Bovenhuis, André M. Hidalgo, Johan A. M. van Arendonk and John W. M. Bastiaansen
Animal Breeding and Genomics Centre, Wageningen University, 6708 PB Wageningen, The Netherlands

Matteo Bergamaschi, Alessio Cecchinato and Giovanni Bittante
Department of Agronomy, Food, Natural Resources, Animals and Environment (DAFNAE), University of Padua, Viale dell'Università 16, 35020 Legnaro, PD, Italy

Franco Biasioli and Flavia Gasperi
Department of Food Quality and Nutrition, Research and Innovation Centre, Fondazione Edmund Mach (FEM), Via E. Mach 1, 38010 San Michele all'Adige, TN, Italy

Bruno Martin
INRA, UMR Herbivores, 63122 Saint-Genès
Champanelle, France
Clermont Université, VetAgro Sup, BP 10448, 63000
Clermont-Ferrand, France

Oscar O. M. Iheshiulor and Theo H. E. Meuwissen
Department of Animal and Aquacultural Sciences,
Norwegian University of Life Sciences, PO Box
5003, 1432 Ås, Norway

John A. Woolliams
Department of Animal and Aquacultural Sciences,
Norwegian University of Life Sciences, PO Box
5003, 1432 Ås, Norway
The Roslin Institute (Edinburgh), Royal (Dick)
School of Veterinary Studies, University of
Edinburgh, Midlothian EH25 9RG, Scotland, UK

Morten Svendsen and Trygve Solberg
GENO SA, Holsegata 22, 2317 Hamar, Norway

Jörn Bennewitz and Robin Wellmann
Institute of Animal Science, University of
Hohenheim, 70593 Stuttgart, Germany

Christian Edel
Institute of Animal Breeding, Bavarian State
Research Center for Agriculture, 85580 Grub,
Germany

Ruedi Fries
Chair of Animal Breeding, Technical University of
Munich, Liesel-Beckmann-Strasse 1, 85354 Freising,
Germany

Theo H. E. Meuwissen
Institute of Animal and Aquacultural Science,
Norwegian University of Life Science, 1450 Aas,
Norway

Index

www.ingramcontent.com/pod-product-compliance
Lightning Source LLC
Chambersburg PA
CBHW061241190326
41458CB00011B/3544